# NEURAL NETWORKS FOR SIGNAL PROCESSING

## PROCEEDINGS OF THE 1991 IEEE WORKSHOP

Edited by

**B. H. Juang**
*AT&T Bell Laboratories*

**S. Y. Kung**
*Princeton University*

**Candace A. Kamm**
*Bellcore*

Published under the sponsorship of the
IEEE Signal Processing Society
(in cooperation with the IEEE Neural Networks Council)

**Neural Networks for Signal Processing - Proceedings of the 1991 IEEE Workshop**

Abstracting is permitted with credit to the source. Libraries are permitted to photocopy beyond the limits of U.S. copyright law for private use of patrons those articles in this volume that carry a code at the bottom of the first page, provided the per-copy fee indicated in the code is paid through the Copyright Clearance Center, 29 Congress Street, Salem, MA 01970. Instructors are permitted to photocopy isolated articles for noncommercial classroom use without fee. For other copying, reprint, or republication permission, write to the Staff Director of Publishing Services at the IEEE, 345 East 47th Street, New York, NY 10017-2394. All rights reserved. Copyright © 1991 by The Institute of Electrical and Electronics Engineers, Inc.

IEEE Catalog Number: 91TH0385-5

ISBN Casebound: 0-7803-0118-8

Michofiche: 0-7803-0119-6

Library of Congress Number: 91-72653

Additional copies of this publication are available from

IEEE Service Center
445 Hoes Lane
Piscataway, NJ 08854-4150

1-800-678-IEEE

# Contents

**Preface**

**Part 1: Theory and Modeling**

Note on Generalization, Weight Decay, and Architecture
Selection in Nonlinear Learning Systems ..... 1
*J. E. Moody*

Discriminative Multi-Layer Feed-Forward Networks ..... 11
*S. Katagiri, C. H. Lee, and B. H. Juang*

Efficient Training Procedures for Adaptive
Kernel Classifiers ..... 21
*S. V. Chakravarthy, J. Ghosh, L. Deuser, and S. Beck*

Concept Formation and Statistical Learning in
Nonhomogeneous Neural Nets ..... 30
*R. L. Tutwiler and L. H. Sibul*

An Alternative Proof of Convergence for
Kung-Diamantaras APEX Algorithm ..... 40
*H. Chen and R. Liu*

Neural Networks for Extracting Unsymmetric
Principal Components ..... 50
*S. Y. Kung and K. I. Diamantaras*

The Outlier Process ..... 60
*D. Geiger and R. A. M. Pereira*

A Mapping Approach for Designing Neural Sub-Nets ..... 70
*K. Rohani, M. S. Chen, and M. T. Manry*

Three-Dimensional Structured Networks for
Matrix Equation Solving ..... 80
*L. X. Wang and J. M. Mendel*

Improving Learning Rate of Neural Tree Networks
Using Thermal Perceptrons  90
*A. Sankar and R. Mammone*

Adaline with Adaptive Recursive Memory  101
*B. De Vries, J. C. Principe, and P. Guedes de Oliveira*

Learned Representation Normalization:
Attention Focusing with Multiple Input Modules  111
*M. L. Rossen*

A Parallel Learning Filter System That Learns
the KL-Expansion from Examples  121
*R. Lenz and M. Osterberg*

Restricted Learning Algorithm and Its Application
to Neural Network Training  131
*T. Miyamura, I. Yamada, and K. Sakaniwa*

Multiply Descent Cost Competitive Neural Networks
with Cooperation and Categorization  141
*Y. Matsuyama*

Nonlinear Adaptive Filtering of Systems with
Hysteresis by Quantized Mean Field Annealing  151
*R. A. Nobakht, S. H. Ardalan, and D. E. Van den Bout*

An Outer Product Neural Net for Extracting
Principal Components from a Time Series  161
*L. E. Russo*

## Part 2: Pattern Recognition

Pattern Recognition Properties of Neural Networks  173
*J. Makhoul*

Edge Detection for Optical Image Metrology
Using Unsupervised Neural Network Learning  188
*H. K. Aghajan, C. D. Schaper, and T. Kailath*

| | |
|---|---|
| Improving Generalization Performance in Character Recognition<br>H. Drucker and Y. Le Cun | 198 |
| Neural Networks for Sidescan SONAR Automatic Target Detection<br>M. J. LeBlanc and E. S. Manolakos | 208 |
| An Effection Method for Visual Pattern Recognition<br>I. N. M. Papadakis | 217 |
| Fingerprint Recognition Using Neural Network<br>W. F. Leung, S. H. Leung, W. H. Lau, and A. Luk | 226 |
| A Comparison of Second-Order Neural Networks to Transform-Based Methods for Translation- and Orientation-Invariant Object Recognition<br>R. Duren and B. Peikari | 236 |
| Shape Recognition with Nearest Neighbor Isomorphic Network<br>H. C. Yau and M. T. Manry | 246 |
| Dimensionality Reduction of Dynamical Patterns Using a Neural Network<br>S. Nakagawa, Y. Ono, and Y. Hirata | 256 |
| A Critical Overview of Neural Network Pattern Classifiers<br>R. Lippmann | 266 |

## Part 3: Speech Processing

| | |
|---|---|
| Workstation-Based Phonetic Typewriter<br>T. Kohonen | 279 |
| Word Recognition with the Feature Finding Neural Network (FFNN)<br>T. Gramss | 289 |

New Discriminative Training Algorithms Based
on the Generalized Probabilistic Descent Method ... 299
*S. Katagiri, C. H. Lee, and B. H. Juang*

Probability Estimation by Feed-Forward Networks
in Continuous Speech Recognition ... 309
*S. Renals, N. Morgan, and H. Bourlard*

Nonlinear Resampling Transformation
for Automatic Speech Recognition ... 319
*Y. D. Liu, Y. C. Lee, H. H. Chen, and G. Z. Sun*

Speech Recognition by Combining Pairwise
Discriminant Time-Delay Neural Networks and
Predictive LR-Parser ... 327
*J. Takami, A. Kai, and S. Sagayama*

Speech Recognition Using Time-Warping Neural
Networks ... 337
*K. Aikawa*

A Hybrid Continuous Speech Recognition System
Using Segmental Neural Nets with Hidden Markov
Models ... 347
*S. Austin, G. Zavaliagkos, J. Makhoul, and R. Schwartz*

Connectionist Speaker Normalization and Its
Application to Speech Recognition ... 357
*X. D. Huang, K. F. Lee, and A. Waibel*

A Time-Derivative Neural Net Architecture -
An Alternative to the Time-Delay Neural Net
Architecture ... 367
*K. K. Paliwal*

Word Recognition Based on the Combination
of a Sequential Neural Network and
the GPDM Discriminative Training Algorithm ... 376
*W. Y. Chen and S. H. Chen*

A Space-Perturbance/Time-Delay Neural Network
for Speech Recognition ... 385
*M. Ji, H. H. Chen, and Z. K. Shen*

Non-Linear Prediction of Speech Signals Using
Memory Neuron Networks   395
P. Poddar and K. P. Unnikrishnan

Experiments with Temporal Resolution for
Continuous Speech Recognition with
Multi-Layer Perceptrons   405
N. Morgan, C. Wooters, and H. Hermansky

Neural-Network Architecture for Linear and
Nonlinear Predictive Hidden Markov Models:
Application to Speech Recognition   411
L. Deng, K. Hassanein, and M. Elmasry

On Adaptive Acquisition of Spoken Language   422
A. L. Gorin, S. E. Levinson, L. G. Miller & A. N. Gertner

Vector Quantisation with a Codebook-Excited
Neural Network   432
L. Wu and F. Fallside

Segment-Based Speaker Adaptation by Neural
Network   440
K. Fukuzawa, H. Sawai, and M. Sugiyama

A Simple Word-Recognition Network with the
Ability to Choose Its Own Decision Criteria   452
K. A. Fischer and H. W. Strube

Supervised and Unsupervised Feature Extraction
from a Cochlear Model for Speech Recognition   460
N. Intrator and G. Tajchman

**Part 4: Signal Processing**

A Relaxation Neural Network Model for Optimal
Multi-Level Image Representation   473
by Local-Parallel Computations
N. Sonehara

Lithofacies Determination from Wire-Line Log
Data Using a Distributed Neural Network  483
M. Smith, N. Carmichael, I. Reid & C. Bruce

Improved Structures Based on Neural Networks
for Image Compression  493
S. Carrato, G. Ramponi, A. Premoli, and G. L. Sicuranza

Adaptive Neural Filters  503
L. Yin, J. Astola, and Y. Neuvo

A Surface Reconstruction Neural Network
for Absolute Orientation Problems  513
J. N. Hwang and H. Li

Recursive Neural Networks for Signal Processing
and Control  523
D. Hush, C. Abdallah, and B. Horne

A Neural Architecture for Nonlinear Adaptive
Filtering of Time Series  533
N. Hoffmann and J. Larsen

Ordered Neural Maps and Their Applications
to Data Compression  543
E. A. Riskin, L. E. Atlas, and S. R. Lay

Vector Quantization of Images Using Neural
Networks and Simulated Annealing  552
M. Lech and Y. Hua

A Multilayer Perceptron Feature Extractor
for Reading Sequenced DNA Autoradiograms  562
M. Murdock, N. Cotter, and R. Gesteland

Configuring Stack Filters by the LMS Algorithm  570
N. Ansari, Y. Huang, and J. H. Lin

A Neural Network Pre-Processor for Multi-Tone
Detection and Estimation  580
S. S. Rao and S. Sethuraman

Fuzzy Tracking of Multiple Objects  589
L. I. Perlovsky

**Part 5: System Implementation**

Neural Nets for Signal/Image Processing
Using the Princeton Engine Multiprocessor  595
N. Binenbaum, L. Dias, P. Hsieh, J. Ju, S. Markel,
J. C. Pearson, and H. Taylor, Jr.

Design of a Digital VLSI Neuroprocessor for Signal
and Image Processing  606
C. F. Chang and B. Sheu

Tutorial: Digital Neurocomputing for Signal/Image
Processing  616
S. Y. Kung

**Author Index**  645

# Preface

This book contains papers presented at the IEEE Workshop on Neural Networks for Signal Processing (NNSP-91) at Princeton, New Jersey, USA on September 30 - October 2, 1991. This is the first workshop on the subject sponsored by the IEEE Signal Processing Society, in cooperation with the IEEE Neural Networks Council.

The workshop, organized by the Neural Network Technical Committee of the IEEE Signal Processing Society, is designed to serve as a regular forum for researchers from universities and industry who are interested in interdisciplinary research on neural networks for signal processing applications. In the present scope, the workshop encompasses up-to-date results in several key areas, including learning theory, neural models, speech processing, signal processing, image processing, pattern recognition, and system implementation. This Conference Proceedings is crafted to be an archival reference in the rapidly growing field of Neural Networks for Signal Processing.

Our deep appreciation is extended to Professor Teuvo Kohonen, Helsinki University of Technology, Helsinki, Finland, for his keynote address titled "Workstation-Based Phonetic Typewriter", and to Dr. J. Makhoul, BBN Systems & Technologies, Cambridge, MA., USA, for his keynote address titled "Pattern Recognition Properties of Neural Networks". Our sincere thanks go to all the authors for their timely contributions and to all the members of the Program Committee for the outstanding and high-quality program. Also, we would like to express our gratitude to Dr. John Vlontzos for taking care of the local arrangements, to Dr. Gary Kuhn for providing the workshop publicity, to Dr. Bastiaan Kleijn for handling the tedious finance and registration matters, and to all the session chairs for their help in making the workshop a success. Finally, we are indebted to Ms. Susan Gafgen and Ms. Kim Hegelbach of Princeton University for their invaluable assistance in organizing the workshop.

*B. H. Juang*

*S. Y. Kung*

*Candace A. Kamm*

# Neural Networks for Signal Processing
# Proceedings of the 1991 IEEE Workshop

The chapters in this book are based on presentations given at the IEEE Signal Processing Society Workshop on Neural Networks for Signal Processing held on September 30 - October 2, 1991 at the Nassau Inn at Princeton, New Jersey.

The Technical Program Committee consisted of:

**Rama Chellappa**, University of Maryland

**Bradley Dickinson**, Princeton University

**Tariq Durrani**, University of Strathclyde

**Frank Fallside**, Cambridge University

**Kunihiko Fukushima**, Osaka University

**Lee Giles**, NEC

**Esther Levin**, AT&T Bell Laboratories

**Richard Lippmann**, MIT Lincoln Laboratories

**John Makhoul**, BBN

**Yasuo Matsuyama**, Ibaraki University

**John Moody**, Yale University

**Erkki Oja**, Tokyo Institute of Technology

**Wojtek Pryztula**, Hughes

**Yoh'ichi Tohkura**, ATR Auditory & Visual Perception Research Laboratories

**Christian Wellekens**, L&H Speechproducts

# Part 1:

# Theory and Modeling

# Note on Generalization, Regularization, and Architecture Selection in Nonlinear Learning Systems

## John E. Moody
Department of Computer Science, Yale University
P.O. Box 2158 Yale Station, New Haven, CT 06520-2158
Internet: moody@cs.yale.edu, Phone: (203)432-1200

### Abstract

I propose a new estimate of generalization performance for nonlinear learning systems called the Generalized Prediction Error ($GPE$) which is based upon the notion of the effective number of parameters $p_{eff}(\lambda)$. $GPE$ does not required the use of a test set or computationally intensive cross validation and generalizes previously proposed model selection criteria (such as $GCV$, $FPE$, $AIC$, and $PSE$) in that it is formulated to include biased, nonlinear models (such as back propagation networks) which may incorporate weight decay or other regularizers. The effective number of parameters $p_{eff}(\lambda)$ depends upon the amount of bias and smoothness (as determined by the regularization parameter $\lambda$) in the model, but generally differs from the number of weights $p$. Construction of an optimal architecture thus requires not just finding the weights $\widehat{w}_\lambda^*$ which minimize the training function $U(\lambda, w)$ but also the $\lambda$ which minimizes $GPE(\lambda)$.

## 1 Background and Motivation

Many of the nonlinear learning systems of current interest for adaptive control, adaptive signal processing, and timeseries prediction, are supervised learning systems of the regression type.[1] Estimating the generalization performance of such systems is of crucial importance. Being able to estimate generalization not only allows one to know how good a given model is, but also allows one to select the best among a set of models of various functional forms and varying sizes and allows one to choose optimal regularization (such as weight decay) parameters. We will take the *prediction risk*, which is the expected error of a model in fitting or predicting future data, as our measure of generalization performance.

---

[1] Supervised learning systems can be classified as either regression or classification systems, depending upon whether the response variables are respectively real-valued or categorical. This paper addresses only regression systems.

## 1.1 Learning from Examples

The problem of learning from examples can be defined as follows. Consider a set of real-valued input/output data pairs $\xi = \{\xi^i = (x^i, y^i); i = 1, \ldots, n\}$ generated according to the "signal plus noise" model[2]

$$y^i = \mu(x^i) + \epsilon^i \tag{1}$$

where $y^i$ is the observed response (dependent variable), $x^i$ is the independent variable sampled with probability density $\Omega(x)$, $\epsilon^i$ is independent, identically-distributed (iid) noise sampled with density $\Psi(\epsilon)$ having mean 0 and variance $\sigma^2$,[3] and $\mu(x)$ is an unknown function. The learning problem is then to find an estimate $\hat{\mu}(x)$ of $\mu(x)$ on the basis of the training set $\xi$.

In many real world problems, few *a priori* assumptions can be made about the functional form of $\mu(x)$. Since a parametric function class is usually not known, one must resort to a *nonparametric regression* approach, whereby one constructs an estimate $\hat{\mu}(x) = f(x)$ for $\mu(x)$ from a large class of functions $\mathcal{F}$ known to have good approximation properties (for example, all possible radial basis function networks and multilayer perceptrons). The class of approximation functions usually contains a countable set of subclasses[4] $f(w, x) \subset \mathcal{F}$ for which the elements of each subclass $f(w, x)$ are continuously parametrized by a set of $p$ weights[5] $w = \{w^\alpha; \alpha = 1, \ldots, p\}$. The task of finding the estimate $f(x)$ thus consists of two problems: choosing the best architecture or functional form $f(w, x)$ and choosing the best set of weights $\hat{w}^*$ given the architecture. We shall use estimates of the prediction risk as an objective guide to finding good solutions to both of these problems.

The first problem (finding the architecture) requires a search over possible archiractures (*e.g.* network sizes and topologies), usually starting with small architectures and then considering larger ones. The search is usually not exhaustive and utilizes heuristics to reduce search complexity. For each specific architecture $f(w, x)$, an optimization is done to choose a good set of weights $\hat{w}^*_\lambda$. These weights locally minimize an objective function of the form:

$$U(\lambda, w, \xi) = n\mathcal{E}_{train}(w, \xi) + \lambda S(w) \tag{2}$$

---

[2]The assumption of additive noise $\epsilon$ which is independent of $x$ is a standard assumption and is not overly restrictive. Many other conceivable signal/noise models can be transformed into this form. For example, the multiplicative model $y = \mu(x)(1 + \epsilon)$ becomes $y' = \mu'(x) + \epsilon'$ for the transformed variable $y' = \log(y)$.

[3]Note that we have made only a minimal assumption about the noise $\epsilon$, that it is has finite variance $\sigma^2$ independent of $x$. Specifically, we do not need to make the assumption that the noise has a known density (*e.g.* gaussian) for the following development.

[4]A function subclasses is usually referred to as a network architecture, for example a "fully connected two layer perceptron with five internal units".

[5]The number of weights $p$ varies with the architecture.

Here, $\mathcal{E}_{train}(w,\xi)$ is an error measuring the "distance" between the target response values $y^i$ and the fitted values $f(w,x^i)$:

$$\mathcal{E}_{train}(w,\xi) = \frac{1}{n}\sum_{i=1}^{n}\mathcal{E}[y^i, f(w,x^i)] \,, \tag{3}$$

and $S(w)$ is a regularization or weight-decay function which biases the solution toward functions with *a priori* "desirable" characteristics, such as smoothness. The parameter $\lambda \geq 0$ is the regularization or weight decay parameter and must itself be optimized.[6]

The most familiar example of an objective function uses the squared error[7] $\mathcal{E}[y^i, f(w,x^i)] = [y^i - f(w,x^i)]^2$ and a weight decay term:

$$U(\lambda, w, \xi) = \sum_{i=1}^{n}(y^i - f(w,x^i))^2 + \lambda \sum_{\alpha=1}^{p} g(w^\alpha) \,. \tag{4}$$

The first term is thus the sum of squared errors ($SSE$) of the model $f(w,x)$ with respect to the training data, while the second term penalizes either small, medium, or large weights, depending on the form of $g(w^\alpha)$. Two common examples of weight decay functions are the ridge regression form $g(w^\alpha) = (w^\alpha)^2$ (which penalizes large weights) and the Rumelhart form $g(w^\alpha) = (w^\alpha)^2/[(w^0)^2 + (w^\alpha)^2]$ (which penalizes weights of intermediate values near $w^0$).

An example of a regularizer which is not explicitly a weight decay term is:

$$S(w) = \int_x dx \Omega(x) \|\partial_{xx} f(w,x)\|^2 \,. \tag{5}$$

This is a smoothing term which penalizes functional fits with high curvature.

## 1.2 Prediction Risk, Bias & Variance, and Model Size

With $\hat{\mu}(x)$ is the "learned" estimate of the true regression function $\mu(x)$, then the prediction risk $P$ is the expected error $\mathcal{E}[y(x), \hat{\mu}(x)]$ on future data. This expectation must be taken over all possible training sets of size $n$ and all possible test sets.[8]

---

[6] The optimization of $\lambda$ will be discussed in a later section.

[7] Other error functions, such as those used in generalized linear models (see for example McCullagh and Nelder 1983) or robust statistics (see for example Huber 1981) are more appropriate than the squared error if the noise is known to be non-gaussian or the data contains many outliers.

[8] Recall that since $\hat{\mu}(x)$ is estimated from a finite, noisy training set, it is an implicit function of the random variables $\{\epsilon^i; i = 1, \ldots, n\}$. If we consider all possible $\hat{\mu}(x)$ which can be estimated from all possible randomly-sampled training sets of size $n$, we obtain a conditional probability density characterized by conditional mean $\langle\hat{\mu}(x)\rangle$ and conditional variance $\langle(\hat{\mu}(x) - \langle\hat{\mu}(x)\rangle)^2\rangle$.

We define prediction risk in the context of the "signal plus noise" data model described above. For this model, the joint density $\Xi(\xi)$ for the input/output data pairs $\xi$ is

$$\Xi(x,y) = \Omega(x) \int d\epsilon\, \Psi(\epsilon) \delta[\mu(x) + \epsilon - y] \; . \tag{6}$$

For the squared error $\mathcal{E}[y(x), \widehat{\mu}(x)] = [y(x) - \widehat{\mu}(x)]^2$, the prediction risk is:[9]

$$P = \int dx\, dy\, \Xi(x,y) \, \langle \mathcal{E}[y(x), \widehat{\mu}(x)] \rangle \tag{7}$$

$$= \int dx\, d\epsilon\, \Omega(x) \Psi(\epsilon) \langle [\mu(x) + \epsilon - \widehat{\mu}(x)]^2 \rangle \tag{8}$$

$$= \sigma^2 + \int dx\, \Omega(x) \langle [\mu(x) - \widehat{\mu}(x)]^2 \rangle \; . \tag{9}$$

The last integral measures the expected squared distance between the functions $\widehat{\mu}$ and $\mu$. The residual noise $\sigma^2$ determines the lower bound on the value of $P$.

The prediction risk for the squared error can be further decomposed into three terms:

$$P = \sigma^2 \text{ residual noise} \tag{10}$$

$$+ \int dx\, \Omega(x) [\mu(x) - \langle \widehat{\mu}(x) \rangle]^2 \text{ mean squared bias} \tag{11}$$

$$+ \int dx\, \Omega(x) \langle [\mu(x) - \langle \widehat{\mu}(x) \rangle]^2 \rangle \text{ mean model variance} \; . \tag{12}$$

The mean squared bias is the expected approximation error between $\widehat{\mu}(x)$ and $\mu(x)$, while the mean model variance measures the estimation error, the expected fluctuations in our model due to random sampling of training sets.

Although training error typically decreases with model size $p$, the prediction risk typically first drops and then begins to rise. This is because the mean squared biase tends to drop with model size, while the mean model variance increases with model size. Given a sequence of models with increasing $p$, some model of optimal size will minimize the prediction risk.

## 1.3 Estimates of Prediction Risk

Since the joint density of data pairs $\Xi(x, y)$ is generally not known, the prediction risk $P$ can not be calculated directly and must be estimated from available data.

---

[9]The angled brackets $\langle \rangle$ are used here to denote an expectation value for $\widehat{\mu}(x)$ taken over all possible training sets of size $n$.

The most standard method of estimating prediction risk is test set validation. For $\mathcal{E}[y(x), \hat{\mu}(x)]$ taken to be the squared error, the test set estimate of prediction risk is the average squared error (ASE) on a set of test data:

$$\mathcal{E}_{test} = \frac{1}{n'} \sum_{j=1}^{n'} (y^j - \hat{\mu}(x^j))^2 . \tag{13}$$

Here, the data pairs $\zeta^j = (x^j, y^j)$ are data which were not used in the estimation of $\hat{\mu}(x)$. As the size $n'$ of the test set gets large, the expected value of $\mathcal{E}_{test}$ approaches the prediction risk:

$$\lim_{n' \to \infty} E(\mathcal{E}_{test}) = P . \tag{14}$$

For many important problems, data is scarce or expensive, making test set validation impractical or impossible. In such situations, cross validation[10] (a training sample re-use method) is often used to estimate prediction risk. Cross validation is computationally intensive, however, since it requires training multiple models on data subsets. This limits its usefulness.

Two recent theoretical approaches for analyzing generalization performance for nonlinear neural network models are based on statistical physics [essentially a Bayesian approach] (Tishby et al. 1989, Levin et al. 1990) and VC dimension (see Abu Mostafa 1989) While both of these approaches are extremely interesting, they are not necessarily practical or convenient. The statistical physics approach does not provide a straightforward method of computing the generalization performance for nonlinear models. Indeed, Levin et al. are unable to compute the generalization performance exactly using their methods even for a linear regression model.[11] Similarly, the determination of the VC dimension required for the VC estimates is in general difficult, if not impossible. Even if the VC dimension for a problem is known, the VC results provide only worst-case bounds on the sample sizes required to obtain good generalization. These worst-case bounds are useless for most practical situations.

---

[10] Although cross validation has been reinvented many times over the last five decades or so, the first published description can be found in Mosteller and Tukey (1954); see Stone (1978) for a relatively recent review. For an illustrative application in the context of neural nets, see Moody and Utans (1991) and Utans and Moody (1991) who use the method to select among various network architectures for a financial application.

[11] The exact expression for prediction risk in linear regression is a standard result in statistics, while the calculation leading to equation (23) of Levin et al. (1990) is only approximate. The statistical physics approach becomes even more unmanageable for nonlinear problems: in section 4B of Levin et al., the authors abandon their approach entirely and resort to conventional cross validation for estimating prediction risk.

Ideally, one would like a simple, algebraic formula for estimating prediction risk, so that a test set is not needed and cross validation can be avoided. The statistical physics and VC approaches have not provided this.

However, the main-stream statistics literature contains a number of reliable and computationally cheap estimates of prediction risk for linear models and unbiased nonlinear models. These estimates can be expressed in the general form:

Prediction Risk Estimate = Training Set Error + Model Complexity Penalty    (15)

These estimates include Mallow's (1973) $CP$, the generalized cross-validation $(GCV)$ formula (Craven and Wahba 1979), Akaike's (1970) final prediction error $(FPE)$, and Akaike's (1973) information criterion $(AIC)$.[12] A generalization of these estimates with better statistical properties is Barron's (1984) predicted squared error $(PSE)$. However, no extension of these types of estimates to general *biased* nonlinear systems (such as those often used in neural networks and other nonparametric regression models) has previously been formulated.

My proposed estimate of prediction risk is called the *Generalized Prediction Error* $(GPE)$ which is valid for general nonlinear learning systems which may be biased and may include weight decay or other regularizers. Many of the nonparametric learning systems of current interest (particularly neural networks) fall into this category.

## 2   The Generalized Prediction Error Estimate

In this short paper, there is not space to present a derivation of GPE or examples of its empirical use. The derivation and examples will be presented in a longer paper (Moody 1991). This section thus serves as a cursory overview of GPE.

The prediction risk $P$, and therefore the form of $GPE$, depends on the model complexity as measured by $p_{eff}(\lambda)$, the *effective number of parameters* in the system. If we denote the training error for a data set of size $n$ by $\mathcal{E}_{train}(n)$ (for example, the average squared error $(ASE)$ or some other error measure) then the prediction risk is:

$$P(\lambda) = E(\mathcal{E}_{train}(n)) + 2\sigma^2 \frac{p_{eff}(\lambda)}{n} \qquad (16)$$

Note that this expression is of the general form (15), but differs from previous estimates (such as $AIC$, $PSE$, etc.) due to the appearance of $p_{eff}$ rather than $p$. $GPE$ estimates (16) as follows:

$$GPE(\lambda) = \widehat{P}_{GPE} = \mathcal{E}_{train}(n) + 2\widehat{\sigma}^2 \frac{\widehat{p}_{eff}(\lambda)}{n} \qquad (17)$$

---

[12] See also Akaike (1974).

The regularization parameter $\lambda$ controls the effective number of parameters $p_{eff}(\lambda)$ of the solution. Generally, as $p_{eff}(\lambda)$ is increased, the training set error decreases (eventually leading to overfitting), but the model complexity term in $P$ and $GPE$ increases to compensate. For a given problem, there is an optimal value of $\lambda > 0$ which balances the training set error and model complexity term and thus minimizes $GPE$.

Thus, the approach to learning which optimizes generalization performance requires optimizing $\lambda$ in addition to finding the best weights $\widehat{w}_\lambda$ for a given $\lambda$.

## 2.1 Computation of GPE

The computation of $GPE$ according to (17) requires the estimation of $p_{eff}(\lambda)$ and $\sigma^2$.

The form of $p_{eff}$ is:

$$p_{eff}(\lambda) \equiv \operatorname{tr} G(\lambda) \qquad (18)$$

Here, $G(\lambda)$ is the *generalized influence matrix*, which is not directly observable, but can be estimated. An estimate for $G$ is:

$$\widehat{G}^{ij}(\lambda) = \frac{n}{2}(\mathcal{E}_{train})_{i\alpha}(\frac{1}{n}U(\lambda)_{\alpha\beta})^{-1}(\mathcal{E}_{train})_{\beta j} \; . \qquad (19)$$

Here, subscripts denote derivatives ( $(\mathcal{E}_{train})_{i\alpha} \equiv \partial^2(\mathcal{E}_{train})/\partial y^i \partial w^\alpha$, and so on) and there is an implied summation over repeated indexes on the right hand side (this is the Einstein summation convention).

For an objective function of the form (4), evaluating $\widehat{G}$ requires computation of the following quantities:

The residuals: $r^i = y^i - f(\widehat{w}_\lambda^*, x^i)$.

The gradient matrix: $F_{i\alpha} = \partial f(\widehat{w}_\lambda^*, x^i)/\partial w^\alpha$.

The bias-weighted curvature matrix: $C_{\alpha\beta} = \frac{1}{n}\sum_{i=1}^{n} r^i \partial^2 f(\widehat{w}_\lambda^*, x^i)/\partial w^\alpha \partial w^\beta$.

The diagonal weight decay hessian: $S_{\alpha\alpha} = \partial^2 g(\widehat{w}_\lambda^{\alpha*})/(\partial w^\alpha)^2$ .

All quantities are easily computed from the model $f$ and the training data $\xi$. This ease of computation extends to other, more general objective functions (2).

Several estimates of $\sigma^2$ are available, the most standard being:

$$\widehat{\sigma}^2 = \frac{n}{n-p} ASE(w_\lambda^*) \; . \qquad (20)$$

This estimator is unbiased for linear models and unbiased nonlinear models without regularization or weight decay. An estimator which is more accurate for biased nonlinear models is

$$\widehat{\sigma}^2 = \frac{n}{n-\widehat{p}_{eff}(\lambda)} ASE(w_\lambda^*) \; . \qquad (21)$$

For this choice, the expression for $GPE$ becomes

$$GPE(\lambda) = \frac{n + \widehat{p}_{eff}(\lambda)}{n - \widehat{p}_{eff}(\lambda)} ASE(w_\lambda^*) \ . \tag{22}$$

Note that this form of $GPE$ is simplified in certain limits. If the model is unbiased or the model is linear, then the bias-weighted curvature $C_{\alpha\beta}$ is zero. If in addition, weight decay is not used ($\lambda = 0$), then $p_{eff} = p$ and $GPE$ reduces to Akaike's $FPE$:

$$FPE = \frac{n + p}{n - p} ASE(w_\lambda^*) \ , \tag{23}$$

If regularization or weight decay is used, or if the model is nonlinear or biased, then the value of $GPE$ may be either larger or smaller than in (23).

As emphasized by Barron (1984), however, biased estimators for $\sigma^2$ which do not depend upon the $ASE$ for the model at hand can offer lower variance in the estimate of prediction risk. Hence, it is desirable to to use an estimate such as

$$\widehat{\sigma}^2 = \frac{n}{n - \widehat{p}_{eff}(\text{smaller model})} ASE(\text{smaller model}) \ . \tag{24}$$

obtained by fitting a smaller model to the data.

## 2.2  Sample Size Bounds and Architecture Selection

An issue of critical concern for constructing a model for a given application problem is to know how much data $n$ will be required to attain good generalization for a fixed architecture $f(w, x)$ with a number of parameters $p$. The general form (17) of $GPE$ provides this information explicitly. The *mean sample size* $n$ required to achieve a desired *mean relative excess prediction error* $\epsilon \equiv [GPE(\lambda) - \mathcal{E}_{train}(w_\lambda^*)]/\widehat{\sigma}^2$ is:

$$n \approx 2\frac{\widehat{p}_{eff}(\lambda)}{\epsilon} \ . \tag{25}$$

Note that unlike the VC worst-case bounds, this result contains no logarithmic factors or large coefficients.

More commonly, however, the network designer has a fixed dataset and varies the architecture. $GPE$ is extremely useful as an objective guide for the following architecture selection problems:

1. Choosing between various classes of models such as multilayer perceptrons, projection pursuit models, or radial basis function networks for a specific problem.

2. Determining the optimal number of internal units in a network architecture.

3. Determining which input variables to remove from a network.

4. Determining how many weights to remove using the optimal brain damage procedure.

5. Finding the best value of the regularization coefficient $\lambda$.

An application of using prediction risk estimates for architecture selection problems 1-4 is presented in Moody and Utans (1991) and Utans and Moody (1991). They use both cross validation and FPE (a special case of GPE) to determine an optimal architecture for corporate bond rating.

## 2.3 Weight Decay and Architecture Selection

In the neural network community, the most commonly used regularizers are weight decay functions. The use of weight decay is motivated by the intuitive notion that it removes unnecessary weights from the model. An analysis of $GPE$ with weight decay ($\lambda > 0$) confirms this intuitive notion. Furthermore, some amount of weight decay generally lowers the generalization error. This is because weight decay methods yield models with lower model variance, even though they are biased.

For nonlinear unbiased or linear models with quadratic weight decay $g(w^\alpha) = (w^\alpha)^2$ (this is the ridge regression form), the exact expression for $p_{eff}(\lambda)$ is

$$p_{eff}(\lambda) = \sum_{\alpha=1}^{p} \frac{\kappa^\alpha}{\kappa^\alpha + \lambda} , \qquad (26)$$

Here, $\kappa_\alpha$ is the $\alpha^{th}$ eigenvalue of the $p \times p$ matrix $K = F^T F$, with $F$ the gradient matrix defined above.

The form of $p_{eff}(\lambda)$ can be computed easily for other weight decay functions, such as the Rumelhart form $g(w^\alpha) = (w^\alpha)^2/[(w^0)^2+(w^\alpha)^2]$. The basic result for all weight decay functions, however, is that $p_{eff}(\lambda)$ is a decreasing function of $\lambda$ with $p_{eff}(0) = p$ and $p_{eff}(\infty) = 0$, as is evident in the special case (26). If the model is nonlinear and biased, then $p_{eff}(0)$ generally differs from $p$.

Regardless of the form of the regularizer $S(w)$, an optimal value of $\lambda$ can be found. This is done by minimizing $GPE(\lambda)$ with respect to $\lambda$. This is a constrained optimization problem, since $GPE(\lambda)$ must be evaluated with weight values $\widehat{w}_\lambda^* = \text{argmin } U(\lambda, w, \xi)$ .

## References

Y. Abu Mostafa. (1989) The Vapnik-Chervonenkis Dimension: Information versus Complexity in Learning. *Neural Computation* 1:312-317.

H. Akaike. (1970) Statistical predictor identification. *Ann. Inst. Statist. Math.*, **22**:203-217.

H. Akaike. (1973) Information theory and an extension of the maximum likelihood principle. In *2nd International Symposium on Information Theory*, Akademia Kiado, Budapest, 267-281.

H. Akaike. (1974) A new look at the statistical model identification. *IEEE Trans. Auto. Control*, **19**:716-723.

A. Barron. (1984) Predicted squared error: a criterion for automatic model selection. In *Self-Organizing Methods in Modeling*, S. Farlow, ed., Marcel Dekker, New York.

P. Craven and G. Wahba. (1979) Smoothing noisy data with spline functions: estimating the correct degree of smoothing by the method of generalized cross-validation. *Numer. Math.*, **31**:377-403.

P. J. Huber. (1981) *Robust Statistics*. Wiley, New York.

E. Levin, N. Tishby, and S. Solla. (1990) A statistical approach to learning and generalization in layered neural networks. *Proceedings of the IEEE*, **78**10:1568-1574.

P. McCullagh and J.A. Nelder. (1983) *Generalized Linear Models*. Chapman and Hall, New York.

J. Moody. (1991) Generalization, Regularization, and Architecture Selection in Nonlinear Learning Systems. In preparation.

J. Moody and J. Utans. (1991) Manuscript in preparation.

F. Mosteller and J. W. Tukey. (1954) Data analysis, including statistics. In G. Lindzey and E. Aronson, editors, *Handbook of Social Psychology, Vol. 2*. Addison-Wesley, 1968 (first edition 1954).

J. Rissanen. (1989) *Stochastic Complexity in Statistical Inquiry*. World Scientific, Singapore.

M. Stone. (1978) Cross-validation: A review. *Math. Operationsforsch. Statist., Ser. Statistics*, 9(1).

N. Tishby, E. Levin, and S. Solla.(1989) Consistent Inference of Probabilities in Layered Networks: Predictions and Generalization. *International Joint Conference on Neural Networks*, IEEE Press, Piscataway, NJ.

J. Utans and J. Moody. (1991) Selecting Neural Network Architectures via the Prediction Risk: Application to Corporate Bond Rating Prediction. *First International Conference on Artificial Intelligence Applications and Wall Street*. IEEE Computer Society Press, Los Alamitos, CA.

# DISCRIMINATIVE MULTI-LAYER FEED-FORWARD NETWORKS

Shigeru Katagiri*, Chin-Hui Lee** and Biing-Hwang Juang**

*ATR Auditory and Visual
Perception Research Laboratories
Sanpeidani, Inuidani
Seika-cho, Soraku-gun
Kyoto 619-02, Japan

**AT&T Bell Laboratories
600 Mountain Avenue
Murray Hill
NJ 07974, USA

Abstract - In this paper, we propose a new family of multi-layer, feed-forward network (FFN) architectures. This framework allows us to examine several feed-forward networks, including the well-known *multi-layer perceptron* (MLP) *network*, the *likelihood network* (LNET) and the *distance network* (DNET), in a unified manner. We then introduce a novel formulation which embeds network parameters into a functional form of the classifier design objective so that the network's parameters can be adjusted by gradient search algorithms, such as the *generalized probabilistic descent* (GPD) method. We evaluate several discriminative three-layer networks by performing a pattern classification task. We demonstrate that the performance of a network can be significantly improved when discriminative formulations are incorporated into the design of the pattern classification networks.

## 1. INTRODUCTION

In the past few years, several algorithms based on learning vector quantization (LVQ) have been successfully applied to a number of classifier designs [1-2]. In fact, it has been shown that even a simple *multi-reference distance classifier* trained with LVQ can achieve very high discriminative power, mostly equivalent to that by a classifier based on an MLP [2-3]. Furthermore, several other interesting classifiers having a three-layer structure have been independently proposed; e.g., a *radial-basis function network* (e.g. [4-5]). The structural similarity of these networks has been investigated by one of the authors [6]. However due to the lack of theoretical foundations in formulating discriminative rules, a number of questions related to learning algorithms and convergence properties of these algorithms remain unanswered.

In this paper, we propose a new family of multi-layer FFN architectures. This framework allows us to examine several feed-forward networks, including the LNET, the DNET, and the well-known MLP network, in a unified manner. We then introduce a novel formulation which embeds network parameters into a functional form of the classifier design objective so that the parameters can be adjusted to increase the network's discriminative capability using the GPD method [6-7]. Based on this formulation, the remaining questions on theoretical justification of discriminative networks can be addressed. We also evaluate several discriminative three-layer networks by performing a common pattern classification task. Experimental results demonstrate that the performance of a network can be improved when discriminative formulations are incorporated into the design of the classification networks.

## 2. MULTI-LAYER FEED-FORWARD NETWORKS

Consider an FFN structure with $K$ output nodes (at the $N$-th layer, or the output layer), $M$ input nodes (at the first layer, or the input layer) and ($N$-2) hidden layers. There are $S_n$ nodes on the $n$-th layer. We denote the $j$-th node at the $n$-th layer by $\Psi_{nj}$, and the input and output at $\Psi_{nj}$ by $i_{nj}$ and $o_{nj}$, respectively. We also assume that $i_{nj}$ is a function of the parameter vector $\underline{\lambda}_{nj}$ associated with node $\Psi_{nj}$ and the output vector $\mathbf{o}_{n-1}$ from the immediately lower layer, i.e., $i_{nj} = h_{nj}(\underline{\lambda}_{nj}, \mathbf{o}_{n-1})$. We call $h_{nj}(\cdot)$ the *input function* at node $\Psi_{nj}$. We also assume that the output $o_{nj}$ is a function of only $i_{nj}$, i.e., $o_{nj} = f_{nj}(i_{nj})$. We call $f_{nj}(\cdot)$ the *output function* at node $\Psi_{nj}$. In Figure 1, we show the computation performed at the node $\Psi_{nj}$. In Figure 2, we illustrate a simple three-layer network.

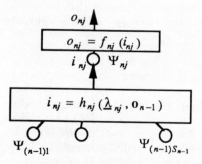

Figure 1. The computation at a node.

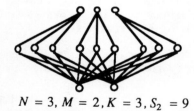

$N = 3, M = 2, K = 3, S_2 = 9$

Figure 2. An example of a partially connected network. The hidden nodes grouped by connection are assigned to one class.

In the remainder of this section, we describe three FFN examples, namely the popular MLP network, the likelihood network and the distance network [6].

### 2.1 Multi-layer Perceptron Networks

A multi-layer perceptron network is defined as follows. We assume here that $\underline{\lambda}_{nj} = \mathbf{w}_{nj}$ with $w_{njl}$ being the connection weight for the connection between the $j$-th node of the $n$-th layer and the $l$-th node of the ($n$-1)-th layer.

The output function is usually a sigmoid function (e.g. [3]) and the input function $h_{nj}$ is a simple dot product of $\lambda_{nj}$ and $\mathbf{o}_{n-1}$, i.e., $i_{nj} = \mathbf{w}_{nj} \cdot \mathbf{o}_{n-1}$.

## 2.2 Likelihood Networks

It was noted that a three-layer network structure is adequate in representing most of the essential characteristics of FFN's [8]. Moreover, the three-layer distance network was shown to have a close structural relation to an LVQ classifier [6]. In this light, we introduce a likelihood network having the three-layer structure. In a three-layer LNET, the input function at each hidden node is defined as the likelihood $G_l(\mathbf{x}|\lambda_{2l})$ of observing the input vector $\mathbf{x} = \mathbf{o}_1$. We also assume that, at each node $\Psi_{3j}$ on the output (3rd) layer, a conditional class probability of the form of a mixture distribution is evaluated, i.e.,

$$o_{3j} = p_\Lambda(\mathbf{x}|C_j) = \sum_l w_{3jl} G_l(\mathbf{x}|\lambda_{2l}) \qquad (1)$$

where $w_{3jl}$ is the mixture weight assigned to the connection between node $\Psi_{3j}$ and node $\Psi_{2l}$. We often impose the constraint $\sum_l w_{3jl} = 1$ to make the network output satisfy the probability constraints. Depending on the network structure, $G_l(\cdot)$ can be of any distribution function. The most popular choice is the Gaussian kernel. In that sense, the conditional class distribution is similar to the output of either the classical Parzen density estimation algorithm (e.g., [9]) or the regularization network [5].

Different connection strategies between the output layer and the hidden layer lead to a variety of LNET's. We here assume that the connection weight is zero if the corresponding nodes are not connected. Three examples are worth noting. The first is a fully-connected network which implies that a common set of kernels at the hidden layer is shared by all the output nodes. This is referred to as the *tied-mixture connection*. The second type of connection assumes that each hidden node is connected to only one output node. Therefore each conditional class probability is characterized by an independent set of kernels. We refer to this strategy as the *independent mixture connection*. The third type of connection assumes that the set of output nodes are partitioned into a fixed number of groups. The set of hidden nodes are also partitioned into the same number of groups. Only the output nodes and hidden nodes with the same group label are full connected. We call this the *group-dependent tied-mixture connection*. Similar strategies have also been applied to estimating parameters of hidden Markov models (HMM's) for context-dependent phone modeling. The discussed three strategies are similar to three types of HMM frameworks, namely the tied-mixture HMM (also known as a semi-continuous HMM, e.g [10]), the mixture Gaussian density HMM [11] and the phone-dependent tied-mixture HMM approach [11]. The connection and modeling strategies are closely related to efficiency of training and the capacity of the network or of the models.

On top of the LNET, a fourth layer can also be added to compute, at each class output, the mutual information $I(\mathbf{x}, C_j)$ between the input vector $\mathbf{x}$ and the class $C_j$, i.e.

$$I(\mathbf{x}, C_j) = \frac{p_\Lambda(\mathbf{x}|C_j)}{\sum_i p_\Lambda(\mathbf{x}|C_i) \cdot p(C_i)} \qquad (2)$$

We call this network a *mutual information network* (MINET).

## 2.3 Distance Networks

The above LNET appears to be an ideal choice for classifier design because the output of the network evaluates the conditional class probability which can readily be used to achieve the minimum Bayes risk in classification. However in practice, we are usually given only a small number of training samples. Under such circumstance, not all the network parameters can be reliably estimated. Many simplified likelihood-based distances, such as *Maharanobis distance* and *Euclidean distance*, have also been used successfully. Therefore the above LNET can be modified to perform classification using a weighted distance. We refer to this network, original proposed in [4], as a distance network (DNET). Consider a three-layer network. At each hidden node $\Psi_{2q}$, a distance, between the input vector x and the parameter $\underline{\lambda}_{2q}$ characterizing the node, is computed, i.e.,

$$i_{3j} = \sum_{q=1}^{S_2} w_{3jq} o_{2q} \text{ and } i_{2q} = h_{2q}(\underline{\lambda}_{2q}, \mathbf{x}) = D_{2q}(\underline{\lambda}_{2q}, \mathbf{x}) \qquad (3)$$

where $D_{2q}$ is a general distance function with a parameter $\underline{\lambda}_{2q}$. Since the network outputs are no longer probabilities, the constraints on the weighting vectors such as $\sum_l w_{3jl} = 1$ can be removed in the training process.

## 3. DISCRIMINATIVE FFN'S

We show in this section how to incorporate discriminative training formulations [8-9] into the networks we discussed in Section 2. We first present a GPD method which allows us to perform training of network parameters.

### 3.1 Generalized Probabilistic Descent Method

GPD is a general adaptive learning algorithm applicable to a variety of classifier designs including the design of classification networks. There are two key concepts contribute to the effectiveness of GPD formulation, namely: (1) the classification factors are entirely embedded in a smooth objective function suitable for numerical optimization based on existing gradient search methods; and (2) *dynamic* (of variable and unspecified dimension) patterns as well as *static* (of fixed dimension) patterns can be handled properly. In this study, we focus our discussion on network structures for static pattern classification.

Consider the task of designing a classifier to classify a given input pattern $\mathbf{x} \in \Re^M$ into one of K competing classes $\{C_j; j = 1, \ldots, K\}$. A classifier consists of a set of parameters $\Lambda = \{\underline{\lambda}_j; j = 1,...,K\}$ and a classification rule $C(\cdot)$ which operates on the input pattern and produces a class label as a decision. Assume that we are given a set of training samples $\mathbf{X} = \{\mathbf{x}_1,...,\mathbf{x}_R\}$. The purpose of training is to find a classifier parameter $\Lambda$ and a classification rule, based on $\mathbf{X}$, so that the error on classifying all tokens in $\mathbf{X}$ is minimized.

To embed the classifier structure into a function form, a three-step procedure is required in the GPD formulation. First, we define a *discriminant function* $g_j(\mathbf{x}; \Lambda)$ to measure the degree to which the input pattern $\mathbf{x}$ belongs to class $C_j$. We then define a *misclassification measure* $d_k(\mathbf{x}; \Lambda)$ which measures the distance between the distance between the correct class and other competing classes for a given input $\mathbf{x} \in C_k$. Finally, we define a *loss (cost) function* $\ell_k(\mathbf{x}; \Lambda) = \ell_k(d_k(\mathbf{x}; \Lambda))$ which is usually a smooth 0-1 function. The overall average cost (or error count) is then approximated as

$$L(\Lambda) = \frac{1}{N} \sum_i \sum_k \ell_k(d_k(\mathbf{x}_i; \Lambda)) 1(\mathbf{x}_i \in C_k). \quad (4)$$

There are many possible combinations of selections of the functions and parameters which lead to a wide variety of implementation examples. The GPD adjustment rule then defines [6-7]

$$\delta \Lambda(\mathbf{x}, C_k, \Lambda) = -\varepsilon \mathbf{U} \nabla \ell_k(\mathbf{x}; \Lambda) = -\varepsilon \mathbf{U} \nabla \ell'_k(\mathbf{x}; \Lambda) \frac{\partial d_k(\mathbf{x}; \Lambda)}{\partial \Lambda} \quad (5)$$

where $\mathbf{x}$ is the training vector for adaptation, $\varepsilon$ is a small positive number for controlling the adjustment step size and $\mathbf{U}$ is a positive-definite matrix. The above rule is applied to all the training samples sequentially over and over until convergence. In practice, a small number of iterations (or epochs) is enough to give a good solution $\Lambda$. We then use estimated $\Lambda$ in the following classification rule,

$$C(\mathbf{x}) = C_i \quad \text{if} \quad i = \underset{j}{\arg\max}\{g_j(\mathbf{x}; \Lambda)\} \quad \text{or} \quad i = \underset{j}{\arg\min}\{g_j(\mathbf{x}; \Lambda)\}. \quad (6)$$

According to the GPD formulation, the adjustment rule in Eq. (5) results in a classifier Eq. (6) that achieves the minimum error in a probabilistic sense [6-7].

## 3.2 Discriminative Multi-Layer FFN's

Using the above discriminative training formulation we can derive GPD-based adjustment rules for the multi-layer FFN's discussed in Section 2.1. At the $n$-th layer, define the following three network adjustment vectors

$$\underline{\alpha}_{nl} = \frac{\partial i_{nl}}{\partial \underline{\lambda}_{nl}}, \quad \underline{\beta}_{nl} = \frac{\partial i_{nl}}{\partial o_{n-1}} \quad \text{and} \quad \underline{\gamma}_{nj} = \frac{\partial d_j(\mathbf{x}; \Lambda)}{\partial i_n} \quad (7)$$

then it can be shown that $\gamma_{njl}$, l-th element of $\underline{\gamma}_{nj}$, can be computed recursively,

$$\gamma_{njl} = \left[\sum_{q=1}^{S_{n+1}} \gamma_{(n+1)jq}\beta_{(n+1)ql}\right] f'_{nl}(i_{nl}), \qquad 1 \le n < N. \tag{8}$$

The adjustment rule in Eq. (5) can now be evaluated as follows

$$\frac{\partial d_j(\mathbf{x};\Lambda)}{\partial \underline{\lambda}_{nl}} = \gamma_{njl}\underline{\alpha}_{nl}, \qquad 1 \le n \le N. \tag{9}$$

In order to evaluate Eq. (8), we need the adjustment factors at the $N$-th layer which depends on the definitions of the discriminant function and the misclassification measure. In this Section, we assume that the discriminant function is defined as the network output at node $\psi_{Nj}$, i.e., $g_j(\mathbf{x};\Lambda) = o_{Nj}$.

For a training token $\mathbf{x} \in C_k$, we also assume the following three types of misclassification measures for MLP networks, LNET's and DNET's respectively, i.e.

$$d_k(\mathbf{x};\Lambda) = -g_k(\mathbf{x};\Lambda) + \left[\frac{1}{K-1}\sum_{i,i \ne k}\{g_i(\mathbf{x};\Lambda)\}^\eta\right]^{1/\eta}, \tag{10}$$

$$d_k(\mathbf{x};\Lambda) = -g_k(\mathbf{x};\Lambda) + \ln\left[\frac{1}{K-1}\sum_{i,i \ne k}\exp\{\eta g_i(\mathbf{x};\Lambda)\}\right]^{1/\eta}, \tag{11}$$

$$d_k(\mathbf{x};\Lambda) = g_k(\mathbf{x};\Lambda) - \left[\frac{1}{K-1}\sum_{i,i \ne k}\{g_i(\mathbf{x};\Lambda)\}^{-\eta}\right]^{-1/\eta}. \tag{12}$$

Now, $\gamma_{Nkj}$ is evaluated as follows. For $j = k$,

$$\gamma_{Nkj} = \begin{cases} -f'_{Nj}(i_{Nj}), & \text{for MLP and LNET,} \\ f'_{Nj}(i_{Nj}), & \text{for DNET.} \end{cases} \tag{13}$$

For $j \ne k$, there are three possibilities, i.e.

$$\gamma_{Nkj} = \begin{cases} \dfrac{1}{K-1}\left[\dfrac{1}{K-1}\sum_{i,i \ne k}\left\{\dfrac{g_i(\mathbf{x};\Lambda)}{g_j(\mathbf{x};\Lambda)}\right\}^\eta\right]^{\frac{1}{\eta}-1} f'_{Nj}(i_{Nj}), & \text{for MLP,} \\ \left[\sum_{i,i \ne k}\exp\{\eta g_i(\mathbf{x};\Lambda) - \eta g_j(\mathbf{x};\Lambda)\}\right]^{-1} f'_{Nj}(i_{Nj}), & \text{for LNET,} \\ -\dfrac{1}{K-1}\left[\dfrac{1}{K-1}\sum_{i,i \ne k}\left\{\dfrac{g_i(\mathbf{x};\Lambda)}{g_j(\mathbf{x};\Lambda)}\right\}^{-\eta}\right]^{-\frac{1}{\eta}-1} f'_{Nj}(i_{Nj}), & \text{for DNET.} \end{cases} \tag{14}$$

## 3.3 Discriminative MLP's, LNET's and DNET's

For MLP networks, the input function is simply the inner product, therefore the two vectors $\underline{\alpha}_{nl}$ and $\underline{\beta}_{nl}$ can be easily computed.

$$\underline{\alpha}_{nl} = \mathbf{0}_{n-1} \text{ and } \underline{\beta}_{nl} = \mathbf{w}_{nl} \tag{15}$$

and the adjustment rules in Eqs. (7)-(15) can now be used to adjusted parameters (weights) of the discriminative MLP networks.

For designing discriminative three-layer LNET's, the reader is referred to [7] for a detailed description. It is noted that the weights in LNET's need to satisfy the probability constraint discussed in Section 2.2. One way to accomplish this is to normalize the weights after they are adjusted. It can be shown [6-7] that the GPD convergence properties can still be preserved.

We now discuss discriminative DNET's. Using the notations in Section 2.3, we can now evaluate the three vectors in Eq.(7). Assume that

$$\Lambda = \left\{ (\mathbf{w}_{3j}, \mathbf{m}_{2q}); j = 1, \ldots, K, q = 1, \ldots, S_2 \right\}. \tag{16}$$

Then,

$$\underline{\alpha}_{3j} = \frac{\partial i_{3j}}{\partial \mathbf{w}_{3j}} = \mathbf{0}_2 \text{ and } \underline{\beta}_{3j} = \frac{\partial i_{3j}}{\partial \mathbf{o}_2} = \mathbf{w}_{3j}. \tag{17}$$

$\gamma_{3j}$ is evaluated in the same way as in Eqs. (8), (13), and (14).

If the Euclidean distance at each hidden node is used, then

$$\underline{\alpha}_{2q} = 2(\mathbf{m}_{2q} - \mathbf{x}) \text{ and } \underline{\beta}_{2q} = -2(\mathbf{m}_{2q} - \mathbf{x}). \tag{18}$$

Again Eqs. (8), (13) and (14) can be used to evaluate $\underline{\gamma}_{2j}$.

## 3.4 Relations with LVQ Classifiers

Considering using a three-layer DNET to implement a multi-reference distance classifier [4]. Assume each output node for class $C_j$ is connected to a set $Q_j$ of $L_j$ hidden nodes. This is similar to the group-dependent connection strategy we discussed in Section 2.2. We also assume that the input function at each output node is defined by

$$i_{3j} = \left\{ \frac{1}{L_j} \sum_{i \in Q_j} (o_{2i})^{-\xi} \right\}^{-1/\xi} \tag{19}$$

where $\xi$ is a positive real number. Then there is no parameter on the third layer, and the network adjustment vectors are evaluated as

$$\underline{\alpha}_{3j} = \frac{\partial i_{3j}}{\partial w_{3j}} = 0 \quad \text{and} \quad \beta_{3jl} = \frac{1}{L_j}\left\{\frac{1}{L_j}\Sigma\left(\frac{o_{2i}}{o_{2l}}\right)^{-\xi}\right\}^{\frac{1}{\xi}-1}. \qquad (20)$$

We can now use Eqs. (8) and (9) to compute the adjustment factors, i.e.,

$$\frac{\partial d_k(\mathbf{x};\Lambda)}{\partial \mathbf{m}_{2q}} = 2\gamma_{2kq}(\mathbf{m}_{2q}-\mathbf{x}) = 2\left[\sum_{l=1}^{K}\gamma_{3kl}\beta_{3lq}\right]f'_{2q}(i_{2q})(\mathbf{m}_{2q}-\mathbf{x}). \qquad (21)$$

If we let both $\eta \to \infty$ and $\xi \to \infty$ and let the output function at all nodes be equal to the identity function, i.e., $f'_{nj}(\cdot) = 1$ for each node $\Psi_{nj}$, then

$$\begin{cases}\gamma_{3kj} = -1, & \text{for } j = k, \\ \gamma_{3kj} = \delta(j-i) & i = \underset{r \neq k}{\operatorname{argmin}}(o_{3r}), \quad \text{for } j \neq k.\end{cases} \qquad (22)$$

Also

$$\beta_{3lq} = \delta(q-t), \quad t = \underset{p \in Q_l}{\operatorname{argmin}}(o_{2p}) \qquad (23)$$

$$\begin{cases}\frac{\partial d_k(\mathbf{x};\Lambda)}{\partial \mathbf{m}_{2q}} = -2(\mathbf{m}_{2q}-\mathbf{x}) & \text{if } q \in Q_k \text{ and } q = \underset{p \in Q_k}{\operatorname{argmin}}(o_{2p}) \\ \frac{\partial d_k(\mathbf{x};\Lambda)}{\partial \mathbf{m}_{2q}} = 2(\mathbf{m}_{2q}-\mathbf{x}) & \text{if } j \neq k, j = \underset{r \neq k}{\operatorname{argmin}}(o_{3r}) \text{ and } q = \underset{p \in Q_j}{\operatorname{argmin}}(o_{2p}) \\ 0 & \text{otherwise}\end{cases} \qquad (24)$$

Now, if we assume the cost function $\ell_k(\mathbf{x};\Lambda) = d_k(\mathbf{x};\Lambda)$ and also use a likelihood-based distance in place of the Euclidean distance evaluated at the hidden nodes, we have a training algorithm similar to the modified LVQ2-L algorithm proposed by two of the authors [13]. If we define the loss function

$$\ell_k(d_k(\mathbf{x};\Lambda)) = \begin{cases}1 & \theta_1 < d_k(\mathbf{x};\Lambda) \\ d_k(\mathbf{x};\Lambda) & \theta_2 \leq d_k(\mathbf{x};\Lambda) \leq \theta_1 \\ 0 & d_k(\mathbf{x};\Lambda) < \theta_2\end{cases} \qquad (25)$$

where $\theta_1$ and $\theta_2$ are positive constants, each near 0, then

$$\ell_k'(d_k(\mathbf{x};\Lambda)) = \begin{cases}1, & \theta_2 \leq d_k(\mathbf{x};\Lambda) \leq \theta_1, \\ 0, & \text{otherwise.}\end{cases} \qquad (26)$$

This special case corresponds to the modified LVQ2 proposed in [14], in which the adjustment is made only when the training vector falls within a small window region near the corresponding reference vectors.

## 4. EXPERIMENTS

To show the capability of the proposed networks, we conducted a set of experiments using the following network structures: E1) a linear classifier trained with the perceptron cost (e.g. [15]); E2) a conventional three-layer MLPN trained with the minimum mean squared error (MMSE); E3) a three-layer discriminative (GPD-trained) MLPN; E4) a three-layer discriminative DNET; E5) an LVQ-E (using the Euclidean distance), and E6) an LVQ-L (using the uncorrelated Gaussian likelihood distance) [13]. The data set used for evaluation was the famous Fisher's data consisting of the three-class iris tokens, each token being of four dimensions (e.g., see [16]). Fifty training tokens were used for each class. Independent testing data were not prepared.

In the cases of E2) through E4), we used the fully connected three-layer structure of FFN's, each having fifteen hidden nodes. The LVQ-trained multi-reference distance classifiers in E5) and E6) used five references per class. We also performed four training runs for each classifier, and evaluated the average recognition rates of these four runs. Each training consisted of 100 epochs. Table 1 shows the average recognition rates. The results very clearly demonstrate the high discriminative power of the proposed DNET and its simplified versions, the LVQ systems. It also shows that the significant improvement of GPD training over the popular MMSE training (E2 vs. E3).

| E1) linear classifier + perceptron; | 86.0% |
|---|---|
| E2) 3-layer MLPN + MMSE; | 89.5% |
| E3) 3-layer MLPN + GPD; | 97.8% |
| E4) 3-layer DNET + GPD; | 98.7% |
| E5) LVQ-E; | 98.7% |
| E6) LVQ-L; | 100.0% |

TABLE 1. RECOGNITION RATES ON THE IRIS DATA.

## 5. SUMMARY

We have proposed a new family of multi-layer feed-forward networks. The network has a structure similar to that of the conventional multi-layer perceptron. The difference lies in that this network uses a more general input function at each of the network nodes to replace the inner product computation at each MLP node. Therefore, gradient based search methods, such as the error back-propagation algorithm, can still be used to design a wide variety of networks, e.g. the proposed likelihood networks and the distance networks discussed earlier. We have also introduced a new discriminative training formulation to embed the network structure into a functional form of design objective so that a numerical optimization can be carried out directly. The proposed GPD method is then applied to adjust the network parameters based on a given set of training data.

For pattern classification problems, the learned parameters can be used to design classifiers which achieve a minimum error rate on the training data in a probabilistic sense. We perform a common classification task using Fisher's iris data set to test the effectiveness of the proposed networks and the discriminative training formulations. The results clearly demonstrated that the newly proposed DNET's are more effective than the conventional MLP networks. When comparing the same network structure, such as the MLP, the experimental results also showed that the proposed discriminative training algorithm, based on the minimum error objective, performed much better compared to the conventional algorithm based on the MMSE optimization criterion.

So far, we have only addressed the issue of designing static feed-forward networks. Our future work includes investigation of dynamic FFN's and classification of dynamic patterns using static and dynamic FFN's.

## REFERENCES

[1] T. Kohonen; The self-organizing map, IEEE Proc., pp.1464-1480 (1990. 9).
[2] E. McDermott, and S. Katagiri; Shift-invariant, multi-category phoneme recognition using Kohonen's LVQ, IEEE, ICASSP, vol. 1, pp.81-84 (1989. 5).
[3] A. Waibel, et. al; Phoneme recognition: Neural networks vs. hidden Markov models, IEEE, ICASSP, vol. 1, pp.107-110 (1988. 4).
[4] S. Katagiri; Systematic explanation of learning vector quantization and multi-layer perceptron, IEICE, MBE88-72, (1988. 10) (in Japanese).
[5] T. Poggio, and F. Girosi; Networks for approximation and learning, IEEE Proc., pp.1481-1497 (1990. 9).
[6] S. Katagiri, C.-H. Lee, B.-H. Juang; A generalized probabilistic descent method, ASJ Fall Meeting, (1990. 9).
[7] S. Katagiri, C.-H. Lee,, and B.-H. Juang; New discriminative training algorithms based on the generalized probabilistic descent method, to appear in Neural networks for signal processing - IEEE-SP workshop - (1991. 9).
[8] K. Funahashi; On the approximate realization of continuous mappings by neural networks, Neural Networks, vol. 2, pp.183-192, Pergamon Press (1989).
[9] K. Fukunaga; Introduction to statistical pattern recognition, Academic Press, pp.166-177 (1972).
[10] X.D. Hwang, et. al; "Improved hidden Markov modeling for speaker independent continuous speech recognition," DARPA Workshop (1990. 6).
[11] C.-H. Lee, et. al; "Acoustic modeling for large vocabulary speech recognition," Computer Speech and Language, Vol.4, pp.127-165 (1990).
[12] B.-H. Juang, and S. Katagiri; Discriminative learning for minimum error classification, in preparation.
[13] S. Katagiri, and C.-H. Lee; A new HMM/LVQ hybrid algorithm for speech recognition, IEEE, GLOBECOM, 608.2, pp.1032-1036 (1990. 12).
[14] E. McDermott; LVQ3 for phoneme recognition, ASJ, Spring Conf., pp.151-152 (1990 3).
[15] R. Duda, and P. Hart; Pattern classification and scene analysis, A Wiley-interscience publication, pp.130-188 (1973).
[16] M. James; Classification algorithms, John Wiley & Sons (1985).

# EFFICIENT TRAINING PROCEDURES FOR ADAPTIVE KERNEL CLASSIFIERS

Srinivasa V. Chakravarthy and Joydeep Ghosh
*Department of Electrical and Computer Engineering*
*The University of Texas at Austin*
*Austin, TX 78712-1084*

Larry Deuser and Steven Beck
*Tracor Applied Sciences Inc.*
*6500 Tracor Lane, Austin, TX 78725*

Abstract - We investigate two training schemes for adapting the locations and receptive field widths of the centroids in Radial Basis Function Classifiers. The Adaptive Kernel Classifier is able to adjust the responses of the hidden units during training using an extension of the Delta rule, thus leading to improved performance and reduced network size. The Rapid Kernel Classifier, on the other hand, uses the faster Learned Vector Quantization algorithm to adapt the centroids. This network shows a remarkable reduction in training time with little compromise in accuracy. The performance of these two networks is evaluated using underwater acoustic transient signals.

## BACKGROUND AND MOTIVATION

Neural network approaches to problems in the field of pattern recognition and signal processing have led to the development of various static "neural" classifiers using feed-forward networks [14]. These include the Multi-Layer Perceptron (MLP) as well as kernel based classifiers such as those employing Radial Basis Functions (RBFs) [2,15]. A second group of neural-like schemes such as Learning Vector Quantization (LVQ) have also gained considerable attention [13]. These are adaptive, exemplar-based classifiers that are closer in spirit to the classical K-nearest neighbor method.

The strength of both groups of classifiers lies in their applicability to problems involving arbitrary distributions. Most neural network classifiers do not require simultaneous availability of all training data and frequently yield error rates comparable to Bayesian methods without needing *apriori* information. Moreover, it has been recently shown that training these networks using a mean square error criterion and a 0/1 teaching function yields network outputs that approximate *aposteriori* class probabilities [7]. A good review of static classifiers discussing the relative merit of various schemes is available in [12,14].

We observe that LVQ and its variants such as LVQ2, LVQ2.1 and "conscience learning" [4] need somewhat less training time than an MLP-based classifier for comparable performance [12]. The memory requirements are similar but their performance is more sensitive to initial choice of reference vectors. Networks using RBFs, on the other hand need much shorter training times at the expense of additional memory as compared to the MLP. These networks are primarily aimed at multivariate function interpolation or function approximation, and have been used successfully for problems such as prediction of chaotic time series [15]. They serve as universal aproximators using only a single hidden layer [11]. However, they can also be used for classification. For example, Niranjan and Fallside were able to achieve good results on voice and digit speech categorization by using one "centroid" for each training vector [16]. The results are robust with respect to variations in the class distributions.

This motivates us to investigate hybrid networks that combine the best features of LVQ and RBF based classifiers so that an accurate classifier is obtained that requires less training time and is not memory intensive. We now summarize the LVQ and RBF procedures and then introduce the two hybrid networks.

LVQ is an adaptive version of the classical Vector Quantization algorithm whose aim is to represent a set of input vectors by a smaller set of codebook vectors or reference vectors (RVs) so as to minimize an error functional. Typically the algorithm consists of the following steps: The RVs are initialized by a random selection from, or by performing K-means clustering on the training set. These vectors are now adjusted iteratively by moving them closer to or further away from training inputs depending on whether the closest RV is of the same class as the input vector or not. Unknown inputs are assigned to the class of the nearest RV.

Let the n$th$ training pattern vector x(n) belong to class $C_r$, and $m_c$, the reference vector closest to x(n), belong to class $C_s$. At the presentation of the input the RVs are adapted as follows.

$$m_c(n+1) = m_c(n) + \alpha(n) [x(n) - m_c(n)] \quad \text{if} \quad C_r = C_s$$

$$m_c(n+1) = m_c(n) - \alpha(n) [x(n) - m_c(n)] \quad \text{if} \quad C_r \neq C_s$$

$$m_i(n+1) = m_i(n) \quad \text{for all other RV's} \qquad (0)$$

where $\alpha(n)$ is a learning factor that decreases monotonically with time. Details of LVQ-type learning procedures can be found in [4,5,13].

RBF networks are a class of single hidden-layer feedforward networks in which radially symmetric basis functions are used as the activation functions for the hidden layer units. The output of the $j$th hidden/ kernel node, $R_j(x)$, and that of $i$ th output node, $f_i(x)$ are given by:

$$f_i(x) = \sum_j w_{ij} R_j(x) \qquad (1)$$

$$R_j(x) = R(\| x - x_j \| / \sigma_j) \qquad (2)$$

where $R(x)$ is a radially symmetric function such as a Gaussian. We call $x_j$ the $j$ th centroid, $\sigma_j$ its width and, $w_{ij}$ the weight connecting the $j$ th kernel/hidden node to the $i$ th output node.

## HYBRID KERNEL CLASSIFIERS

Both LVQ and RBF involve construction of a representative set of the training data - the centroids of the RBF network and the reference vectors of LVQ - which determine the final decision. In [15] the centroids in an RBF network were determined using heuristics such as performing k-means clustering on the input set, and widths were held fixed during training. Alternatively, we can vary both the centroid locations and associated widths of the receptive fields by performing gradient descent on the mean square error of the output. This leads to the Adaptive Kernel Classifier (AKC).

### Adaptive Kernel Classifier

Consider a quadratic error function, $E = \Sigma E_p$ where
$$E_p = 1/2 \sum_i (t_i - f_i(x_p))^2 .$$
Here $t_i$ is the target function for the $p$ th input, $x_p$, and $f_i(.)$ is as defined in (1). For the $p$ th pattern the update rules for various network parameters can be derived by performing gradient descent on the expected mean square error, $E_p$, as follows:

$$\Delta w_{ij} = \eta_1 (t_i - f_i(x)) R_j(x_p) \qquad (3)$$

$$\Delta x_{jk} = \eta_2 R_j(x_p) ((x_{pk} - x_{jk})/\sigma_j^2) (\sum_i (t_i - f_i(x_p)).w_{ij}) \qquad (4)$$

$$\Delta \sigma_j = \eta_3 (\| x_p - x_j \|^2 / \sigma_j^3) R_j(x_p) (\sum_i (t_i - f_i(x_p)).w_{ij}) \qquad (5)$$

where $\| . \|$ denotes the Euclidean norm, and $R(.)$ is a Gaussian. These constitute the learning scheme for the Adaptive Kernel Classifier.

A similar scheme, called the *Gaussian Potential Function Network* (GPFN), which involves segmentation of the input domain into several potential fields in form of Gaussians, was proposed by Lee and Kil [10]. The Gaussian potential functions of this scheme need not be radially symmetric functions. Instead, the sizes of these potential fields are determined by a *correlation matrix*. The network parameters are computed by gradient descent as in the case of AKC. More general schemes like the *regularization networks* have been studied by Poggio et al. [18]. Though more complex decision regions can be shaped out of potential fields that are not radially symmetric, receptive fields of radial functions can achieve universal approximation even if each kernel node has the same smoothing factor, $\sigma$ [17].

The AKC is able to perform the same level of approximation as RBF using fewer hidden units. However, training time is increased since centroids and center widths are also adapted using the generalized delta rule, which is a slower procedure.

## Rapid Kernel Classifier

Hybrid schemes which combine unsupervised and supervised learning in a single network have been proposed in [8], [9]. For instance, the *hierarchical feature map classifier* of Huang and Lippmann [9] consists of a feature map stage followed by a stage of adaptive weights. The central idea behind these approaches is to have a layer that is trained in an unsupervised way followed by a layer that will be trained using the delta rule. Since backpropagation of error through multiple layers is avoided these hybrid networks yield remarkable speedups in computation. Such methods are particularly useful when there are a large number of training samples. It must be noted that these hybrid schemes are not optimized in the manner of backpropagation since the first stage parameters are not optimized with respect to the output performance. A question that immediately arises is: how does the location of centroids obtained by hybrid training compare with that obtained by strict gradient descent?

Keeping in view the similarities in form between the equations describing update of centroids by the delta rule and the LVQ algorithm, one can replace the former equation with the latter in the AKC training scheme. This results in the Rapid Kernel Classifier (RKC) which requires shorter training times with little change in performance as compared to the Adaptive Kernel Classifier. In the hybrid procedures mentioned above various layers of the network are trained sequentially. The first stage parameters are first trained in an unsupervised way and are held fixed during the second stage training. On the contrary, in the RKC scheme we let the LVQ algorithm run in parallel with the training of the second layer.

The distinguishing features of some of the training schemes introduced are summarized below:

| Type | Network Parameters | | |
|------|----------|------------|--------|
|      | Centroid | Weights    | Widths |
| RBF  | K-means/fixed | Delta rule | Fixed |
| AKC  | Delta rule | Delta rule | Delta rule |
| RKC  | LVQ | Delta rule | Delta rule |
| GPFN | Delta rule | Delta rule | Delta rule (vectorized widths) |

Interestingly, in all our experiments so far with RKC, the mean square error decreases monotonically as in the case when all the parameters are adapted by backpropagation. This indicates that adjusting the centroids using LVQ might amount to performing gradient descent on the "centroid-location space". While proving this seems difficult, we are able to derive a result for a simplified version of the RBF, which is similar to a characterization of the Self-Organizing Feature Map of Kohonen. To this end, consider a situation where the RBF network is trained to map a one-dimensional input $X$, of a known distribution $P(x)$, onto a constant $a$. Let the weights of all links from the kernel nodes to the output be held at $w$, and each centroid have a constant width σ. Moreover, let the centroids of the network be adapted as in eqn.(4). The aim is to study the relation between distribution of the centroids thus obtained and that of the input. A formula for the distribution of the centroids for the instance described above can be derived in the form of the differential equation:

$$\frac{dg}{dy} = P(g)^{-2/3} \exp(-\frac{2}{3}\gamma y)  \tag{6}$$

where $y$ is a continuous variable representing the kernel node index, $g(y)$ is the centroid at $y$, and $\gamma$ is the ratio of the constant weight, $w$ and the target output $a$. Equation (6) can be solved for $g(.)$ if the form of $P(x)$ is known. For example, let $X$ come from a uniform distribution lying between 0 and $c$. Then $g(y)$ falls exponentially from the origin to $c$ and vanishes for all other $y$. That is, node density is lesser at higher magnitudes of the centroid since in this case $dy/dg$ is inversely proportional to $g$ between 0 and $c$. A similar result for the density of the output nodes in Kohonen's Feature Map has been derived by Ritter and Schulten [RiSc86]. It is shown that, for the one-dimensional case, the density of the output nodes of the Feature Map is proportional to $P(x)^{2/3}$. On comparing our formula with the result of [19] it is apparent that for the special situation described above, the distribution of the centroids approaches that of the Feature Map as a limit, i.e. when $\gamma$ goes to zero. Moreover, if an input distribution lies within a range around the origin, the RBF network would require fewer number of hidden nodes compared to the Feature Map for the same level of representation.

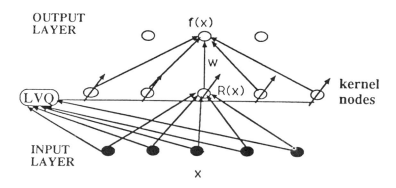

Figure 1: A Rapid Kernel Classifier.

## DATA DESCRIPTION AND REPRESENTATION

The various kernel classifiers have been evaluated for their ability to classify short duration acoustic signals. These signals are extracted from DARPA Data Set I which consists of digital time records of a training and a test set of 6 signal types propagated over short oceanic data paths with variable source and background noise levels. Only those portions of the records that contain the signals are used. As shown in Table 1, the 6 classes are denoted by the letters A through F with durations from a few msec to 8 seconds, and bandwidths from less that 1 Hz to several kHz. The signal to noise ratios of the records varies over 24 dB. There were 42 training samples. Of the test data, 179 samples were similar enough to the training data to constitute a good test set for classification comparisons. An additional 19 test samples were added selectively to observe, without retraining, the robustness of the techniques to other deterministic signals. These 19 additional records had duration and frequency characteristics that deviated significantly from the training data. Overall, the data included simultaneously occurring transient signals, sequentially occurring signals requiring event association and also signals that are similar to one of the training samples but with one prominently missing feature.

TABLE 1: DESCRIPTION OF DARPA DATA SET I

| | Signal Classes | Number of Samples | | |
|---|---|---|---|---|
| Class | Description | Training | Test 179 | Additional Test 19 Deviant Signals |
| A | Broadband, 15 msec pulse | 7 | 53 | 8 |
| B | Two 4 msec pulses, 27 msec separation | 7 | 54 | 6 |
| C | 3 kHz tonal, 10 msec duration | 8 | 31 | 1 |
| D | 3 kHz tonal, 100 msec duration | 9 | 14 | 0 |
| E | 150 Hz tonal, 1 sec duration | 6 | 19 | 4 |
| F | 250 Hz tonal, 8 sec duration | 5 | 8 | 0 |
| | TOTALS | 42 | 179 | 19 |

It is observed in our previous study [1,6] that discriminant parameters obtained using wavelet transforms yield better performance than using autoreggressive coefficients or spectral coefficients. A constant Q prototype wavelet with 24 coefficients is used, in addition to time duration to characterize the signals in our current work. The particular wavelet transform used represents a signal $X(t)$ by shifted and dilated versions of an analyzing waveform [3].

$$T_x(\tau, a) = a^{-m/2} \int_{-\infty}^{\infty} x(t) \, h^* \left( \frac{t - \tau}{a^m} \right) dt$$

where the prototype wavelet is given by

$$h(t) = a^{-m/2} \, h\left( \frac{t - \tau}{a^m} \right)$$

Here $a^m$ is a scaling factor and * indicates the complex conjugate. The choice of the most appropriate prototype function is still an open issue. To date, octave frequency spacing has proved quite successful. Besides the wavelet coefficients, signal duration is also used as an important classification parameter. In fact, it is the only discriminant between classes C and D.

## PERFORMANCE EVALUATION

Both AKC and RKC have been used to train an RBF network with data extracted from the preprocessed underwater signals described above. The performance of these two schemes is compared with RBF network trained in the traditional way. The classification results are given in Figure 2 in the form of confusion matrices. The correct class is given by horizontal row label and only misclassifications are entered. The primed labels represent the results for the 19 deviant signals. For the 179 regular test signals, the accuracy is from 98.9% (RBF and RKC) to 100% (AKC), while for the deviant signals, only 4 of the 19 samples got labels in agreement with the provided ground truth for all three networks.

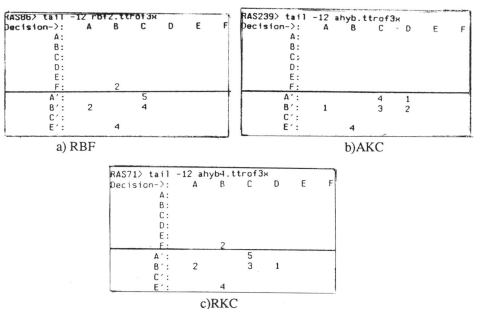

Figure 2: Confusion matrices showing classification results for a) RBF, b) Adaptive Kernel Classifier, and c) Rapid Kernel Classifier. The top half of each display shows the confusion matrix of mis-classifications for the 179 regular signals. The primed class labels shown at the bottom half of each display denote additional errors due to the 19 deviant signals.

We note that the AKC is able to classify all regular test signals correctly (100%). The drastic contrast in performance for the additional 19 test signals indicates that the neural nets were sharply tuned to the training set and not amenable to grossly deviant signals. This conclusion is further reinforced by observing that the cell outputs have very low values for most of the deviant signals.

Besides classification accuracy, training and testing times and memory requirements are also of concern, particularly for real-time implementations. The AKC took the longest training time, while the RKC was quicker than AKC while providing comparable results. More accurate timing estimates will require careful separation of CPU and input/output times, and is part of ongoing research.

## DISCUSSION

We have conducted some experiments comparing classification capabilities of a two-layer MLP with the AKC and RKC. In all cases, the training time for AKC is much shorter than that for the MLP, and RKC is much faster than AKC. The reason for this is easily seen. There is no good initialization procedure for the MLP, whereas for the AKC, first stage weights can be initialized by simply selecting representatives from the input set.

One of the curious features of the RKC is that for all our experiments, the MSE has been observed to decrease monotonically as if the network were performing strict gradient descent. Oscillations or other anomalous behavior have not been observed. While we are yet to fully explain this phenomenon analytically, the results presented in the second section indicate a deeper connection between adaptation of centroids using a self-organizing scheme, and adaptation based on gradient-descent on the mean squared error. We are presently studying how well the distributions constructed by the hybrid algorithms comply with the Bayesian philosophy.

**Acknowledgements**: This research was supported by contract N00014-89-C-0298 with Dr.Thomas McKenna (ONR) and Dr. Barbara Yoon (DARPA) as Government cognizants.

REFERENCES:
[1] Beck, S., Deuser, R., Still, R. & Whiteley, J., "A Hybrid Neural Network Classifier of Short Duration Acoustic Signals", *Proc. IJCNN*, Seattle, July 1991.
[2] Broomhead, D.S. & Lowe, D., "Multivariable Functional Interpolation and Adaptive Networks," *Complex Systems*, 2, 1988, pp. 321-355.
[3] Combes, J.M., Grossman, A. & Tchamitchian, Ph. (Eds.), *Wavelets: Time-Frequency Methods and Phase Space*, Springer Verlag, 1989.
[4] DeSieno, D., "Adding Conscience to Competitive Learning", *IEEE Int'l Conf. on Neural Networks*, 1117-1124, 1988.
[5] Geva, S. & Sitte, J., "Adaptive Nearest Neighbor Classification", *IEEE Trans. Neural Networks*, Vol.2, No. 2, pp 318-322, 1991.
[6] Ghosh, J., Deuser, L. & Beck, S., "Impact of Feature Vector Selection on Static Classification of Acoustic Transient Signals", *Govt. Neural Network Application Workshop - NOSC*, San Diego, CA, August 29, 1990.

[7] Gish, H., "A probablistic approach to understanding and training of neural network classifiers", *Proc. Int'l Conf. on ASSP*, Albuquerque, NM, April 1990, pp. 1361-64.

[8] Hecht-Nielsen, R., " Counterpropagation Networks", *Applied Optics* **26**, pp. 4979-4984.

[9] Huang, W.Y. & Lippmann, R.P., " Neural Network and Traditional Classifiers", In *Neural Information Processing Systems* ( Denver 1987), ed. D.Z.Anderson, 387-396. New York: Amer. Inst. of Physics.

[10] Lee, S. & Kil, R., "Multilayer Feedforward Potential Function Network", Proc. IJCNN, San Diego.

[11] Kowalski, J., Hartman, E. & and Keeler, J., "Layered Neural Networks with Gaussian Hidden Units as Universal Approximators", *Neural Computation*, 1990, 2, pp. 210-215.

[12] Kohonen, T., Barna, G., & Chrisley, R., "Statistical Pattern Recognition with Neural Networks: Benchmarking Studies," *IEEE Annual Int'l Conf on Neural Networks*, San Diego, July 1988.

[13] Kohonen, T., *Self-Organization and Associative Memory*, 3rd ed., Berlin, Heidelberg : Springer-Verlag, 1989.

[14] Lippmann, R.P., "Pattern Classification using Neural Networks", *IEEE Communications Magazine*, November 1989, pp. 47-64.

[15] Moody, J. & Darken, C., " Fast Learning in Networks of Locally-Tuned Processing Units", *Neural Computation*, V.1, No.2, pp. 281-294.

[16] Niranjan, M., & Fallside, F., "Neural Networks and Radial Basis Functions in Classifying Static Speech Patterns", *Tech. Rep. CUED/FINFENG/TR22*, Cambridge University Engineering Dept., 1988.

[17] Park, J. & Sandberg, I., "Universal Approximation Using Radial Basis Function Networks", *Neural Computation*, Vol. **3**, No. 2, Summer 1991, pp. 246-257.

[18] Poggio, T. & Girosi, F., "Regularization Algorithms for Learning that are equivalent to Multilayer Networks", *Science* **247**, 978-982, 1990.

[19] Ritter, H. & Schulten, K., "On the Stationary State of Kohonen's Self-Organizing Sensory Mapping", *Biological Cybernetics* **54**, pp. 99-106.

# CONCEPT FORMATION AND STATISTICAL LEARNING IN NONHOMOGENEOUS NEURAL NETS

Richard L. Tutwiler          Leon H. Sibul
Pennsylvania State University
Applied Research Laboratory
Post Office Box 30
State College, PA 16804 USA
Phone: (814) 863-2188
Fax: (814) 863-7841
Email: rlt@psuarl.bitnet

Abstract: This paper presents an analysis of complex nonhomogeneous neural nets, an adaptive statistical learning algorithm, and the potential use of these types of systems to perform a general sensor fusion problem. The three main contributions of this paper are the following. First, an extension to the theory of *Statistical Neurodynamics* is introduced to include the analysis of complex nonhomogeneous neuron pools consisting of three subnets. Second, a statistical learning algorithm is developed based on the differential geometric theory of statistical inference for the adaptive updating of the synaptic interconnection weights. The statistical learning algorithm is merged with the subnets of nonhomogeneous nets and it is shown how these ensembles of nets can be applied to solve a general sensor fusion problem.

## STATISTICAL NEURODYNAMICS

The theory of *Statistical Neurodynamics* developed by Amari [1, 2] treats an ensemble of neural nets as a probability space whose sample space is a function of the dimensionality of the synaptic interconnection matrix and the threshold vector. Therefore, the weight matrix and the threshold vector are treated as random variables. The probability measure is defined on the Borel sets of the space. In our case we assume that they are both normally distributed and i.i.d. A complex net is composed of a series of simple nets which are called subnets. Amari's theory treats these networks from a macroscopic point of view. Our analysis derives the state equations describing the dynamic behavior of the complex net consisting of three subnets. Figure 1 illustrates a general block diagram of the complex net consisting of three subnets. The *microstate transition law* for the nets is given by

$$u_i = \sum_{j=1}^{n} w_{ij} x_j - h_i \quad (1)$$

$$i = 1,...,n$$

$$\tilde{X}_i = sgn(u_i) \quad (2)$$

The *state transition operator* is defined by

$$\tilde{X}_i = T_\omega X_i = sgn(u_i) \quad (3)$$

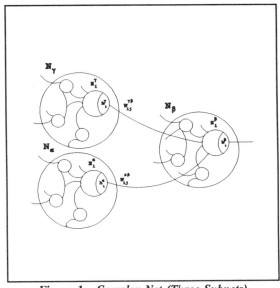

**Figure 1.** *Complex Net (Three Subnets)*

$T_\omega$ is a nonlinear operator depending on $\omega$ that maps a state to its next state. If S = {set of all states}, then $T_\omega : S \to S$. The *activity level* of the net yields the *macrostate transition equation*

$$Z = a(X) = \frac{1}{n} \sum_{i=1}^{n} \tilde{X}_i = \frac{1}{n} \sum_{i=1}^{n} sgn(u_i), -1 \leq Z \leq 1 \quad (4)$$

The explicit form of the macroscopic transition function is expressed through the ensemble mean $\varphi(Z)$, which for n large (n = number of neurons) maps to

$$\varphi(Z) = \Phi(WZ - H)$$

$$W = \frac{nE\{w_{ij}\}}{[n\sigma_w^2 + \sigma_h^2]^{1/2}}, \quad H = \frac{E\{h_i\}}{[n\sigma_w^2 + \sigma_h^2]^{1/2}} \quad (5)$$

with

$$\Phi(u) = 2 \int_0^u \frac{1}{\sqrt{2\pi}} e^{-v^2/2} \, dv \quad (6)$$

Referring to Figure 1 (and letting $\alpha = 1$, $\gamma = 2$, $\beta = 3$) the *equilibrium* macrostate transition equations for the three subnet case are given by

$$Z_1 = \varphi_1(Z_1,Z_2,Z_3) = \Phi(W_{11}Z_1 + W_{12}Z_2 + W_{13}Z_3 - H_1) = \Phi(\eta) = 0$$
$$Z_2 = \varphi_2(Z_1,Z_2,Z_3) = \Phi(W_{21}Z_1 + W_{22}Z_2 + W_{23}Z_3 - H_2) = \Phi(\phi) = 0 \quad (7)$$
$$Z_3 = \varphi_3(Z_1,Z_2,Z_3) = \Phi(W_{31}Z_1 + W_{32}Z_2 + W_{33}Z_3 - H_3) = \Phi(\psi) = 0$$

The stability of the equilibrium states is conducted by evaluating the generalized Jacobian matrix J at the solutions of the simultaneous equilibrium equations given above. The generalized Jacobian is given by

$$J = \sqrt{\frac{2}{\pi}} \begin{bmatrix} W_{11} e^{-\eta} & W_{21} e^{-\eta} & W_{31} e^{-\eta} \\ W_{21} e^{-\phi} & W_{22} e^{-\phi} & W_{32} e^{-\phi} \\ W_{31} e^{-\psi} & W_{23} e^{-\psi} & W_{33} e^{-\psi} \end{bmatrix} \quad (8)$$

Decomposing J by standard eigensystem analysis yields

$$J = U\Lambda U^T$$
$$|\lambda_i| > 1 \rightarrow Z_i \text{ is UNSTABLE} \quad (9)$$
$$|\lambda_i| \leq 1 \rightarrow Z_i \text{ is STABLE}$$

## STATISTICAL LEARNING

The statistical learning algorithm [3, 4, 5, 6] is essentially a nonlinear recursive density estimation procedure. Differential geometry of statistical inference treats the probability space as a statistical manifold. The objective of the estimation procedure is to project the true density orthogonally onto a prespecified approximation family. The differential geometric analogy is the alignment of parallel tangent spaces. The approximating density is chosen such that the projection of the tangent vectors coincides. When this is accomplished the divergence also known as the Kullback-Leibler distance is minimized. The main goal is to continuously adapt the interconnection matrix to train the network to identify specific conditional statistical estimates.

The assumption taken is that the n-dimensional manifold includes all the possible posterior densities together. We consider the family of normal distributions $N(\mu, \sigma^2)$, the family of densities given by

$$p(x; \mu, \sigma) = \frac{1}{\sqrt{2\pi\sigma^2}} e^{-\{\frac{(x-\mu)^2}{2\sigma^2}\}} \qquad (10)$$

with respect to the Lebesque measure on $\mathbb{R}$. The *Univariate Gaussian Manifold* is defined as

$$\mathcal{F} = \{p(\theta; w) | W \in \mathbb{R}^n\} \qquad (11)$$

The submanifolds of approximating distributions are given by

$$A = \{p(\theta; s) | s \in \mathbb{R}^n\} \subset \mathcal{F} \qquad (12)$$

If $W = (w_1, \ldots, w_N)$ is a coordinate system in $\mathcal{F}$ and $s = (s_1, \ldots, s_M)$ is a coordinate system in $A$, a relationship between coordinate systems is defined by the mapping $W = w(s)$. If the mapping is *differentiable* and regular, then $A$ is *embedded* in $\mathcal{F}$. Therefore, the submanifolds are *foliations* of $\mathcal{F}$ as shown in Figure 2a. where

$$\mathcal{F} = \bigcup_{u \in A} A(u) \qquad (13)$$

A *foliation* is a partition of $\mathcal{F}$ into submanifolds $A(u)$ of dimension N-M. The tangent space of $\mathcal{F}$ at point $p$ is a vector space obtained by local linearization around $p$. This space is composed of tangent vectors of all smooth curves passing through $p$. The *natural basis* of the tangent space $T_w$, is formed by $\frac{\partial}{\partial W_i}$ ($i = 1, N$), as illustrated in Figure 2b. The concepts of *affine connection* and *geodesic*, shown in Figure 2c, specifies the way of transferring the tangent space over the manifold $\mathcal{F}$. The *affine connection* describes the relationship between adjacent tangent spaces. In addition, if the tangent vector is displaced parallelly along the curve $c$ (shown in Figure 2c), the curve forms a *geodesic* with respect to the *affine connection*. A detailed explanation of these concepts is beyond the scope of this paper and the interested reader is referred to references [3]-[6]. The principle of approximation defines geodesics connecting; (1) the prior density $p(\theta) \in A$ and the true posterior density $p(\theta;w_x) \in \mathcal{F}$, and (2), the prior density $p(\theta) \in A$ and the approximating density $p(\theta;\hat{s}) \in A$.. The approximating density $p(\theta;\hat{s})$ is chosen such that the *projection* of the tangent vectors of the *geodesics* defined by (1) and (2) onto the tangent space of $A$ at point $p(\theta)$ coincide as illustrated in Fig. 2d.

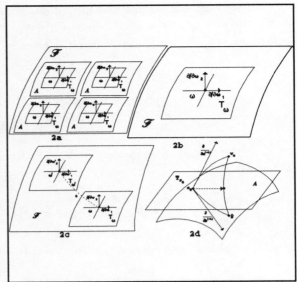

Figure 2. *Manifold Illustrations*

The *Parallel Projection Algorithm (PPA)*, shown in Figure 3, places a grid over both $\mathcal{F}$ and all *foliations* of $\mathcal{F}$. The projection algorithm is computed in parallel and the decision network searches for the minimum projection function which is the true *orthogonal projection* of $A$ onto $\mathcal{F}$. The calculations performed in parallel at each processing stage are the following

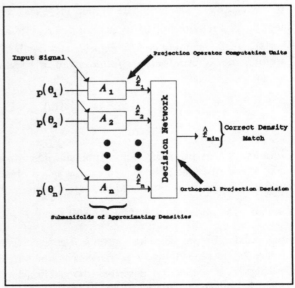

Figure 3. *Parallel Projection Algorithm*

**Divergence Measure**

$$D = \int p(\theta;s) \ln\left(\frac{p(\theta;\hat{s})}{p(\theta;w_x)}\right) d\lambda(\theta) \tag{14}$$

## Inner Product Calculation of Tangent Vectors

$$\left\langle \frac{\partial}{\partial b^{(+1)}}, \frac{\partial}{\partial s_i} \right\rangle = E\left\{ \left( \ell n \left( \frac{p(\theta;\hat{s})}{p(\theta)} \right) \right) \left( \frac{\partial}{\partial s_i} \ell n(p(\theta;s)) \right) \right\} \quad (15)$$

$$\left\langle \frac{\partial}{\partial a^{(+1)}}, \frac{\partial}{\partial s_i} \right\rangle = E\left\{ \left( \ell n \left( \frac{p(\theta;w_x)}{p(\theta)} \right) \right) \left( \frac{\partial}{\partial s_i} \ell n(p(\theta;s)) \right) \right\}$$

## Angle between the Tangent Vectors

$$\cos(\alpha) = \frac{\frac{\partial}{\partial b^{(+1)}} \cdot \frac{\partial}{\partial a^{(+1)}}}{\left| \frac{\partial}{\partial b^{(+1)}} \right| \left| \frac{\partial}{\partial a^{(+1)}} \right|} \quad (16)$$

Therefore, at convergence the following conditions would hold i.) $D = 0$ in equation (14), ii) The inner product calculations in equation (15) would be equal, and (iii) $\cos(\alpha) = 1$ in equation (16).

## Classification Procedure

The objective behind the recognition scheme is to train each subnet to recognize specific segments of the distribution function as shown in Figure 4.

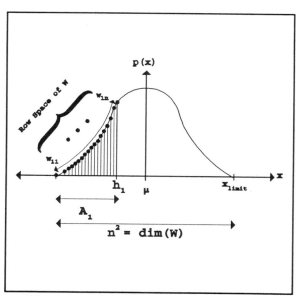

Assuming a density estimate has been calculated according to the *PPA* described in the previous section, the synaptic weights are given by the quantized levels of the density function. The number of quantization levels is proportioned to the dimension of the synaptic interconnection matrix. Given that $dim(\underline{W}) \sim n^2$, choose $x_{limit}$ such that $p(x) \approx 0$.

**Figure 4.** *Density Matching Functions*

To accomplish this set equation (10) equal to $\epsilon$ and solve for $x_{limit}$. The value of $\epsilon$ would be equal to the machine precision of the hardware used. The stepsize $\Delta$ for the weight increment is given by $\Delta = \dfrac{|2 x_{limit}|}{n^2}$. Knowing these values $\underline{W}$ can be calculated and mapped to the neuron plane. This implies that the row space of $\underline{W}$ corresponds to specific segments of $p(x)$ needed to be classified. Since $\sum_{j=1}^{n} w_{ij} x_j$ approximates the area under the curve of $p(x)$, the thresholds $h_i$'s can be precomputed based on standard probability theory. From Figure 4

$$h_1 = P_r(X \le x_1) = A_1 = \int_{-\infty}^{x_1} f(t)\, dt = \int_{-\infty}^{x_1} \dfrac{1}{\sqrt{2\pi\sigma^2}} e^{-\left\{\dfrac{(x-\mu)^2}{2\sigma^2}\right\}} dx$$

where the value $x_1$ corresponds to the value of $x$ used to calculate $w_{1n}$. Therefore, knowing $n$, $x_{limit}$, $\Delta$, $\underline{w}$, one can precompute the $h_i$'s and index them via a lookup table for each subnet. Correct classification implies that the macrostate activity level given by equation (4) will be equal to one.

## APPLICATION TO GENERAL SENSOR FUSION

The objective of fusion in estimation theory is to combine various estimates to obtain a good estimate. The basic concept related in [7] is that there exists a Borel measurable function that performs the mapping $g: \mathbb{R}^n \to \mathbb{R}$. This mapping allows conditional estimates from $n$ type sensors to be fused into a total conditional estimate. If the random variables are Gaussian as ours are by assumption, then further simplifications may be made as mentioned in [7].

Let $Z$ be the *activity level* of the net with $Prob(Z=1) = \dfrac{1}{2}$ and $Prob(Z=-1) = \dfrac{1}{2}$. Then

$$E\{X|Z=z\} = \tanh(Z/\sigma_i^2) \qquad (17)$$

This implies that

$$E\{X|Z_1,\cdots,Z_n\} = \tanh\left(\sum_{i=1}^{n} \tanh^{-1}(E\{X|Z_i\})\right) \qquad (18)$$

Figure 5 illustrates the concepts behind the nonhomogeneous network ensembles for the general sensor fusion problem. The *classification net* performs the functions mentioned in the previous section for correct density matching and the adjustment net performs the density estimation via the *parallel projection algorithm*. The inputs to the net are the conditional estimates $E\{X|Z_1\}$ and $E\{X|Z_2\}$ from these respective ensembles. The conditional estimates are given by equation (17). However, the output mapping function for the macrostate activity level is given by equation (6). The main concept of *Statistical Neurodynamics* treats both the $\underline{W}$ and $\underline{h}$ as normally distributed random variables that are i.i.d. This assumption implies that

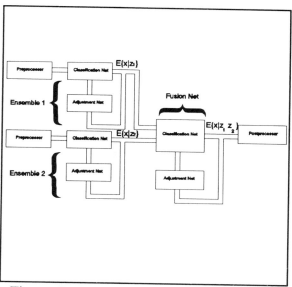

**Figure 5. Network Ensembles for General Sensor Fusion**

$$P_{\underline{W}\underline{h}}(x, y) = P_{\underline{W}}(x)P_{\underline{h}}(y) \tag{19}$$

and

$$\mu_{\underline{W}} = \mu_{\underline{h}} \quad \sigma_W^2 = \sigma_h^2 \tag{20}$$

Therefore, $WZ-H$ reduces to the following

$$WZ-H = \frac{\overline{w}(nZ-1)}{\sqrt{(n+1)\sigma_w^2}} \tag{21}$$

Furthermore, assuming $n \gg 1$ ($n = \#$ of neurons) equation (21) reduces to

$$WZ-H = \frac{\sqrt{n}\overline{w}Z}{\sigma_w} \tag{22}$$

Since $n$, $\overline{w}$, and $\sigma_w^2$ are known we can compute a scale factor (S.F.) equal to $\dfrac{1}{\sqrt{n}\overline{w}\sigma_w}$, which applied to equation (22) yields

$$WZ - H = S.F. \frac{\sqrt{n\bar{w}}Z}{\sigma_w} = \frac{Z}{\sigma_w^2} \qquad (23)$$

which is the same argument as in equation (17). Equations (6) and (17) are both $S$ shaped curves. Transcendental solutions for $erf(x) - \tanh(x) = 0$ exist only for $x = 0$. However, since we know the dimension of the system ahead of time, we can compute a polynomial fit of $\tanh(x)$ to $erf(x)$ using a *Linear Mean Square Error* estimation technique. This implies that if $\hat{y} = erf(x)$ and $y = \tanh(x)$, then $\hat{y} = \underline{c}\, y$ where $\underline{c}$ is the vector of polynomial coefficients. This illustrates the following point. If $\tilde{T}_\omega: S \to S$ is the *macrostate transition operator* defined by equation (6), and $g: \mathbb{R}^n \to \mathbb{R}$ is the Borel measurable fusion function defined by equation (17), then $\tilde{T}_\omega$ is *isomorphic* to $g$ to within linear scaling constant. The isomorphism allows the *fusion* function to be *imbedded* in the *nonlinear output mapping* function of the neural network.

The fusion network takes the conditional estimates from each subnet and fuzes the estimates. The fusion net equations are given by

$$u_i = \sum_{j=1}^{n} w_{ij} x_j - h_i \qquad i = 1, m$$

with $\underline{\underline{W}} = I_{n \times n}$ and $\underline{h} = 0 \;\forall\, i$. In our particular case, $n = 2$ and $m = 1$. Therefore, given $u_i$

$$\tilde{X}_1 = \Pi(u_1)$$

$$\Pi(u_i) = \{1 \cdot u_i \;\forall\, i\}$$

$$Z = \frac{1}{m} \sum_{i=1}^{m} \tilde{X}_i = \tilde{X}_1 \qquad m = 1$$

The output mapping is given by equation (6) with $\Phi(WZ - H) = \Phi(WZ)$ since $\underline{h} = 0 \;\forall\, i$. Letting $W = \frac{\sqrt{n\bar{w}}Z}{\sigma_w}$, and defining the scale factor to be $S.F. = \frac{\sigma_w}{\sqrt{n\bar{w}}}$ $\Phi(WZ)$ reduces to $\Phi(Z)$ and equation (6) is used to calculate the fuzed estimate $E\{X | Z_1 Z_2\}$.

## CONCLUSIONS

An analysis of a complex nonhomogeneous neural net structure applicable to statistical inference procedures has been described. The three main points elucidated in this paper are the extension of the theory of *Statistical Neurodynamics* to include the analysis of complex nonhomogeneous neuron pools consisting of three subnets, the development of a statistical learning algorithm based on the orthogonal alignment of tangent spaces with respect to the submanifold of approximating probability density to the statistical manifold containing the true probability density, and the application of these structures to a general sensor fusion problem.

## REFERENCES

[1] S. Amari," A Mathematical Theory of Nerve Nets", *Advan. in Biophys.*, Vol. 6, pp. 75-120, 1974.

[2] S. Amari, K. Yoshida, and K. Kanatani," A Mathematical Foundation for Statistical Neurodynamics", *SIAM J. Appl. Math.*, Vol. 33, No. 1, pp. 95-126, July 1977.

[3] S. Amari, *Differential-Geometrical Methods in Statistics*, Springer-Verlag, 1990.

[4] S. Amari, O.E. Barndorff-Nielsen, R.E. Kass, S.L. Lauritzen, and C.R. Rao, *Differential Geometry and Statistical Inference*, Institute of Mathematical Statistics Lecture Notes - Monograph Series, S.S. Gupta (Series Editor), 1987.

[5] R. Kulhavy," Recursive Nonlinear Estimation: A Geometric Approach", *Automatica*, Vol. 26, No. 3, pp. 545-555, 1990.

[6] R. Kulhavy, "Differential Geometry of Recursive Estimation," *11th IFAC World Congress*, Tallinn, 1990.

[7] E.B. Hall, A. E. Wessel, and G.L. Wise," Some Aspects of Fusion in Estimation Theory", *IEEE Trans. Information Theory*, Vol. 37, No.2, March 1983.

# AN ALTERNATIVE PROOF OF CONVERGENCE FOR KUNG-DIAMANTARAS APEX ALGORITHM

H. Chen and R. Liu
Department of Electrical Engineering
University of Notre Dame, Notre Dame, IN 46556

Abstract—The problem of adaptive principal components extraction (APEX) has gained much interest recently [1] [2] [3] [4] [5]. In 1990, a new neuro-computation algorithm for this purpose was proposed by Kung and Diamautaras[3]. In this paper an alternative proof is presented to illustrate that the K-D algorithm is in fact richer than what has been proved in [3] [4] [5]. Unlike in [3], our proof shows that neural network will converge and the principal components can be extracted, without assuming that some of projections of synaptic weight vectors have diminished to zero. In addition, we show that the K-D algorithm converges exponentially.

## 1 INTRODUCTION

The computation of part or all of the principal components of a vector stochastic process is a fundamental issue and is frequently encountered in various fields of engineering. For example, in signal processing (especially sensor array and time series problems), a signal subspace approach is often used to effectively reduce signal-to-noise ratio and thereby improve performances. In many cases, principal components of the input data sequence are computed to obtain the spectral modes of the data which can be divided into signal and noise subspaces. Each significant spectral mode is described by one of the principal components.

Other important applications of computing principal components include data compression, pattern recognition and neural network itself, where a mapping from an input space of higher dimensionality to a representation space of lower dimensionality is often used. Principal components analysis(PCA) provides an ideal tool for such applications, optimal in the sense that the mean square distance between the estimated inputs (given the lower-dimensioned representation) and the actual inputs are minimized. In short, the mapping is in the form of an orthogonal transformation $P$, where rows of matrix $P$ are the computed principal components.

Recently, Kung and Diamautaras proposed a new APEX model and algorithm, which will be summarized in the following section.

## 2 KUNG-DIAMAUTARAS ALGORITHM

Let $\mathbf{x}(t) \in \mathcal{R}^n$ be a stationary vector stochastic process, and

$$R = E\{\mathbf{x}(t)\mathbf{x}^T(t)\} \tag{1}$$

its covariance matrix, where E denotes the expectation operator, and $T$ denotes transpose. The objective is to present a robust neuro-computation method for finding the eigenvalues and eigenvectors of $R$.

For simplicity, let its eigenvalues be distinct, positive and arranged in the descending order, $\lambda_1 > \lambda_2 > \cdots > \lambda_m > 0$. Denote $(\lambda_i, \mathbf{e}_i)$ the pair of the $i$th eigenvalue and eigenvector. The neuro-computation follows the iterative procedure of first computing $(\lambda_1, \mathbf{e}_1)$, then computing $(\lambda_2, \mathbf{e}_2)$, so on and so forth. When $(\lambda_i, \mathbf{e}_i)$, $i = 1, 2, \ldots, m-1$, are known, the neural network model used by this neuro-computation method to compute $(\lambda_m, \mathbf{e}_m)$ is shown in Fig. 1.

In Fig. 1, $\mathbf{x} = [x_1, x_2, \cdots, x_n]^T$ is the input vector, with $x_i$ the $i$th input; $\mathbf{y} = [y_1, y_2, \cdots, y_m]^T$ is the output vector, with $y_j$ the $j$th output; $\mathbf{p}_i = [p_{i1}, p_{i2}, \cdots, p_{in}]$ is a *row* vector, with $p_{ij}$ the weight connecting $j$th input to the $i$th neuron; $\mathbf{w} = [w_1, w_2, \cdots, w_{m-1}]$ is a *row* vector, with $w_i$ connecting the $i$th output to the $m$ neuron. It will be shown that $\mathbf{w}$ will asymptotically approach zero as the network converges.

This is a linear model and when computing $m$th eigenvector and eigenvalue, we clearly have

$$\mathbf{y} = P\mathbf{x} \tag{2}$$

$$y_m = \mathbf{p}_m \mathbf{x} + \mathbf{w}\mathbf{y} \tag{3}$$

where matrix $P = [\mathbf{p}_1^T, \mathbf{p}_2^T, \cdots, \mathbf{p}_n^T]^T$.

The Kung-Diamautaras Learning Algorithm(K-D Algorithm) proposed in [3] is as follows

$$\triangle \mathbf{p}_m = \beta(y_m \mathbf{x}^T - y_m^2 \mathbf{p}_m) \quad \text{(hebbian learning)}, \tag{4}$$

$$\triangle \mathbf{w} = -\gamma(y_m \mathbf{y}^T + y_m^2 \mathbf{w}) \quad \text{(anti-hebbian learning)}, \tag{5}$$

where $\beta$ and $\gamma$ are two learning rates, and $\mathbf{p}_m$ is the $m$th row of $P$.

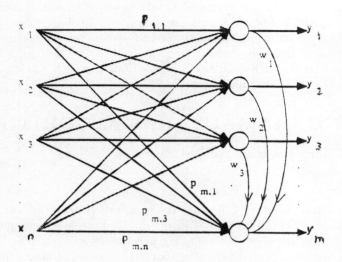

Figure 1: The neural-like network model used in this method.

## 3 ANALYSIS OF K-D ALGORITHM

Let the $n$ orthonormal (since $R$ is a real symmetric matrix) eigenvectors (principal components) associated with $R$ be denoted $e_1, e_2, \cdots, e_n$. The purpose of the Kung-Diamautaras Algorithm is to find part or all of these eigenvectors and their corresponding eigenvalues without using $R$.

Suppose that we want to compute the $m$th eigenvector $p_m$. Let $k$ denote the number of training cycles for the neural network, and the projection of $p_m(k)$ on $e_i$ be denoted by $\theta_i(k)$, then

$$p_m(k) = \sum_{i=1}^{n} \theta_i(k) e_i^T. \qquad (6)$$

We shall present an alternative but rigorous proof of convergence, taking into account for the actual situation when the principal components(p.c.'s) are being extracted, that is the projections of synaptic weights vector $p_m$ will *simultaneously* approach to the $m$th p.c., in contrast to assuming some of these projections are already zeros when proving convergences of rest of them. Firstly, it can be shown [3] that the analysis of convergence properties of this algorithm (except for the case of $\theta_1$) can be reduced to the problem of studying the following dynamic system

$$\begin{bmatrix} \theta_i(k+1) \\ w_i(k+1) \end{bmatrix} = \begin{bmatrix} 1 + \beta'(\lambda_i - \sigma(k)) & \beta' \lambda_i \\ -\gamma' \lambda_i & 1 - \gamma'(\lambda_i + \sigma(k)) \end{bmatrix} \begin{bmatrix} \theta_i(k) \\ w_i(k) \end{bmatrix} \qquad (7)$$

$$\sigma(k) = \mathrm{E}\{y_m^2(k)\} = \sum_{j=m}^{n} \lambda_j \theta_j^2(k) + 2 \sum_{j=1}^{m-1} \lambda_j (\theta_j(k) + w_j(k))^2 \tag{8}$$

for $i = 2, 3, \cdots$, and where $w_i(k) \equiv 0$ for $i \geq m$ and $\beta' = N\beta, \gamma' = N\gamma$, and $N$ is the the number of iterations per cycle. We want to investigate the convergence property of this *nonlinear* discrete-time dynamical system. Note that in equations (7) and (8), *we do not assume that $\theta_i(k)$ and $w_i(k)$ have reached their steady states*, for $i = 1, 2, \cdots, m - 1$. Hence, they are not zeros, as is implicitly assumed in the proofs in [3] [4].

## 3.1 Proof of Convergence for K-D Algorithm

**Theorem 1** *If $\theta_m(0) \neq 0$ and $|\theta_i(0)| \leq 1$ for $i \neq m$, then $\theta_i(k) \to 0$ for $i \neq m$ and $|\theta_m(k)| \to 1, \sigma(k) \to \lambda_m$, all exponentially, as $k \to \infty$. This implies that $\mathbf{p}_m(k) \to e_m$, as $k \to \infty$.*

*Comment.* The condition that all $|\theta_i(0)| \leq 1$ can be ensured by initializing the weights $p_{m,i}$ with small values.

*Proof.* For the case that $i < m$, the proof is similar to that given in [3]. The important cases for $i = m$ and for $i > m$ are to be proved by the following Lemmas.

**Lemma 1** *There hold*

$$\theta_i(k+1) = [1 + \beta'(\lambda_i - \sigma(k))]\theta_i(k), \tag{9}$$

$$\sigma(k) = \sum_{j=m}^{n} \lambda_j \theta_j^2(k) + \sum_{j=1}^{m-1} \lambda_j (\theta_j(k) + w_j(k))^2 \tag{10}$$

*for $i = m, \cdots, n$, and*

$$\begin{bmatrix} \theta_i(k+1) \\ w_i(k+1) \end{bmatrix} = \begin{bmatrix} 1 + \beta'(\lambda_i - \sigma(k)) & \beta' \lambda_i \\ -\gamma' \lambda_i & 1 - \gamma'(\lambda_i + \sigma(k)) \end{bmatrix} \begin{bmatrix} \theta_i(k) \\ w_i(k) \end{bmatrix} \tag{11}$$

*for $i = 1, \cdots, m - 1$.*

Clearly this is a re-statement of (7) and (8), their dynamics can now be further explored as follows.

**Lemma 2** *For any $i \in \{1, 2, \cdots, n\}$, if $|\theta_i(0)| \leq 1$ then $|\theta_i(k)| \leq 1$, where $k = 1, 2, \cdots$.*

*Proof.* For $i \in \{1, 2, \cdots, m-1\}$, the assertion can be shown similarly to the one given in [3].

For $i \in \{m, m+1, \cdots, n\}$, we can discuss it by two cases:
**CASE ONE:**
If $0 \leq \theta_i(k) \leq 1$, $(i > m)$, then $0 \leq \theta_i(k+1) \leq 1$.

In fact, since $\beta'$ is usually chosen to be sufficiently small, by (9) we clearly have $\theta_i(k+1) \geq 0$ and moreover we also have $\theta_i(k+1) \leq 1$, since

$$\begin{aligned}
&1 - \theta_i(k+1) \\
&= 1 - [1 + \beta'(\lambda_i - \sigma(k))]\theta_i(k) \\
&= 1 - \theta_i(k) - \beta'\theta_i(k)[\lambda_i - \lambda_i\theta_i^2(k) - \sum_{j=m}^{i-1}\lambda_j\theta_j^2(k) - \sum_{j=i+1}^{n}\lambda_j\theta_j^2(k) \\
&\quad - \sum_{j=1}^{m-1}\lambda_j(\theta_j(k) + w_j(k))^2] \\
&= [1 - \theta_i(k)][1 - \beta'\theta_i(k)\lambda_i(1 + \theta_i(k))] + \beta'\theta_i(k)[\sum_{j=m}^{i-1}\lambda_j\theta_j^2(k) \\
&\quad + \sum_{j=i+1}^{n}\lambda_j\theta_j^2(k) + \sum_{j=1}^{m-1}\lambda_j(\theta_j(k) + w_j(k))^2] \quad (12) \\
&\geq [1 - \theta_i(k)][1 - 2\beta'\lambda_i] \\
&\geq 0
\end{aligned}$$

which holds as long as $\beta' \leq \frac{1}{2\lambda_1}$.

**CASE TWO:**
If $-1 \leq \theta_i(k) < 0$ $(i > m)$, then $-1 \leq \theta_i(k+1) \leq 0$.
In fact, since $\beta'$ is chosen small, by (9) we have $\theta_i(k+1) \leq 0$ and moreover we also have $\theta_i(k+1) \geq -1$, since

$$\begin{aligned}
&1 + \theta_i(k+1) \\
&= 1 + \theta_i(k)[1 + \beta'(\lambda_i - \sigma(k))] \\
&= 1 + \theta_i(k) + \beta'\theta_i(k)[\lambda_i - \lambda_i\theta_i^2(k) - \sum_{j=m}^{i-1}\lambda_j\theta_j^2(k) - \sum_{j=i+1}^{n}\lambda_j\theta_j^2(k) \\
&\quad - \sum_{j=1}^{m-1}\lambda_j(\theta_j(k) + w_j(k))^2] \\
&= [1 + \theta_i(k)][1 + \beta'\theta_i(k)\lambda_i(1 - \theta_i(k))] - \beta'\theta_i(k)[\sum_{j=m}^{i-1}\lambda_j\theta_j^2(k) \\
&\quad + \sum_{j=i+1}^{n}\lambda_j\theta_j^2(k) + \sum_{j=1}^{m-1}\lambda_j(\theta_j(k) + w_j(k))^2] \quad (13) \\
&\geq [1 + \theta_i(k)][1 - 2\beta'\lambda_i] \\
&\geq 0
\end{aligned}$$

which again holds as long as $\beta' \leq \frac{1}{2\lambda_1}$.

In summary of the above two cases, we have shown that for any $k$, if $|\theta_i(0)| \leq 1$, then for all $k \geq 1$, $|\theta_i(k)| \leq 1$.

**Lemma 3** *If $\theta_m(0) \neq 0$, then $\theta_i(k)$ tends to zero exponentially, as $k \to \infty$, for $i = 1, 2, \cdots, n$, and $w_i(k)$ tends to zero exponentially, as $k \to \infty$, for $i = 1, 2, \cdots, m-1$.*

*Comment.* If we initialize $\mathbf{p}_m$ with small random values, then $\theta_m(0) \neq 0$ with probability 1.

*Proof.* For $i < m$, the proof is similar to the one given in [3].
For $i > m$, by (9)

$$\theta_i(k+1) = \{1 + \beta'[\lambda_i - \sigma(k)]\}\theta_i(k), \tag{14}$$

$$\theta_m(k+1) = \{1 + \beta'[\lambda_m - \sigma(k)]\}\theta_m(k). \tag{15}$$

Now, assume that $\theta_m(0) \neq 0$ (with probability 1), then

$$\frac{\theta_i(k)}{\theta_m(k)} = \prod_{j=0}^{j=k-1} \frac{1 + \beta'[\lambda_i - \sigma(j)]}{1 + \beta'[\lambda_m - \sigma(j)]} \cdot \frac{\theta_i(0)}{\theta_m(0)} \tag{16}$$

Under the condition $\lambda_m > \lambda_i$ for $i > m$, we can further show that

$$0 < \frac{1 + \beta'[\lambda_i - \sigma(j)]}{1 + \beta'[\lambda_m - \sigma(j)]} \leq \frac{1 + \beta'\lambda_i}{1 + \beta'\lambda_m}. \tag{17}$$

Therefore,

$$|\theta_i(k)| \leq \left[\frac{1 + \beta'\lambda_i}{1 + \beta'\lambda_m}\right]^k |\frac{\theta_i(0)}{\theta_m(0)} \theta_m(k)|$$

$$\leq \left[\frac{1 + \beta'\lambda_{m+1}}{1 + \beta'\lambda_m}\right]^k |\frac{\theta_i(0)}{\theta_m(0)} \theta_m(k)|$$

$$\leq \left[\frac{1 + \beta'\lambda_{m+1}}{1 + \beta'\lambda_m}\right]^k |\frac{1}{\theta_m(0)}|. \tag{18}$$

Letting $d \equiv \frac{1+\beta'\lambda_{m+1}}{1+\beta'\lambda_m}$ and $c \equiv \frac{1}{\theta_m(0)}$, then

$$\theta_i(k) < cd^k \tag{19}$$

where $|d| < 1$ and $k = m+1, m+2 \cdots$.

**Lemma 4** *If $\theta_m(0) \neq 0$, then $\theta_m(k)$ tends to 1 or $-1$ exponentially, and $\sigma(k)$ tends to $\lambda_m$ exponentially, as $k \to \infty$.*

*Proof.* By (9), recall that

$$\theta_m(k+1) = [1 + \beta'(\lambda_m - \sigma(k))]\theta_m(k). \tag{20}$$

By Lemma 2, we also have $|\theta_m(k)| \leq 1$ for all $k$, if $|\theta_m(0)| \leq 1$, therefore we conclude that for some $k_0$ large enough, $\sigma(k_0) > \frac{\lambda_m}{2}$. Otherwise, $\theta_m(k) \geq \theta_m(0)[1 + \beta'(\lambda_m - \frac{\lambda_m}{2})]^k$, which is in contradiction with Theorem 3. Recall from Theorem 3 that for $i \neq m, \theta_i(k) \to 0$ as $k \to \infty$. Combining the consideration of (10), we conclude that there exists some $k_0'$, such that for any $k > k_0'$, $1 \geq |\theta_m(k)| > \frac{1}{2}$.

Based on the randomness of the initial condition, we shall have the following two cases to discuss

**CASE ONE:**
If $0 < \theta_m(0) < 1$, then from (13)

$$1 - \theta_m(k+1)$$
$$= [1 - \theta_m(k)][1 - \beta'\theta_m(k)\lambda_m(1 + \theta_m(k))]$$
$$+ \beta'\theta_m(k)[\sum_{j=m+1}^{n} \lambda_j \theta_j^2(k) + \sum_{j=1}^{m-1} \lambda_j (\theta_j(k) + w_j(k))^2] \tag{21}$$
$$\equiv I_1(k) + I_2(k), \tag{22}$$

where

$$I_1(k) \equiv [1 - \theta_m(k)][1 - \beta'\theta_m(k)\lambda_m(1 + \theta_m(k))], \tag{23}$$

and

$$I_2(k) \equiv \beta'\theta_m(k)[\sum_{j=m+1}^{n} \lambda_j \theta_j^2(k) + \sum_{j=1}^{m-1} \lambda_j (\theta_j(k) + w_j(k))^2]. \tag{24}$$

By Lemma 3, we can write $0 < |I_2| \leq cd^{2k}$, where $c$ and $d$ are some positive constants, independent of $i$ and $k$, and $|d| < 1$.

Considering the quadratic term in $I_1(k)$ and studying the parabola curve (now that it is known that there exist some $k_0$, such that for any $k > k_0$, $\theta_m(k) \in (\frac{1}{2}, 1]$), we get

$$1 - 2\beta'\lambda_m \leq 1 - \beta'\theta_m(k)\lambda_m(1 + \theta_m(k)) < 1 - \frac{3}{4}\beta'\lambda_m, \tag{25}$$

hence

$$0 \leq I_1(k) \leq [1 + \theta_m(k)](1 - \frac{3}{4}\beta'\lambda_m). \tag{26}$$

Choose $\delta$ small enough such that $(1 - \frac{3}{4}\beta'\lambda_m)(1+\delta) < 1$. (In fact, we can pick $\delta = \frac{1}{2}[\frac{2}{1-\frac{3}{4}\beta'\lambda_m} - 1]$, where [ ] is the integer truncation function. This way $\delta$ can be close to 1, since $\beta'$ is small.) In between, there always exists some positive constant $e$, such that $(1 - \frac{3}{4}\beta'\lambda_m)(1+\delta) < e < 1$.

1. If $I_2(k) > \delta I_1(k)$, then
$$0 \leq 1 - \theta_m(k+1) \leq (1+\frac{1}{\delta})I_2(k) \leq c(1+\frac{1}{\delta})d^{2k} \qquad (27)$$

2. If $I_2(k) \leq \delta I_1(k)$, then
$$\begin{aligned} 0 &\leq 1 - \theta_m(k+1) \leq (1+\delta)I_1 \\ &\leq (1+\delta)(1-\theta_m(k))(1-\frac{3}{4}\beta'\lambda_m(k)) \\ &\leq e[1-\theta_m(k)]. \end{aligned}$$

Let $\mathcal{K} = \{k_1 < k_2 < \cdots < k_l < \cdots\}$ be the set of indices ($\mathcal{K}$ can be empty) such that for $k \in \mathcal{K}$, $I_2(k) > \delta I_1(k)$, hence
$$1 - \theta_m(k_l+1) \leq c(1+\frac{1}{\delta})d^{2k_l}. \qquad (28)$$

For the rest of integers, that is $k \in \mathcal{K}^c$ (complementary set of $\mathcal{K}$), assuming that $k_l < k \leq k_{l+1}$, then
$$1 - \theta_m(k+1) < c(1+\frac{1}{\delta})d^{2k_l}e^{k-k_l} \leq c'f^k \qquad (29)$$

where $c'$ and $f$ are some positive constants, and $|f| < 1$.

These two cases together show that if $0 < \theta_m(0) < 1$, then $\theta_m(k) \to 0$ exponentially, as $k \to \infty$, and hence by Lemma 3,
$$\lambda_m - \sigma(k) = \lambda_m(1-\theta_m^2(k)) + \sum_{j=m+1}^{n}\lambda_j\theta_j^2(k) + \sum_{j=1}^{m-1}\lambda_j(\theta_j(k)+w_j(k))^2 \qquad (30)$$

we also see that $\sigma(k)$ tends to $\lambda_m$ exponentially.

**CASE TWO:**
If $-1 < \theta_m(0) < 0$: then from (14)
$$\begin{aligned} &1 + \theta_m(k+1) \\ &= [1+\theta_m(k)][1+\beta'\theta_m(k)\lambda_m(1-\theta_m(k))] \\ &\quad - \beta'\theta_m(k)[\sum_{j=m+1}^{n}\lambda_j\theta_j^2(k) + \sum_{j=1}^{m-1}\lambda_j(\theta_j(k)+w_j(k))^2] \qquad (31) \\ &\equiv I_1(k) + I_2(k) \qquad (32) \end{aligned}$$

where
$$I_1(k) \equiv [1+\theta_m(k)][1+\beta'\theta_m(k)\lambda_m(1-\theta_m(k))], \qquad (33)$$

and
$$I_2(k) \equiv -\beta'\theta_m(k)[\sum_{j=m+1}^{n}\lambda_j\theta_j^2(k) + \sum_{j=1}^{m-1}\lambda_j(\theta_j(k)+w_j(k))^2]. \qquad (34)$$

By Lemma 3, we can write $0 < |I_2| \leq cd^{2k}$, where $c$ and $d$ are some positive constants, independent of $i$ and $k$, and $|d| < 1$.

Considering the quadratic term in $I_1(k)$ and studying the parabola curve (now that it is known that there exist some $k_0$, such that for any $k > k_0$, $\theta_m(k) \in [-1, -\frac{1}{2})$), we can get

$$1 - 2\beta'\lambda_m \leq 1 + \beta'\theta_m(k)\lambda_m(1 - \theta_m(k)) < 1 - \frac{3}{4}\beta'\lambda_m, \quad (35)$$

hence

$$0 \leq I_1(k) \leq [1 + \theta_m(k)](1 - \frac{3}{4}\beta'\lambda_m). \quad (36)$$

Choose $\delta$ small enough such that $(1 - \frac{3}{4}\beta'\lambda_m)(1 + \delta) < 1$. (In fact, we can pick $\delta = \frac{1}{2}[\frac{2}{1 - \frac{3}{4}\beta'\lambda_m} - 1]$, where [ ] is the integer truncation function. This way $\delta$ can be close to 1, since $\beta'$ is small.) In between, there always exists some positive constant $e$, such that $(1 - \frac{3}{4}\beta'\lambda_m)(1 + \delta) < e < 1$.

1. If $I_2(k) > \delta I_1(k)$, then

$$0 \leq 1 + \theta_m(k+1) \leq (1 + \frac{1}{\delta})I_2(k) \leq c(1 + \frac{1}{\delta})d^{2k}. \quad (37)$$

2. If $I_2(k) \leq \delta I_1(k)$, then

$$\begin{aligned} 0 &\leq 1 + \theta_m(k+1) \leq (1+\delta)I_1 \\ &\leq (1+\delta)(1 + \theta_m(k))(1 - \frac{3}{4}\beta'\lambda_m(k)) \\ &\leq e[1 - \theta_m(k)]. \end{aligned}$$

Let $\mathcal{K} = \{k_1 < k_2 < \cdots < k_l < \cdots\}$ be the set of indices ($\mathcal{K}$ can be empty) such that for $k \in \mathcal{K}$, $I_2(k) > \delta I_1(k)$, hence

$$1 - \theta_m(k_l + 1) \leq c(1 + \frac{1}{\delta})d^{2k_l}. \quad (38)$$

For the rest of integers, that is $k \in \mathcal{K}^c$ (complementary set of $\mathcal{K}$), assuming that $k_l < k \leq k_{l+1}$, then

$$1 - \theta_m(k+1) < c(1 + \frac{1}{\delta})d^{2k_l}e^{k-k_l} \leq c'f^k \quad (39)$$

where $c'$ and $f$ are some positive constants, and $|f| < 1$.

These two cases together show that when $-1 \leq \theta_m(0) < 0$, $\theta_m(k) \to 0$ exponentially, as $k \to \infty$, and hence by Lemma 3,

$$\lambda_m - \sigma(k) = \lambda_m(1 - \theta_m^2(k)) + \sum_{j=m+1}^{n} \lambda_j \theta_j^2(k) + \sum_{j=1}^{m-1} \lambda_j(\theta_j(k) + w_j(k))^2, \quad (40)$$

therefore we also see that $\sigma(k)$ tends to $\lambda_m$ exponentially.

## 4 CONCLUSION

A proof showing that the neural network for APEX algorithm converges *exponentially* has been presented. Its technicality is necessary to ensure that all transitional non-steady states, namely $\theta_i$ and $w_i$ as explained in the context be considered. This takes into account of the more reasonable description of the behavior of Kung-Diamautaras algorithm.

**Acknowledgement:** The authors wish to thank Professor Tianping Chen of the Department of Mathematics at Fudan University, Shanghai, China for his many interesting and helpful suggestions.

## References

[1] E. Oja, "A Simplified Neuron Model as a Principal Component Analyzer", *J. Math. Biology*, Vol. 15, pp. 267-273, 1982.

[2] T. D. Sanger, "An Optimality Principle for Unsupervised Learning", in *Advances in Neural Information Processing Systems*, Vol. 1, pp.11-19, (D. S. Touretzky, editor), 1989.

[3] S. Y. Kung and K. I. Diamantaras, "A Neural Network Learning Algorithm For Adaptive Principal Component Extraction(APEX)", *ICASSP*, pp. 861-864, 1990.

[4] S. Y. Kung, "Constrained Principal Component Analysis via an Orthogonal Learning Network", *ICCAS*, pp. 719-722, 1990.

[5] S. Y. Kung, "Orthogonal Learning Network for Constrained Principal Component Problem", *IJCNN*, San Diego, 1990.

[6] E. Oja, "On Stochastic Approximation of the Eigenvectors and Eigenvalues of the Expectation of a Random Matrix", *J. of Mathematical Analysis and Applications*, 106, pp. 69-84, 1985.

# Neural Networks for Extracting Unsymmetric Principal Components

S. Y. Kung and K. I. Diamantaras
Department of Electrical Engineering
Princeton University, Princeton, NJ 08544, USA*

Abstract - In this paper we introduce two forms of *Unsymmetric Principal Component Analysis* (UPCA), namely the *cross-correlation UPCA* and the *linear approximation UPCA problem*. Both are concerned with the SVD of the input-teacher crosscorrelation matrix itself (first problem) or after prewhitening (second problem). The second problem is also equivalent to reduced-rank Wiener filtering. For the former problem, we propose an unsymmetric linear model for extracting one or more components using lateral inhibition connections in the hidden layer. The numerical convergence properties of the model are theoretically established. For the *linear approximation UPCA problem*, we can apply Back-Propagation extended either using a straightforward deflation procedure or with the use of lateral orthogonalizing connections in the hidden layer. All proposed models were tested and the simulation results confirm the theoretical expectations.

## 1 Introduction

Principal Component Analysis (PCA), is a statistical method for extracting the most representative low-dimensional subspace for a signal of high dimensionality, namely, it is a method for finding those subspace directions that contain (in a way) the maximum signal information. Such an extraction process has many useful applications in signal/image filtering, restoration, pattern classification, recognition, etc [3][8]. In this paper we distinguish two basic categories of PCA problems: the symmetric and the unsymmetric PCA problem. Different Neural Network models can be used to solve these problems as described below.

---

*This research was supported in part by Air Force Office of Scientific Research under Grant AFOSR-89-0501A.

1. **Symmetric (ordinary) PCA and Self-supervised networks**

   *Symmetric* principal components analysis (PCA) is concerned with the Singular Value Decomposition(SVD) of the autocorrelation matrix $R_x$ where $\{x\}$ is a vector stochastic process. It has been shown that self-supervised two-layer networks with linear units, trained by back propagation produce the optimal subspace spanned by the eigen-components [2]. Other neural models have been proposed as well for extracting the components themselves rather than their subspaces [5][6][4].

2. **Unsymmetric PCA and Hetero-Supervised networks**

   The *Unsymmetric PCA* (UPCA) problem involves two (at least wide sense stationary) vector stochastic processes: the input process $\{x\}$, and the teacher process $\{y\}$, with autocorrelation matrices $R_x$ and $R_y$ respectively, and cross-correlation matrix $R_{xy}$. Two cases are defined to belong to the Unsymmetric PCA problem: the *linear approximation UPCA problem* and the *cross-correlation UPCA problem*.

   •**Linear Approximation UPCA:**

   Find the most representative components of $\{x\}$ so that they may be used to best recover $\{y\}$. As shown in Section 2.1 if we use the least-squares-error as the optimality criterion, then the UPCA problem can be reduced to the problem of finding the principal left (right) singular vectors of the matrix $R_x^{-1/2} R_{xy}$. The problem is also equivalent to *reduced rank Wiener filtering* [7].

   •**Cross-correlation UPCA:**

   Find the left and right singular vectors for $R_{xy}$, corresponding to the largest $k$ singular values, namely produce the optimal $k$-rank approximation of the cross-correlation matrix.

It's worth mentioning that if $y = x$ then the ordinary PCA and the two UPCA problems discussed above become the same.

# 2 Principal Components for the Linear Approximation Problem

## 2.1 From Least-Squares Solution to UPCA

In this paper (except when explicitly noted) we are dealing with linear two-layer networks. We shall denote by $\overline{W}$ and $\underline{W}^T$, the upper and lower layer weight-matrices respectively. Obviously, the input-output mapping realized by the two-layer net is $W = \overline{W}\underline{W}^T$. The number of input and output units is $n$ and $m$ respectively, and we'll define $p \equiv \min\{m, n\}$. The number of the hidden units $k$, simply guards the rank of the overall map $W$.

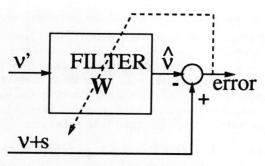

Figure 1: If $\nu(t)$ is low-rank then a low-rank Wiener filter may be prefered.

**Full Rank Wiener Filtering**  The least squares error criterion for linear approximation is

$$J = E\{\|\mathbf{y} - \mathbf{W}\mathbf{x}\|^2\} = E\{\|\mathbf{y} - \overline{\mathbf{W}}\underline{\mathbf{W}}^T\mathbf{x}\|^2\}$$

If the network has enough hidden units then the full-rank optimal solution $\mathbf{W}^* = R_{xy}^T R_x^{-1}$ can be achieved, which is the classical optimal Wiener Filter. We note that although the matrix $\mathbf{W}^*$ is unique, the optimal solution for the individual matrices $\overline{\mathbf{W}}$ and $\underline{\mathbf{W}}$ is not.

**Rank-Deficient Wiener Filtering: Reduced Hidden Units**  In the case where the rank $k$ of the overall mapping is less than full (i.e. $k < p$) we are looking for the optimal vectors $\overline{\mathbf{w}}_i$ and $\underline{\mathbf{w}}_i$, $i = 1, \ldots, k$, that minimize:

$$J_k = E\{\|\mathbf{y} - \overline{\mathbf{W}}\underline{\mathbf{W}}^T\mathbf{x}\|^2\} = E\{\|\mathbf{y} - \sum_{i=1}^{k}\overline{\mathbf{w}}_i\underline{\mathbf{w}}_i^T\mathbf{x}\|^2\} \tag{1}$$

A possible application example could be an echo-cancellation problem where the noise or interference $\nu(t)$ that corrupts the signal $s(t)$, has inherent low-rankness and it is correlated with some observed noise $\nu'(t)$. The low-rankness of $\nu$ can be, for example, due to a small number of interference sources the combination of which produces $\nu$. In that case there is reason to prefer a low-rank approximation filter rather than the standard Wiener filter for estimating $\nu$ (Figure 1). Nevertheless, no experiments have been conducted so far to prove (or disprove) the advantage of the low-rank filter over the full-rank one.

Now, let us return to the solution of the above approximation problem. Let $\mathbf{U}\Sigma\mathbf{V}^T$ be the Singular Value Decomposition of the matrix $R_x^{-1/2}R_{xy}$, where $\mathbf{U} = [\mathbf{u}_1 \ldots \mathbf{u}_n]$, $\mathbf{V} = [\mathbf{v}_1 \ldots \mathbf{v}_m]$ are square orthonormal matrices, and $\Sigma = \begin{bmatrix} \mathbf{D} \\ \mathbf{0} \end{bmatrix}$ or $\Sigma = [\mathbf{D} \mid \mathbf{0}]$, where $\mathbf{D} = \text{diag}[\sigma_1 \ldots \sigma_p]$, and $\sigma_1 > \sigma_2 > \ldots > \sigma_p > 0$. It has been proved [1] that $J_k$ is minimized if $R_x^{1/2}\underline{\mathbf{w}}_i = \mu_i \mathbf{u}_i$, $\overline{\mathbf{w}}_i = \rho_i \mathbf{v}_i$ and $\mu_i \rho_i = \sigma_i$, for $i = 1, \ldots, k$. Of course the minimizing solution is not

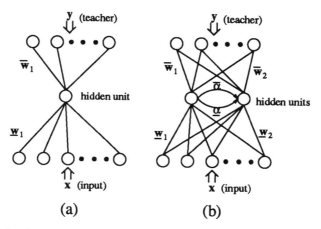

Figure 2: (a) The linear BP network with 1 hidden unit. (b) The proposed model for many hidden units and two Lateral Orthogonalization Networks.

unique since whenever $(\overline{\mathbf{W}}, \underline{\mathbf{W}}^T)$ is a solution then so is $(\overline{\mathbf{W}}\mathbf{C}, \mathbf{C}^{-1}\underline{\mathbf{W}}^T)$, for any invertible matrix $\mathbf{C}$.

## 2.2 Neural Models for Extracting the First PC

**The Linear BP Model** The linear back-propagation (BP) method for a two-layer network with one hidden unit (Figure 2a) minimizes $J_1$, therefore is suitable for extracting the first component. In this case the standard BP method becomes

$$\Delta\overline{w}(t) = \beta \left[y(t) - \overline{w}(t)a(t)\right] a(t) \qquad (2)$$
$$\Delta\underline{w}(t) = \beta\, x(t) \left[b(t) - \|\overline{w}(t)\|^2 a(t)\right] \qquad (3)$$

where $a(t) = \underline{w}(t)^T x(t)$ and $b(t) = \overline{w}(t)^T y(t)$. According to [1] the problem has no local minima, while the global minimum is achieved at the points $(\underline{w}, \overline{w}) = (\mu_1 R_x^{-1/2}\mathbf{u}_1, \rho_1 \mathbf{v}_1)$ where $\mu_1 \rho_1 = \sigma_1$.

## 2.3 Neural Models for Extracting Multiple PCs

**Definition of Deflation** *Let $U\Sigma V^T$ be the SVD of a matrix $A$, where $\Sigma = \begin{bmatrix} D \\ 0 \end{bmatrix}$ or $\Sigma = [D \mid 0]$, and $D = diag[\sigma_1 \ldots \sigma_p]$, where $\sigma_1 > \sigma_2 > \ldots > \sigma_p > 0$. We say that $A$ is deflated if it is transformed by one of the following ways*

$$A_1 = (I - \mathbf{u}_1\mathbf{u}_1^T)\, A \qquad (4)$$
$$A_2 = A\, (I - \mathbf{v}_1\mathbf{v}_1^T) \qquad (5)$$

$$\begin{align}
A_3 &= (I - u_1 u_1^T)\, A\, (I - v_1 v_1^T) \tag{6}\\
A_4 &= A - \sigma_1 u_1 v_1^T \tag{7}
\end{align}$$

It is easy to show that all four transformations are equivalent, namely

$$A_1 = A_2 = A_3 = A_4 = U\hat{\Sigma} V^T \tag{8}$$

where $\hat{\Sigma} = \begin{bmatrix} \hat{D} \\ 0 \end{bmatrix}$ or $\hat{\Sigma} = [\hat{D} \mid 0]$, and $\hat{D} = \text{diag}[0\ \sigma_2 \ldots \sigma_p]$. Thus, after deflation, the second component becomes dominant so it can be extracted using exactly the same learning rule used for the first component, applied now on the deflated data. Note that the recursive nature of the above argument lends itself easily to the extension to the $k$th component case. In the following we discuss two ways to implement deflation: the straightforward method and the Lateral Orthogonalization Net (LON).

**Straightforward deflation** For the $k$th neuron we assume that the previous $k-1$ principal components have already been extracted, so that $R_x^{1/2}\underline{w}_i = \mu_i u_i$, $\overline{w}_i = \rho_i v_i$ and $\mu_i \rho_i = \sigma_i$. Then the modified BP rule for this neuron is

$$\Delta \overline{w}_k(t) = \beta \left[ y(t) - \sum_{i=1}^{k-1} \overline{w}_i a_i(t) - \overline{w}_k(t) a_k(t) \right] a_k(t) \tag{9}$$

$$\Delta \underline{w}_k(t) = \beta x(t) \left[ y(t) - \sum_{i=1}^{k-1} \overline{w}_i a_i(t) - \overline{w}_k(t) a_k(t) \right]^T \overline{w}_k(t) \tag{10}$$

where $a_i(t) = \underline{w}_i^T x(t)$, $i = 1, \ldots, k$. The above rule is equivalent to a single-hidden-unit BP rule with a modified teacher $\tilde{y}(t) = y(t) - \sum_{i=1}^{k-1} \overline{w}_i \underline{w}_i^T x(t)$, therefore the algorithm will convergence to the principal singular vectors of the matrix $R_x^{-1/2} R_{x\tilde{y}}$. It is easy to show that

$$R_x^{-1/2} R_{x\tilde{y}} = R_x^{-1/2} R_{xy} - \sum_{i=1}^{k-1} R_x^{1/2} \underline{w}_i \overline{w}_i^T \tag{11}$$

Under the initial assumptions it follows that the matrix $\tilde{A} = R_x^{-1/2} R_{x\tilde{y}}$ is a $(k-1)$-times multiple deflation of the fourth type of the matrix $A = R_x^{-1/2} R_{xy}$, hence $\underline{w}_2$ and $\overline{w}_2$ will converge to $\mu_k R_x^{-1/2} u_k$ and $\rho_k v_k$ with $\mu_k \rho_k = \sigma_k$.

**The Lateral Orthogonalization Net** A lateral connection network among hidden neurons may be used to implement the deflation process in an adaptive manner. Two kinds of learning rules can be used to train this network:

**Local Orthogonalization Rule** The goal is to orthogonalize the activations $a_1(t)$ and $a_2(t)$, of a pair of connected neurons. That means we want the "inner product" $< a_1, a_2' > \equiv \sum_t a_1(t) a_2'(t)$, to become zero, where $a_2'$ is the modified activation of the second neuron

$$a_2'(t) = a_2(t) - \alpha a_1(t) \tag{12}$$

The local orthogonalization rule is then defined as follows

$$\Delta \alpha(t) = \beta \left[ a_1(t) a_2(t) - \alpha(t) a_1^2(t) \right] \tag{13}$$

Assuming that $a_1(t)$, $a_2(t)$ are bounded and stationary as $t \to \infty$, and $\beta$ is small enough, then $\alpha$ will converge to the value

$$\alpha = \frac{< a_1, a_2 >}{\|a_1\|^2} \tag{14}$$

where $\|a_1\|^2 = < a_1, a_1 >$. It follows then that $< a_1, a_2' > = 0$, which is the desired orthogonality condition.

**Dynamic (Case-Dependent) Orthogonalization Rule** In some cases the lateral weights can be trained by a specialized rule which depends on the dynamics of the specific system, and which achieves orthogonalization. No general equation can be prescribed for such a rule since it is usually different for each different system it is applied to. One example is the APEX lateral orthogonalization network described in [4].

**Using the LON** In the case of the linear approximation model we will need two lateral connections for each pair of hidden neurons: one for controlling the orthogonality of the upper-layer weights and one for the lower-layer weights (see Figure 2b). Assuming that the first $k-1$ components have already been extracted the proposed model for the $k$th neuron is defined below (compare with equations (2)-(3))

$$\Delta \overline{\mathbf{w}}_k(t) = \beta \left[ y(t) a_k'(t) - \overline{\mathbf{w}}_k(t) a_k^2(t) \right] \tag{15}$$
$$\Delta \underline{\mathbf{w}}_k(t) = \beta \, \mathbf{x}(t) \left[ b_k'(t) - \|\overline{\mathbf{w}}_k(t)\|^2 a_k(t) \right] \tag{16}$$

where

$$a_i(t) = \underline{\mathbf{w}}_i^T \mathbf{x}(t), \quad i=1,\ldots,k; \qquad a_k'(t) = a_k(t) - \sum_{i<k} \underline{\alpha}_{ki} a_i(t) \tag{17}$$

$$b_i(t) = \overline{\mathbf{w}}_i^T \mathbf{y}(t), \quad i=1,\ldots,k; \qquad b_k'(t) = b_k(t) - \sum_{i<k} \overline{\alpha}_{ki} b_i(t) \tag{18}$$

$$\Delta \underline{\alpha}_{ki}(t) = \beta \left( a_i(t) a_k(t) - \underline{\alpha}_{ki}(t) a_i^2(t) \right) \tag{19}$$
$$\Delta \overline{\alpha}_{ki}(t) = \beta \left( \overline{\mathbf{w}}_i^T \overline{\mathbf{w}}_k(t) - \overline{\alpha}_{ki}(t) \|\overline{\mathbf{w}}_i\|^2 \right) \tag{20}$$

The updating rules for the lateral weights are of the local type. Furthermore, $\overline{\alpha}_{ki}$ and $\underline{\alpha}_{ki}$ track the values $\frac{\overline{\mathbf{w}}_i^T \overline{\mathbf{w}}_k(t)}{\|\overline{\mathbf{w}}_i\|^2}$ and $\frac{\mathbf{w}_i^T R_x \mathbf{w}_k(t)}{\|R_x^{1/2}\mathbf{w}_i\|^2}$ respectively. If these values were put into (15) and (16) then the resulting system of equations would be similar to the single-hidden-unit BP system but with $(k-1)$-times deflated data. Therefore, the $k$th component would have been extracted. Even though the tracking of these values introduces additional error, it turns out that this does not affect the convergence of the system to the desired steady-state solution. Due to lack of space we will postpone further discussion on the theoretical aspects of this model for a later report.

### 2.3.1 Simulation Results

Figure 3a shows the simulation studies on multiple components. The network uses the straightforward deflation rule to extract the first two components in sequence (only the second is shown in the Figure). The $y$-axis in the plot corresponds to the inner product of the normalized vector $\overline{\mathbf{w}}_i$ (or $R_x^{1/2}\mathbf{w}_i$) with the corresponding principal singular vector $\mathbf{v}_i$ (or $\mathbf{u}_i$). The $x$-axis corresponds to the number of iterations, (i.e. updating steps). The data used are a single block of 200 ($\mathbf{x}$, $\mathbf{y}$) pairs, repeated cyclicly in sweeps (so each sweep has 200 iterations). The value of $\beta$ used in these experiments is 0.005. Figure 3b shows the convergence of the second neuron to the second component of the same experiment using BP with the LON. The convergence speed of the second neuron is very similar to Fig. 3a. Comparing the two plots, notice that it is not important whether the aforementioned inner products converge to 1 or $-1$ as long as they have the same sign.

## 3 Neural Nets for the Cross-Correlation UPCA Problem

As before we assume that the system to be determined receives two at least wide sense stationary signals $\mathbf{x}(t)$ (input) and $\mathbf{y}(t)$ (teacher). In the following, let $U\Sigma V^T$ be the SVD of the cross-correlation matrix $R_{xy} = E\{\mathbf{xy}^T\}$ the principal singular vectors of which we are interested in extracting.

**Extracting the First PC** The proposed model has one hidden unit (Figure 4a) and two neuron values associated with it: the forward value $a(t) = \underline{\mathbf{w}}(t)^T \mathbf{x}(t)$, and the backward value $b(t) = \overline{\mathbf{w}}(t)^T \mathbf{y}(t)$. The weights are updated by the following rule

$$\Delta \overline{\mathbf{w}}(t) = \beta \left[ \mathbf{y}(t) - \overline{\mathbf{w}}(t)b(t) \right] a(t) \qquad (21)$$

$$\Delta \underline{\mathbf{w}}(t) = \beta \left[ \mathbf{x}(t) - \underline{\mathbf{w}}(t)a(t) \right] b(t) \qquad (22)$$

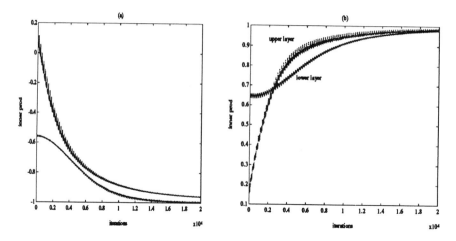

Figure 3: (a) Convergence of the linear approximation model with 2 hidden units using straightforward deflation. (b) Convergence of model with 2 hidden units using LON.

The method will converge if the following initial conditions are met (see proof in later report):

$$\sum_t \underline{\mathbf{w}} \mathbf{x}(t) \mathbf{y}(t)^T \overline{\mathbf{w}} > 0 \qquad (23)$$

$$\|\overline{\mathbf{w}}\| < 1, \quad \|\underline{\mathbf{w}}\| < 1 \qquad (24)$$

These conditions can be easily satisfied in practice by controlling the sign and/or the magnitude of $\underline{\mathbf{w}}$ and $\overline{\mathbf{w}}$ before the updating starts. One advantage of the algorithm is that it produces normalized weights $\underline{\mathbf{w}}$ and $\overline{\mathbf{w}}$ in the steady state which obviates the need for extra normalization whenever a deflation procedure is required for the extraction of the remaining components.

**Extracting the $k$th PC** Assume that the first $k-1$ components have already been extracted, namely, $\underline{\mathbf{w}}_i = \mathbf{u}_i$ and $\overline{\mathbf{w}}_i = \mathbf{v}_i$. The objective now is to find $\mathbf{u}_k$ and $\mathbf{v}_k$. Just like the APEX network, the inhibition connections in the hidden layer are unidirectional. The connections and their corresponding parameters are shown in Figure 4b. The updating equations for the $k$th neuron have the form

$$\Delta \overline{\mathbf{w}}_k(t) = \beta \left[ \mathbf{y}(t) - \overline{\mathbf{w}}_k(t) b_k(t) \right] a'_k(t) \qquad (25)$$
$$\Delta \underline{\mathbf{w}}_k(t) = \beta \left[ \mathbf{x}(t) - \underline{\mathbf{w}}_k(t) a_k(t) \right] b'_k(t) \qquad (26)$$

where

$$a_i(t) \equiv \underline{\mathbf{w}}_i^T \mathbf{x}(t), \quad i = 1, \ldots, k; \qquad a'_k(t) \equiv a_k(t) - \sum_{i=1}^{k-1} \underline{\alpha}_{ik}(t) a_i(t) \qquad (27)$$

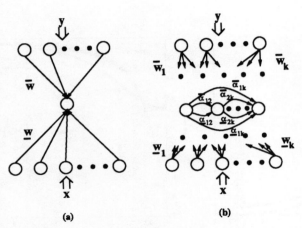

Figure 4: (a) A single hidden unit cross-correlation network. (b) A multiple hidden units cross-correlation network with lateral inhibition parameters.

$$b_i(t) \equiv \overline{\mathbf{w}}_i^T \mathbf{y}(t), \quad i=1,\ldots,k; \qquad b'_k(t) \equiv b_k(t) - \sum_{i=1}^{k-1} \overline{\alpha}_{ik}(t) b_i(t) \quad (28)$$

$$\overline{\alpha}_{ik}(t) = \overline{\mathbf{w}}_i^T \overline{\mathbf{w}}_k(t) \tag{29}$$
$$\underline{\alpha}_{ik}(t) = \underline{\mathbf{w}}_i^T \underline{\mathbf{w}}_k(t) \tag{30}$$

It can be shown that (25)-(26) are the same as the single-component equations (21)-(22) applied on $(k-1)$-times deflated data. Therefore, as in the approximation case, the $k$th component is dominant and it is going to be extracted by the $k$th neuron. In addition, using the LON idea one can define the following adaptive (tracking) formulas for the inhibition parameters

$$\Delta\underline{\alpha}_{ik}(t) = \beta \left[ a_i(t) - \underline{\alpha}_{ik}(t) a_k(t) \right] b'_k(t) \tag{31}$$
$$\Delta\overline{\alpha}_{ik}(t) = \beta \left[ b_i(t) - \overline{\alpha}_{ik}(t) b_k(t) \right] a'_k(t) \tag{32}$$

These equations are of the dynamic orthogonalization type. Again, due to the restricted space we will further elaborate on that in a future report.

## 3.1 Simulation Results

Figure 5 shows the convergence of the network for the first and the second component. The $y$-axis corresponds to the inner product of $\underline{\mathbf{w}}_i$ and $\overline{\mathbf{w}}_i$ with $\mathbf{u}_i$ and $\mathbf{v}_i$ respectively. Notice that the inner product converges to 1 exponentially fast which implies that the norms of the weight vectors also converge to 1.

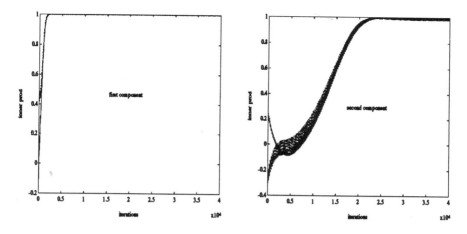

Figure 5: Simulation results for the cross-correlation network. The learning rate $\beta$ is 0.01 for both examples.

# References

[1] P. Baldi and K. Hornik. Neural Networks for Principal Component Analysis : Learning from Examples Without Local Minima. *Neural Networks*, 2:53–58, 1989.

[2] H. Bourlard and Y. Kamp. Auto-association by multilayer perceptrons and singular value decomposition. *Biological Cybernetics*, 59:291–294, 1988.

[3] E. F. Deprettere, editor. *SVD and Signal Processing*. Elsevier Science Publishers B.V. (North Holland), 1988.

[4] S. Y. Kung and K. I. Diamantaras. A Neural Network Learning Algorithm for Adaptive Principal Component EXtraction (APEX). In *Proc. ICASSP*, pages 861–864, Albuquerque, April 1990.

[5] E. Oja. A Simplified Neuron Model as a Principal Component Analyzer. *J. Math. Biology*, 15:267–273, 1982.

[6] T. D. Sanger. Optimal Unsupervised Learning in a Single-Layer Linear Feedforward Neural Network. *Neural Networks*, 2(6):459–473, 1989.

[7] L. L. Scharf. The SVD and Reduced-Rank Signal Processing. In R. Vaccaro, editor, *SVD and Signal Processing II*, pages 3–32. Elsevier, Amsterdam, Netherlands, 1991.

[8] R. Vaccaro, editor. *SVD and Signal Processing, II: Algorithms, Analysis and Applications*. Elsevier Science Publishers B.V., Amsterdam, Netherlands, 1991.

# The outlier process

*Davi Geiger*
Siemens Corporate Research, Inc.
755 College East Road
Princeton, NJ 08540
USA

*Ricardo Alberto Marques Pereira*
Istituto di Informatica
Universita di Trento
Via Antonio Rosmini 42
Trento, TN 38100 ITALY

**Abstract** : We discuss the problem of detecting outliers from a set of surface data. We start from the Bayes approach and the assumption that surfaces are (i) piecewise smooth and (ii) corrupted by a combination of white Gaussian and salt and pepper noise. We show that we can model such surfaces by introducing an outlier process that is capable of "throwing away" data.

We make use of mean field techniques to finally obtain a deterministic network. The experimental results with real images support the model.

## 1 Introduction

We discuss the problem of detecting outliers i.e, data points that are corrupted by salt and pepper noise, from a set of surface data. Our starting point is the Bayes approach and the weak membrane (WM) model of surface reconstruction. The literature on the subject is extensive, see for instance [BlaZis87], [GemGem84], [Marroquin87],[KocMarYui85], [GeiGir90], [MumSha85], [Ambr88] and [GeiYui90].

### 1.1 Bayes and the weak membrane model

Using Bayes theorem and assuming Gaussian noise and piecewise smooth surfaces with borders of discontinuity, the posterior probability is then given by the following coupled Markov random field

$$P(f,l|g) = \frac{1}{Z}e^{-\beta \sum_{i,j}[\frac{(f_{ij}-g_{ij})^2}{2\sigma^2}+\mu'(\Delta_i^2 f+\Delta_j^2 f)(1-l_{ij})+\gamma'_{ij}l_{ij}]} = \frac{1}{Z}e^{-\beta \frac{1}{2\sigma^2}V(f,l)} \tag{1}$$

where we represented surfaces by $f_{ij}$ at pixel $(i,j)$ and discontinuities by $l_{ij}$. $\Delta_i f$ is some form of horizontal gradient of $f$, and $\Delta_i^2 f$ is short for $(\Delta_i f)^2$. $Z$ is a normalization constant called partition function, $\beta$ controls the model uncertainty and is inspired in thermodynamics, $\sigma$ is the standard deviation of the noise (to be estimated) and $\mu'$ and $\gamma'_{ij}$ are parameters to be estimated. The first term is the Gaussian white noise, the second one imposes first order smoothness except at discontinuities and the last term is the cost to create discontinuities, otherwise lines would be created everywhere. This model, when used for image segmentation, has been shown to give a good pattern of discontinuities and eliminate the noise.

The parameter $\sigma$ can be absorved by the parameter $\beta$ and the visual cost function of the weak membrane is given by

$$V(f,l) = \sum_{i,j}[(f_{ij}-g_{ij})^2 + \mu(\Delta_i^2 f+\Delta_j^2 f)(1-l_{ij}) + \gamma_{ij}l_{ij}]. \tag{2}$$

where $\gamma_{ij} = 2\sigma^2\gamma'_{ij}$ and $\mu = 2\sigma^2\mu'$.

We consider the following discretization of the gradient

$$\Delta_i f = K_{ij} + M_{ij} \quad \text{and} \quad \Delta_j f = K_{ij} - M_{ij}$$
$$K_{ij} = \frac{1}{2}(f_{i,j} - f_{i-1,j-1}) \quad \text{and} \quad M_{ij} = \frac{1}{2}(f_{i,j-1} - f_{i-1,j}). \tag{3}$$

Notice that $\Delta_i^2 f + \Delta_j^2 f = 2(K_{ij}^2 + M_{ij}^2)$ and the localization of $K_{ij}$ coincides with $M_{ij}$ to attain consistency with a single line process [GeiYui90]. The problem of this discretization is that at places where $l_{ij} = 0$ the lattice is decoupled into two lattices, each one with half of the size, like a checkerboard. We are now considering other discretizations.

## 2 The outlier process and noise

The proposal here is to extend the WM model by including an additional binary process - the *outlier process* - interacting with the data and capable of supressing it for large amounts of noise. The noise could come from the sensor in which the data was acquired or from some other source. The outlier process $s$ take binary values 0 (data on) and 1 (data off), and the visual cost (2) becomes

$$V(f,l,s) = \sum_{i,j}[(1-s_{ij})(f_{ij}-g_{ij})^2 + \mu(K_{ij}^2 + M_{ij}^2)(1-l_{ij}) + \eta_{ij}s_{ij} + \gamma_{ij}l_{ij}]$$

(4)

where we have introduced the term $\eta_{ij}s_{ij}$ to keep some data, otherwise $s_{ij} = 1$ everywhere. In this model $\eta_{ij}$ is directly related to the "tolerable" amount of noise. More precisely, if the data is very noisy (salt and pepper noise), such as $(f_{ij}-g_{ij}) \geq \eta_{ij}$, then $s_{ij} = 1$ will lower the cost.

In the visual cost (4) above the field $s$ stands to the data term as the line process $l$ does to the smoothness term. Thus if the data is too noisy the field $s = 1$ can suppress it.

## 2.1 The competition between $l$ and $s$

We notice that there is a competition between the line process and the outlier process. Both processes detect changes from the smoothness assumption. The outlier process detects outliers while the line process detects " legal " discontinuities. More precisely, if two neighbor data points have very different values there is a competition for either create a line process and not smooth or to activate the outlier process, suppress one of the data points and, consequently, smooth over. Let us make this argument more precise, for the one dimensional case, with the configuration shown in figure 1 of three pixels where $g_{i-1} = g_{i+1} = b$ and $g_i = a$, what does the model prefers, to create two discontinuities, and keep all the data, or to simply detect an outlier ? Let us consider that $f_{i-1} = g_{i-1}$ and $f_{i+1} = g_{i+1}$, therefore, $s_{i-1} = s_{i+1} = 0$. As for the parameters we assume $\mu(a-b)^2 > \gamma$ and $(a-b)^2 > \eta_i$. So the model either prefers (1) to create two lines ($l_{i-1} = l_i = 1$) and set $f_i = g_i$ ($s_i = s_{i-1} = s_{i+1} = 0$) or (2) to select out $g_i$ ($s_i = 1$) and set $f_i = b$. The cost to create two lines ($f_i = a$) is $V_{(1)} = 2\gamma$ and the cost for selecting out $g_i$ ($f_i = b$) is $V_{(2)} = \eta_i$. In this case the competition between the line process and the outlier process is such that $\eta_i < 2\gamma \Rightarrow V_{(2)} < V(1)$ and the data is selected out, otherwise, edges are created. This analyses also give us some bounds on how to estimate the parameter $\eta_i$ as to detect outliers.

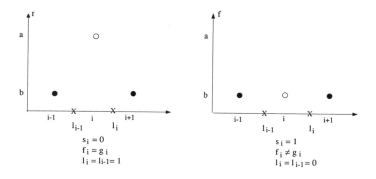

Figure 1: *The data is given by $g_i = a$ and $g_{i-1} = g_{i+1} = b$. The competition between (a) detecting two edges ($l_i = l_{i-1} = 1$) or (b) selecting out the data ($s_i = 1$). The second case is prefered if and only if $\eta_i < 2\gamma$.*

## 2.2 Averaging out the outlier process

Following [GeiGir90] we can sum over the outlier process. From (1) and the model (4) we obtain

$$\begin{aligned}
P(f, l|g) &= \sum_{s=\{0,1\}^{N^2}} P(f, l, s|g) \\
&= \sum_s \frac{1}{Z} e^{-\beta \sum_{ij}[(1-s_{ij})(f_{ij}-g_{ij})^2 + \mu(K_{ij}^2 + M_{ij}^2)(1-l_{ij}) + \eta_{ij}s_{ij} + \gamma_{ij}l_{ij}]} \\
&= \frac{1}{Z} \prod_{i,j} (e^{-\beta(f_{ij}-g_{ij})^2} + e^{-\beta\eta_{ij}}) e^{-\beta[\mu(K_{ij}^2+M_{ij}^2)(1-l_{ij}) + \gamma_{ij}l_{ij}]} \quad (5)
\end{aligned}$$

where $\sum_s$ represents the sum over all configurations of the outlier process $s$. We notice that (5) is very similar to (1) but with the noise model modified. This model accounts for a greater variety of noise (other than just Gaussian noise), including the salt and pepper noise. Notice that for $\eta \to \infty$ we recover the original pure Gaussian noise case.

### 2.2.1 Alternative derivation:

Girosi et al. [GirPogCap90] obtained this model of the noise by considering the a priori distribution for the standard deviation of the noise to be

$$P(\vec{x}) = \prod_{ij}[(1-\epsilon)\delta(x_{ij} - \frac{1}{2\sigma^2}) + \epsilon\delta(x_{ij})] \quad (6)$$

where $\frac{1}{2\sigma^2} = \beta$, $\vec{x} = (x_{00}, x_{0,1}, ..., x_{N-1,N}, x_{N,N})$, and $\epsilon$ controls the percentage of "outliers" (infinite standard deviation). By combining the Gaussian white noise with the *a priori* (6) they obtained

$$\begin{aligned}
P(g|f) &= \int_0^\infty d^{N^2}x P(g|f,\vec{x})P(\vec{x}) \\
&= \prod_{ij} \int_0^\infty dx_{ij}[(1-\epsilon)\delta(x_{ij} - \frac{1}{2\sigma^2}) + \epsilon\delta(x_{ij})]\frac{1}{C_2}e^{-x_{ij}(f_{ij}-g_{ij})^2} \\
&= \prod_{ij} \frac{1}{C_2}[(1-\epsilon)e^{-\frac{1}{2\sigma^2}(f_{ij}-g_{ij})^2} + \epsilon]
\end{aligned} \quad (7)$$

This distribution is the same as obtained in (5), with $e^{-\beta\eta_{ij}} = \frac{\epsilon}{(1-\epsilon)}$ or $\epsilon = \frac{1}{1+e^{\beta\eta_{ij}}}$ and so $\eta_{ij}$ is directly related to the percentage of "outliers". In a more general approach, known as *Robust Statistics* [Hub81], this modelization for the noise distribution is just a special case.

## 3 The network and mean field approach

Given a probability distribution we estimate the average values since it is robust (minimizes the Bayes variance estimator, see for example [GeiGir90])

$$\begin{aligned}
\bar{l}_{ij} &= \sum_{f,l,s} l_{ij} P(f,l,s|g) \quad, \\
\bar{s}_{ij} &= \sum_{f,l,s} s_{ij} P(f,l,s|g) \quad, \\
\bar{f}_{ij} &= \sum_{f,l,s} f_{ij} P(f,l,s|g)
\end{aligned} \quad (8)$$

It is not difficult to check that for the outlier model (4) the equations (9) becomes

$$\begin{aligned}
\bar{l}_{ij} &= -\frac{1}{\beta}\frac{\partial lnZ}{\partial \gamma_{ij}} \\
\bar{s}_{ij} &= -\frac{1}{\beta}\frac{\partial lnZ}{\partial \eta_{ij}} \\
\overline{(1-s_{ij})(f_{ij}-g_{ij})} &= \frac{1}{2\beta}\frac{\partial lnZ}{\partial g_{ij}}
\end{aligned} \quad (9)$$

We can not compute exactly the partition function, $Z = \sum_{f,l,s} e^{-\beta V(f,l,s)}$, but under the saddle point approximation [Parisi88] we obtain

$$Z \approx Cmax_f \sum_{l,s=(0,1)^{N^2}} e^{-\beta V(f,l,s)} \qquad (10)$$
$$= Cmax_f \prod_{ij}(e^{-\beta(f_{ij}-g_{ij})^2} + e^{-\beta\eta_{ij}})(e^{-\beta[\mu(K_{ij}^2+M_{ij}^2)]} + e^{-\beta\gamma_{ij}})$$

where $C$ is a scaling constant and $max_f h(f)$ represents the maximum value that $h(f)$ can assume.

## 3.1 Mean field equations

Using (9) and (11) we obtain the following mean field equations

$$\bar{l}_{ij} = \frac{1}{1+e^{\beta[\gamma_{ij}-\mu(K_{ij}^2+M_{ij}^2)]}}$$
$$\bar{s}_{ij} = \frac{1}{1+e^{\beta[\eta_{ij}-(f_{ij}-g_{ij})^2]}}$$
$$(1-\bar{s}_{ij})(\bar{f}_{ij}-g_{ij}) = -\mu[\bar{K}_{ij}(1-\bar{l}_{ij})-\bar{K}_{i+1j+1}(1-\bar{l}_{i+1j+1})$$
$$+\bar{M}_{ij+1}(1-\bar{l}_{ij+1})-\bar{M}_{i+1j}(1-\bar{l}_{i+1j})] \qquad (11)$$

We notice that the mean field values of the binary variables become continuous and in the limit of low temperature ($lim\beta \to \infty$) we have again the binary values, $0, 1$. The last equation, the one for $f$, is a hard one to solve. We then consider a strategy to solve (11).

## 3.2 Continuation and Jacobi's method

In order to solve the set of coupled equation (11) we start by exactly solving for **a specific** set of parameters $\beta, \mu, \gamma$. We then continuously change the parameters and at the same time solve the set of equations for the update parameters. The idea is that although it is hard to solve the equation for the field $f$, we can use iterative methods like the Jacobi's method starting with a good initial condition given by the solution obtained with the previous set of parameters. This is the essence of continuation methods [Was73] in which a deterministic annealing is a special case.

We have already discussed the competition between the line process and the outlier process. It is therefore important to start with an exact solution. We notice that an initial state of a blurried surface data (with no detection of discontinuities or outliers) is better than simply using the initial data, since the initial data is corrupted by some salt and pepper noise. So we start with a blurried surface, i.e. large $\mu$ and very large $\gamma$, which is easy to obtain (we

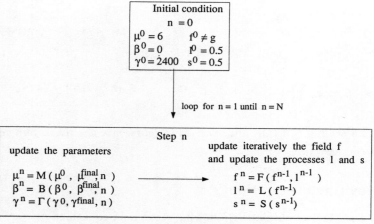

Figure 2: *A sketch of the method.*

can use a Gaussian blur for speed up the computation of the initial state). We also start with $\beta = 0$ (high temperature) and the line and the outlier process are $l_{ij} = s_{ij} = \frac{1}{2}$.

We then gradually **decrease** the values of $\mu$, and $\gamma$, increase the value of $\beta$ and at the same time update the equations (11), using Jacobi's method for the equation on $f$. In figure 2 we sketch the method.

## 4 Result

We have concentrated in the application of this model for image restoration. The schedule for updating the parameters was linear,

$param^{p+1} = param_0 + \frac{p}{N}(param_{final} - param_0)$

where $p$ is an integer that indicates the step of the iteration procedure, $N$ is the number of iterations. We show below the results to one of the images that was artificially corrupted with salt and pepper noise at 100 pixels. The outlier process detected 94% of the noisy data points.

## 5 Theoretical comparison to median filter

Results similar to those shown above could also be obtained by median filtering. Indeed, our nonlinear filter has some similarities to the median filter. The median filter selects some median value within some window according to a criteria that can be the highest value, second highest or the nth-highest. Consequently, the contaminated data is substituted by the median value within

Figure 3: a. An 8-bit image of 256 X 256 pixels. b. "Artificially' added salt and pepper noise (randomly chosen 100 pixels with values 0). c. The initial condition for the smoothed image: $\mu = 8$, $\gamma_{ij} = \eta_{ij} = \infty$ (just smoothing for 50 iterations). d. The final smoothed image, $\mu = 1$, $\gamma_{ij} = 144$, $\eta_{ij} = 22500$ and 200 iterations. e. The corresponding line process f. The corresponding outlier process (94% correct).

the window. The analogous substitute in our model is obtained from the piecewise smoothness assumption.

For places where the distinction between salt and pepper noise data and thin edges are difficult to detect, we expect our approach to outperform the median filter. This is because, unlike the median filtering, our algorithm inherits a model of piecewise smooth surfaces and the presence of Gaussian and salt and pepper noise. In particular, the network has been derived from explicitly modeling the competition between the line process and the outlier process to optimize a global cost function.

In this way, while the median filter is *a valid* local procedure to eliminate the contamination of salt and pepper noise, our alternative proposal is (i) a global model of surfaces and (ii) a *derived* local procedure (network, distributed algorithm) to optimize the model and consequently detect the outliers (and detect edges and smooth the surface).

# Bibliography

[Ambr88]  L. Ambrosio, *Variational problems in SBV*, Technical Report Lectures Notes, M.I.T., 1988.

[BlaZis87]  A. Blake and A. Zisserman, *Visual Reconstruction*, Cambridge, Mass: MIT Press, 1987.

[GeiGir90]  D. Geiger and F. Girosi, "Parallel and deterministic algorithms for mrfs: surface reconstruction and integration," in O. Faugeras, ed., *Lecture Notes in Computer Science*, pp. 89–98, Springer-Verlag, Berlin, 1990.

[GeiYui90]  D. Geiger and A. Yuille, "A common framework for image segmentation," in *ICPR*, June 1990.

[GemGem84]  S. Geman and D. Geman, "Stochastic relaxation, gibbs distributions, and the bayesian restoration of images," *IEEE Transactions on Pattern Analysis and Machine Intelligence*, vol. PAMI-6, pp. 721–741, 1984.

[GirPogCap90]  F. Girosi, T. Poggio, and B. Caprile, *Extensions of a theory of Networks for Approximation and Learning: outliers and negative examples*, Technical Report, Artificial Intelligence Laboratory, Massachusetts Institute of Technology, 1989.

[Hub81]  P.J. Huber, *Robust Statistics*, New York: John Wiley and Sons, 1981.

[KocMarYui85]  C. Koch, J. Marroquin, and A. Yuille, "Analog 'neuronal' networks in early vision," *Proc. Natl. Acad. Sci.*, vol. 83, pp. 4263–4267, 1985.

[Marroquin87]  J. L. Marroquin, "Deterministic Bayesian estimation of Markovian random fields with applications to computational vision," in *Proceedings of the International Conference on Computer Vision*, IEEE, Washington, DC, London, England, June 1987.

[MumSha85]  D. Mumford and J. Shah, "Boundary detection by minimizing functionals, i," in *Proc. IEEE Conf. on Computer Vision & Pattern Recognition*, San Francisco, CA, 1985.

[Parisi88]  G. Parisi, *Statistical Field Theory*, Reading, Massachusets: Addison-Wesley, 1988.

[Was73]  E. Wasserstrom, "Nume.al solutions by the continuation method," *SIAM Review*, no. 15, pp. 89–119, 1973.

# A MAPPING APPROACH FOR DESIGNING NEURAL SUB-NETS

[1]Kamyar Rohani, [2]Mu-Song Chen, and [2]Michael T. Manry
[1]Motorola Inc., 5555 N Beach, Ft. Worth, TX 76137
[2]Dept. of EE, U. of Texas, Arlington, TX 76019

Abstract-Several investigators have constructed back-propagation (BP) neural networks by assembling smaller, pre-trained building blocks, [1,2]. This approach leads to faster training and provides a known topology for the network. Here, we carry this process down one additional level, by describing methods for mapping given functions to sub-blocks. First, polynomial approximations to the desired function are found. Then the polynomial is mapped to a BP network, using an extension of a constructive proof to universal approximation. Examples are given to illustrate the method.

## INTRODUCTION

Many signal and image processing algorithms can be represented as a cascade (composition) of several linear and nonlinear operations. Example linear operations include convolution, inner product, and derivative. Also, example nonlinear operations, which include absolute value, $n^{th}$ power, and $n^{th}$ root, are basically power operations ( $f(x)=x^v$, $v \in R$ ) or combinations of power operations. Here, we develop techniques for mapping such operations to small feedforward networks called building blocks or sub-nets.

The operations of an L-layer feedforward neural network with $N_k$ units (neurons) in each layer can be summarized as

$$u_i^{(k+1)} = \sum_{j=1}^{N_k} W_{ij}^{(k+1)} s_j^{(k)} + \theta_i^{(k+1)} \quad ; \quad s_i^{(k+1)} = g(u_i^{(k+1)}), \quad 1 \leq i \leq N_{k+1} \quad (1)$$

where inputs $s_j$ to each unit are first weighted by the corresponding connection weights, $W_{ij}$. Then the sum of these inputs, $u_i$, is fed into a nonlinear activation $g(x)$ to produce the final output. Each unit has an associated bias denoted by $\theta_i$. Thus, free parameters associated with a network realization are $W=\{ W_{ij}^{(k)}, 1\leq j\leq N_{k-1}, 1\leq i\leq N_k, 1\leq k\leq L \}$, and $\theta=\{ \theta_i^{(k)}, 1\leq i\leq N_k, 1\leq k\leq L \}$.

The problem of learning in the context of neural networks can be defined as follows: Find an approximation to an unknown multi-dimentional mapping $\overline{y} = F(\overline{x})$; this approximation is captured by the network weights $W$. In the case of the well known back-propagation (BP) training, the actual mapping is unknown; however, input-output spaces can be sampled in order to derive a set of training examples. The $N_p$ training examples are defined as $\rho = \{ \rho_p, \ 1 \leq p \leq N_p \}$ where

$$\rho_p \equiv (\overline{x_p}, \overline{y_p}); \ \overline{x_p} \in X \subset R^n, \ \overline{y_p} \in Y \subset R^m \tag{2}$$

Thus, the overall objective is to determine the optimal set of weights, $W_{opt}$, such that a close approximation can be made to the unknown mapping,

$$\overline{y_p} = \hat{F}(\overline{x_p}) \ \forall \ 1 \leq p \leq N_p \ ; \quad \hat{F}: R^n \times W \to R^m \tag{3}$$

Note that the number of free parameters associated with a network realization are

$$D = \sum_{k=1}^{L} N_k(N_{k-1}+1) \tag{4}$$

Hence, a solution point $W_{opt}$ must be determined in the $D$-dimensional Euclidean space $R^D$. It should be clear that closed form solutions cannot be found except for special cases. BP training is a generalized LMS algorithm. Thus, in general, some problems with this method are: optimal solutions are not guaranteed, convergence is dependent on the initial starting point, and convergence rate depends on the training examples.

The universal approximation property of neural networks, having a single hidden layer of sigmoidal units, has been demonstrated by several authors including [4-7]. Such networks are capable of approximating any continuous multi-dimensional mapping defined on a finite support,

$$F: X \in R^n \to Y \in R^m \tag{5}$$

Given this powerful theorem, the next logical step is to find ways to efficiently map known functions, to BP networks. Note that this is a different viewpoint than that of BP training.

In this paper, we outline the steps in mapping a given polynomial function onto a network structure, by extending the constructive proof of universal approximation seen in [6]. Such a methodology should prove highly useful, since it leads to good initial weights, results in a known network topology, gives us useful insights into the interworkings of the network units, and yields a useful final network. Some simulation examples are included in order to clarify the proposed algorithm.

# MAPPING ALGORITHM

Nonlinear operators can often be approximated using power series in terms of their input variables. Consequently, in this section, we devise a method for designing nonlinear sub-nets, by mapping a given $N^{th}$ order polynomial function $p(x)$ onto a neural network. Most of the analysis in this section is directed towards single-input/single-output networks. However, this analysis can be extended to multi-input/multi-output networks. The mapping problem can be formulated as follows: Given a $N^{th}$ order polynomial function $p(x)$, find an equivalent network which approximates the mapping, i.e. determine the network weights. Thus, the objective is to find the network weights and constants (refer to Figure 1),

$$\overline{K} = [k_1, k_2, \ldots, k_m]^T \; ; \; \overline{B} = [b_1, b_2, \ldots, b_m]^T \; ; \; \overline{W} = [w_1, w_2, \ldots, w_{m+2}]^T \qquad (6)$$

where m=N-1 is the number of hidden units, $\overline{K}$ stores input connection weights, $\overline{B}$ stores the hidden unit thresholds, and $\overline{W}$ stores the output connection weights.

Assuming that the hidden units have analytic activation functions, their outputs can be modeled as polynomial functions of their net inputs, as in [6]. Thus, for a given radius of convergence $R_i$ and point of expansion $\eta_i$, we can approximate the activation output $s_i$ of the $i^{th}$ unit with the polynomial

$$s_i \approx \sum_{n=0}^{N} \alpha_i(n) \, (u_i - \eta_i)^n \qquad (7)$$

where $u_i$ denotes the net function and $\alpha_i(n)$ are the polynomial expansion coefficients. Based on this terminology, the network's output is a linear combination of the polynomials formed by the hidden units. Ideally, $p(x)$ is composed of a linear combination of orthogonal kernels which span its output space

$$p(x) = \sum_{k=0}^{\infty} c_k \, \psi_k(x) \qquad (8)$$

On the other hand, $p(x)$ can be approximated as follows

$$p(x) \approx \sum_{k=0}^{N} c_k \, p_k(x) \qquad (9)$$

where $p_k(x)$ denote the polynomial approximation to the output of the $k^{th}$ sigmoidal unit. Clearly, this approximation is not compact and many solutions exist to the above approximation problem. Our objective in the following algorithm is to devise a technique which minimizes the overall system error in a mean-square sense.

Now, we construct an algorithm which automatically maps a given $p(x)$ to a network. A detailed description of the procedure is given below:

**Step 1.** Given an input/output nonlinear mapping, our first task is to describe this mapping in terms of power series

$$p(x) = \sum_{n=0}^{N} c_n x^n \quad (10)$$

In the experiments that follow, we perform this task by computing a polynomial fit in a least-square sense. This is accomplished by sampling the nonlinear function at distinct points, $x_i$, in the range $[x_{min}, x_{max}]$ and forming the matrix

$$\begin{bmatrix} 1 & x_1 & x_1^2 & \cdots & x_1^{N_p} \\ 1 & x_2 & x_2^2 & \cdots & x_2^{N_p} \\ \vdots & \vdots & \vdots & \ddots & \vdots \\ 1 & x_M & x_M^2 & \cdots & x_M^{N_p} \end{bmatrix} \begin{bmatrix} c_0 \\ c_1 \\ \vdots \\ c_{N_p} \end{bmatrix} = \begin{bmatrix} p(x_1) \\ p(x_2) \\ \vdots \\ p(x_M) \end{bmatrix} \quad (11)$$

or in compact form

$$X \overline{C} = \overline{Y} \quad (12)$$

The least-square solution to this matrix equation is

$$\overline{C} = (X^T X)^{-1} X^T \overline{Y} \quad (13)$$

Although the equation is ill-conditioned, it is solvable in practice through the well known LU decomposition and other techniques.

**Step 2.** In order to generate an arbitrary desired output polynomial, (N-1) sigmoidal units are sufficient. The output of the first unit is a $N^{th}$ order polynomial; the second units output is a $(N-1)^{th}$ order polynomial; the third unit approximates a $(N-2)^{th}$ order polynomial, and so on. By arranging the sigmoidal units in this fashion, $i^{th}$ unit corrects the $(N-i+1)^{th}$ degree terms generated by the preceding units.

For the $i^{th}$ unit, we choose an arbitrary point of expansion, $\eta_i$. Given this point, the radius of convergence, $R_i$, is determined such that the approximating polynomial has the desired degree. In our implementation, points of expansion are chosen in regular intervals according to $y_o(i) = \beta(i-1)$ where $\beta = 0.25$. The corresponding radius of convergence is found by increasing the interval until the resulting polynomial approximation is $i^{th}$ degree.

**Step 3.** Once the points of expansion and radii of convergence are known for all hidden units, the input weights and thresholds can be computed from

$$\quad (14)$$

$$b_i + k_i x_{max} = \eta_i + R_i \quad ; \quad b_i + k_i x_{min} = \eta_i - R_i$$

or after simplifying

$$k_i = \frac{2R_i}{x_{max} - x_{min}} \quad ; \quad b_i = \eta_i + R_i - k_i + x_{max} \quad (15)$$

**Step 4.** Up to this point, all network parameters are known except for the output connection weights. The output weights, $W$, are determined such that the mean-squared output error

$$E = \sum_n \left[ p(x_n) - \sum_{i=1}^{N+1} w_i \; s_i(n) \right]^2 \tag{16}$$

is minimized, where $s_i(n)$ denotes output of the $i^{th}$ unit for the $n^{th}$ input sample $x_n$, $s_N(n) = x_n$, $s_{N+1}(n) = 1$, and $p(x_n) = \{p(x) \text{ for } x = x_n\}$. We then set the derivative of the error with respect to each weight,

$$\frac{\partial E}{\partial w_j} = -2 \sum_n \left[ p(x_n) - \sum_{i=1}^{N+1} w_i \; s_i(n) \right] s_j(n) \tag{17}$$

to zero, yielding

$$\sum_{i=1}^{N+1} w_i \sum_n s_i(n) \; s_j(n) = \sum_n p(x_n) \; s_j(n) \tag{18}$$

This can be easily written in matrix form as

$$A_w \; W = \overline{B_w} \tag{19}$$

where elements of $A_w$ and $\overline{B}_w$ are

$$a_{ij} = \sum_n s_i(n) s_j(n) \quad ; \quad b_j = \sum_n p(x_n) s_j(n) \quad 1 \leq i,j \leq N+1 \tag{20}$$

This completes our polynomial mapping procedure. The next section contains examples which illustrate the usefulness of this technique. At this point, two important notes are in order: (1) The method described in this section can be extended to multi-input multi-output mapping. However, most signal processing algorithms require only single-input single-output operators (or their composites). Thus the algorithm described here is sufficient for most signal processing tasks, [2]. (2) The point of expansion in step 2 was arbitrarily chosen at regular intervals which performs well for most mapping described here. One should note, however, that better points of expansion can be determined by searching the entire input space (e.g. via a brute force method).

## SIMULATION RESULTS

To illustrate the validity of our mapping algorithm, this section contains results of three simulation examples.

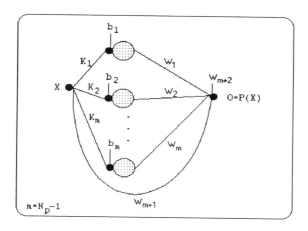

**Figure 1** Mapping Polynomial to Network.

## Squaring Network

Squaring sub-nets were designed via BP training and using the mapping procedure described above. Since $y = x^2$ is a second order polynomial, these sub-nets each had one hidden unit (1-1-1 sub-nets). The 600 input training numbers were randomly chosen, with a uniform probability density, between -4 and 4. The desired and actual outputs versus input are shown in Figures 2-3. The mapped sub-net has a clear advantage in performance and speed of design. The failure of the randomly-initialized sub-net to learn, when we know the topology is appropriate, is a phenomena that can occur with any size BP network.

We investigated the problem with BP training by examining the error surface. Three dimensional plots of mean-square error,

$$E = \sum_n \left[ x_n^2 - \hat{y}_n \right]^2 \; ; \; \hat{y}_n = W_{32} \left[ \frac{1}{1+e^{-(W_{21} x_n + \theta_1)}} \right] + W_{31} x_n + \theta_2 \quad (21)$$

as a function of free parameters $W_{21}$ and $\theta_1$ are given in Figures 4-5. The global minimum corresponds to a sudden sharp drop in the magnitude of the mean-square error (MSE) with a minimum value of 0.007. On the other hand, the local minima corresponds to a shallow minimum. Once, the BP training algorithm is trapped within the bounds of this minimum point, it remains at this point and cannot detect the "non-optimality" of the solution.

## Scalar Inversion Network

A difficulty with the scalar inversion operation is that it contains a singular point at zero. Thus, approximations have to be limited to intervals not containing this point. A 7th order polynomial approximation to the inversion operator was obtained via the least-squares analysis where $y = 1/x$ ; $x:[0.25, 4]$. Next, a 1-6-1

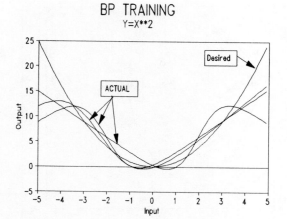

**Figure 2** Output Versus Input for Trained Squaring Sub-Nets.

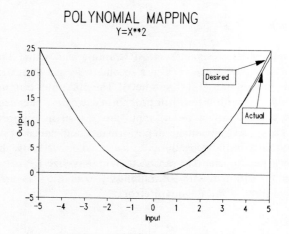

**Figure 3** Output Versus Input for Mapped Squaring Sub-Net.

network was designed via the algorithm described in the previous section. The output of this network is shown in Figure 6. Clearly, the network output closely approximates the desired polynomial as expected.

## Square Root Network

To approximate the square-root operation in the range [0,1], a 9th order polynomial was derived using the least-squares method. Next, a 1-8-1 network was designed using the analysis of the previous section. The output of this network is shown in Figure 7. Note that the network output follows the desired polynomial closely.

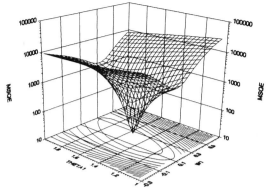

**Figure 4** Error Surface at Absolute Minima for $y = x^2$ Training (Minimum Point is Below XY-Plane at MSE=0.007).

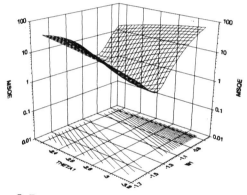

**Figure 5** Error Surface at a Local Minima for $y = x^2$ Training.

## CONCLUSIONS

In this paper, we have described an approach for mapping simple mathematical operations to BP sub-networks, via polynomial approximation. This approach, which follows from the constructive universal approximation proof of [6], leads to a known topology for the sub-net. For clarity, most of the analysis presented in this paper is limited to the case of single-input single-output networks. However, it should be mentioned that the same strategy can be extended to the case of multi-input multi-output systems.

## ACKNOWLEGMENT

The contributions to this work by Mu-Song Chen and Michael T. Manry were funded by the State of Texas through the Advanced Technology Program.

## REFERENCES

[1]  A. Waibel, H. Sawai, K. Shikano, "Consonant Recognition by Modular Construction of Large Phonemic Time-Delay Neural Networks", International Conf. on Acoust, Speech, and Signal Processing, ICASSP-89, Vol. S3.9, pp.112-115, May 1990.

[2]  K. Rohani, M. Manry, "The Design of Multi-Layer Perceptron using Building Blocks", Submitted to International Joint Conf. on Neural Net, IJCNN'91.

[3]  D.E. Rumelhart, G.E. Hinton, and R.J. Williams, Learning internal representations by error propagation,. In D.E. Rumelhart and J.L. McClelland (Eds.). Parallel distributed processing, Vol. I, Cambridge, Massachusetts: The MIT Press, 1986.

[4]  G. Cybenko, "Approximations by Superpositions of a Sigmoidal Function," Math. Contrl., Signals, Syst., vol. 2, 1989, pp. 303-314.

[5]  K. Hornik, M. Stinchcombe, H. White, "Multilayer Feedforward Networks Are Universal Approximators," Neural Networks, vol. 2, 1989, pp. 359-366.

[6]  M. Chen, M. Manry, "Back-Propagation Representation Theorem using Power Series", International Joint Conf. on Neural Net, IJCNN'90, Vol.I, pp.643-648, San Diego, California, June 17-21, 1990.

[7]  I. Sandberg, "Approximation Theorems for Discrete-Time Systems," IEEE Trans on Circuits and Systems, Vol.38, No.5, May 1991.

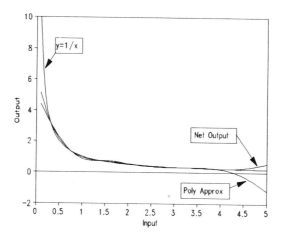

**Figure 6** Neural Network Approx. to $y = 1/x$; $0.25 < x < 4.0$.

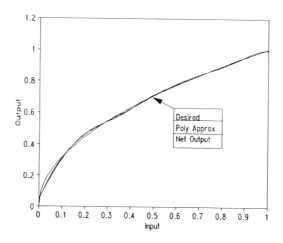

**Figure 7** Neural Network Approx. to square-root operation.

# THREE-DIMENSIONAL STRUCTURED NETWORKS FOR MATRIX EQUATION SOLVING

Li-Xin Wang and Jerry M. Mendel

Signal and Image Processing Institute
Department of Electrical Engineering-Systems
University of Southern California
Los Angeles, CA, 90089-2560

## Abstract

Structured networks are feedforward neural networks with linear neurons that use special training algorithms. In this paper, two three-dimensional (3-D) structured networks are developed for solving linear equations and the Lyapunov equation.' The basic idea of the structured network approaches is: first, represent a given equation-solving problem by a 3-D structured network so that if the network matches a desired pattern array, the weights of the linear neurons give the solution to the problem; then, train the 3-D structured network to match the desired pattern array using some training algorithms; finally, obtain the solution to the specific problem from the converged weights of the network. The training algorithms for the two 3-D structured networks are proved to converge exponentially fast to the correct solutions.

## 1 INTRODUCTION

During the last two decades, many parallel methods for matrix algebra problems have been developed, analysed, and implemented on specific architectures [1-10]. Most of these studies concentrated on parallelizing existing serial algorithms, or developing new algorithms which admit a high degree of parallelization. Due to the difficulty of developing a general and easy-to-use software system for parallel computers, and other reasons [11], practical applications of parallel algorithms for matrix algebra problems are limited. With the rapid developments of artificial neural networks and the advances of neural chips [12,13], one may ask: Is it possible to use artificial neural networks to solve matrix algebra problems ?

Recently, *structured networks* were developed for solving a wide variety of matrix algebra problems in a massively parallel fashion [14-17]. Structured networks are feedforward neural networks with linear neurons that use special training algorithms. The basic idea of the structured network approaches is: first, represent a given matrix algebra problem by a structured network so that if the network matches a set of desired patterns, the weights of the linear neurons give the solution to the problem; then, train the structured network to match the desired patterns using some training algorithms; and, finally, obtain the solution to the problem from the converged weights of the network.

The structured networks of [14-17] are two-dimensional (2-D); hence, the massively parallel processing advantage of structured networks is not fully utilized. Also, because the architectures are 2-D, a set of patterns (the number of which usually equals the dimension of the matrix involved in the matrix algebra problem) must be matched, which makes it difficult to analyse the convergence properties of the training algorithms for these 2-D structured networks. At present, convergence analysis has only been done for linear problems [17].

In this paper, we extend the methods of [14-17] to three-dimensional (3-D) structured networks. A 3-D structured network is a combination of $n$ 2-D structured networks, where $n$ is the dimension of the (square) matrix involved in the matrix algebra problem to be solved. The basic idea remains the same, i.e., we first represent a given problem by a 3-D structured network architecture, then train the 3-D network to match a pattern array, and finally obtain the solution to the problem from the converged weights. The difference is that 3-D structured networks are required to match only one pattern, whereas 2-D structured networks are required to match $n$ patterns. 3-D structured networks introduce more parallelism into the processing, hence the massively parallel processing advantage of structured networks is further utilized. However, 3-D structured networks are more complicated than 2-D structured networks.

In Section 2, a detailed 3-D structured network for solving linear equations is constructed, and a new adaptive steepest descent training algorithm is developed for this specific network. Convergence properties of the training algorithm are analyzed. In Section 3, we construct a 3-D structured network for solving the Lyapunov equation, and develop a new adaptive steepest descent training algorithm for this network. Convergence properties of the training algorithm are analyzed. Conclusions are given in Section 4.

# 2 LINEAR EQUATION SOLVER

1 . *Network Architecture and Desired Pattern*

Consider the problem of solving the linear equation

$$Ax = b \qquad (1)$$

where $A \in R^{n \times n}$ is invertible. We solve this equation by first determining the inverse of $A$, i.e., we first find a matrix $B \in R^{n \times n}$ such that

$$AB = I_{n \times n}; \tag{2}$$

then the solution to (1) is obtained as

$$x = A^{-1}b = Bb. \tag{3}$$

We construct a 3-D structured network for this linear equation solving problem in the following way: first, we construct a 2-D 2-layer structured network whose lower-layer performs transformation $B$ and whose upper-layer performs transformation $A$; then, we combine $n$ of these 2-D structured networks into a 3-D structured network, as shown in Fig.1 for the case of $n = 3$. In Fig.1, the input pattern array is denoted by $n \times n$ matrix $X = (x_1, x_2, ..., x_n)$, where $x_i$ is the input vector to the $i$'th 2-D sub-network ( the ordering directions, i.e., how we order $i$ from 1 to $n$, are shown in the figure ); and, the actual output pattern array is denoted by $n \times n$ matrix $Y = (y_1, y_2, ..., y_n)$, where $y_i$ is the actual output pattern vector of the i'th 2-D sub-network, and $y_i = ABx_i$.

From this construction we see that all $n$ 2-D sub-networks are identical; hence, the training algorithm for the 3-D structured network must meet this constraint. The horizontal directed lines in the 3-D network of Fig.1 represent sharing of weight values among the 2-D sub-networks.

THEOREM 1: If the 3-D structured network of Fig.1 matches the pattern pair $(X, Y) = (I_{n \times n}, I_{n \times n})$, then the lower-layer weight matrix $B$ of any 2-D sub-network satisfies $AB = I_{n \times n}$.

Proofs of the theorems in this paper can be found in [18].

2 . *Training Algorithm and Convergence Analysis*

We propose the following adaptive steepest descent (ASD) training algorithm for the Fig.1 3-D structured network.

Step 1: Set all the upper-layer weights of the $n$ 2-D sub-networks equal to the corresponding elements of $A$ such that all these upper layers perform transformation $A$. These weights are never changed during the training procedure. Let $b_{ij}^p$ denote the j'th weight of the i'th neuron of the lower-layer of the p'th 2-D sub-network, where the ordering directions are shown in Fig.1. Set the initial lower-layer weights of the $n$ 2-D sub-networks, $b_{ij}^p(0)$, equal to any value under the constraint that $b_{ij}^{p_1}(0) = b_{ij}^{p_2}(0)$ for all $1 \leq i, j, p_1, p_2 \leq n$. Set the input pattern array $X$ equal to the desired pattern array $I_{n \times n}$.

Step 2: Run the 3-D structured network in the forward direction with the weights $b_{ij}^p(k)$ to obtain $y_{ij}(k)$, where $1 \leq i, j, p \leq n$ and $k = 0, 1, 2, ...$.

Step 3: Update the weights $b_{ij}^j$ by

$$b_{ij}^j(k+1) = b_{ij}^j(k) + \alpha_j(k)[d_{ij} - y_{ij}(k)], \tag{4}$$

where $1 \leq i, j \leq n$, $k = 0, 1, 2, ...$; $b_{ij}^j(k)$ is the weight $b_{ij}^j$ at the k'th training cycle; $d_{ij}$ is the $(i, j)$'th element of the desired output pattern array $I_{n \times n}$, i.e., $d_{ij} = e_{ij}$; $y_{ij}(k)$ is the $(i,j)$'th element of the actual output pattern array at the k'th training cycle; and,

$$\alpha_j(k) = \sum_{i=1}^n (d_{ij} - y_{ij}(k))^2 / \sum_{i=1}^n \sum_{q=1}^n (d_{ij} - y_{ij}(k))a_{iq}(d_{qj} - y_{qj}(k)). \quad (5)$$

Then, set
$$b_{ij}^p(k+1) = b_{ij}^j(k+1) \quad (6)$$
for all $1 \leq i,j,p \leq n$ and $p \neq j$.

Step 4: Set $k = k+1$, and repeat by going to Step 2 for another training cycle until $E(k) = \sum_{i=1}^n \sum_{j=1}^n (d_{ij} - y_{ij}(k))^2$ is less than a prespecified threshold $\epsilon$.

Comment 1: This training algorithm updates the $n^3$ weights $b_{ij}^p$ ( $1 \leq i,j,p \leq n$ ) in two stages: first, it updates the $n^2$ weights $b_{ij}^j$ for $1 \leq i,j \leq n$ using Eq.(4), where only $n$ weights are updated in each of the $n$ 2-D sub-networks ( we call these $n^2$ weights *essential* weights ); then, the remaining $n^3 - n^2$ weights, $b_{ij}^p$ for $1 \leq i,j,p \leq n$ and $p \neq j$, are set equal to the corresponding essential weights according to Eq.(6) (the horizontal directed lines in Fig.1 indicate these settings). In this way, all the lower-layers of the $n$ 2-D sub-networks remain identical, i.e., our ASD training algorithm guarantees that the resulting 3-D structured network meets the constraint that all the 2-D sub-networks must be identical.

Comment 2: The stepsizes $\alpha_j(k)$ change from one training cycle to another and are different for weights in different 2-D sub-networks. This is a major difference between this ASD training algorithm and the error back-propagation algorithm [20-22] or the Least-Mean-Squares algorithm [23].

THEOREM 2: Suppose $A \in R^{n \times n}$ is positive-definite, and there exist positive $m$ and $M$ such that

$$m = \inf_{x \neq \theta} \frac{(x, Ax)}{(x,x)}, \qquad M = \sup_{x \neq \theta} \frac{(x, Ax)}{(x,x)} \quad (7)$$

where $(x,y) = x^T y$ denotes the real inner-product of vectors $x$ and $y$, and $\theta$ is a zero vector. Then the $n \times n$ matrix $B(k) = [b_{ij}^{p_0}(k)]$ ( i.e., the $(i,j)$'th element of $B(k)$ is $b_{ij}^{p_0}(k)$ where $1 \leq i,j \leq n$ and $p_0$ is any fixed integer $1 \leq p_0 \leq n$; since Eq.(6) guarantees that $b_{ij}^{p_0}(k) = b_{ij}^p(k)$ for any $1 \leq p_0, p \leq n$, we do not need an index $p_0$ for $B(k)$ ) obtained from the ASD training algorithm converges to $B_0$ such that $AB_0 = I_{n \times n}$. Additionally, let $B_0 = [b_1^0, b_2^0, ..., b_n^0]$, $B(k) = [b_1(k), b_2(k), ..., b_n(k)]$, and define

$$F[b_j(k)] = (b_j(k) - b_j^0, A(b_j(k) - b_j^0)); \quad (8)$$

then the rate of convergence of the ASD training algorithm is characterized by
$$F[b_j(k)] \leq (1 - \frac{m}{M})^k F[b_j(0)] \quad (9)$$
for $j = 1, 2, ..., n$.

Simulations of the training algorithms in this paper can be found in [18].

# 3 LYAPUNOV EQUATION SOLVER

1. *Network Architecture and Desired Pattern*

Consider the problem of solving the following Lyapunov equation

$$AP + PA^T = W, \qquad (10)$$

where $A, W \in R^{n \times n}$ are given, $P \in R^{n \times n}$ is unknown, and $W$ and $P$ are symmetric. The Lyapunov equation is a very important and basic equation in stability analyses of systems and lots of other applications. The basic idea of the 3-D structured network approach to this problem is to view the problem of solving Eq.(10) as a matrix decomposition problem, i.e., to decompose the given matrix $W$ into the form $AP + PA^T$, where $P$ is constrained to be symmetric. Consequently, we first construct a 3-D structured network which consists of $n$ 2-D sub-networks in which each 2-D sub-network performs the transformation $AP + PA^T$; then, we train this 3-D structured network to match a desired pattern array, with the constraint that $P$ must be symmetric. The solution to the Lyapunov equation comes from the final converged weights of the 3-D structured network.

The 3-D structured network for Lyapunov equation solving is shown in Fig.2 for the $n = 2$ case. This 3-D structured network consists of $n$ sub-networks where each 2-D sub-network is a three-layer feedforward network which performs transformation the $AP + PA^T$. These $n$ 2-D sub-networks are required to be identical. The input and output pattern arrays of the 3-D structured network are denoted by $X = (x_1, x_2, ..., x_n)$ and $Y = (y_1, y_2, ..., y_n)$, respectively; and, $x_j$ and $y_j$ are the input and output vectors for the j'th 2-D sub-network, respectively. We see from Fig.2 that $y_j = (AP + PA^T)x_j$.

THEOREM 3: If the 3-D structured network of Fig.2 matches the pattern array pair $(X, Y) = (I_{n \times n}, W)$, then the weight matrix $P$ of any 2-D sub-network satisfies $AP + PA^T = W$.

2. *Training Algorithm and Convergence Analysis*

We propose the following adaptive steepest descent (ASD) algorithm to train the 3-D structured network of Fig.2.

Step 1: Set the weights corresponding to $A$ in the 3-D structured network equal to the $A$ of a given Lyapunov equation. These weights are never changed during the training procedure. Let $p_{ij}^q$ denote the $(i,j)$'th element of the $P$ matrix in the q'th 2-D sub-network. Set the initial $p_{ij}^q(0)$ equal to any value under the conditions that $p_{ij}^{q_1}(0) = p_{ij}^{q_2}(0)$ for all $1 \leq i, j, q_1, q_2 \leq n$ and $p_{ij}^q(0) = p_{ji}^q(0)$ for all $1 \leq i, j, q \leq n$. Set the input pattern array $X = I_{n \times n}$.

Step 2: Run the 3-D structured network in the forward direction with the weights $p_{ij}^q(k)$ to obtain $y_{ij}(k)$, where $1 \leq i, j, q \leq n$ and $k = 0, 1, 2, ...$.

Step 3: Update the weight $p_{ij}^j$ by

$$p_{ij}^j(k+1) = p_{ij}^j(k) + \alpha(k)[d_{ij} - y_{ij}(k)] \qquad (11)$$

where $1 \leq i, j \leq n$, $k = 0, 1, 2, ...$; $d_{ij}$ and $y_{ij}(k)$ are the $(i,j)$'th elements of the desired and actual output pattern arrays at the k'th training cycle, respectively; and,

$$\alpha(k) = \sum_{i=1}^{n}\sum_{j=1}^{n}(d_{ij}-y_{ij}(k))^2 / \sum_{i=1}^{n}\sum_{j=1}^{n}\sum_{p=1}^{n}\sum_{q=1}^{n}(d_{pi}-y_{pi}(k))(a_{pq}+a_{ij})(d_{qj}-y_{qj}(k)). \quad (12)$$

Then, set
$$p_{ij}^q(k+1) = p_{ij}^j(k+1) \quad (13)$$

for $1 \leq i,j,q \leq n$ and $q \neq j$.

Step 4: Set $k = k+1$, and return to Step 2 for another training cycle until the error $E(k) = \sum_{i=1}^{n}\sum_{j=1}^{n}(d_{ij} - y_{ij}(k))^2$ is less than a threshold $\epsilon$.

THEOREM 4: Suppose the $A$ in the Lyapunov equation (10) is such that the $n^2 \times n^2$ matrix
$$G = I_{n \times n} \bigotimes A + A \bigotimes I_{n \times n} \quad (14)$$

is positive-definite, and there exist $0 < m_1 \leq M_1$ such that
$$m_1 = \inf_{x \neq \theta} \frac{(x, Gx)}{(x, x)}, \quad M_1 = \sup_{x \neq \theta} \frac{(x, Gx)}{(x, x)}, \quad (15)$$

where $\bigotimes$ denotes the Kronecker product [25]. Then the $n \times n$ matrix $P(k) = [p_{ij}^{q_0}(k)]$ (Eq.(13) guarantees that this $P(k)$ is identical for all $1 \leq q_0 \leq n$) obtained from the ASD training algorithm for the Lyapunov equation solving network converges to such a $P_0$ that $AP_0 + P_0 A^T = W$. Additionally, define
$$F[P(k)] = (vec(P(k) - P_0), Gvec(P(k) - P_0)); \quad (16)$$

then the rate of convergence is characterized by
$$F[P(k)] \leq (1 - \frac{m_1}{M_1})^k F[P(0)]. \quad (17)$$

## 4 CONCLUSIONS

In this paper two three-dimensional (3-D) structured networks were developed for solving linear and Lyapunov equations, respectively. These 3-D structured networks preserve the most important advantage of the previous 2-D structured networks [14-17], i.e., massively parallel processing. New training algorithms were proposed which guarantee that the network weights converge exponentially fast to the correct problem solution under some conditions.

An advantage of the structured network approaches is that different training algorithms can be used for the same network architecture, because our prime purpose is to make the network match a desired pattern (see Theorems 1 and 3); hence, users can choose either a simpler but possibly less accurate training algorithm or a more complicated but more accurate training algorithm without changing the network architecture. In this paper, we proposed only one training algorithm for each network. The error back-propagation algorithm [20-22] can also be used to train the 3-D structured networks in this paper. In fact, the 2-D structured networks of [14-16] used the back-propagation algorithm. The main advantage of the back-propagation algorithm over the training algorithms in this paper is its simplicity. The main reason for this simplicity is that the back-propagation algorithm uses a constant stepsize $\alpha$, while the training algorithms in this paper use very complicated adaptive rules for choosing the stepsizes (see Eqs.(5) and (12)). However, due to a constant stepsize, convergence analysis of the back-propagation algorithm is difficult, and its convergence rate is slow. On the other hand, due to the adaptive stepsizes, the training algorithms in this paper converge exponentially fast to the correct solutions. Hence, the back-propagation algorithm and the algorithms in this paper may be viewed as two extremes: one is simple but slow, the other is complicated but fast. Some algorithms in between may be developed, e.g., by approximating Eqs.(5) and (12) with simpler expressions.

## 5 REFERENCES

[1] D. P. Bertsekas and J. N. Tsitsiklis, "*Parallel and Distributed Computation,*" Prentice-Hall, Inc., Englewood Cliff, NJ, 1989.

[2] D. Heller, "A Survey of Parallel Algorithms in Numerical Linear Algebra," *SIAM Rev.*, 20:740-777, 1978.

[3] S. L. Johnson, "Solving Tridiagonal Systems on Ensemble Architectures," *SIAM J. Sci. Stat. Comput.*, 8:345-392, 1987.

[4] A. Bojanczyk, R. P. Brent and H. T. Kung, "Numerically Stable Solution of Linear Equations Using Mesh-Connected Processors," *SIAM J. Sci. Stat. Comput.*, 5:95-104, 1984.

[5] S. Y. Kung, *VLSI Array Processors*, Prentice-Hall, Inc., Englewood Cliff, NJ, 1988.

[6] H. T. Kung, B. Sproul and G. Steele, *VLSI Systems and Computations*, Computer Science Press, Rockville, MD, 1981.

[7] L. Csanky, "Fast Parallel Matrix Inversion Algorithms," *SIAM J. Comput.*, 5:618-623, 1976.

[8] A. Bojanczy, "Complexity of Solving Linear Systems in Different Models of Computation," *J. ACM*, 32:792-803, 1984.

[9] R. Barlow, D. Evans and J. Shanehchi, "Parallel Multisection Applied to the Eigenvalue Problem," *The Computer Journal*, 26:6-9, 1983.

[10] E. Jessup and D. Sorensen, "A Multiprocessor Scheme for the Singular Value Decomposition," in *Parallel Processing for Scientific Computing*, pp.61-66, 1989.

[11] A. Brenner, etc., "Panel: How Do We Make Parallel Processing a Reality ?: Bridging the Gap Between Theory and Practice," *Proc. 5th International Parallel Processing Symposium*, pp. 648-653, 1991.

[12] N. Morgan (ed.), *Artificial Neural Networks Electronic Implementations*, IEEE Computer Society Press, 1990.

[13] U. Ramacher and U. Ruckert (ed.), *VLSI Design of Neural Networks*, Klumer Academic Publishers, 1991.

·[14] L. X. Wang and J. M. Mendel, "Structured Trainable Networks for Matrix Algebra," *Proc. of 1990 International Joint Conf. on Neural Networks*, Vol.2, pp.II125-II132, 1990.

[15] L. X. Wang and J. M. Mendel, "Matrix Computations and Equation Solving Using Structured Networks and Training," *Proc. of IEEE 1990 Conf. on Decision and Control*, pp.1747-1750, 1990.

[16] L. X. Wang and J. M. Mendel, "Parallel Structured Networks for Solving a Wide Variety of Matrix Algebra Problems," submitted to *J. of Parallel and Distributed Processing*, 1991.

[17] M. M. Polycarpou and P. A. Ioannou, "Learning and Convergence Analysis of Neural-Type Structured Networks," *IEEE Trans. on Neural Networks*, to appear, 1990.

[18] L. X. Wang and J. M. Mendel, "Three-Dimensional Structured Networks for Matrix Equation Sloving," *IEEE Trans. on Computers*, Dec., 1991.

[19] L. X. Wang and J. M. Mendel, "Cumulant-Based Parameter Estimation Using Structured Networks," *IEEE Trans. on Neural Networks*, Vol. 2, No. 1, pp. 73-83, 1991.

[20] P. Werbos, "New Tools for Predictions and Analysis in the Behavioral Science," Ph.D. Thesis, Harvard U. Committee on Applied Mathematics, 1974.

[21] D. E. Rumelhart, G. E. Hinton and R. J. Williams, "Learning Internal Representation by Error Propagation," in *Parallel Distributed Processing I*, edited by D.E.Rumelhart and J.L.McClelland, MIT Press, 1986.

[22] D. B. Parker, "Learning Logic," Technical Report TR-47, MIT Center for Computational Economics and Statistics, 1985.

[23] B. Widrow and S. D. Stearns, *Adaptive Signal Processing*, Prentice-Hall, New Jersey, Englewood Cliff, NJ, 1985.

[24] D. G. Luenberger, *Optimization by Vector Space Methods*, John Wiley, New York, 1969.

[25] J. W. Brewer, "Kronecker Products and Matrix Calculus in System Theory," *IEEE Trans. on Circuits and Systems*, Vol.CAS-25, No.9, pp.772-781, 1978.

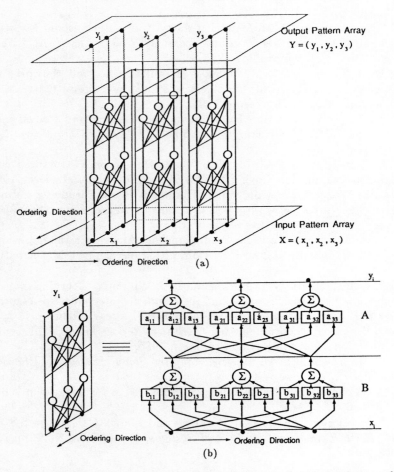

Figure 1: Three-dimensional structured network for linear equation solving for the case of $n = 3$. (a): the 3-D network. (b): a detailed 2-D sub-network.

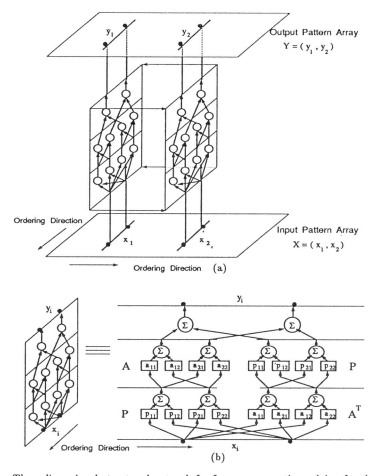

Figure 2: Three-dimensional structured network for Lyapunov equation solving for the case of $n = 3$. (a): the 3-D network. (b): a detailed 2-D sub-network.

# IMPROVING LEARNING RATE OF NEURAL TREE NETWORKS USING THERMAL PERCEPTRONS

Ananth Sankar[1]     Richard J. Mammone
CAIP Center, Rutgers University
Brett and Bowser Roads
P.O. Box 1390
Piscataway, NJ 08855-1390

### Abstract

Recently we proposed a new neural network called Neural Tree Networks (NTN). The NTN is a combination of decision trees and multi-layer perceptrons (MLP). The NTN grows the network as opposed to MLPs. The learning algorithm for growing NTNs is more efficient than standard decision tree algorithms. Simulation results have shown that the NTN is superior in performance to both decision trees and MLPs. In this paper a new NTN learning algorithm is proposed based on the thermal perceptron algorithm. It is shown that the new algorithm greatly increases the speed of learning of the NTN and attains similar classification performance as the previously used algorithm.

## 1  INTRODUCTION

Neural networks, particularly multi-layer perceptrons (MLP) have received a lot of attention recently. MLPs have been successfully used in a variety of pattern recognition tasks [1, 2, 3, 4, 5]. The most popular learning algorithm for MLPs is backpropagation [1]. This is essentially a gradient descent algorithm and hence has problems associated with local minima. Another problem with backpropagation is that the MLP architecture must be specified before learning can begin. This is a problem since there is no guarantee that backpropagation will find a solution for a given network architecture. The MLP architecture is usually chosen by trial and error, which results in very slow learning.

Another approach to pattern recognition is decision trees [6, 7]. Decision trees use a sequential decision making strategy to classify an input feature vector. At each internal node of the decision tree a hyperplane test is used to route the feature vector to one

---
[1] Currently with AT&T Bell Labs., Murray Hill, NJ

of the child nodes. This process is repeated for each successive child node until the feature vector reaches a leaf node. The leaf nodes classify the feature vector. The hyperplanes at the internal decision tree nodes are formed by a training algorithm. Most decision tree training algorithms like CART [6] and ID3 [7] generate a set of candidate hyperplanes and then use a distortion metric to search for the best hyperplane. This process is done for each internal decision tree node. Since the candidate hyperplanes are arbitrarily generated, these algorithms may not result in an optimal hyperplane. Furthermore, most decision tree algorithms use hyperplanes that are perpendicular to the feature space axes. This results in an unnecessarily large number of nodes. CART [6] allows for slanted hyperplanes but the training algorithm used is computationally expensive.

Neural Tree Networks (NTN) use a decision tree structure with a neural network at each internal node [8, 9]. The NTN can be implemented very efficiently in hardware [8]. The NTN is grown by the learning algorithm as opposed to backpropagation, where the network must be known before learning begins [8]. The NTN always converges to a solution. Simulations have indicated that learning is much faster than backpropagation [8]. The performance of the NTN was shown to be superior to both backpropagation and CART, a popular decision tree approach [8, 9].

In this paper we propose a new learning algorithm for the NTN based on a variant of the perceptron learning algorithm called the thermal perceptron algorithm [10]. The new approach greatly increases the learning speed of the NTN. In section 2 the NTN is briefly reviewed. The new learning algorithm is presented in section 3 and simulation results are presented in section 4. The conclusion is given in section 5.

## 2 NEURAL TREE NETWORKS (NTN)

The NTN works very much like decision trees by using a sequential decision making strategy. The NTN uses a single layer perceptron network at each internal decision tree node to make decisions as to which path to follow to the leaf nodes. The NTN recursively partitions the feature space until, finally, all sub-regions correspond to a single class. These single class sub-regions correspond to the leaf nodes of the NTN. For example, figure 1 shows a 3-class pattern recognition problem and an NTN that solves the problem. A single layer perceptron network is used at each internal node of the NTN. The perceptron network at the root node divides the feature space into three regions corresponding to three child nodes. One of these regions (region 2) has only one class in it and so the corresponding node is a leaf node. The other two regions are further subdivided using a perceptron network at each of the other two child nodes. The number of neurons used in the perceptron network at any particular NTN node, $t$, is equal to the number of classes, $N_t$, in the subset of training data corresponding to node $t$. Thus the sub-region, $R_t$, corresponding to node, $t$, is partitioned into $N_t$ regions, where $N_t$ is the number of classes in the training set contained in $R_t$. Since all the $N_t$ regions may not necessarily contain training data, the training set corresponding to node $t$ is split into *at most* $N_t$ subsets. If the number of classes is 2 then this results in a binary tree structure such as in the CART decision tree [6]. For a multi-class problem, the NTN has an arbitrary tree structure in that the number of child nodes does not have to remain constant for all internal tree nodes. The perceptron network at the internal NTN nodes use a winner take all rule to classify an input pattern. Thus, $\mathbf{X}$ is classified as class $i$ iff

$$O_i(\mathbf{X}) \geq O_j(\mathbf{X}) \quad \forall j \neq i, 1 \leq i, j \leq N_t \tag{1}$$

where $O_i$ denotes the output of neuron $i$ for input $\mathbf{X}$, and is given by

$$O_i(\mathbf{X}) = \frac{1}{1 + e^{\mathbf{W}_i \cdot \mathbf{X} + \theta_i}} \tag{2}$$

where $\mathbf{W}_i$ is the weight vector for neuron $i$ and $\theta_i$ is a bias term. The perceptron network at node $t$ splits the subregion, $R_t$, into $N_t$ regions by assigning a feature vector to region $i$ if it is classified as class $i$. Each region $i$ is then assigned to a different child node.

We have described the architecture of the NTN. In order to derive a training algorithm to grow the NTN, we must first specify the conditions the algorithm must satisfy. It is required to find

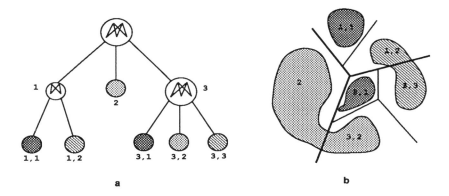

Figure 1: (a) Two level NTN with neural nets in internal nodes. Leaf nodes shaded according to appropriate class. (b) Feature space regions: first number refers to first level split of feature space, and second number to a second level split. First level region boundaries are shown by thick lines.

the smallest possible NTN that correctly classifies all the training data. The smallest possible NTN is required since using too many nodes may result in "memorizing" the training data, producing an NTN that performs poorly on the test data. It has been shown that this problem is NP-complete [11]. Thus the only feasible solutions are heuristic solutions. The heuristic that we have used is to train the perceptron network at node $t$ so that it makes the minimum number of classification errors on the training set contained in $R_t$. It is important to realize that a network that minimizes some error metric such as the mean squared error as is used in backpropagation may not minimize the number of misclassifications. In fact minimizing the squared error or L2 norm metric results in poor classification performance in the nonlinearly separable case.

Consider node $t$ of the NTN. The $N_t$ classes in node $t$ are labeled by a local encoding scheme. Thus class $k$ is labeled by the $N_t$-dimensional binary basis vector whose $k$th component is 1 and the rest of the components are 0. When pattern **X** is presented to the perceptron network, the weights of the $i$th neuron are updated by using a gradient descent technique that minimizes the L1 norm of the errors rather than the L2 norm. This method gives good performance even in the case of non-linearly separable data [8].

The weight update rule for neuron $i$ is given by

$$w_{j,i}^{n+1} = w_{j,i}^n - \eta y_i(1-y_i)\text{sgn}(t_i - y_i)x_j \quad 1 \leq j \leq d \quad (3)$$

where $w_{j,i}$ is the weight from input $j$ to neuron $i$, $t_i$ is the required output for neuron $i$, $y_i$ is the output of neuron $i$, $x_j$ is the $j$th component of the input vector $\mathbf{X}$ and $\eta$ is a step size. The sgn function gives the sign of its argument and is, in fact, the derivative of the L1 norm. The weights of the perceptron network are updated using this rule until some stopping condition is satisfied. This could be when the change in error is below some threshold. The perceptron network now splits the training set into at most $N_t$ subsets as discussed earlier. Each subset is assigned to a different child node and the algorithm is repeated for each child node. The algorithm stops when each subset contains only one class. After growing the NTN, it is pruned to form a sequence of pruned NTNs. Each Pruned NTN is then evaluated on an independent test set to get the best NTN [12, 9].

In this section we discussed the L1 norm algorithm for minimum misclassifications. However, there are many other possible algorithms such as that of Amari [13], Frean [10], and Juang [14]. In this paper we use Frean's approach called thermal perceptrons. The next section gives a review of the thermal perceptron algorithm.

## 3 THERMAL PERCEPTRONS

The thermal perceptron algorithm is a modification of the perceptron algorithm [15]. For the two class case, suppose one class is labeled by 1 and the second class is labeled by 0. The perceptron algorithm works by updating the weights according to the following rule:

$$\mathbf{W}^{n+1} = \mathbf{W}^n + \eta(t-y)\mathbf{X} \quad (4)$$

where $t$ is the required output and $y$ is the actual output of the neuron. Since a hard limiter is used, $y$ can be only 0 or 1. This rule can easily be extended to the multi-class case [16]. It can be shown that this rule will find an optimum weight vector if the classes are linearly separable [16]. However, in the nonlinearly separable case, the perceptron algorithm is unstable and does not give good results. One reason is that when a error is made where the input vector **X** is far away from the hyperplane, **W**, the updated weight vector is prone to misclassify patterns that were previously correctly classified. This is because the amount by which the weight vector is updated is large. The thermal perceptron algorithm works by biasing the updates in favor of the errors where the input vector is close to the hyperplane. The distance of the vector, **X**, to the hyperplane, **W**, can be measured by **W.X**. The update rule for the thermal perceptron is given by:

$$\mathbf{W}^{n+1} = \mathbf{W}^n + \eta(t-y)e^{-\mathbf{W}.\mathbf{X}/T}\mathbf{X} \tag{5}$$

where $T$ is a temperature such as is used in simulated annealing [17] and controls the amount by which the weights are changed. When $T$ is large, the thermal perceptron algorithm behaves very much like the perceptron algorithm, i.e., the weight change is essentially independent of the distance between the input vector and the hyperplane. In practice the algorithm is started with a large value of $T$ which is gradually annealed over the iterations. The gain term can also be annealed over time.

The thermal perceptron algorithm as described above can easily be generalized to the multi-class case. Suppose the input pattern **X** belongs to class $i$ but the maximum output neuron is $j$. Then the following updates are made:

$$\mathbf{W}_i^{n+1} = \mathbf{W}_i^n + \eta e^{-\mathbf{W}_i.\mathbf{X}/T}\mathbf{X} \tag{6}$$
$$\mathbf{W}_j^{n+1} = \mathbf{W}_j^n - \eta e^{-\mathbf{W}_j.\mathbf{X}/T}\mathbf{X} \tag{7}$$

where $\mathbf{W}_i^n$ is the weight vector for neuron $i$ at the $n$th iteration. The temperature, $T$, and the gain, $\eta$, can be annealed.

The modification to the existing NTN algorithm is obtained by replacing the gradient descent over an L1 norm error surface with the thermal perceptron algorithm. In the gradient descent technique, all $N$ neurons are updated for each input pattern **X**, but in the thermal perceptron algorithm only two neurons are updated. Thus the thermal perceptron algorithm is much faster when the number of classes is large.

| Classifier | Number of Neurons | Percent Correct |
|---|---|---|
| MLP | 88 | 51 |
| MLP | 22 | 45 |
| MLP | 11 | 44 |
| RBF | 528 | 53 |
| RBF | 88 | 48 |
| GN | 528 | 55 |
| GN | 88 | 53 |
| CART | - | 44 |
| NTN using L1 Norm | 59 [2] | 54 |

Table 1: Results on Speaker Independent Vowel Recognition

## 4 Simulation Results

In this section, we present simulation results on a speaker independent vowel recognition problem. This database is maintained as a benchmark by Scott Fahlman at CMU. The training set consists of 528 10-dimensional log area feature vectors representing 11 vowel sounds. An independent test set of 462 feature vectors is used to test the recognition performance. Results on this data set using MLPs with various type of nodes such as Guassian Nodes and Radial Basis Function nodes are reported in [18]. The performance of CART [6] on this data set has been reported in [19]. Table 1 summarizes these results. The NTN performs much better than CART, but is similar in performance to Multi-Layer Perceptrons (MLP) that have been optimized by using the appropriate number of Guassian Nodes (GN) or Radial Basis Function (RBF) nodes. The NTN algorithm, however, grows the network and finds the correct number of nodes as opposed to MLPs.

As mentioned before, the step size, $\eta(n)$, and the temperature, $T(n)$, of the thermal perceptron algorithm decrease with time. This corresponds to the annealing schedule of simulated annealing [17, 20]. In this section, a guassian annealing function is used

---

[2] The neurons in the NTN are not hidden neurons. Rather this number refers to the total number of neurons used in the NTN

| anneal $\eta$? | yes | no | yes | no |
|---|---|---|---|---|
| anneal $T$? | yes | yes | no | no |
| percent correct | 52 | 52 | 48 | 48 |

Table 2: Speaker Independent Vowel Recognition using Thermal Perceptron for the NTN

for the step size and the temperature. Thus the step size and temperature are given by

$$\eta(n) = \eta_0 \exp(-\tfrac{an^2}{n_{max}^2})$$
$$T(n) = T_0 \exp(-\tfrac{an^2}{n_{max}^2}) \quad (8)$$

where $\eta_0$ is the initial step size, $T_0$ is the initial temperature, $a$ is a constant that controls the rate at which $\eta$ and $T$ decrease, and $n_{max}$ is the maximum number of epochs of training. It is possible to anneal just the temperature or the step size, or both the temperature and the step size together.

The thermal perceptron algorithm was used to grow the NTN for the speaker independent vowel recognition problem. Different annealing schemes were tried corresponding to annealing the temperature and step size independently and annealing them together. Table 2 shows the classification performance of the NTN for different annealing schemes. For all the experiments, we used $\eta_0 = 0.1$ and $T_0 = 2.0$. The value of $a$ and $n_{max}$ was 7 and 1500 respectively. The results are averaged over 20 runs of the algorithm. From table 2 it can be seen that the best performance is 52%. This is slightly worse than the L1 norm NTN algorithm which had 54% classification accuracy (see table 1). However, it is possible that a better choice of initial values for the step size and the temperature or a better annealing schedule may improve the performance of the thermal perceptron NTN algorithm.

It was also observed that the thermal perceptron NTN did not need much pruning. In most of the runs of the algorithm, the best performance was achieved by the NTN grown by the learning algorithm with no pruning used. The thermal perceptron algorithm also grows the NTN faster than the L1 norm algorithm. This is because at each internal node of the NTN, the thermal perceptron only updates two neurons for every misclassified pattern. However

| L1 Norm Algo. | Thermal Perceptron |
|---|---|
| 10756 | 2794 |

Table 3: Number of neuron updates for NTN

the L1 norm algorithm updates all the neurons for every pattern presented. In the simulations of the NTN algorithm, the data set at each NTN node was presented 1500 times in random order to the perceptron network while training. Table 3 shows the number of neuron updates required for the NTN using the L1 norm algorithm and the thermal perceptron algorithm. From this table, we see that the thermal perceptron algorithm is around 5 times faster than the L1 norm algorithm.

## 5 Conclusion

In this paper, we presented a new algorithm for the NTN based on the thermal perceptron algorithm. The new algorithm learns much faster than the previous method of training the NTN based on the L1 norm. The new method does not need pruning as was the case in the L1 norm algorithm. The classification performance of the new method was found to be comparable to MLPs and the NTN trained by the L1 norm method.

## References

[1] D. Rumelhart and J. McClelland, *Parallel Distributed Processing*. MIT Cambridge Press, 1986.

[2] A. Waibel, "Modular Construction of Time Delay Neural Networks for Speech Recognition," *Neural Computation*, vol. 1, March 1989.

[3] K. Lang, A. Waibel, and G. Hinton, "A Time Delay Neural Network Architecture for Isolated Word Recognition," *Neural Networks*, vol. 3, no. 1, pp. 23–43, 1990.

[4] A. Rajavelu, M. Musavi, and M. Shirvaikar, "A Neural Network Approach to Character Recognition," *Neural Networks*, vol. 2, no. 5, pp. 387–394, 1989.

[5] Y. L. Cun *et al.*, "Handwritten Zip Code Recognition with Multilayer Networks," in *Proceedings of the 10th International Conference on Pattern Recognition*, pp. vol. 2, 35–40, 1990.

[6] L. Breiman, J. Friedman, R. Olshen, and C. Stone, *Classification and Regression Trees*. Belmont,CA: Wadsworth international group, 1984.

[7] J. Quinlan, "Induction of Decision Trees," *Machine Learning*, vol. 1, pp. 81–106, 1986.

[8] A. Sankar and R. Mammone, "Neural Tree Networks," in *Neural Networks: Theory and Applications* (R. Mammone and Y. Zeevi, eds.), pp. 281–302, Academic Press, 1991.

[9] A. Sankar and R. Mammone, "Speaker Independent Vowel Recognition using Neural Tree Networks," in *Proceedings of the IJCNN*, July 1991.

[10] M. Frean, *Small Nets and Short Paths: Optimising Neural Computation*. PhD thesis, University of Edinburgh, 1990.

[11] L. Hyafil and R. Rivest, "Constructing Optimal Decision Trees is NP-Complete," *Information Processing Letters*, vol. 5, no. 1, pp. 15–17, 1976.

[12] A. Sankar and R. Mammone, "Pruning Neural Tree Networks for Good Generalization." in preparation.

[13] S. Amari, "A Theory of Adaptive Pattern Classifiers," *IEEE Transactions on Electronic Computers*, vol. EC-16, pp. 299–307, June 1967.

[14] B. Juang. Personal communication.

[15] F. Rosenblatt, *Principles of Neurodynamics*. New York: Spartan, 1962.

[16] N. J. Nilsson, *Learning Machines*. McGraw-Hill, 1965.

[17] S. Kirkpatrick, C. Gelatt(Jr.), and M. Vecchi, "Optimization by simulated annealing," *Science*, vol. 220, May 1983.

[18] A. Robinson, *Dynamic Error Propagation Networks*. PhD thesis, Cambridge University Engineering Department, 1989.

[19] A. C. Tsoi and R. Pearson, "Comparison of Three Classification Techniques, CART, C4.5, and Multi-Layer Perceptrons," in *Advances in Neural Information Processing Systems 3* (D. Touretzky, ed.), San Mateo, CA: Morgan Kaufmann, 1990.

[20] S. Kirkpatrick, "Optimization by simulated annealing: Quantitative studies," *Journal of Statistical Physics*, vol. 34, no. 5 & 6, pp. 975–986, 1984.

# ADALINE WITH ADAPTIVE RECURSIVE MEMORY

Bert De Vries, Jose C. Principe, and Pedro Guedes de Oliveira*

Department of Electrical Engineering
University of Florida
Gainesville, FL 32611
principe@brain.ee.ufl.edu

*Departamento Eletronica/INESC
Universidade de Aveiro
Aveiro, Portugal

**Abstract** - We present a generalization of Widrow's adaptive linear combiner with an adaptive recursive memory. Expressions for memory depth and resolution are derived. The LMS procedure is extended to adapt the memory depth and resolution so as to match the signal characteristics. The particular memory structure, gamma memory, was originally developed as part of a neural net model for temporal processing.

## INTRODUCTION

Although infinite impulse response (IIR) systems are more powerful for modeling, identification, and signal processing, the subclass of finite impulse response (FIR) filters is almost exclusively utilized in adaptive signal processing. The main reason is centered in the difficulty of ensuring stability during adaptation of IIR systems. Moreover, gradient descent update rules are not guaranteed to find global minima in the non-convex error surfaces of IIR systems.

Feedforward systems, on the other hand, are always stable. Yet the tapped delay line memory structure implies that the depth of memory is always of the same order as the number of adaptive weights in an FIR structure. For some real world signal environments, in particular biological signals, this is a severe modeling drawback. Thus, the choice between a recurrent or feedforward processing or modeling system can be a difficult one.

In this paper we present the gamma memory - a structure that is characterized by a restricted IIR architecture. The uncoupling of memory depth from filter order is inherited from recurrent systems. Yet, the stability condition for the gamma memory will prove to be trivial.

Consider the filter network defined by -

$$y(n) = \sum_{k=0}^{K} w_k \sum_{m=0}^{\infty} e(m) g_k(n-m)$$
$$= \sum_{k=0}^{K} w_k [g_k(n) \bullet e(n)] \qquad (1)$$
$$= \sum_{k=0}^{K} w_k x_k(n)$$

where $e(n)$ is an input sequence, $y(n)$ is an output sequence, the tap variables $x_k(n)$, $k = 0,..,K$, are the convolution of $e(n)$ and a delay kernel $g_k(n)$, $w_k$ is a set of adaptive parameters and $\bullet$ denotes the convolution operator. We assume that the kernels $g_k(n)$ are *causal* and *normalized*, that is, $g_k(n) = 0$ for $n < 0$ and $\sum_{n=0}^{\infty} g_k(n) = 1$. When additionally the kernels admit a *recurrence* relation of the form -

$$g_k(n) = g(n) \bullet g_{k-1}(n), k = 1,..,K, \qquad (2)$$

then it is possible to generate the tap variables $x_k(n)$ recursively by $x_k(n) = g(n) \bullet x_{k-1}(n)$ or in the z-domain by -

$$X_k(z) = G(z) X_{k-1}(z). \qquad (3)$$

We refer to the structure described by (1) and (3) as the *generalized feedforward filter*. The tap-to-tap transfer function $G(z)$ is the (*generalized*) *delay operator*. This structure is depicted in Figure 1.

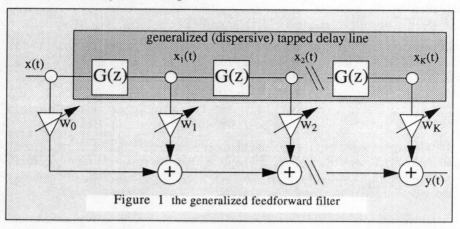

Figure 1 the generalized feedforward filter

Note that we can write the transfer function $H(z) \equiv \dfrac{Y(z)}{X(z)}$ of the generalized feedforward filter as follows -

$$H(z) = \sum_{k=0}^{K} w_k [G(z)]^k. \qquad (4)$$

It follows from (4) that $H(z)$ is stable whenever $G(z)$ is stable.

For $g_k(n) = \delta(n-k)$, this structure reduces to Widrow's *Adaline* [1]. The past of $x(t)$ is represented in the tap variables $x_k(t)$. Although conventional digital signal processing structures are built around the tapped delay line ($G(z) = z^{-1}$), we have observed that alternative delay operators may lead to better filter performance. In general, the optimal memory structure $G(z)$ is a function of the input signal characteristics as well as the goal of the filter operation. This observation has led us to consider *adaptive* delay operators $G(z)$.

In this paper we consider *gamma delay kernels* as defined by -

$$g_k(n) = \binom{n-1}{k-1} \mu^k (1-\mu)^{n-k} U(n-k), \ k = 1,..,K, \qquad (5)$$

where $U(n)$ is the unit step function. The gamma delay kernels were originally developed in continuous time as part of a neural net model for temporal processing [2]. We showed that - by the transformation $s = \dfrac{z-1}{T_s}$ - the impulse response of the continuous time gamma filter can be written as $h(t) = \sum_{k=0}^{K} w_k g_k(t)$, where

$$g_k(t) = \frac{\mu^k}{(k-1)!} t^{k-1} e^{-\mu t}, \ k = 1,...,K, \text{ and } g_0(t) = \delta(t).$$ The functions $g_k(t)$ are the integrands of the (normalized) gamma function. Hence the name gamma model for structures that utilize tap variables of type $x_k(t) = (g_k \bullet x)(t)$ to store the past of x(t).

The Z-transform of (5) yields -

$$G_k(z) = \left(\frac{\mu}{z-(1-\mu)}\right)^k \qquad (6)$$

We define the *gamma delay operator* as $G(z) = \dfrac{\mu}{z-(1-\mu)}$. The gamma delay operator can be interpreted as a leaky integrator with loop gain $1-\mu$. The parameter $\mu$ controls the memory depth and resolution. Note that stability is guaranteed when $0 < \mu < 2$.

In the following, we refer to an adaline structure with gamma memory as adaline($\mu$). Note that Widrow's adaline is equivalent to adaline(1). Adaline($\mu$) is shown in Figure 2.

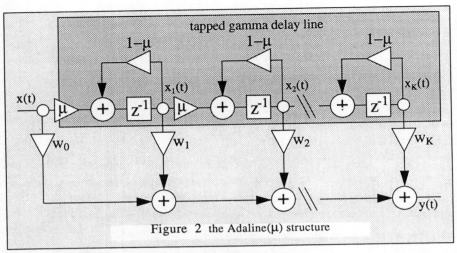

Figure 2 the Adaline($\mu$) structure

## CHARACTERISTICS OF GAMMA MEMORY

An extensive account of the properties of the gamma memory structure has been presented in [3]. Here we summarize a discussion regarding the memory depth and we present an LMS learning procedure for adaline($\mu$).

**Memory Depth versus Filter Order**

Although the impulse response $g_k(n) = Z^{-1}\{G_k(z)\}$ of the $k$th tap of the gamma memory extends to infinite time for $0 < \mu < 1$, it is possible to formulate a mean memory depth for a given memory structure $g_k(n)$. Let us define the *mean sampling time* $n_k$ for the $k$th tap as -

$$n_k = \sum_{n=0}^{\infty} n g_k(n) = Z\{n g_k(n)\}|_{z=1} = -z \frac{dG_k(z)}{dz}\bigg|_{z=1} = \frac{k}{\mu}. \quad (7)$$

We also define the *mean sampling period* $\Delta n_k$ (at tap $k$) as $\Delta n_k = n_k - n_{k-1} = \frac{1}{\mu}$. The *mean memory depth* $D_k$ for a gamma memory of order $k$ then becomes -

$$D_k = \sum_{i=1}^{k} \Delta n_i = n_k - n_0 = \frac{k}{\mu}. \quad (8)$$

If we define the *resolution* $R_k$ as $R_k = \frac{1}{\Delta n_k} = \mu$, the following formula arises which is of fundamental importance for the characterization of the gamma memory structure[1] -

$$K = D \times R \qquad (9)$$

Equation (9) reflects the possible trade-off of resolution versus memory depth in a memory structure for fixed dimensionality $K$. Such a trade-off is not possible in a non-dispersive tapped delay line, since the fixed choice of $\mu = 1$ sets the depth and resolution to $D = K$ and $R = 1$ respectively. However, in the gamma memory, depth and resolution can be adapted by variation of $\mu$. The choice $\mu = 1$ represents a memory structure with maximal resolution and minimal depth. For this case, the order $K$ and depth $D$ of the memory are equal. In the adaline structure, the number of adaptive parameters $w_k$ equals o the number of taps ($K+1$). Thus, when $\mu = 1$, the number of weights equals the memory depth. Very often this coupling leads to overfitting of the data set (using parameters to model the noise). The parameter $\mu$ provides a means *to uncouple the memory order and depth.*

As an example, assume a signal whose dynamics are described by a system with 5 parameters and maximal delay 10, that is, $y(t) = f(x(t - n_i), w_i)$ where $i = 1,...,5$, and $max_i(n_i) = 10$. If we try to model this signal with an adaline structure, the choice $K = 10$ leads to overfitting while $K < 10$ leaves the network unable to incorporate the influence of $x(t -10)$. In an adaline with gamma memory network, the choice $K = 5$ and $\mu = 0.5$ leads to 5 free network parameters and mean memory depth of 10, obviously a better compromise.

## LMS Adaptation

In this section we present the least mean square (LMS) adaptation update rules for the parameters $w_k$ and $\mu$. Let the performance of the system be measured by the *total error E*, defined as -

$$E \equiv \sum_{n=0}^{T} E_n = \sum_{n=0}^{T} \frac{1}{2}\varepsilon^2(n) = \sum_{n=0}^{T} \frac{1}{2}(d(n) - y(n))^2 \qquad (10)$$

where $d(n)$ is a *target signal*. We first expand for $w_k$, yielding-

$$\Delta w_k(n) = -\eta \frac{\partial E_n}{\partial w_k} = \eta \varepsilon(n) \frac{\partial y(n)}{\partial w_k} = \eta \varepsilon(n) x_k(n) \qquad (11)$$

Similarly, the update equation for $\mu$ evaluates to -

$$\Delta \mu = -\eta \frac{\partial E_n}{\partial \mu} = \eta \varepsilon(n) \sum_k w_k \alpha_k(n) \qquad (12)$$

---

1. We dropped the subscript $k$ when $k = K$.

where $\alpha_k(n) \equiv \dfrac{\partial x_k(n)}{\partial \mu}$. The gradient signal $\alpha_k(n)$ can be computed on-line by differentiating (Eq.6) leading to -

$$\alpha_0(n) = 0$$

$$\begin{aligned}\alpha_k(n) = (1-\mu)\alpha_k(n-1) + \mu\alpha_{k-1}(n-1) \\ + [x_{k-1}(n-1) - x_k(n-1)]\end{aligned}, \text{ for } k = 1,..,K. \quad (13)$$

## A DSP INTERPRETATION OF ADALINE($\mu$)

In order to quantify the signal processing effects of the gamma kernels, we define an auxiliary variable $\gamma$ as -

$$\gamma = \frac{z - (1-\mu)}{\mu} \quad (14)$$

Substitution of (14) in the z-transform of $x(n)$ yields the following expression for the $\gamma$-transform -

$$X(\gamma) = \sum_{n=0}^{\infty} \mu^{-n} x[n] \left\{\gamma - \frac{\mu-1}{\mu}\right\}^{-n}, \quad (15)$$

that is, the (one-sided) Laurent series of the sequence $\mu^{-n}x(n)$ evaluated at $\gamma_0 = \dfrac{\mu-1}{\mu}$. The relation between the z- and $\gamma$-plane is depicted in Figure 3. Note that adaline($\mu$) is a *FIR* filter constructed around ideal delays $\gamma^{-1}$ in the $\gamma$-domain.

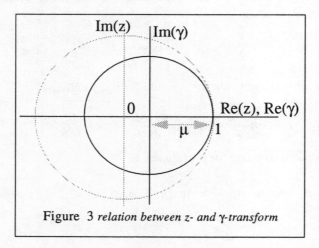

Figure 3 *relation between z- and $\gamma$-transform*

Therefore it is straightforward to analyze and design gamma memory structures in the γ-domain as FIR filters, and obtain the corresponding pole zero maps and impulse responses.

## EXPERIMENTAL DATA

The performance of adaline($\mu$) versus adaline(1) has been tested for various experimental protocols. In this section we present data on a prediction and system identification experiments.

### prediction/noise reduction of sinusoids in noise

We constructed an input signal consisting of a sum of sinusoids, contaminated by gaussian noise. Specifically, e(t) was described by -

$$e(t) = sin(\pi(0.06t + 0.1)) + 3sin(\pi(0.12t + 0.45)) + \\ 1.5sin(\pi(0.2t + 0.34)) + sin(\pi(0.4t + 0.67)) + N(0, 1)$$

(16)

where $N(0,1)$ denotes zero mean gaussian noise with unit variance. This signal is shown in Figure 4a. The desired output signal was the next sample of the sum of sinusoids. Hence, the processing problem involves a combination of prediction and noise reduction. The processing system was an adaline($\mu$). We trained the network using the LMS adaptation rules as derived in this paper. Both the training set and the validation set consisted of 300 samples. The parameters were only adapted during a run on the training data set. The tracks of the memory parameter $\mu$ are shown in Figure 4c. After each run on the training data, we ran the system on the validation data set and measured the normalized performance index $E = \frac{var[\varepsilon(n)]}{var[d(n)]}$. A run over the training data set followed by a run over the validation data is called an epoch. If the performance index for the validation data set increases for 4 consecutive epochs, we assume that the system parameters have converged and thus we stop training (Figure 4d). In Figure 4b we plot the normalized performance index after convergence as a function of memory parameter $\mu$. $\mu$ was parametrized over the domain [0,1] using a step size $\Delta\mu = 0.1$. The system performance is measured for memory orders $K = 1$ through $K = 5$. Note that $\mu = 1$ refers to Widrow's adaline structure. The graph clearly shows that adaline($\mu$) outperforms adaline(1). Even for low order $K = 1$, adaline(0.1) performs better than a fifth order adaline(1).

### Elliptic Filter System Identification

In this section we present numerical simulation results when adaline($\mu$) is used in a system identification configuration. The system to be identified is the third order elliptic low pass filter described by[1] -

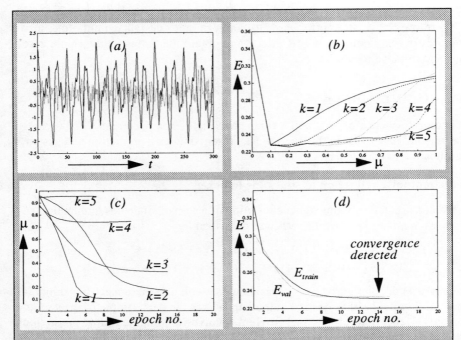

Figure 4 results for prediction/noise removal experiment. The processing system is adaline(µ). The number of prediction steps is 1. (a) input signal consists of a sum of four sinusoids contaminated with gaussian noise. SNR = 10 dB. (b) Normalized error criterion (mse) after convergence versus µ. (c) The tracks of µ during training. (d) The performance index over the training set ($E_{train}$) and the validation set ($E_{val}$) versus the epoch number.

$$H(z) = \frac{0.0563 - 0.0009z^{-1} - 0.0009z^{-2} + 0.0563z^{-3}}{1 - 2.1291z^{-1} + 1.7834z^{-2} - 0.5435z^{-3}} \quad (17)$$

In Figure 5 we show the performance index after convergence. Note that the optimal memory depth $D_{opt} \equiv \frac{K}{\mu_{opt}} \approx 5$ is constant for different memory orders. The graph shows that it is irrelevant to use memory orders higher than $K = 5$ for this identification problem. In fact, when $K = 5$, adaline(1) performs as well as $K = 3$ for adaline(0.6). However, we still prefer $K = 3$, since this structure has 5 free parameters ($K+1$ weights $w_k$ plus µ) whereas adaline uses 6 parameters.

---

1. This filter has been described in [4] on pg.226

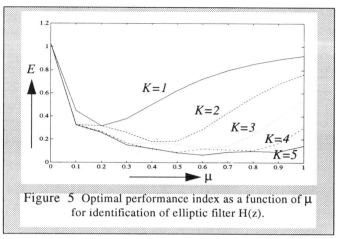

Figure 5 Optimal performance index as a function of $\mu$ for identification of elliptic filter H(z).

Parsimony in the number of free parameters provides adaline(0.6) with better modeling (generalization) characteristics.

We have experimented with several signals (sinusoids in noise, Feigenbaum map, electroencephalogram (EEG)) for various processing protocols (prediction, system identification, classification). Invariably the optimal memory structure[1] was obtained for $\mu < 1$.

## CONCLUSIONS

We presented a new memory structure initially developed for a neural network, the gamma memory, and applied it in the adaptive linear combiner. The gamma memory structure is characterized by two properties in comparison to fully recurrent networks (IIR filters). First, the loops in the gamma filter are local with respect to the taps. As a result, stability is easily controlled in the gamma memory. Secondly, the memory parameter $\mu$ is global with respect to the memory taps. This is advantageous in an adaptive environment, since a single parameter controls the trade-off between memory depth and resolution. We showed that the gamma filter is more efficient than the adaline (an FIR filter) in two experimental settings, system identification and prediction in noisy environments.

It is worth mentioning that the application of gamma memory is not limited to the adaptive linear combiner. This principle for short term memory has been initially developed for a neural network model, the gamma model (DeVries and Principe, 1991), which extends the time delay neural network (TDNN) (Lang et al, 1990), and also provides a fully neural framework for the concentration in time network of Tank and Hopfield (Tank and Hopfield, 1987). The study of the linear

---

1. The optimal memory structure is defined as the structure of lowest dimensionality that minimizes the performance index $E$.

case provided insight into the functioning of the memory mechanism and helped us analyze and understand better the properties of the nonlinear neural network.

## Acknowledgments

This work was partially supported by NSF grant ECS-8915218. The stay of the third author (P.G.O.) at the University of Florida has been partially supported by JNICT.

## References

[1] B. Widrow and S.D. Stearns, <u>Adaptive Signal Processing</u>. Prentice Hall, Inc., Englewood Cliffs, NJ, 1985.

[2] B. De Vries and J.C. Principe, A theory for neural nets with time delays. in <u>Proceedings of NIPS90</u>, Denver, CO, 1991.

[3] J.C. Principe and B. De Vries, The Gamma Filter - A New Class of Adaptive IIR Filters with Restricted Feedback, Submitted to <u>IEEE Transactions on Signal Processing</u>, June 1991.

[4] A. Oppenheim and R.J. Schafer, <u>Digital Signal Processing</u>, Prentice Hall, Inc., Englewood Cliffs, NJ, 1975.

[5] Lang K., Waibel A.H. and Hinton G.E., A time-delay neural network architecture for isolated word recognition, <u>Neural Networks</u>, vol.3.(1), pp. 23-44, 1990.

[6] Tank D.W. and Hopfield, Concentrating information in time: analog neural network with applications to speech recognition problems, in <u>IEEE 1st IJCNN</u>, vol IV, pp.465-468, 1987.

# LEARNED REPRESENTATION NORMALIZATION: ATTENTION FOCUSING WITH MULTIPLE INPUT MODULES

Michael L. Rossen
HNC, Inc.
5501 Oberlin Dr.
San Diego, CA  92121
rossen@amos.ucsd.edu

*Abstract*—A large, multi-modular neural network can be envisioned for use in a complex, multi-task application. The optimum data representation for each sub-task of such an application is often unknown and different from the optimum data representation for the other sub-tasks. A method is needed that allows a network that contains several alternate input representations to learn to focus its attention on the best representation(s) for each sub-task to be learned, without *a priori* information on best representation–sub-task combinations.

An adaptive attention focusing method is introduced that addresses this issue. The method involves training recurrent connections for each input module to selectively attenuate input to that module that causes training error in a final target module. The method is shown to have similarities with both gating networks and anti-Hebbian learning. A task scenario is proposed for which adaptive attention focusing provides superior classification performance relative to standard training methods.

## OVERVIEW

Attention can be thought of as the method by which perceptual data is admitted into consciousness[4]. To pay good attention, a person must simultaneously consider data from a multitude of perceptual centers while performing a task. The person must choose which among the incoming percepts are relevant for performing a particular task. This choice must be efficient, or else the person will expend so much effort on sifting perceptual data that insufficient processing capacity is left for actual task performance. Thus the quality of attention depends on how well relevant input perceptions are chosen for a particular task and how much processing power is taken up by this choice. Attention focusing is a method by which attention is dynamically shifted as a function of changing task conditions. Focusing can be exclusive, where certain perceptual inputs are completely ignored, or soft, where some perceptual inputs are given less attention than others.

Neural networks need attention focusing for the same reason that people do. The analogy is imperfect since the quantities limited by the processing capacity of the brain, whatever they are, most probably bear little resemblance to the those resources that are limited in current computer technology. Nevertheless, it is a good analogy to keep in mind as tasks of increasing complexity are attempted with neural network technology.

Two purposes form the context for the investigations discussed in this paper. The first is to develop a method of attention focusing for a modular neural network. Attention focusing can be implemented in a neural network as a supervisory network that operates on input from a set of sub-networks. In the *representation normalization* method of attention focusing described here, the supervisory network is decentralized, consisting of a set of recurrent synapses associated with the input module of each sub-network. The recurrent synapses are trained with a variant of anti-Hebbian learning.

When an input representation is well-suited for classification of activity patterns only in certain sub-regions of its activity space, but ill-suited for classification of data from other sub-regions, this algorithm is designed to recognize the patterns from the 'good' sub-regions and to propagate them forward unchanged. Conversely, activity patterns from the 'bad' sub-regions are attenuated to zero.

Alternate methods of attention focusing exist, notably those using a gating paradigm. Comparisons of these techniques with representation normalization are reserved for the discussion section.

The second purpose addressed in this paper is to come to an understanding of the nature of the tasks for which attention focusing provides neural networks with a performance advantage. Of particular interest are situations in which multiple representations of data are available, yet it is not clear *a priori* which representation is best for a given sub-task of an application. Neural network classification error rates can be very sensitive to input representation. This is exemplified in speech research where large error rate variations have been found as a function of input representation[7]. Combining several candidate representations into a single 'multiple input module' representation can help, but as we intend to show in this paper, this simple solution has its limitations.

Several studies have indicated that attention focusing (using the gating paradigm) provides *faster* training than do standard neural network techniques on a particular task (e.g., [6]). To our knowledge, only in [5] have results been presented suggesting that attention focusing *reduces error rates* in a particular task relative to error rates using standard neural network techniques. Even in this latter study, in which a vowel classification task was considered, it was not clear what properties of the task and/or data necessitated attention focusing for optimal performance.

We focus on an application involving multiple input modules with identical representation, but receiving data from distinct channels differentially corrupted by noise. An example of this would be microphones in different locations of a big hall. This application is simple enough to allow a simulation with artificial data that is easy to analyze, yet non-trivial enough to provide an example in which attention focusing yields a performance advantage.

In the next section, a description is provided for the architecture and dynamics of representation normalization. Following this, a data scenario is described, involving two channels of noisy data, where attention focusing might be used to best advantage. Some preliminary experimental results are then discussed. Finally, we compare representation normalization with other models in the literature and discuss possible model extensions.

## THE TRAINING RULE

The representation normalization algorithm for attention focusing involves training in two stages. In the first training stage, two sub-networks are independently trained with standard backpropagation on the same classification task. As shown in Figure 1, each of the upper and lower sub-networks has an input module, a hidden module and a target module. The processing elements (PEs) in each module have standard bi-polar logistic activation functions, except for input module PEs, which are linear. The training data for each input module comes from the same set of events. However, as mentioned above, the input data vector for a training presentation is different for each sub-network due to distinct noise sources acting on the separate channels supplying data to each module. For the second training stage, recurrent synapses[1] are added

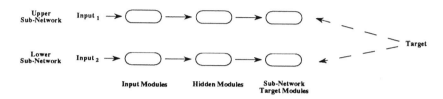

Figure 1: Architecture of two independent classifier networks.

to each input module, and all the target modules are connected to a final target module, as shown in Figure 2. The new synapses are trained while

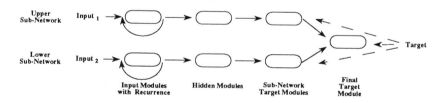

Figure 2: Architecture of two classifier networks connected to implement attention focusing dynamics.

synapses trained during the first training stage are frozen. The new synapses from intermediate target modules to the final target module could be trained with a standard delta rule. Alternatively (and this is what we assume in the present discussion), these weights could be fixed so that the final target module simply averages the answers from the intermediate target modules.[2]

---

[1] In our implementation, the recurrent synapses have zero delay, and so are not dynamically recurrent since they do not encode time history. In an analog system, these synapses would have a finite time-delay that was much shorter than the delay of the feed-forward synapses.

[2] More precisely, each synaptic weight matrix from a sub-network target module to the final target module is an identity matrix, scaled by the number of sub-networks in the system.

The recurrent synapses implement the attention focusing ability of the network. Let $i$ index the sub-networks and $j$ index the stimuli of a data set. Let $\mathbf{m}_{I_i}$, $\mathbf{m}_{H_i}$, and $\mathbf{m}_{T_i}$ represent the input, hidden, and target modules of the sub-network $i$. If $\mathcal{A}_{sub-net_i}$ is the recurrent synaptic weight matrix associated with $\mathbf{m}_{I_i}$, then for a given input activity vector $\mathbf{f}_{sub-net_i}^{(j)}$ to $\mathbf{m}_{I_i}$, the output activity of $\mathbf{m}_{I_i}$ is:

$$\mathbf{x}_{sub-net_i}^{(j)} = (\mathcal{A}_{sub-net_i}\, \mathbf{f}_{sub-net_i}^{(j)}) + \mathbf{f}_{sub-net_i}^{(j)}. \quad (1)$$

The training rule for the recurrent synaptic weight matrix $\mathcal{A}_{sub-net_i}$ follows a modified delta rule:[3]

$$\Delta \mathcal{A}_{sub-net_i} = \alpha\, \sigma(\varepsilon_{finaltgt}^{(j)}) \\ (\mu_{sub-net_i}^{(j)}\, \mathbf{f}_{sub-net_i}^{(j)} - \mathbf{x}_{sub-net_i}^{(j)})\, \mathbf{f}_{sub-net_i}^{*(j)}; \quad (2)$$

where:

$\mu_{sub-net_i}^{(j)} = 1 - \sigma(\varepsilon_{finaltgt}^{(j)})\sigma(\epsilon_{sub-net_i}^{(j)})$,
$\alpha$ : learning parameter,
$\varepsilon_{finaltgt}^{(j)}$ : error parameter of final target for stimulus $j$,
$\epsilon_{sub-net_i}^{(j)}$ : error parameter of target for sub-network $i$ and stimulus $j$,
$\sigma$ : sigmoid function, with range $(0,1)$,
$\mathbf{f}_{sub-net_i}^{*(j)}$ : transpose of $\mathbf{f}_{sub-net_i}^{(j)}$.

The final target error parameter $\varepsilon_{finaltgt}^{(j)}$ is equal to the final target MSE (squared error averaged over target PEs) for stimulus $j$ divided by a running average of final target MSE, with the running average accumulating throughout the second training stage. The sigmoid $\sigma$ that operates upon $\varepsilon_{finaltgt}^{(j)}$ has a steep slope and thus serves to push values above (or below) the running average towards the extreme value of 1 (or 0). Note that this $\sigma$ is distinct from the *bi-polar* logistic which serves as activation function for each PE in the network. Similarly, each sub-network target error parameter $\epsilon_{sub-net_i}^{(j)}$ is equal to that sub-network's MSE for stimulus $j$ divided by its running average. The sigmoid $\sigma$ serves the same function for $\epsilon_{sub-net_i}^{(j)}$ as it does for $\varepsilon_{finaltgt}^{(j)}$.[4]

The central purpose of the update algorithm in eq. (2) is to train the recurrent weight matrix $\mathcal{A}_{sub-net_i}$ of each sub-network to *selectively attenuate activity vectors that cause both its sub-network target error and the final target error to be large*.[5] After the second training stage the weight matrix $\mathcal{A}_{sub-net_i}$

---

[3] The algorithm is best viewed as modified auto-associative training. This would be clear if we had adopted the convention that external input is not part of the total activity, so that the module output activity term in eq. (1) would be $\mathcal{A}_{sub-net_i}\, \mathbf{f}_{sub-net_i}^{(j)}$.

[4] We have also experimented with using the binary categorization for a target module as the error parameter for that target module. Preliminary results suggest this might be the better choice.

[5] Since the PE activation functions are bi-polar logistics, 0 output from an input module of a sub-network causes 0 output from that sub-network's target module which then has no affect on final target module activity.

carries out the selective attenuation for each sub-network; but during the second training stage, for each presentation of stimulus $j$, the parameters $\varepsilon^{(j)}_{finaltgt}$ and $\mu^{(j)}_{sub-net_i}$ control whether the recurrent synapses are updated and how much attenuation is caused by the update so as to adapt $\mathcal{A}_{sub-net_i}$ for its role of selective attenuation.[6]

Table 1: Update magnitudes and amount of attenuation for different combinations of sub-network target error and final target error.

| Sub-Network Target Error | Final Target Error Large | Final Target Error Small |
|---|---|---|
| Large | $\mu^{(j)}_{sub-net_i} \to 0$ (full attenuation) | $\mu^{(j)}_{sub-net_i} \to 1$ (no attenuation) |
|  | $\sigma(\varepsilon^{(j)}_{finaltgt}) \to 1$ (yes update) | $\sigma(\varepsilon^{(j)}_{finaltgt}) \to 0$ (no update) |
| Small | $\mu^{(j)}_{sub-net_i} \to 1$ (no attenuation) | $\mu^{(j)}_{sub-net_i} \to 1$ (no attenuation) |
|  | $\sigma(\varepsilon^{(j)}_{finaltgt}) \to 1$ (yes update) | $\sigma(\varepsilon^{(j)}_{finaltgt}) \to 0$ (no update) |

Table 1 provides a qualitative picture of how $\varepsilon^{(j)}_{finaltgt}$ and $\mu^{(j)}_{sub-net_i}$ control recurrent synapse training in the $i^{th}$ sub-network for the $j^{th}$ stimulus. Since $\sigma(\varepsilon^{(j)}_{finaltgt})$ multiplies the learning parameter $\alpha$, synaptic update for any sub-network only occurs if there is final target error. The result is that, regardless of sub-network target error, matrix weights are not updated if the final target module generates no error.

If there is final target error, then the degree of attenuation that the weight matrix $\mathcal{A}_{sub-net_i}$ is adapted to produce is controlled by $\mu^{(j)}_{sub-net_i}$ since it is a factor in the desired output of the input module (i.e., $\mu^{(j)}_{sub-net_i} \mathbf{f}^{(j)}_{sub-net_i}$), as specified in eq. (2). For a given value of $\mu^{(j)}_{sub-net_i}$, the weight matrix $\mathcal{A}_{sub-net_i}$ is trained so that the output of module $\mathbf{m}_{I_i}$ is equal to its external input $\mathbf{f}^{(j)}_{sub-net_i}$ attenuated by the factor $\mu^{(j)}_{sub-net_i}$. If input training vectors were orthogonal, this training would be equivalent to adjusting $\mathcal{A}_{sub-net_i}$ to have an eigenvector in the direction of $\mathbf{f}^{(j)}_{sub-net_i}$ with associated eigenvalue equal to $(-1 + \mu^{(j)}_{sub-net_i})$.

Since the desired output of the $i^{th}$ input module contains the factor $\mu^{(j)}_{sub-net_i}$, the smaller $\mu^{(j)}_{sub-net_i}$, the more severe the attenuation that training imposes.

---

[6]$\mathcal{A}_{sub-net_i}$ could be replaced by a multi-layer non-linear sub-network, if more complex mapping were needed.

Table 1 indicates that attenuation occurs only when both final target error and sub-network target error are large. This is because the sigmoid used has a very sharp slope and, as mentioned above, each error parameter is scaled by its running average. Thus, if either $\varepsilon^{(j)}_{finaltgt}$ or $\epsilon^{(j)}_{sub-net_i}$ is much below its running average value, $\mu^{(j)}_{sub-net_i} \to 1$, and training imposes no attenuation.

## AN EXPERIMENT WITH ARTIFICIAL DATA

### Stimuli

We now consider a data scenario in which the input modules of two sub-networks receive the same base input representation of the signal. For each stimulus, the final target module and the two intermediate target modules always receive the same target category. However, the data channels to the two input modules are corrupted by different noise sources. Data is generated in three categories, $ctg_a$, $ctg_b$, and $ctg_c$. When $ctg_c$ or $ctg_a$ is present, there is never any noise. When $ctg_b$ is present, there can be noise in one of the two channels such that the input data from that channel belongs to $ctg_a$, even though the 'correct' training vector and the input to the other channel indicate $ctg_b$. This method of noise generation allows strict control over the regions in input space *in which a noise signal may be present*.

For simplicity, let 'up' and 'low' index the upper and lower sub-networks in Figure 2, so that, for example, $\mathbf{m}_{I_{up}}$ refers to the input module of the upper sub-network and $\mathbf{m}_{I_{low}}$ refers to the input module of the lower sub-network. We denote the activity space of module $\mathbf{m}_{I_{up}}$ as $\mathbf{I}_{up}$. $\mathbf{I}_{up}(a)$ is that portion of $\mathbf{m}_{I_{up}}$ for which the *true* target category is always $ctg_a$. Let $\mathbf{N}_{up}(a)$ be a small, localized region of $\mathbf{I}_{up}(a)$ that is corrupted by noise, as described above, so that the *actual* target module training vector associated with a particular input vector in $\mathbf{N}_{up}(a)$ varies between two categories, being category $ctg_a$ (correctly) 30% of the time and category $ctg_b$ (incorrectly) 70% of the time. For this reason, the input activity space region $\mathbf{N}_{up}(a)$ is called the noise region for $\mathbf{m}_{up}$. Throughout the rest of input activity space $\mathbf{I}_{up}$, an incorrect target is specified 0% of the time.

Assume the same situation for data entering the lower sub-network and that the probability of activity simultaneously occurring in $\mathbf{N}_{up}(a)$ of module $\mathbf{m}_{I_{up}}$ and in $\mathbf{N}_{low}(a)$ of $\mathbf{m}_{I_{low}}$ is zero.

### Method

The experiment was performed under four conditions:

1. Train each sub-network independently, as shown in Figure 1.
2. Take the best results from condition 1. Extend the system architecture to include connections between the intermediate target modules and the final target module, as shown in Figure 2. However, *do not include the recurrent synapses*. With the feed-forward synapses frozen, train the new synapses using the standard delta rule.
3. Take the best results from condition 1. Extend the system architecture to implement representation normalization, as shown in Figure 2. Then, with the feed-forward synapses frozen, train the recurrent synapses following eq. (2). Synapses to final target module are set to average the outputs of the intermediate target modules.

4. Repeat condition 3, but instead of training the recurrent synapses, hard-wire the system to attenuate to zero all input to the upper sub-network from $\mathbf{N}_{up}(a)$, and all input to the lower sub-network from $\mathbf{N}_{low}(a)$.

In each condition, three stimulus sets, each containing 400 stimuli per category, are used. The first and second stimulus sets contain random stimuli generated according to the noise conditions described above. The first stimulus set is used for training. System performance on the second stimulus set is monitored to decide when training is completed. The final system for a given condition is the one with minimum error rate on the second stimulus set. The third stimulus set is noise-free. It is used to grade the performance of the final system.

The best results we can expect in condition 1, where each sub-network is trained as an independent classifier, is to obtain a correct answer 30% of the time from a sub-network when data occurs in the noise region of that sub-network.

Intuition might lead one to believe that the network in condition 2 could do better than 30% correct when data from $\mathbf{N}_{up}(a)$ occurred in module $\mathbf{m}_{I_{up}}$ and/or data from $\mathbf{N}_{low}(a)$ occurred in module $\mathbf{m}_{I_{low}}$. The reasoning would be that when 'bad' data occurs in one of the input modules, 'good' data will occur most of the time in the other input module, so the network will learn to ignore the 'bad' data.

However, the above reasoning is not correct. Consider a training presentation in which data from $\mathbf{N}_{up}(a)$ has occurred in module $\mathbf{m}_{I_{up}}$. Since data occurred within $\mathbf{N}_{up}(a)$, there is a 30% chance of the correct training target category $ctg_a$, and a 70% chance of the incorrect target category $ctg_b$. Whichever category it is, there is a 100% chance that it is the correct category for the data in module $\mathbf{m}_{I_{low}}$. Assume the most advantageous combination: that the category is $ctg_b$, wrong for data in $\mathbf{m}_{I_{up}}$ and right for data in $\mathbf{m}_{I_{low}}$. The network will not learn to ignore data in module $\mathbf{m}_{I_{up}}$, since no target error is generated in the upper sub-network.[7]

There is no magic method of telling the network to ignore a noisy region of input space when the target is *wrong*, if ground truth about the right target is not consistently available because your data is permanently noisy. However, it is possible that attention focusing can provide a way to tell a sub-network to ignore (attenuate) a noisy region of input space because it is associated with an *inconsistent* target.

Consider the full network shown in Figure 2, as implemented in condition 3. After the first training stage, each of the two sub-networks will ideally obtain 100% classification performance outside its noise region, and 30% correct classification performance inside its noise region. During the second training stage, the contribution of $\mu_{sub-net_i}^{(j)}$ is to attenuate the output activity of the input modules in their respective noise regions, $\mathbf{N}_{up}(a)$ and $\mathbf{N}_{low}(a)$. This can be seen by considering again a training presentation in which data from $\mathbf{N}_{up}(a)$ has occurred in module $\mathbf{m}_{I_{up}}$, but the data in $\mathbf{m}_{I_{low}}$ is from *outside* $\mathbf{N}_{low}(a)$.

---

[7] We note that it is important that the noise in both the sub-networks occurs in the same category. If they occurred in different categories, then target output in the noisy category would be of smaller magnitude, thus allowing a final target module that performed simple averaging to boost performance.

We consider the two possible cases: (1) the target associated with $\mathbf{N}_{up}(a)$ is correct (30% probability), and (2) the target associated with $\mathbf{N}_{up}(a)$ is incorrect (70% probability).

The target training vector is correct for $\mathbf{N}_{up}(a)$ when it corresponds to $ctg_a$. However, during the second learning stage, this target will result in a large target error in the top sub-network. This is because during the first learning stage, module $\mathbf{m}_{T_{up}}$, the target module of the top sub-network, learned to generate $ctg_b$ in response to activity in $\mathbf{N}_{up}(a)$ since this combination occurred 70% of the time. The final target module will also generate some target error since it simply averages the contributions of the two intermediate target modules. Referring to Table 1, it can be seen that a large target error combined with a large intermediate target error for the upper sub-network leads to a small value of attenuation parameter $\mu_{sub-net_{up}}^{(j)}$ and a non-vanishing learning parameter. Thus, during this training presentation, the recurrent synapses of the top sub-network will learn to attenuate activity in the $\mathbf{N}_{up}(a)$ activity region. Because the intermediate target error for the bottom sub-network is zero, essentially no attenuation will occur in the associated recurrent synapses.

The target training vector is incorrect for $\mathbf{N}_{up}(a)$ when it corresponds to $ctg_b$. Because of first stage training, this will result in no intermediate target error for the top sub-network, and, since we still assume that the target for the bottom sub-network is correct, no final target error will occur either. Therefore, no update of $\mathcal{A}_{sub-net_{up}}$, the recurrent weight matrix for the upper sub-network input module, occurs when the target training vector is *incorrect* for $\mathbf{N}_{up}(a)$.

Update only occurs when the target training vector is *correct*. The unexpected result is thus that attenuation occurs in the noise regions because the target is occasionally *correct* for data in these regions. In other words, attention focusing should help not because the algorithm can divine when the target is right or wrong – it cannot – but because it can use the available information that the target in noise region $\mathbf{N}_{up}(a)$ is *inconsistent*.

The final condition, condition 4, attenuates all data occurring in noise regions $\mathbf{N}_{up}(a)$ and $\mathbf{N}_{low}(a)$. This essentially implements (by 'cheating') the ideal end result of attention focusing training. The purpose of this condition is both to provide an upper limit for performance of the representation normalization algorithm on the present task, and to see whether the artificial data scenario devised for this task is a model of a situation where attention focusing can improve classification performance at all.

## RESULTS

At this time we have only preliminary results from the experiment described above. We have trained a system according to condition 1 where the sub-networks perform at about a 20% overall error rate (only error rates from stimulus set 3, the 'clean data,' are given). The error rate is about 38% for data in $ctg_a$, and about 10% for data outside $ctg_a$. Condition 2, where only intermediate to final target synapses are trained, did not improve performance. On the other hand, the error rate is reduced in condition 4 by about 7%. This was the condition in which the system is hardwired to attenuates all data occurring in noise regions.

Unfortunately, the error rate is not reduced in condition 3, where representation normalization is implemented. Selective damping of input module activity in noise region areas was recorded. However, the damping was apparently not sufficiently strong to significantly reduce the error rate.

## DISCUSSION

The success of condition 4 in reducing the error rate, together with the negative result in condition 2, suggest that the data scenario constructed for the experiment correctly reflects an application-type in which attention focusing is necessary for optimal neural network performance. The negative result in condition 3 suggests that, as presently implemented representation normalization is not sufficiently powerful to train the system to successfully perform attention focusing for the data in the experimental task being considered. However, this does not constitute a fundamental limitation of the method, as discussed later in this section.

The *dynamics* of the representation normalization technique of attention focusing, as expressed in eq. (2), are similar to Hebbian anti-learning as discussed in [1]. Although adaptive control of the degree of anti-learning was not discussed in this work, it is interesting to note that when $\mu^{(j)}_{sub-net_i} = 0$, eq. (2) is in effect identical to the expression for anti-learning in [1].

The *spirit* of the representation normalization technique bears similarity to several previously proposed methods of attention focusing in neural networks [8, 6, 2, 3, 5]. Each of these methods employs a variation of a *gating* network: a trained supervisory network whose outputs are used to to selectively gate the output of each of the sub-networks. Gating network outputs are analogous to the attenuation parameters ($\mu^{(j)}_{sub-net_i}$) with $i$ indexing the gating output for each sub-network for stimulus $j$. The gating paradigm of attention focusing differs from representation normalization in several ways. Representation normalization requires no multiplicative (gating) connections. ($\mu^{(j)}_{sub-net_i}$) appears explicitly only to control training. It also differs in that the locus of attention focusing is earlier in the system, operating directly on the input module of each sub-network. On the other hand, representation normalization will in general require more synapses than a gating network for a particular application. The importance of these differences is related to the capacity limitations of the machine implementing the algorithms. If synapses are expensive, then a gating paradigm would be desirable. However, if the number of *active* synapses are more important than the total number of synapses, then the representation normalization paradigm would be more attractive. Our method is not explicitly biologically-based, in contrast with some other recently proposed attention focusing techniques (e.g., [9]). However, the capacity limitation on active synapses, which favors the representation normalization paradigm, would be relevant in an analog machine where power consumption is an expensive quantity, and possibly in the brain, where synapses abound but the glucose supply is limited.

Relative processing power is also an issue. As depicted in Figure 2, each set of recurrent synapses constitutes a one-layer associative net, which has limited processing power. This may be contrasted with the architecture of some of the gating paradigms that have been studied in which the supervisory gating network contains hidden layers. However, each one-layer associative network of the representation normalization architecture can be extended to a multi-layer associative network with hidden layers. Sets of synapses can also be added between input modules. No fundamental changes would be required to the update algorithm in eq. (2) to train these additional synapses. With these extensions, there is no apparent difference in processing power between representation normalization network and gating network realizations of attention focusing.

Processing power could be further increased by allowing the associative/recurrent synapses associated with the input layers to have resonant dynamics, similar to those discussed in [1]. If the time scale of the resonance were small relative to the feed-forward delay of the system, then an input to a sub-network for which attenuation was only partially learned by the recurrent synapses would be fully attenuated by the dynamical system.

Thanks are due to Todd Gutschow for thoughtful reading and criticism of this paper. This work was supported in part by the DARPA contract MDA904-90-C-5260.

## References

[1] J.A. Anderson, J.W. Silverstein, S.A. Ritz, & R.S. Jones, 'Distinctive features, categorical perception, and probability learning: Some applications of a neural model,' *Psychological Review*, vol. 84, pp. 413-451, 1977.

[2] J.B. Hampshire II, & A.H. Waibel, 'The Meta-Pi network: Building distributed knowledge representations for robust pattern recognition,' *Tech. Report CMU-CS-89-166-B*, CMU, August 1989.

[3] R.A. Jacobs, M.I. Jordan, & A.G. Barto, 'Task decomposition through competition in a modular connectionist architecture: The what and where vision tasks,' *COINS Tech. Report 90-27*, U. of Massachusetts, Amherst, March 1990.

[4] W.A. Johnston, & S.P. Heinz, 'Flexibility and Capacity Demands of Attention,' *Journal of Experimental Psychology: General*, vol. 107(4), pp. 420-435, 1978.

[5] S.J. Nowlan, 'Competing experts: An experimental investigation of associative mixture models,' *Tech. Report CRG-TR-90-5*, U. of Toronto, September 1990.

[6] J.B. Pollack, 'Cascaded back-propagation on dynamic connectionist networks,' *Proceedings of the Ninth Annual Conference of the Cognitive Science Society*, 1987, pp. 391-404.

[7] M.L. Rossen, & J.A. Anderson, 'Representational Issues in an Neural Network Model of Syllable Recognition,' in *Proceedings of the Third Annual International Conference on Neural Networks*, vol. I, 1989, pp. 19-25.

[8] D.E. Rumelhart, G.E. Hinton, & R.J. Williams, 'Learning internal representations by error propagation,' in D.E. Rumelhart and J.L. McClelland (Eds.), *Parallel Distributed Processing: Explorations in the Microstructure of Cognition: Vol 1*, Cambridge:MIT Bradford, 1986, ch. 8, pp. 319-362.

[9] D. Wang, J. Buhmann, & C. von der Malsburg, 'Pattern segmentation in associative memory,' *Neural Computation*, vol. 2, pp. 94-106, 1990.

# A parallel learning filter system that learns the KL-expansion from examples

Reiner Lenz and Mats Österberg
Linköping University
S-58183 Linköping Sweden

### Abstract

A new method for learning in a single-layer linear neural network is investigated. It is based on an optimality criterion that maximizes the information in the outputs and simultaneously concentrates the outputs. The system consists of a number of so-called basic units and it is shown that the stable states of these basic units correspond to the (pure) eigenvectors of the input correlation matrix. We show that the basic units learn in parallel and that the communication between the units is kept to a minimum.

We discuss two different implementations of the learning rule, a heuristical one and one based on the Newton-rule. We demonstrate the properties of the system with the help of two classes of examples: waveform analysis and simple OCR-reading. In the waveform-analysis case we know the eigenfunctions of the systems from our group-theoretical studies and we show that the system indeed stabilizes in these states.

## Introduction

Unsupervised learning neural networks try to describe the structure of the space of input signals. One type of unsupervised neural networks is closely related to the traditional statistical method of principal component analysis. One of the earliest attempts to compute the eigenfunction of a class of input signals is the principal component analyzer developed by Oja (see [9] for the original description and [10] for a recent overview). Oja's analyzer is based on the idea that the system is essentially a similarity detector or a correlation unit. This leads naturally to a Hebbian learning rule which updates the correlation coefficients so that the output of the analyzer is maximized. Oja showed that this analyzer could learn the first eigenvector of the input process. Recently, Sanger (see [13]) generalized Oja's result by showing how different one-dimensional analyzers could be used to compute a number of different eigenfunctions. The system proposed by Sanger consists of a number of analyzers of the type introduced by Oja which are connected in a serial fashion: The first analyzer works on the original input signal, then the contribution from the first analyzer is subtracted from the original signal and the modified signal is fed into the next analyzer. Analyzer number $i$ is thus trained with that part of the signal which is unexplained by the previous analyzers of the system.

In this paper we propose a system that learns the principal components of the input process in parallel. The basic parts of the system are essentially the same correlation units as those used by Oja and Sanger. In the following we will call one such correlation unit a basic unit. The fundamental difference between our system (see also [7]) and the earlier systems lies in the learning rule that is used to update the basic units. In this paper we will show that our new learning rule makes it possible to train the basic units in parallel. We will also show that the communication between the basic units is minimal. By this we mean that the only information that is send from unit $i$ to unit $j$ is the response of unit $i$ to the actual input pattern. A given unit has therefore no information about the internal coefficients of the other units in the system. This leads to a clear modular structure of the overall system.

The learning rule proposed in this paper is based on a combination of three principles: the structure preserving principle, the maximum information extraction principle and the maximum concentration principle.

The structure preserving or invariance principle is one of the oldest design principles in pattern recognition (see, for example [2], [3] and [4]). It is based on the idea that the measurement- or feature extraction process should preserve the structures in pattern space. The standard example is rotation-invariant edge-detection in image processing. Here the pattern space is structured in the sense that patterns differing only by their orientations are considered to be equal. An edge detector should thus produce features that are independent of the orientation of the edge in the image.

The maximum information principle is by now well-known in neural network research. This principle states that the space of computed output signals should contain as much information about the space of input signals as possible. This maximum information principle was recognized as a basic principle for neural network design by Linsker (see [8]) and it was shown to be of fundamental importance in the group-theoretical filter design strategy (described in [4]). In [4] it was also shown that the invariance principle and the maximum information principle are closely related and that they lead to identical solutions. The amount of extracted information is however much easier to handle algorithmically and we will therefore base our design on the requirement of maximum information extraction. In our analytical studies we showed that the solutions obtained with this approach can be described in terms of eigenfunctions of the input covariance function.

It can be shown that these two principles are not strong enough to force the system into a unique stable state. Using only the maximum information principle usually leads to basic units that correspond to arbitrary (orthogonal) mixtures of pure eigenfunctions. Therefore we propose to combine the maximum information principle with the maximum concentration principle.

The idea behind this principle is the following: Assume we have two systems $\mathcal{C}$ and $\mathcal{S}$ that extract an equal amount of information from the input process. Assume further that system $\mathcal{C}$ tries to concentrate this information into as few output channels as possible, whereas system $\mathcal{S}$ tries to spread this information equally over all channels. A single output channel of system $\mathcal{C}$ will (in the mean) have either very strong or very weak output signals whereas a typical output channel in system $\mathcal{S}$ will produce medium strength signals. If the different units in the systems communicate via noisy communication channels then it should be clear that system $\mathcal{C}$ is much less vulnerable to communication errors than system $\mathcal{S}$. We will therefore require that our system should extract a maximum amount of information from the input signals and that it should concentrate this information as much as possible.

We will first introduce the quality function that we use to measure the quality of the current set of filter coefficients. Then we describe the update rule that optimizes these coefficients by performing a gradient search. We show that this update rule leads to a parallel filter system with a minimal amount of internal communications. In the theoretical part of the paper we will show that the stable states of the system correspond to those eigenfunctions of the input process that have the largest eigenvalues. Then we describe some aspects of the implementation of the system and finally we describe some experiments where we used the system for 1-D signal analysis and OCR-recognition.

## The quality function and its properties

Based on the heuristical considerations described in the introduction can now be formalized in the following construction.

We propose a quality function of the form:

$$Q(W) = \frac{Q_{Information}(W)}{Q_{Concentration}(W)} \quad (1)$$

where $W$ is the matrix of the weight coefficients of the system.

We assume further that the pattern vectors have $K$ components and that the system consists of $N$ basic units. The coefficients of the system are collected in the $K \times N$ matrix $W$. We require that all columns of this matrix have unit length, i.e. the units are represented by unit vectors. If $p$ denotes the input vector and $o$ the output vector then we have: $o = p \times W$, where $\times$ is the product of a vector and a matrix. The components of $o$ will be denoted by $o_i$ : $o_i$ is thus the output of the unit number $i$. The output covariance matrix $S$ is defined as the matrix of all products $o_i \cdot o_j$.

Using these notations we introduce the quality function $Q$ defined as:

$$Q(W) = \frac{Q_V(W)}{Q_C(W)} \quad (2)$$

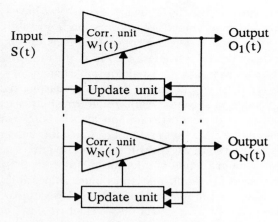

Figure 1: The Learning Filter System

where $Q_V(W) = \det S$ measures the amount of extracted information by the variation of the output signals. The function $Q_C(W)$ is defined as $Q_C(W) = \sum_{i=1}^{N} o_i^2(1 - o_i^2)$. It measures the concentration within the output vectors.

Using gradient based learning rules derived from this quality function, it can be shown, that the learning rule can indeed be implemented in a parallel system of the form shown in Figure 1. This is possible since the increment for a weight factor depends only on the output values of all units and the current pattern vector.

Summarizing we can describe the operation of the system as follows:

- At each iteration all basic units in the system receive the same signal

- The correlation part of a basic unit convolves the signal with the internal state and sends the result to the output line and to all update units in the system.

- From the input signal and all the responses from the correlation units the update unit computes the increment vector that increases the quality function.

- The increment vector is added to the current state vector to produce the new state vector of the unit.

Finally we want to mention that we also experimented with quality functions where the concentration of the feature vectors was measured by the entropy-like function $Q_C(W) = -\sum_{i=1}^{N} o_i^2 \log(o_i^2)$ and we found that both systems had essentially the same performance. In the rest of the paper we will therefore investigate mainly systems based on the first version of the quality function. Some of the experiments are however done with the entropy-like concentration measure.

# Stable States of the System

The stable states of the system are the points where the quality function $Q$ has a maximum. To find a characterization of these points we assume that the system has already stabilized so that the weight coefficients do not change anymore. This constant weight matrix will be denoted by $W$. The output vector at time $t$ is then computed as: $o(t) = p(t) \times W$ and the covariance matrix of the output values is given by:

$$\begin{aligned} S &= \text{mean}_t o'(t) o(t) \\ &= \text{mean}_t W' p'(t) p(t) W \\ &= W' \left[ \text{mean}_t p'(t) p(t) \right] W = W' T W \end{aligned} \quad (3)$$

where $T$ is the covariance matrix of the input patterns: $t_{ij} = \text{mean}_t p_i(t) p_j(t)$. We assume that we know the first and second order statistical properties of the pattern process: especially that the mean vector is zero and that the covariance matrix $T$ is given. Our goal is to find the matrices $W$ which have centered and normed columns and which are maximum points of the quality function: $Q(W) \stackrel{!}{=} \max$.

Our analysis is based on the singular value decomposition of the weight matrix $X$. This decomposition can be described as follows (see [1] and [12]):

**Theorem 1** Assume $X$ is an $K \times N$ matrix with $K > N$. Then we can find orthogonal matrices $U$ and $V$ of size $K$ and $N$ respectively and a matrix $D$ of size $K \times N$ such that

$$X = UDV. \quad (4)$$

For the diagonal elements $d_{kk}$ of $D$ we have $d_{11} \geq d_{22} \geq ... \geq d_{NN} \geq 0$ and the other elements in $D$ are all zero. The decomposition $X = UDV$ is called the *singular value decomposition* or the *SVD* of $X$.

In the following we will use the term diagonal matrix for all matrices that have zero entries outside the diagonal. The matrix $D$ in the previous theorem is therefore a diagonal matrix although it is rectangular. The $N \times N$ unit matrix will be denoted by $E_N$. But we will also use $E_N$ for rectangular matrices if they are essentially equal to the unit matrix, that is if $d_{ii} = 1$ for all $i = 1..N$ and $d_{ij} = 0$ for all $i \neq j$. This should not lead to any confusion since the size of the matrices are always clear from the context.

Using these notations we can summarize our main result in the following theorem (see [6]):

**Theorem 2** Assume that $W$ is a filter system with columns of unit length. Then we found that:

1. The variation value is maximal for all filter systems of the form

$$W = U_0 E_N V, \quad (5)$$

where $U_0$ is the eigenvector matrix of the input covariance matrix and $V$ is an arbitrary orthogonal matrix.

2. For all filter systems with maximum variation value the concentration value is minimal for the filter system consisting of the pure eigenvectors:

$$Q_C(W) = Q_C(U_0 E_N V) \geq Q_C(U_0 E_N E_N) \tag{6}$$

# Implementation and Experiments

In our implementation we initialize the system with random coefficients and use gradient based optimization techniques to update the basic units of the system. Computing the gradient $G$ of the quality function $Q$ one finds that the update rule for the state function $w_n(t)$ is of the form:

$$w_n(t+1) = w_n(t) + \lambda(t,n) G(w_n(t), p(t), o_1(t), ..., o_N(t))$$

Where $\lambda(t,n)$ is the step length function.

The status of the basic unit at iteration $t+1$ is thus a function of its status at the previous iteration $t$, the incoming pattern $p(t)$ and the *output values* of the other basic units in the filter system.

In our first implementation we used the step-length parameter $\lambda$ defined as:

$$\lambda(t,n) = \lambda(o_n(t)) = \begin{cases} \text{abs}(o_n(t)) & 0 \leq \text{abs}(o_n(t)) \leq \frac{1}{2} \\ 1 - \text{abs}(o_n(t)) & \frac{1}{2} \leq \text{abs}(o_n(t)) \leq 1 \end{cases}$$

This definition was motivated by the goal to have more or less binary component values in the final feature vectors. A basic unit that has made up its mind and selects or rejects a given pattern does not need to be updated.

Later we used the Newton method to compute the update rule (see for example [11]). Using this strategy the update rule becomes now:

$$w_n^{(j)}(t+1) = w_n^{(j)}(t) - \frac{g_n^{(j)}}{g_n'^{(j)}}$$

where $w_n^{(j)}$ is the $j$-th coefficient in the $n$-th unit, $g_n^{(j)}$ is a function of the first partial derivative $\frac{\partial Q}{\partial w_n^{(j)}(t)}$ and $g_n'^{(j)}$ is a function of the second partial derivative $\frac{\partial^2 Q}{\partial w_n^{(j)}(t)^2}$ of the quality function.

In a number of experiments we used the system to analyze 1-D signals. Most of these experiments were motivated by the edge and line detection problem from image processing. In this case we can compute the optimal filter functions analytically using group theoretical methods.

The setup for these experiments is the following: Consider the gray value distributions along the dotted circles in figure 2A. These two patterns represent two edge- and line-like gray value distributions on the disk. The one-dimensional intensity functions on these circles are given by the square waves

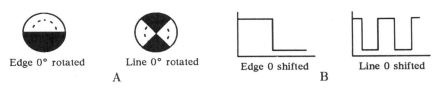

Figure 2: A. 2-D Patterns  B. 1-D Patterns

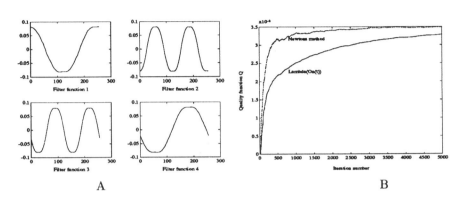

Figure 3: A. Final states of the basic units: Using $Q_E$ and $\lambda(o_n(t))$  B. The step length function $\lambda(o_n(t))$ vs. Newtons method as update rule

in figure 2. Rotated patterns are obtained by circular shifts of the square waves. These shifted (1-D) intensity functions are the input signals to our filter system. Using the group theoretical approach it can be shown that the four best filter functions are: $\cos\phi, \sin\phi$ (for the edges) and $\cos 2\phi, \sin 2\phi$ (for the lines). In all the experiments in this section we use a filter system consisting of four basic units. The input signals are vectors of length 256.

After training a system consisting of four basic units with 5000 different edge and line waves we find a system with final states as shown in figure 3A. One can see immediately that the four filter functions are indeed similar to the first four trigonometric functions. This experiment shows that the filter system converges to the same filter functions as given by group theory. We see also that we could classify the incoming patterns into edge- and line patterns by simply observing which basic unit had the largest response: the basic units one and four react only to edge patterns whereas units two and three react only to lines. This clear separation of the different classes is an effect of the concentration part of our quality function. This feature can not be obtained by systems that are based on principle components properties only.

In another set of experiments with edge and line patterns we compared the performance of the heuristical update rule with the Newton method. We suspect that the learning filter system will converge faster with the New-

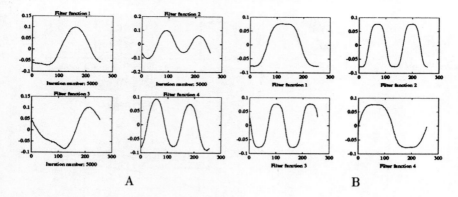

Figure 4: A. Noisy edge and line patterns, step length function $\lambda(o_n(t))$ B. Noisy edge and line patterns, Newton rule

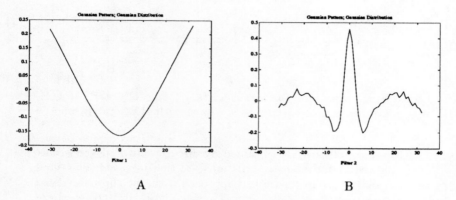

Figure 5: A. Gaussians: Filter 1    B. Gaussians: Filter 2

ton method since it updates every coefficient individually, whereas our first method uses the same step length for all coefficients in one unit. In figure 3 we see that the Newton method really is superior (in both experiments we measured the concentration of the feature vectors with the entropy-like function $Q_E$).

This property of the Newton method was quite expected whereas the result of the next set of experiments was less obvious. In these experiments we compare the performance of the old update rule and the Newton method in the case where the patterns are distorted by added noise. We used exactly the same input sequence in both cases. The results are shown in the figure 4 for the old update rule and the Newton method respectively. We see that the old rule is much more sensitive to noise. This is probably due to the fact that the Newton method is much more flexible since it updates each filter

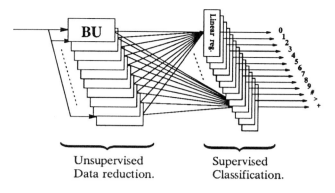

Figure 6: A two level system for OCR-character recognition

coefficient separately.

In the next experiment we trained two filter functions with Gaussian patterns of variable width. The distribution of the width parameters was also a Gaussian. The system consisted of two basic units and their final internal states are shown in the figure 5. They are good approximations of Hermite functions which are known to be optimal solutions in this case (see [5]).

In the last experiment we trained a network with characters read by an OCR-reader. The network consisted of two layers, one for unsupervised data reduction and one for supervised classification, see figure 6. The first layer consisted of ten basic units which was trained with 224 examples of the 13 different OCR-characters 0-9, >, # and +. The second layer consisted of 13 different linear regression functions: one for each character class. First we trained the two level system using the training set, then we tested the recognition rate using a test set consisting of 226 different OCR-characters.

The error rate on the test set was 2/226 which can be compared with the error rate on the training set: 1/224.

# References

[1] G. H. Golub and C. F. Van Loan. *Matrix Computations*. North-Oxford Academic, 1986.

[2] Robert A. Hummel. Feature detection using basis functions. *Computer Graphics and Image Processing*, 9:40–55, 1979.

[3] Reiner Lenz. A group theoretical model of feature extraction. *Journal of the Optical Society of America A*, 6(6):827–834, 1989.

[4] Reiner Lenz. *Group Theoretical Methods in Image Processing.* Lecture Notes in Computer Science (Vol. 413). Springer Verlag, Heidelberg, Berlin, New York, 1990.

[5] Reiner Lenz. On probabilistic invariance. *Neural Networks*, in print.

[6] Reiner Lenz. Computing the karhunen-loeve expansion with a parallel, unsupervised filter system. Submitted for publication.

[7] Reiner Lenz and Mats Österberg. Learning filter systems. In *Proc. Int. Neural Networks Conference, Paris*, 1990.

[8] Ralph Linsker. Self-organization in a perceptual network. *IEEE Computer*, 21(3):105–117, 1988.

[9] E. Oja. A simplified neuron model as a principle component analyser. *Journal of Mathematical Biology*, 15:267–273, 1982.

[10] E. Oja. Neural networks, principal components, and subspaces. *Int. Journal of Neural Systems*, 1:61–68, 1989.

[11] J. M. Ortega and W. C. Rheinboldt. *Iterative Solution of nonlinear equations in several variables*. Academic Press, New York, London, 1970.

[12] W. H. Press, B. P. Flannery, S. A. Teukolsky, and W. T. Vetterling. *Numerical Recipes*. Cambridge University Press, 1986.

[13] Terence D. Sanger. Optimal unsupervised learning in a single-layer linear feedforward neural network. *Neural Networks*, 2(6):459–474, 1989.

# RESTRICTED LEARNING ALGORITHM AND ITS APPLICATION TO NEURAL NETWORK TRAINING

Tsuyoshi MIYAMURA    Isao YAMADA    Kohichi SAKANIWA

Department of Electrical and Electronic Engineering,
Tokyo Institute of Technology
2-12-1 Ookayama, Meguro-ku Tokyo 152 JAPAN

## 1. INTRODUCTION

This paper proposes a new (semi-)optimization algorithm, called the *restricted learning algorithm*, for a nonnegative evaluating function which is 2 times continuously differentiable on a compact set $\Omega$ in $R^N$. The *restricted learning algorithm* utilizes the *maximal excluding regions* which are newly derived in this paper and is shown to converge to the global $\varepsilon$-optimum in $\Omega$.

A most effective application of the proposed algorithm is the training of multi-layered neural networks. In this case, we can estimate the *Lipschitz's constants* for the evaluating function and its derivative very efficiently and thereby we can obtain sufficiently large excluding regions. It is confirmed through numerical examples that the proposed *restricted learning algorithm* provides much better performance than the conventional *back propagation* algorithm and its modified versions.

## 2 RESTRICTED LEARNING ALGORITHM

### 2.1 Preliminaries

In this section, we first introduce some notations which shall be used throughout the paper.

Let $E(w)$ be a nonnegative evaluating function defined in $R^N$ and let $\Omega$ be a compact set in $R^N$. Then the (Semi)-Minimization problem we are concerned with is described as follows:

$$\text{Find } w_{opt} \text{ such that } w_{opt} \in \Omega_{opt}(\varepsilon) \stackrel{\triangle}{=} \{w \mid E(w) \leq \varepsilon, w \in \Omega\}, \quad (1)$$

where $\varepsilon$ is an arbitrary positive constant such that $\min_{w \in \Omega} E(w) < \varepsilon$ ($\varepsilon$ also provides the criterion for the convergence).

In this paper we assume that the evaluating function $E(w)$ is 2 times continuously differentiable on a open set $U(\supset \Omega)$. Hereafter, we denote by $C^i(U)$ the set of functions that are $i$ times continuously differentiable on the open set $U \subset R^N$. Then $E(w)$ and its gradient $\nabla E(w) \stackrel{\triangle}{=} \left(\frac{\partial E(w)}{\partial w_1}, \ldots, \frac{\partial E(w)}{\partial w_N}\right)$ satisfy the following *Lipschitz's conditions*.

C1)    $|E(w_b) - E(w_a)| \leq L \|w_b - w_a\|$

$$\text{for } {}^\exists L(>0) \text{ and } {}^\forall w_a, {}^\forall w_b \in \Omega, \tag{2}$$

C2)  $\|\nabla E(w_b) - \nabla E(w_a)\| \le L_d \|w_b - w_a\|$

$$\text{for } {}^\exists L_d(>0) \text{ and } {}^\forall w_a, {}^\forall w_b \in \Omega, \tag{3}$$

where $\|\cdot\|$ denotes *Euclidean norm*. The constants $L$ and $L_d$ are referred to as the *Lipschitz's constants* for $E(w)$ and $\nabla E(w)$ on $\Omega$, respectively.

The next lemma is obtained by using the *mean value theorem for derivative* and the *Cauchy-Schwartz's inequality*.

**Lemma 1** (*Lipschitz's constant for $E(w)$ and $\nabla E(w)$*)
If $E(w) \in C^2(U)$, every constant $L$ and $L_d$ satisfying

$$\max_{w \in \Omega} \|\nabla E(w)\| \le L \text{ and } \max_{w \in \Omega} \left\{ \sum_{i=1}^{N} \sum_{j=1}^{N} \left( \frac{\partial^2 E(w)}{\partial w_i \partial w_j} \right)^2 \right\}^{\frac{1}{2}} \le L_d \tag{4}$$

are *Lipschitz's constants* for $E(w)$ and $\nabla E(w)$ on $\Omega$, respectively. ¶

## 2.2 Excluding Regions

An *excluding region* is formally defined as follows.

**Definition 1**(*Excluding region*)
A region (open connected set) $\Omega_R(w_*) (\subset R^N)$ is called an *excluding region* for $w_*$, if $w_* \in \Omega_R(w_*)$ and $\Omega_R(w_*) \cap \Omega_{opt}(\varepsilon) = \emptyset$. ¶

Note that if we can find a lower bound $G(w)$ for the evaluating function $E(w)$ in the neighborhood of $w_*$, we can get an *excluding region* by

$$\Omega_R(w_*) = \{w \mid G(w) > \varepsilon\}. \tag{5}$$

In this section, utilizing $E(w_*)$ and $\nabla E(w_*)$ as local information, and *Lipschitz's constants* $L$ and $L_d$ as global information, we derive a couple of lower bounds for $E(w)$ and gives corresponding *excluding regions* by Eq.(5).

It is noted that we shall use in this section a constant $L$ satisfying Eq.(4) as a *Lipschitz's constant* for $E(w)$ while for $\nabla E(w)$ we use a constant satisfying Eq.(3) which is less restrictive.

We shall use the notations

$$B(c;d) \triangleq \{w \mid \|w - c\| < d\} \quad \text{and} \quad \overline{B}(c;d) \triangleq \{w \mid \|w - c\| \le d\}$$

for convenience sake.

**A) Excluding Region (I)**

A well known lower bound for $E(w)$, which is an immediate consequence of Eq.(2), is given by[3]

$$G_1(w) \triangleq E(w_*) - L\|w - w_*\| \quad (\le E(w)).$$

Then we have from Eq.(5) one of the simplest *excluding region* for ${}^\forall w_*$ $(\in \Omega \setminus \Omega_{opt}(\varepsilon))$ as

$$\Omega_1(w_*) \triangleq B\left(w_* ; \frac{E(w_*) - \varepsilon}{L}\right).$$

## B) Excluding Region (II)

If we can use *Lipschitz's constant* for $\nabla E(w)$ in addition to that for $E(w)$, we can get a wider *excluding region* as shown below. We shall use the *Newton Leibnitz formula*[2]

$$E(w) = E(w_*) + \int_0^1 \langle \nabla E(w_* + t(w - w_*)), w - w_* \rangle dt, \tag{6}$$

where $\langle\ ,\ \rangle$ denotes the *inner product*. Applying condition C2)(Eq.(3)) and Lemma 1 to Eq.(6), we obtain after some manipulations the second type of lower bound for $E(w)$ as shown below:

$$G_2(w) \triangleq \begin{cases} E(w_*) + \dfrac{\|\nabla E(w_*)\|^2}{2L_d} - \dfrac{L_d}{2}\left\|w - w_* - \dfrac{\nabla E(w_*)}{L_d}\right\|^2, \\ \qquad \text{for } w \in B\left(w_* + \dfrac{\nabla E(w_*)}{L_d}\ ;\ \dfrac{L}{L_d}\right) \\ E(w_*) + \dfrac{\|\nabla E(w_*)\|^2}{2L_d} + \dfrac{L^2}{L_d} - L\left\|w - w_* - \dfrac{\nabla E(w_*)}{L_d}\right\|, \\ \qquad \text{for } w \in \Omega \backslash B\left(w_* + \dfrac{\nabla E(w_*)}{L_d}\ ;\ \dfrac{L}{L_d}\right). \end{cases} \tag{7}$$

Then it is easy to show that

$$G_1(w) \leq G_2(w), \qquad \text{for }^\forall w \in \Omega$$

holds. Thus the *excluding region* $\Omega_2(w_*)$ obtained by using $G_2(w)$ is always larger than $\Omega_1(w_*)$ and is given by the next theorem.

**Theorem 1** (*Excluding region* $\Omega_2(w_*)$)
*Excluding region* $\Omega_2(w_*)$ for $^\forall w_* (\in \Omega \backslash \Omega_{opt}(\varepsilon))$ is given as follows.

i) If $\sqrt{\dfrac{2(E(w_*) - \varepsilon)}{L_d} + \dfrac{\|\nabla E(w_*)\|^2}{L_d^2}} \leq \dfrac{L}{L_d}$,

$$\Omega_2(w_*) \triangleq B\left(w_* + \dfrac{\nabla E(w_*)}{L_d}\ ;\ \sqrt{\dfrac{2(E(w_*) - \varepsilon)}{L_d} + \dfrac{\|\nabla E(w_*)\|^2}{L_d^2}}\right) \tag{8}$$

ii) If $\sqrt{\dfrac{2(E(w_*) - \varepsilon)}{L_d} + \dfrac{\|\nabla E(w_*)\|^2}{L_d^2}} > \dfrac{L}{L_d}$,

$$\Omega_2(w_*) \triangleq B\left(w_* + \dfrac{\nabla E(w_*)}{L_d}\ ;\ \dfrac{E(w_*) - \varepsilon}{L} + \dfrac{\|\nabla E(w_*)\|^2}{2L \cdot L_d} + \dfrac{L}{2L_d}\right). \tag{9}$$

¶

## C) Excluding region (III)

In the next theorem we show a simple way to modify the lower bounds.
**Theorem 2** (Modification of lower bounds)
Let a lower bound $G(w)$ for $E(w)$ be given as

$$G(w) = g(\|w - c\|) \qquad \text{for }^\exists c \in \Omega \text{ and }^\forall w \in \Omega \tag{10}$$

and suppose that $g(\cdot)$ satisfies

$$g(\theta) \geq 0, \quad \text{for } {}^{\exists}\theta(\geq 0) \text{ s.t. } \overline{B}(c;\theta) \subseteq \Omega. \tag{11}$$

Then for a $\theta$ satisfying Eq.(11), ${}^{\forall}b \geq L_d$ and ${}^{\forall}a \left(\geq \sqrt{2bg(\theta)}\right)$,

$$H(w:g(\theta),a,b) \triangleq \begin{cases} G(w) & \text{for } w \in \overline{B}(c;\theta) \\ g(\theta) - a(\|w-c\| - \theta) + \dfrac{b}{2}(\|w-c\| - \theta)^2 \\ & \text{for } w \in \Omega \setminus \overline{B}(c;\theta) \end{cases}$$

satisfies

$$H(w:g(\theta),a,b) \leq E(w), \quad \text{for } {}^{\forall}w \in \overline{B}\left(c;\theta + \alpha(\theta,a,b)\right), \tag{12}$$

where

$$\alpha(\theta,a,b) \triangleq \dfrac{a - \sqrt{a^2 - 2bg(\theta)}}{b}.$$

¶

By using this theorem, we can improve the lower bound $G_2(w)$ given in Eq.(7). Define the constant vector $c$ in Theorem 2 by

$$c \triangleq w_* + \dfrac{\nabla E(w_*)}{L_d}. \tag{13}$$

Then $G_2(w)$ is expressed as

$$G_2(w) \triangleq g(\|w-c\|), \text{ where}$$

$$g(x) = \begin{cases} E(w_*) + \dfrac{\|\nabla E(w_*)\|^2}{2L_d} - \dfrac{L_d}{2}\|x\|^2, & \text{for } x < \dfrac{L}{L_d} \\ E(w_*) + \dfrac{\|\nabla E(w_*)\|^2}{2L_d} + \dfrac{L^2}{L_d} - L\|x\|, & \text{for } x \geq \dfrac{L}{L_d}, \end{cases} \tag{14}$$

and satisfies the condition Eq.(10). Thus applying Theorem 2 to $G_2(w)$, we obtain a new class of lower bounds for $E(w)$. The next theorem gives a new class of *excluding regions* that are obtained by using these bounds.

**Theorem 3** (*Excluding region* $\Omega_3(w_*:\theta,a,b)$)
Employ $G_2(w)$ as $G(w)$ in Theorem 2 ($c$ and $g(\cdot)$ are given by Eq.(13) and Eq.(14) respectively.). Then for

$${}^{\forall}\theta \quad \text{s.t.} \quad \varepsilon \leq g(\theta) \leq E(w_*) + \dfrac{\|\nabla E(w_*)\|^2}{2L_d},$$

${}^{\forall}b(\geq L_d)$ and ${}^{\forall}a\left(\geq \sqrt{2bg(\theta)}\right)$, the region defined by

$$\Omega_3(w_*:\theta,a,b) \triangleq \{w \mid \varepsilon < H(w:g(\theta),a,b),$$
$$w \in \overline{B}(c;R(w_*,\theta) + \alpha(\theta,a,b))\} \tag{15}$$

is an *excluding region*, where $R(w_*, \theta)$ is given by

$$R(w_*, \theta) \triangleq \begin{cases} \sqrt{\dfrac{2(E(w_*) - g(\theta))}{L_d} + \dfrac{\|\nabla E(w_k)\|^2}{L_d^2}}, \\ \quad \text{for } \sqrt{L_d E(w_*) + \|\nabla E(w_*)\|^2} \le L \\ \dfrac{E(w_*) - g(\theta)}{L} + \dfrac{\|\nabla E(w_k)\|^2}{2L \cdot L_d} + \dfrac{L}{L_d}, \\ \quad \text{for } \sqrt{L_d E(w_*) + \|\nabla E(w_*)\|^2} > L. \end{cases}$$

Since $\Omega_3(w_* : \varepsilon, a, b) = \Omega_2(w_*)$ for $g(\theta) = \varepsilon$, the largest *excluding region* among these derived above is the largest one belonging to $\{\Omega_3(w_* : g(\theta), a, b)\}$.

**Definition 2** (*Maximal excluding region*)
We call the largest *excluding region* in the class $\{\Omega_3(w_* : \theta, a, b)\}$ as the *maximal excluding region* $\Omega_{max}(w_*)$. ¶

The *maximal excluding region* is obtained by finding the optimal $\theta$, $a$ and $b$ which make $\Omega_3(w_* : \theta, a, b)$ the largest and is explicitly given by the next theorem.

**Theorem 4** (*Maximal excluding region* $\Omega_{max}(w_*)$)
For $^\forall w_* (\not\in \Omega_{opt}(\varepsilon))$, the *maximal excluding region* $\Omega_{max}(w_*)$ is given as follows.

1. If $\sqrt{\dfrac{E(w_*)}{L_d} + \dfrac{\|\nabla E(w_*)\|^2}{L_d^2}} \le \dfrac{L}{L_d}$,

   (a) if $\varepsilon \le \dfrac{1}{2}\left(E(w_*) + \dfrac{\|\nabla E(w_*)\|^2}{2L_d}\right)$,

   $$\Omega_{max}(w_*) = B\left(w_* + \dfrac{\nabla E(w_*)}{L_d} \; ; \; 2\sqrt{\dfrac{E(w_*)}{L_d} + \dfrac{\|\nabla E(w_*)\|^2}{2L_d^2}} - \sqrt{\dfrac{2\varepsilon}{L_d}}\right)$$

   (b) if $\dfrac{1}{2}\left(E(w_*) + \dfrac{\|\nabla E(w_*)\|^2}{2L_d}\right) < \varepsilon \le E(w_*) + \dfrac{\|\nabla E(w_*)\|^2}{2L_d}$,

   $$\Omega_{max}(w_*) = B\left(w_* + \dfrac{\nabla E(w_*)}{L_d} \; ; \; \sqrt{\dfrac{2(E(w_*) - \varepsilon)}{L_d} + \dfrac{\|\nabla E(w_*)\|^2}{L_d^2}}\right).$$

2. If $\sqrt{\dfrac{E(w_*)}{L_d} + \dfrac{\|\nabla E(w_*)\|^2}{L_d^2}} > \dfrac{L}{L_d}$,

   (a) if $\varepsilon \le \dfrac{L^2}{2L_d}$,

   $$\Omega_{max}(w_*) = B\left(w_* + \dfrac{\nabla E(w_*)}{L_d} \; ; \; \dfrac{E(w_*)}{L} + \dfrac{\|\nabla E(w_*)\|^2}{2L \cdot L_d} + \dfrac{L}{L_d} - \sqrt{\dfrac{2\varepsilon}{L_d}}\right),$$

(b) if $\dfrac{L^2}{2L_d} < \varepsilon \leq E(w_*) + \dfrac{\|\nabla E(w_*)\|^2}{2L_d}$,

$$\Omega_{max}(w_*) = B\left(w_* + \dfrac{\nabla E(w_*)}{L_d} \; ; \; \dfrac{E(w_*) - \varepsilon}{L} + \dfrac{\|\nabla E(w_*)\|^2}{2L \cdot L_d} + \dfrac{L}{2L_d}\right).$$

¶

## 2.3 Restricted Learning Algorithm

In this section, using the *maximal excluding region* derived in Theorem 4, we describe two (semi-)optimization algorithms newly proposed in this paper. The first one is a strict one and is given as follows.

⟨Algorithm I⟩

**Step 0** Set

$w_0 (\in \Omega)$ : initial point,

$\varepsilon'(> 0)$: criterion for convergence

$k := 0, \delta(k) := \delta_0 (> 0)$

**Step 1** Calculate $\Omega_{max}(w_k)$ according to Theorem 4.

**Step 2** Define sets $Y(\delta(k))$ and $\Omega(\delta(k))$ by

$$Y(\delta(k)) := \bigcup_{i=0}^{k} \Omega_{max}(w_i)$$

$$\Omega(\delta(k)) := \Omega \setminus \left( \bigcup_{x \in [R^N \setminus \Omega] \cup Y(\delta(k))} B(x; \delta(k)) \right).$$

**Step 3** If $\Omega(\delta(k)) = \emptyset$, then $\delta(k) := \dfrac{1}{2}\delta(k)$, and goto Step 1.

**Step 4** Choose $w_{k+1}$ such that

$$w_{k+1} \in \left\{ w_*(\in \Omega(\delta(k))) \; | \; \|w_* - w_k\| = \min_{w \in \Omega(\delta(k))} \|w - w_k\| \right\},$$

$\delta(k+1) := \delta(k)$.

**Step 5** If $E(w_{k+1}) > \varepsilon + \varepsilon'$, then let $k := k+1$ and goto Step 1. else end. ¶

The next theorem guarantees that the sequence $\{w_k\}$ obtained by Algorithm $I$ converges to a point belonging to $\Omega_{opt}(\varepsilon + \varepsilon')$.

**Theorem 5** (Convergence theorem)

For $^\forall \varepsilon (> 0)$ and $^\forall w_0 (\in \Omega)$, the sequence $\{w_k\}$ generated by Algorithm $I$ converges to a point belonging to $\Omega_{opt}(\varepsilon + \varepsilon')$. ¶

Through Algorithm $I$ guarantees the global convergence, it requires to memorize the sequence of *excluding regions* as $Y(\delta(k))$ and may not be practical and efficient in some applications.

The algorithm shown below is a simplified version of Algorithm I and is executed very fast and efficiently while it does not guarantee the global convergence in general.

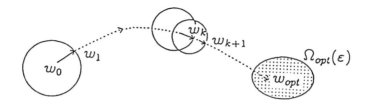

Figure 1: Restricted Learning Algorithm

⟨Algorithm *II* ⟩

**Step 0** Set

$w_0 (\in \Omega)$ : initial point,

$\varepsilon'(> 0)$, $k = 0$.

**Step 1** Calculate $\Omega_{max}(w_k)$ according to Theorem 4.

**Step 2** Choose $w_{k+1}$ such that

$$w_{k+1} \in \left\{ w_*(\in \Omega \setminus \Omega_{max}(w_k)) \mid \|w_* - w_k\| = \min_{w \in \Omega \setminus \Omega_{max}(w_k)} \|w - w_k\| \right\}.$$

**Step 3** If $E(w_{k+1}) > \varepsilon + \varepsilon'$, then let $k := k+1$ and goto Step 1, else end. ¶

Although Algorithm *II* is a simplified version of Algorithm *I*, it still retain the following remarkable properties.

- If $\Omega_{max}(w_k)$ is included in $\Omega$, the innovation vector $w_{k+1} - w_k$ and the gradient vector $-\nabla E(w_k)$ have the same direction, that is, the new point $w_{k+1}$ is in the direction of steepest descent from $w_k$.

- If $w_k$ is a local minimum such that $E(w_k)(> \varepsilon)$, the innovation vector $w_{k+1} - w_k$ can not be 0, that is, the sequence $\{w_k\}$ does not settle at a local minimum.

In Fig.1 is shown the intuitive interpretation of Algorithm *II*. It can be said that this algorithm works just the *steepest descent algorithm* with variable step size which is adjusted by the size of *excluding region* at each iteration.

## 3. APPLICATION TO NEURAL NETWORK TRAINING

In this section, we apply the proposed *restricted learning algorithm* to the training of multi-layered neural networks and compare it with the conventional *back propagation algorithms*.

### 3.1 Neural Network

We consider here the three layered network $\mathcal{N} = (G, w, (P; Q))$.

$G = (I, H, O; E)$ represents the three layered feednext graph where $I$, $H$ and $O$ are the sets of *units(neurons)* in the *input* layer, in the *hidden* layer and in the *output* layer, respectively.
$E = \{(u,v) \mid u \in I, v \in H\} \cup \{(u,v) \mid u \in H, v \in O\}$ denotes the set of connections between the units in *successive* layers. $w (\in R^N)$ represents the vector composed of all weights $w_{(u,v)}$ assigned to each connection $(u,v) \in E$ and the threshold levels $\theta_v$ associated with the output $a_v$ of each unit $v \in H \cup O$. $(P; Q)$ denotes the paired set of input pattern $p = \{p_u\}_{u \in I}$ and the corresponding desired output $q(p) = \{q_v(p)\}_{v \in O}$. The output $a_v$ of a unit $v \in H \cup O$ is assumed to be given as the output of the *sigmoid* function

$$f(x) \triangleq \frac{1}{1 + \exp(-x)}$$

Thus the each output $a_v$ of a unit $v \in H \cup O$, when the input pattern $p$ is applied to $\mathcal{N} = (G, w, (P; Q))$, is recursively given as

$$a_u(p, w) = p_u \quad \text{(for } u \in I\text{)},$$
$$a_v(p, w) = f\left( \sum_{e=(u,v) \in E} w_{(u,v)} a_u(p, w) - \theta_v \right)$$
$$\text{(for } u \in I, v \in H \text{ or } u \in H, v \in O\text{)}.$$

And the evaluating function $E(w)$ for the neural network is defined by

$$E(w) \triangleq \sum_{p \in P} \sum_{v \in O} \{a_v(p, w) - q_v(p)\}^2. \tag{16}$$

## 3.2 Lipschitz's Constants for Neural Networks

The next two theorems are derived by using Lemma 1 and the recursion formula for partial derivatives which is well known and used in the *back propagation algorithm*. These theorems enable us to estimate the *Lipschitz's constants* for $E(w)$ and $\nabla E(w)$ very effectively. In what follows, it is assumed that the value of each element in $w$ of $\mathcal{N} = (G, w, (P; Q))$ is limited to the range $[-T, T]$, that is, $\Omega = [-T, T]^N$.

**Theorem 7** (*Lipschitz's constant $L$ for $E(w)$*)
A *Lipschitz's constant* $L$ for $E(w)$ is given by

$$L = \frac{8}{27} \left[ |O||P| \left\{ |P| + |H| \sum_{p_d \in P} J(p) \right\} \right]^{\frac{1}{2}}$$

where

$$J(p) \triangleq \max \left[ \frac{1}{2^5} \left\{ \xi(p) \left(1 - \frac{8}{\xi(p)}\right)^{\frac{3}{2}} - \frac{8}{\xi(p)} + \xi(p) + 20 \right\}, 1 \right]$$

$$\xi(p) \triangleq |O|T^2 \left( \|p\|^2 + 1 \right)$$

and $|S|$ denotes the cardinality of the set S.

**Theorem 8** (*Lipschitz's constant $L_d$ for $\nabla E(w)$*)
A *Lipschitz's constant $L_d$ for $\nabla E(w)$* is given by

$$L_d = \left[ |H|^2 \sum_{p \in P} \max_{x,y \in [0,1]} \{(1-x)^2 (1-y)^2 \right.$$
$$[\alpha(p) x^2 y^2 + \beta(p) x^2 + \gamma] \}$$
$$\left. + |P| \left[ |H| \max \{(1-\eta)^2 (\eta^2 + 2\gamma), 2\gamma\} + \gamma \right] \right]^{\frac{1}{2}}, \tag{17}$$

where

$$\alpha(p) \triangleq 4 \left( \frac{27}{256} T^2 \right)^2 |P| \sum_{p_u \in p} \sum_{p_v \in p} \{p_u \cdot p_v\}^2$$

$$\beta(p) \triangleq 8|O| \left( \frac{27}{256} T \right)^2 |P| \|p\|^2$$

$$\gamma \triangleq 4|O| \left( \frac{27}{256} \right)^2 |P| \qquad \eta \triangleq \frac{1 + \sqrt{1-16\gamma}}{4}.$$

### 3.3 Numerical Examples

We consider as an example a neural network for the *parity check* problem, i.e., the three layered network that distinguishes whether the number of 1's in the input pattern $p$ (a *binary $|I|$-tuple*) is even or odd.

Fig.2 and Fig.3 show the comparisons of convergence properties of the proposed algorithm (Algorithm $II$) and the conventional ones from the two points of view.

In Fig.2, $\Delta_k(R)$ and $\Delta_k(B)$ denote the innovation $\|w_{k+1} - w_k\|$ for the proposed *restricted learning algorithm II* and that for the conventional *back propagation algorithm* with learning factor 1, respectively. The ratio $\Delta_k(R)/\Delta_k(B)$ is shown for the neural networks whose number of units in each layer is $(|I|, |H|, |O|) = (7, 7, 1)$ and $(10, 10, 1)$, and the number of input patterns is 10, 15 and 20.

Fig.3 shows the behavior of the evaluating function $E(w_k)$ for the network whose number of units in each layer is $(10, 10, 1)$ and the number of input patterns is 20. *Momentum* indicates the modified *back propagation algorithm* with learning factor 0.25 and momentum factor 0.9. It is seen from these figures that the *restricted learning algorithm II* gives a much better performance than the conventional *back propagation algorithms*.

### 4. CONCLUSION

In this paper, we have propose two types of new (semi-)optimization algorithms, called the *restricted learning algorithm*, for a nonnegative evaluating function which is 2 times continuously differentiable on a compact set $\Omega$ in $R^N$. These algorithms utilizes the *maximal excluding regions* newly defined in this paper.

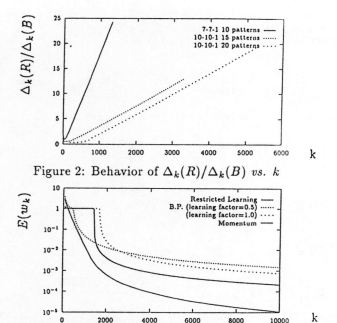

Figure 2: Behavior of $\Delta_k(R)/\Delta_k(B)$ vs. $k$

Figure 3: Convergence of Restricted Learning and Back Propagation

The first one guarantees the convergence to the global $\varepsilon$-optimum in $\Omega$ while it requires large memory size. The second one is a simplified version of the first and supposed to be very effective in the practical applications.

A most effective application of the proposed algorithm is the training of multi-layered neural networks. In this case, it has been confirmed through numerical examples that the proposed *restricted learning algorithm*(the second type) provides much better performance than the conventional *back propagation* algorithm and its modified versions.

## ACKNOWLEDGMENT

The authors are grateful to Prof. S. Tsujii and Prof. K. Kurosawa of Tokyo Institute of Technology for their encouraging suggestions.

## REFERENCES

[1] B. Widrow and M. A. Lehr : "30 Years of Adaptive Neural Networks: Perceptron, Madaline, and Backpropagation", *Proc. IEEE*, vol.78, No.9, 1990.

[2] Y. G. Evtshenko: *Numerical Optimization Techniques*, Optimization Software, Inc. Publications Division, New York, 1985.

[3] A. Törn and A. Žilinskas: *Global Optimization*, Lecture Notes in Computer Science, 350, 1988.

# MULTIPLY DESCENT COST COMPETITIVE NEURAL NETWORKS WITH COOPERATION AND CATEGORIZATION

Yasuo Matsuyama

Department of Computer and Information Science
Ibaraki University
Hitachi-shi, Ibaraki 316, Japan

Abstract - Generalized competitive learning algorithms are described. These algorithms comprise competition handicaps, cooperation and multiply descent cost property. Applications are made on signal processing and combinatorial optimizations. Besides, parallel computation of the presented algorithms is discussed.

## I. INTRODUCTION

A multiply descent cost algorithm is a composition of cost-decreasing mappings. Each mapping decreases a common cost in a distinctive way. In the case of doubly descent cost, the first mapping decreases the cost by generating an optimized grouping on atomic training data. The second mapping decreases the cost by competitive learning.

Competition is a mechanism where only one (winner-take-all), or at most a few (winner-take-quota), qualified neuron can give an output and can update its state in order to decrease the cost. Such a neuron is called *winner*. Nearby neurons of each winner can also update their states. This is a cooperation among locally connected neurons.

There are two learning strategies according to the form of input feeding. The batch mode is the case that all traing data are given to the network and afterwards all winners are updated. In the successive mode, however, only a portion of the training data is shown. For this data piece, a winner is decided and the updates occur. These two modes are equivalent in the sense that a problem solved by one can be solved by the other. We start discussions from the batch mode.

## II. MULTIPLY DESCENT COST COMPETITIVE LEARNING: BATCH MODE

### Training Data and Neural Weight Vectors

Let $\{\mathbf{x}_i\}_{i=0}^{T-1}$ be a fixed finite set of vectors as training data. Each vector $\mathbf{x}_i$ is an element in $\mathbf{R}^M$. The atomic data elements $\mathbf{x}_i$, $(i = 0, \ldots, T-1)$, are grouped so that weight vectors $\mathbf{v}_j$, $(j = 0, \ldots, J-1)$, are generated. That is, $\bigcup_{j=0}^{J-1} \mathbf{v}_j = \bigcup_{i=0}^{T-1} \mathbf{x}_i$, $\mathbf{v}_i \cap \mathbf{v}_j = \emptyset$, $(i \neq j)$. Note that the dimensions of $\mathbf{v}_j$'s are usually different. This grouping to form $\{\mathbf{v}_j\}_{j=0}^{J-1}$ is versatile. Denote the finite class of such grouping by $\mathcal{U}$. If a specific way of grouping, say $u \in \mathcal{U}$, needs to be emphasized, then $\mathbf{v}_j(u)$ is used instead of $\mathbf{v}_j$.

The grouping is done by looking at the cost with respect to a *set of neural weight vectors*, i.e., a *set of standard patterns*: $\mathcal{C} = \{\mathbf{c}_0, \ldots, \mathbf{c}_n, \ldots, \mathbf{c}_{\|N\|-1}\}$.

Here, $c_n \in \mathbf{R}^{KL_0}$. $K$ is the dimension of $\mathbf{x}_i$'s transformed expressions. $L_0$ is the number of elements in a predetermined regular class.

Let $\mathcal{C}^{(q)} = \{c_0^{(q)}, \ldots, c_{n_q}^{(q)}, \ldots, c_{N_q-1}^{(q)}\}$, $(q = 0, \ldots, Q-1)$, with $c_{n_q}^{(q)} \in \mathbf{R}^{K_q L_0}$ and $\sum_{q=0}^{Q-1} K_q = K$. Draw one element from each $\mathcal{C}^{(q)}$, $(q = 0, \ldots, Q-1)$, and form $c_n = \mathrm{col}(c_{n_0}^{(0)}, \ldots, c_{n_q}^{(q)}, \ldots, c_{n_{Q-1}}^{(Q-1)}) \in \mathbf{R}^{KL_0}$. If every $c_n \in \mathcal{C}$ is expressed by this form, then $\mathcal{C}$ is called *product weight set* or *product standard pattern set* since this $\mathcal{C}$ can be expressed by $\mathcal{C} = \prod_{q=0}^{Q-1} \mathcal{C}^{(q)}$ with $\|\mathcal{C}\| = N = \prod_{q=0}^{Q-1} N_q$.

**Supervector's Class**

Let the class for $\mathbf{v}_j$'s be $\mathcal{G}$. A two-dimensional example of the class $\mathcal{G}$ is the convex quadrilateral of pixels[1],[2]. In the case of one dimension, the class $\mathcal{G}$ is the set of variable-length segments. The regular form is restricted to be of a further specific form, say $\mathcal{G}_0$. In the case of two-dimensional data, the class $\mathcal{G}_0$ is often selected to be a square. The class $\mathcal{G}_0$ for one dimensional case is a fixed-length segment.

Since the grouped weight vector $\mathbf{v}_j \in \mathcal{G}$ is compared with a regular form $c_n \in \mathcal{G}_0$, the latter needs to be altered. This is called *pattern reconstruction with warping* or simply *warping*, and is expressed by $\mathbf{w} : \mathcal{G}_0 \times \mathcal{G} \mapsto \mathcal{G}$.

**Cost Function**

A *cost function* defines dissimilarity of two weight vectors. By the warping $\mathbf{w}$, it becomes possible to compare a supervector $\mathbf{v}_j$ with a regular form $c_n$ using the cost function: $d_\mathbf{w}(\mathbf{v}_j, c_n) = d_\mathcal{G}(\mathbf{v}_j, \mathbf{w}(c_n))$. Here, $d_\mathcal{G} : \mathcal{G} \times \mathcal{G} \mapsto [0, \infty)$ is a usual cost function.

**Total Cost.** Let $k_q$, $(q = 0, \ldots, Q-1)$, be loop count indices appearing in the algorithms of the following sections. Let the weight vector set with the loop count indices be $\mathcal{C}[k_0, \ldots, k_{Q-1}] = \prod_{q=0}^{Q-1} \mathcal{C}^{(q)}[k_q] = \prod_{q=0}^{Q-1} \{c_0^{(q)}[k_q], \ldots, c_{N_q-1}^{(q)}[k_q]\}$, and let the grouping pattern be $u_m$. Here, the index $m$ is dependent on $k_q$ by $m = \sum_{q=0}^{Q-1} k_q$. Then, the total cost with respect to the weight vector set is defined as follows:

$$D[k_0, \ldots, k_{Q-1}, u_m] = \sum_{j=0}^{J-1} d_\mathbf{w}(\mathbf{v}_j(u_m), \mathcal{C}[k_0, \ldots, k_{Q-1}])$$
$$= \sum_{j=0}^{J-1} \min_{\substack{0 \le n_p < N_p \\ 0 \le p < Q}} d_\mathcal{G}(\mathbf{v}_j(u_m), \mathbf{w}(\prod_{q=0}^{Q-1} c_{n_q}^{(q)}[k_q], \mathbf{v}_j(u_m))).$$

**Descent Cost Grouping.** A descent cost grouping $\phi$ with respect to the weight vector set $\prod_{q=0}^{Q-1} \mathcal{C}^{(q)}[k_q]$ is to find $u'_m$ such that

$$D(\{\mathbf{x}_i\}_{i=0}^{T-1}, \prod_{q=0}^{Q-1} \mathcal{C}^{(q)}[k_q] \mid u'_m) \le D(\{\mathbf{x}_i\}_{i=0}^{T-1}, \prod_{q=0}^{Q-1} \mathcal{C}^{(q)}[k_q] \mid u_m).$$

Such a mapping is denoted by $\phi(u_m, \prod_{q=0}^{Q-1} \mathcal{C}^{(q)}[k_q]) = u'_m$.

**Descent Cost Partial Update of Weight Vectors.** A descent cost *partial* update of weight vectors with respect to the $p$-th vector set $\mathcal{C}^{(p)}[k_p]$ is to find $\prod_{q=0}^{Q-1} \mathcal{C}^{(q)}[k'_q]$ with

$$D(\{\mathbf{x}_i\}_{i=0}^{T-1}, \prod_{q=0}^{Q-1} \mathcal{C}^{(q)}[k'_q] \mid u'_m) \le D(\{\mathbf{x}_i\}_{i=0}^{T-1}, \prod_{q=0}^{Q-1} \mathcal{C}^{(q)}[k_q] \mid u'_m)$$

and $k'_q = k_q + \delta_{pq}$. Here, $\delta_{pq}$ is the Kronecker's delta. Denote such a mapping for the update of the $p$-th vector set by

$$\psi_p(\prod_{q=0}^{Q-1} \mathcal{C}^{(q)}[k_q], u'_m) = \prod_{\substack{q=0 \\ \{p\,:\,\text{updated}\}}}^{Q-1} \mathcal{C}^{(q)}[k'_q].$$

Here, "$p$ : updated" means that $k'_q = k_q + \delta_{pq}$. This update process starts with generating a partition with respect to the $p$-th vector set:

$$B_{n_p}(u'_m) = \left\{ \mathbf{v}_j(u'_m) \mid d_G\left(\mathbf{v}_j(u'_m), \mathbf{w}(\prod_{q=0}^{Q-1} \mathbf{c}_{n_q}^{(q)}[k_q], \mathbf{v}_j(u'_m)\right)\right.$$
$$\left. \leq d_G\left(\mathbf{v}_j(u'_m), \mathbf{w}(\prod_{q=0}^{Q-1} \mathbf{c}_{n'_q}^{(q)}[k_q], \mathbf{v}_j(u'_m)\right), \; n'_p \neq n_p \right\}.$$

Using this partition $\{B_{n_p}(u'_m)\}_{n_p=0}^{N_p-1}$, the mapping $\psi_p$ to update $\mathcal{C}^{(p)}[k_p]$ is to obtain the *generalized centroid* $\mathbf{c}_{n_p}^{(p)}[k_p + 1] \in \mathbf{R}^{K_p L_0}$. Then,

$$\mathcal{C}^{(p)}[k_p + 1] = \{\mathbf{c}_0^{(p)}[k_p + 1], \ldots, \mathbf{c}_{N_p-1}^{(p)}[k_p + 1]\}$$

is the updated weight vector set. There are other suboptimal descent cost mappings. We denote this class by $\Psi_p$. Note that we select the index for the grouping to be $u'_m = u_{m+1}$.

## Basic Multiply Descent Cost Competitive Learning Algorithm

The basic design algorithm comprises four phases; the initial states, the mapping selection for grouping, the stopping check and the mapping selection for weight vector update.

*[Basic Design Algorithm]*

**Mapping Scheduler.** *The mapping scheduler $\mathcal{A}$ knows the finite set $\Phi$ of mappings for grouping and the finite sets $\Psi_q, q \in \{0, \ldots, Q-1\}$ for weight vector update. The mapping scheduler is settled to select every member of $\Phi$ and $\Psi_q$, $q \in \{0, \ldots, Q-1\}$ infinitely often. The scheduler has the following four phases:*

**Initial State.** *The training set $\{\mathbf{x}_i\}_{i=0}^{T-1}$ to be grouped into $J$ supervectors, a positive constant $\epsilon$ and the following initial states are given; a weight vector set $\prod_{q=0}^{Q-1} \mathcal{C}^{(q)}[0]$, a grouping pattern $u_0$ and a cost $D[k_0, \ldots, k_{Q-1}, m] = D[0, \ldots, 0, 0] = D[old] = \infty$.*

**Mapping Selection for Grouping.** *The mapping scheduler picks up $\phi$ from $\Phi$, and then apply this $\phi$ to $(\prod_{q=0}^{(Q-1)} \mathcal{C}^{(q)}[k_q], u_m)$ in order to yield $u'_m = u_{m+1}$. Then, go to Stopping Check.*

**Stopping Check.** *If every element in $\Phi$ and $\Psi_q$, $q \in \{0, \ldots, Q-1\}$ is selected since the previous Stopping Check, then let $D[new] = D[k_0, \ldots, k_{Q-1}]$, which is the current cost. Then, compute and check the inequality*
$$(D[old] - D[new])/D[new] < \epsilon.$$
*If this inequality holds, then exit from the iteration and adopt $\prod_{q=0}^{Q-1} \mathcal{C}^{(q)}[k_q]$ and $u_m$ to be the final $\prod_{q=0}^{Q-1} \mathcal{C}^{(q)}$ and $u$, respectively. If the inequality does not hold, then replace $D[old]$ by $D[new]$, and go to Mapping Selection for Weight Vector Update. If there is still an unused element in $\Phi$ or in $\Psi_q$, $q \in \{0, \ldots Q-1\}$ since the previous Stopping Check, then simply go to Mapping Selection for Weight Vector Update.*

**Mapping Selection for Weight Vector Update.** *A mapping $\psi_p$ is selected from $\Psi_p$, $p \in \{0, \ldots, Q-1\}$ according to the mapping scheduler's rule. Then, this $\psi_p$ is applied to generate*

$$\psi_p(\prod_{q=0}^{Q-1} \mathcal{C}^{(q)}[k_q], u'_m) = \prod_{\substack{q=0 \\ \{p \,:\, \text{updated}\}}}^{Q-1} \mathcal{C}^{(q)}[k'_q].$$

*The index $k_p$ is increased by one to yield $k'_p = k_p + 1$ whenever $\psi_p$ is applied. Then, go back to Mapping Selection for Grouping.*

We note here that, for any $\epsilon > 0$, the basic algorithm terminates after a finite number of stopping checks. That is, both the grouping pattern and the weight vector sets converge.

### Blockwise Additive Cost and Parallel Partial Optimization

In the Basic Design Algorithm, a mapping $\phi$ is always inserted between the weight vector updates of $\psi_p$ and $\psi_q$ so that the descent cost property is maintained. Under specific circumstances, a concurrent weight vector update such as $\psi_p \circ \psi_q$ still satisfies the descent cost property. Of special interest is the case that the application of the composition mapping $\prod_{q=0}^{Q-1} \psi_q \equiv \psi_{Q-1} \circ \cdots \circ \psi_0$ honors the descent cost property. A sufficient condition for assuring this is the *blockwise additive cost*:

$$d_{\mathcal{G}}(\mathbf{v}, \mathbf{w}) = \sum_{q=0}^{Q-1} d_{K_q}(\mathbf{v}^{(q)}, \mathbf{w}^{(q)}), \qquad \mathbf{v}^{(q)}, \mathbf{w}^{(q)} \in \mathbf{R}^{K_q}.$$

There are quite a few cost functions satisfying this property[3].

## III. MULTIPLY DESCENT COST COMPETITIVE LEARNING: SUCESSSIVE MODE

For the case of partial optimization on a blockwise additive cost, competition is made on a piece of training data. This is the successive mode.

*[Successive Training for the Multiply Descent Cost Competitive Learning]*

**Step 1** ($t = 0$). *The initial grouping pattern on the atomic data elements and the initial weight vector set $\mathcal{C}^{old}$ are given. The finite set $\Phi$ of the mappings for the descent cost grouping is given. A rule $\mathcal{A}$ on the scheduling of descent cost mappings for the grouping is defined. The modification rules to the numbers $r$ and $\epsilon(t, r)$ are also given.*

**Step 2** ($t := t + 1$). *A piece of training data is given and a descent cost grouping $\phi$ is selected from the set $\Phi$ by the scheduler $\mathcal{A}$.*

**Step 3 (feature map by partial optimization).** *The mapping $\phi$ generates new $\mathbf{v}_j$'s.*

**Step 4 (standard pattern update).** *For each new $\mathbf{v}_j$, choose the lowest cost weight vector $\mathbf{c}_{\mathbf{V}_j}^{old} \in \mathcal{C}^{old}$. Then, the weight vector set is updeted by*

$$\mathbf{c}_{\mathbf{V}_j}^{new} = \mathbf{c}_{\mathbf{V}_j}^{old} + \epsilon(t, 0)\{\bar{\mathbf{w}}(\mathbf{v}_j) - \mathbf{c}_{\mathbf{V}_j}^{old}\}.$$

*Here, $\bar{\mathbf{w}}$ is a warping.*

**Step 5 (feature map for the weight vectors).** *For the weight vectors $\mathbf{b}_r^{old}$ within $r$-neighborhood of $\mathbf{c}_{\mathbf{V}_j}^{old}$, the following update is made for each new $\mathbf{v}_j$:*

$$\mathbf{b}_r^{new} = \mathbf{b}_r^{old} + \epsilon(t, r)\{\bar{\mathbf{w}}(\mathbf{v}_j) - \mathbf{b}_r^{old}\}.$$

**Step 6.** *If the modifications to the weight vectors become small enough, then stop the iteration. Otherwise, modify $r$ and compute $\epsilon(t+1, r)$ and $\epsilon(t+1, 0)$ by the predefined rule. Then, go to Step 2.*

## IV. APPLICATIONS TO SIGNAL PROCESSING AND OPTIMIZATION

### Image Processing

**Image feature map by the multiply descent cost.** The successive mode algorithm is applied to images here. In this case, $\{\mathbf{x}\}_{i=0}^{T-1}$ is a finite set of pixels as the atomic data for training. These pixels are grouped into convex quadrilateral patches $\{\mathbf{v}_j\}_{j=0}^{J-1}$. Then, for $u(t) \in \mathcal{U}$, the patches $\{\mathbf{v}_j(u(t))\}_{j=0}^{J-1}$ form a partition. This grouping is done by considering the cost function to be minimized over the weight vector c in $\mathcal{C}$.

$$\begin{aligned} d_{\mathbf{w}}(\mathbf{v}_j, \mathbf{c}) &= d_{\mathbf{w}}(\square abcd, \square ABCD) \\ &= \sum_{\text{physical pixel} \in \square abcd} (\text{OriginalPixel} - \text{InterpolatedPixel})^2 \\ &= \|\mathbf{v}_j - \mathbf{w}(\mathbf{c})\|^2. \end{aligned}$$

Here, $\square abcd$ stands for a variable-sized convex quadrilateral. $\square ABCD$ denotes a unit square to be warped. The space warping w is done by the bilinear transformation and the bilinear patch. in $\square abcd$. The *feature map* by the optimization is the final grouping pattern $u$ as a set of quadrilateral vertices obtained by the multiply descent cost competitive learning.

**Supervised movement of vertices.** Obtained feature map is used for generation of new patterns controlled by external information. This synthesis is done by supervised vertex movement. Contiguous vertices are usually adjusted to maintain the convexity of the patches so that the bilinear transformation is well-posed. For this purpose, iterations of the following equation on the vertex position is used.

$$\begin{aligned} \mathbf{p}^{\text{new}} = \mathbf{p}^{\text{old}} + \epsilon(t)\{&\mathbf{f}_n(\mathbf{p}', \text{NearbyVertices}, \text{PixelValues})(\mathbf{n} - \mathbf{p}^{\text{old}}) \\ +&\mathbf{f}_e(\mathbf{p}', \text{NearbyVertices}, \text{PixelValues})(\mathbf{e} - \mathbf{p}^{\text{old}}) \\ +&\mathbf{f}_w(\mathbf{p}', \text{NearbyVertices}, \text{PixelValues})(\mathbf{w} - \mathbf{p}^{\text{old}}) \\ +&\mathbf{f}_s(\mathbf{p}', \text{NearbyVertices}, \text{PixelValues})(\mathbf{s} - \mathbf{p}^{\text{old}})\}. \end{aligned}$$

Here, the points n, e, w and s are North-East-West-South adjacent vertices of $\mathbf{p}^{\text{old}}$.

**Experimental results.** Fig.1a is the original image. Fig.1b is a feature map for this image obtained by the multiply descent cost competitive learning. This feature map is used to generate a new image by adding information from external intelligence. In Fig.1c, such supervisory information is given to the mesh pattern. This information is to move vertices around an eye. Using the resulting mesh pattern, Fig.1d is obtained. One notices that the girl winks. We note here that, in the above feature transformation, manually generated wire-frame models[4] are *not* used.

## Speech Processing

Multiply descent cost competitive learning is also applicable to speech processing. Because of one dimensional nature as a time series, the optimization on the grouping of atomic data generates variable-length segments of speech. Each atomic data in one segment is a vector whose components are parameters to specify an autoregressive model for a short period of speech. The most popular parameters are partial correlation coefficients. The cost function in this case is nonsymmetric. However, the update term of the weight vector can be obtained by using
$$\Delta c[n] = \epsilon(t)(k[n] - c[n])/\{(d+1)p_k[n]\},$$
where $k[n]$ and $c[n]$ are components of a PARCOR vector and a weight vector. $p_k[n]$ is expressed by
$$p_k[n] = \prod_{j=n}^{m}(1 - k^2[n]),$$

and $d$ is
$$d = \sum_{i=1}^{m}(k[i] - c[i])^2/p_k[i].$$
The mapping for adjusting the dimension is simply a time warping for each segment. Thus, the learning by the multiply descent cost competition gives a set of time normalized typical phonemes[1],[2].

## Combinatorial Optimizations

The presented competitive learning algorithms stretches their hands to the field of combinatorial optimizations besides the above pattern processing. We treat extended vehicle routing problems which contain $N$-person TSP and simple TSP as relaxation problems. For such combinatorial optimization problems, the competition is of single descent cost. However, carefully considered handicaps are used.

**Vehicle routing problems.** The *vehicle routing problem* (VRP)[5] is an extension of the *traveling salesperson problem* (TSP).

[**VRP**] Given a set of cities with a common depot, assigned loads to the cities and $N$ vehicles with capacities, find the shortest total route for the vehicles. Cities are located in the Euclidian space in this article.

Further extended problems are of our interest.

[**EVRP1**] When types of cities to accept specific vehicles are assigned, find the shortest routes for the vehicles.

[**EVRP2**] Solve EVRP1 under the suppression of the longest subtour among $N$ vehicles.

[**EVRP3**] Solve EVRP1 with a request for the equalization of loads per one vehicle.

[**EVRP4**] Solve EVRP1 under the joint constraints of EVRP2 and EVRP3.

Thus, the TSP is the simplest case having one vehicle without any city types nor loads.

*[Algorithm for EVRP's]* At the start of learning, the time count $t$ is set to zero. A fixed set of cities, $V = \{\mathbf{v}_i\}_{i=0}^{M-1}$, is given. Here, the city $\mathbf{v}_0$ is the shared depot. Each city has an assigned type which specifies qualified vehicles. The set of neural weight vectors are as follows:
$$U_n(t) = \{\mathbf{u}_{n,0}(t), \ldots, \mathbf{u}_{n,K-1}(t)\}, \quad \mathbf{u}_{n,k}(t) \in \mathbf{R}^2, \quad (n = 0, \ldots, N-1).$$
Here, each $\mathbf{u}_{n,k}(t)$ at $t=0$ is placed on a circle which corresponds to the $n$-th vehicle. The index $k$ is an order on each circle. A city $\mathbf{v}$ is selected at random from $V$. Time count $t$ is increased by one. If $\mathbf{v}$ is a plain city, then the following handicapped competition and updates are made: A winner $\mathbf{u}_{m,\ell}$ satisfying
$$\min_{0 \leq n < N} \min_{0 \leq k < K} [handicap]_n \|\mathbf{u}_{n,k} - \mathbf{v}\|^2$$
is found. Then, the weights $\mathbf{u}_{m,\ell}$ and $\mathbf{u}_{m,k}$ with $|k - \ell| \leq L(t)$ on the circle are updated:
$$\begin{aligned}\mathbf{u}_{m,k}(t+1) = & \ \mathbf{u}_{m,k}(t) + \epsilon(t)f(k-\ell,t)(\mathbf{v} - \mathbf{u}_{m,k}(t)) \\ & + \alpha(t)(\mathbf{u}_{m,k+1}(t) - 2\mathbf{u}_{m,k}(t) + \mathbf{u}_{m,k-1}(t)). \end{aligned} \quad (1)$$
If the city $\mathbf{v}$ is the depot, $N$ winners, $\mathbf{u}_{n,\ell}, (n = 0, \ldots, N-1)$, are found by $\min_{0 \leq k < K} \|\mathbf{u}_{n,k} - \mathbf{v}\|^2$. Then, neurons are updated in the same way as equation (1). If all cities catch distinct winners after iterations, the learning is completed. Else, if the catch percentage is greater than a prespecified vigilance and non-increasing by iterations, an overlapping neuron is split. Then, learning parameters are updated. This completes one cycle and a city $\mathbf{v}$ is selected

again. Note that new handicaps are computed at every $M$ city feedings for the sake of computational load alleviation.

**Handicaps.** Each vehicle's eligibility for the problems of [EVRP1~4] is achieved by the handicap of

$$H_n^{(1)} = \begin{cases} 1, & \text{if the city } \mathbf{v} \text{ admits the visit of the vehicle } n; \\ \infty, & \text{if not admitted.} \end{cases}$$

Other handicaps are further multiplied to $H_n^{(1)}$ for the validity of obtained solutions. Let $y_i, (i = 0, \ldots, N-1)$, be subtour lengths. Then,

$$y_n^2 / \{(\sum_i y_i)(\sum_i y_i - y_n)\}$$

is multiplied to $H_n^{(1)}$. For subtotal demands $z_i, (i = 0, \ldots, N-1)$, a handicap of the same form as this expression is multiplied.

**Evaluations by experiments.** First, performance for the case of TSP was evaluated by 500 sets of 30 cities in a unit square generated by uniform random numbers. For these test data, our method, Durbin and Willshaw's method[6] and the simulated annealing[7] showed similar performances within 5% excess from the optimal tour lengths. All methods generated true optima around 1/5 of the test data. Note that the method of [8] is the case with $\alpha(t) \equiv 0$ is (1).

The above performances, however, are quite different for a larger set from the real world, the USA532 set[9]: [Padberg and Rinaldi: optimum (6h by a supercomputer)] < [our method: 3.74% longer (2.8h by a WS)] < [simulated annealing: 4.44% longer (0.33h by a WS)] << [Durbin and Willshaw: 34.23% longer (170h by a WS)].

Performance for EVRP's by the presented algorithm is the main point to be examined. Fig.1 illustrates the progress and the result of the computation for [EVRP2] with four vehicles and three city categories. Cities marked by ◯ accept only the vehicle 1. The vehicle 2 can visit cities marked by ◯ and △. The vehicles 3 and 4 can visit any cities. Table 1 summarizes the total tour length, maximum tour lengths and maximum total loads. One notices that the handicaps are very effective.

Table 1    Obtained tour lengths and loads for EVRP1 ~ EVRP4

| problems | EVRP1 | EVRP2 | EVRP3 | EVRP4 |
|---|---|---|---|---|
| total length | 10.2421 | 11.4987 | 13.7066 | 13.3596 |
| max length | 4.4200 | 3.3520 | 3.6453 | 3.5326 |
| max load | 1001 | 863 | 668 | 680 |

## V. SYMBIOSIS OF HETEROGENEOUS PARALLELISMS

**Parallel partial optimization.** Neurocomputation is a paradigm of algorithms. Therefore, its execution on a serial computer is still of some value. In fact, all of the illustrated examples are computed by a conventional workstation. However, most of neurocomputation algorithms are controlled by mapping schedulers which reflects a high-level parallelism after massive data processing. Thus, there are at least two distinctive levels on the parallel computation of neuro-algorithms: Fine-grained computation on raw data and coarse-grained administration. This is a coexistence of heterogeneous parallelisms. In order to examine characteristics of such two-level parallelism, a general-purpose emulator for PDP-style computation is implemented by two workstations. The total system of our two-level heterogeneous parallelism is called NeuroCube reflecting its architecture[10].

**Nondeterminism versus randomness.** Experiments of the emulated two-level heterogeneous parallelism are tried for the multiply descent cost competitive learning on test data. Resulting mesh patterns correspond to lower costs than the case of deterministic scan of vertices on a serial computer. This is due to the parallel processing's nondeterminism which assists escapes from bad local minima.

## VI. CONCLUDING REMARKS

The multiply descent cost competitive learning self-organizes an optimized feature map. The method contains dynamic learning by decisions unlike the static method[11] for the artificial cortical map. Obtained feature map is not intended for recognition but generation of new patterns. Adding information to this feature map allows manipulation of the original data. Refining this process more, a human can impersonate others on a computer. A human-at-large such as an anthropoid could express its emotion like a human via computer.

We note here that the competitive learning can treat combinatorial optimizations besides traditional pattern processing.

## ACKNOWLEDGEMENTS

The author is grateful to Messrs. and Misses Y. Kurosawa, P. V. Krishnamraju, T. Furuya and N. Kumagai for programming supports and discussions. Support by the Grant-in-Aid for Scientific Research, Higher-Order Brain Functions, is appreciated.

## REFERENCES

[1] Y. Matsuyama, "Variable region vector quantization, space warping and speech/image compression," Proc. ICASSP, vol. 4, pp. 2201-2204, 1987.
[2] Y. Matsuyama, "Variable region vector quantization," Trans. IEICE, vol. J-70A, pp. 1830-1837, 1987 (Translation: Electro. and Comm. in Japan, Pt. 1, vol. 71, pp. 49-61, 1988).
[3] Y. Matsuyama, "Mismatch robustness of linear prediction and its relationship to coding," Info. and Control, vol. 47, pp. 237-262, 1980.
[4] K. Aizawa, Y. Yamada, H. Harashima and T. Saito, "Model-based synthesis of image coding system - Modelling a person's face and synthesis of facial expressions," Proc. GLOBECOM, vol. 1, pp. 45-49, 1987.
[5] N. Christofides, "Vehicle routing," in E.L. Lauler et al. Eds., The Traveling Salesman Problem, pp. 431-448, John Wiley & Sons, 1985.
[6] R. Durbin and D. Willshaw, "An analogue approach to the travelling salesman problem using an elastic net method," Nature, vol. 326, pp. 689-691, 1987.
[7] W.H. Press et al. "Numerical Recipes in C," pp. 343-352, Cambridge Univ. Press, 1988
[8] B. Angéniol et al., "Self-organizing feature maps and travelling salesman problem," Neural Networks, vol. 1, pp. 289-293, 1988.
[9] M. Padberg and G. Rinaldi, "Optimization of a 532-city symmetric traveling salesman problem by branch and cut, Operations Res. Lett., vol. 6, pp. 1-7, 1987.
[10] Y. Matsuyama, "Neural net self-organization and two-level parallelism," Proc. InfoJapan'90, Pt. 2, pp. 113-120, 1990.
[11] R. Durbin and G. Mitchison, "A dimension reduction framework for understanding cortical maps," Nature, vol. 343, 644-647, 1990.

(a) ISO/CCITT image.

(b) A feature map.

(c) Controlled feature map.

(d) Created image.

Fig.1 Generation of new image by feature map modification.

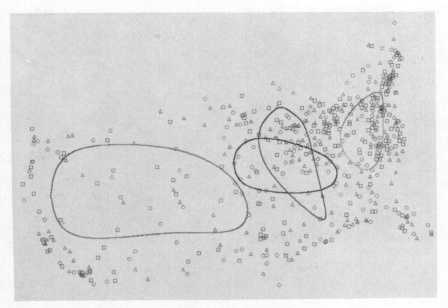

(a) Progress of training.

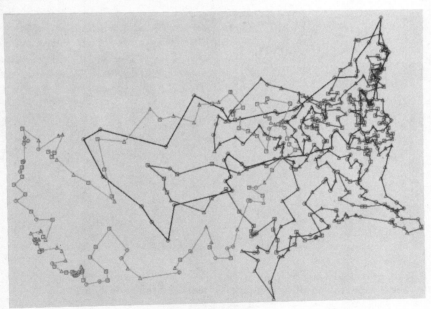

(b) Final tours.
Fig.2 Progress and result of EVRP2.

# NONLINEAR ADAPTIVE FILTERING OF SYSTEMS WITH HYSTERESIS BY QUANTIZED MEAN FIELD ANNEALING

Ramin A. Nobakht*  Sasan H. Ardalan**  David E. Van den Bout**

*International Business Machines Corporation
P.O. Box 12195, C86A/060
Research Triangle Park, NC 27709

**Center for Communications and Signal Processing
Box 7914, Department of Electrical and Computer Engineering
North Carolina State University, Raleigh, NC 27695-7914

## Abstract

In this paper, a technique for nonlinear adaptive filtering of systems with hysteresis has been developed which combines Quantized Mean Field Annealing (QMFA) and conventional RLS/FTF adaptive filtering. Hysteresis is modeled as a nonlinear system with memory. Unlike other methods which rely on Volterra and Wiener models, this technique can efficiently handle large order nonlinearities with or without hysteresis effects. The nonlinear channel is divided into a memory nonlinearity followed by a dispersive linear system. Assuming that the dispersive linear system is stationary during initialization, and the nonlinearity does not change while the dispersive linear system varies with time, QMFA is applied to obtain the coefficients and the order of the memory of the nonlinearity and RLS/FTF is applied to determine the weights of the dispersive linear system. Application of this method to a full duplex digital subscriber loop is made. Simulations show the superior performance of our technique compared to that of ordinary RLS/FTF and steepest-descent algorithms.

## 1 Introduction

Adaptive filter design for nonlinear systems has attracted much attention in the past few years [1, 2, 3]. The existing theory relating to the analysis and modeling of nonlinear systems relies on the Volterra and Wiener models [4]. As the order of nonlinearity increases, such models will become too complex in many practical situations. Also, in addition to such problems, use of the conventional LMS algorithm results in suboptimal solutions.

Within the past several years, two widely applicable new approaches to optimization have appeared which avoid the problems listed above: *simulated annealing* [5] and *neural networks* [6]. Simulated annealing uses random perturbations modulated by a steadily decreasing control parameter (referred to as the *temperature*, $T$) to escape from local optima, but, due to the Monte Carlo nature of this technique, a large number of iterations are required to converge to a good solution. On the other hand, neural networks converge rapidly to good solutions while avoiding local optima, but they require that the function to be optimized be expressed in closed form so that the interaction terms between the neuronal processing elements can be calculated. *Mean Field Annealing* (MFA) [7] combines aspects of both simulated annealing and neural networks to create an algorithm which converges rapidly to near-optimal solutions for a wide variety of problems. The *Quantized Mean Field Annealing* (QMFA) algorithm is an improvement to the MFA algorithm in terms of its efficiency and speed. This is accomplished by using every piece of *a priori* information that can narrow our ignorance of the true object. Hence, it is sensible to adopt consistency with all the pieces of *a priori* knowledge, rather than optimality with respect to some arbitrary standard, as an optimization criterion.

In this paper, a technique for adaptive filtering of nonlinear systems has been developed which combines QMFA and conventional LMS and RLS adaptive filtering. The nonlinear channel is divided into a memory nonlinearity followed by a dispersive linear system (see Figure 1). A dispersive linear system may precede the nonlinearity also [8]. It is assumed that the nonlinearity does not change while the dispersive linear systems vary with time. In applying QMFA to the nonlinear adaptive filter design, the total adaptation time is divided into a training period and a tracking period (Figure 1). Assuming the dispersive linear systems are stationary during initialization (training period), QMFA is applied to obtain the coefficients and the order of the memory of the nonlinearity and RLS/FTF (Fast Transversal Filter) is applied to determine the weights of the dispersive channel. Since it is assumed that the nonlinearity is fixed, the input to the dispersive system can be replicated and a conventional adaptive filtering algorithm can be used to track the dispersive system (tracking period). One of the known problems associated with annealing techniques is the large processing time associated with them. However, the decomposable nature of QMFA combined with the availability of low-cost DSPs makes it practical to achieve near real-time performance with our algorithm through parallel processing. The next section will describe the QMFA technique and the analysis of the adaptive filtering of nonlinear systems with the incorporation of QMFA. In section 3, a realistic example is given which shows the performance of our technique compared to that of ordinary RLS/FTF and steepest-descent (DA) algorithms.

# 2 Adaptive Filtering of Nonlinear Systems

A central objective in engineering is to develop optimal solutions to problems. Unfortunately, many problems of practical value are difficult to solve due to their combinatorial nature — i.e., the quality of their solutions is affected by a large number (possibly millions) of interacting decisions or *degrees of freedom*. Expressed mathematically, a vector $\mathbf{a} = \{a_1, a_2, \cdots, a_N\}$ must be found which minimizes some function of interest, $H(\mathbf{a})$, that depends on $\mathbf{a}$ in some complex, non-linear way. In this paper, $H(\mathbf{a})$ is the root mean square (RMS) of error derived from a conventional RLS/FTF algorithm in an adaptive filtering scheme (see Figure 1), and $\mathbf{a}$ is the set of quantized amplitudes of the $N$ coefficients of the nonlinear element. Adjustments are made to $\mathbf{a}$ based only upon knowledge of the previous and present RMS errors — no other information concerning the system is used. Figure 1 depicts the general system which was used to study the performance of the quantized mean field annealing optimization technique on the adaptive filtering problem. As can be seen from the figure, two optimizations are performed: One to minimize $H(\mathbf{a})$ in order to determine the coefficients and the order of the memory of the nonlinear system (using QMFA), and the other to determine the weights of linear dispersive system (using RLS/FTF).

## 2.1 Nonlinear Subproblem: QMFA

In the quantized mean field annealing algorithm (QMFA), the amplitude range for the $N$ parameters of the system are divided into a contiguous set of $M$ bins, $\{A_1, A_2, \ldots, A_M\}$. At each temperature a randomly selected parameter is stepped through the entire set of $M$ quantized amplitudes (while leaving the other parameters unchanged) and the performance measure at each amplitude, $H_j$ ($1 \leq j \leq M$), is determined either analytically or by simulation. The selected parameter is then set to a weighted average of the quantized amplitudes

$$a_i = \frac{\sum_{j=1}^{M}(A_j)\exp(-H_j/T)}{\sum_{j=1}^{M}\exp(-H_j/T)} \quad (1)$$

where bins with a larger performance degradation contribute less to the average. Let us now assume that the optimal solution which we are in search of actually belongs to a set of distinct quantized values within each and every parameter:

$$S = \{\hat{a}_0, \hat{a}_1, \hat{a}_2, \ldots, \hat{a}_n\} \quad (2)$$

The projection $P_s(a)$ is an $a_p$ in $S$ such that:

$$(\forall i \in \{0, 1, 2, \ldots, n\}) \quad |a_p - a| \leq |\hat{a}_i - a| \quad (3)$$

Therefore the set of all quantized magnitudes of every parameter can be used as a set of *a priori* information to increase the efficiency and the speed of the optimization procedure. Such information can be translated as a quantization factor added to $a_i$ in order to bring its magnitude to the closest $A_i$. The quantization process may be modeled mathematically as

$$\hat{a}_i = a_i + q_i \tag{4}$$

where $\hat{a}_i$ represents the quantized value of $a_i$ and $q_i$ represents the quantization factor. Hence, equation (1) becomes

$$\hat{a}_i = \frac{\sum_{j=1}^{M}(A_j)\exp(-H_j/T)}{\sum_{j=1}^{M}\exp(-H_j/T)} + q_i \tag{5}$$

This procedure is performed for every parameter and, as the temperature decreases, each parameter increasingly avoids amplitudes with a high performance degradation. Since the interactions between the parameters affect the total performance measure, this iterative procedure combined with the temperature adjusted weighting allows the parameters to cooperatively relax to an optimal or near-optimal solution.

In this section we will describe the solution for the problem of adaptive filtering of systems which contain elements of nonlinearity with memory. In addition to the set of $N$ coefficients of the nonlinear system describing $N$ different parameters to be optimized, another parameter defining the set of past memory samples is also added. Similar to other parameters, this memory parameter is also divided into a contiguous set of *bins*, defining the magnitude of the past memory samples.

In applying QMFA to this problem, the amplitude range for the $N$ coefficients of the nonlinear system are divided into a contiguous set of $M$ *bins*, $\{A_1, A_2, \ldots, A_M\}$, and similarly, the memory range for the memory parameter is also divided into a contiguous set of $M$ *bins*, $\{A_1, A_2, \ldots, A_M\}$. Therefore there will be a total of $N+1$ parameters for optimization for this problem. At each temperature a randomly selected parameter is stepped through the entire set of $M$ quantized amplitudes (while leaving the other parameters unchanged) and the RMS error at each amplitude, $H_j$ ($1 \leq j \leq M$), is determined either analytically or by simulation using RLS/FTF methods. According to equation 5, the selected coefficient is then set to a quantized weighted average of the amplitudes, where bins with a larger RMS error contribute less to the average. This procedure is performed for every parameter and, as the temperature decreases, each parameter increasingly avoids amplitudes with a high RMSE. Therefore the QMFA algorithm will not only estimate the magnitude of the nonlinear coefficients, it will also estimate the number of past memory samples which the nonlinear element depends on.

## 2.2 Linear Subproblem: RLS/FTF

For every QMFA iteration (nonlinear parameter update) a new RMS error value related to such update needs to be determined, using *reinitialized* RLS/FTF algorithms. Such error is returned to QMFA for nonlinear minimization. As a result, a new set of dispersive linear system weights will also be determined. This section describes the procedure.

Consider the desired signal $d(n)$ generated by a nonlinear mapping of $x(n)$ convolved by the weight vector $\underline{w}^*$ with elements $w_i^*$ (see Figure 1):

$$d(n) = \sum_{i=0}^{N-1} w_i^* \underline{y}(n-i) = \underline{w}^{*T}\underline{y}(n) \qquad (6)$$

where $\quad \underline{y}(n) = f[\underline{x}(n)]$

For a given fixed input sequence

$$\underline{x}(n) = [x(n), x(n-1), \cdots x(n-N+1)]^T \qquad (7)$$

$f[\underline{x}(n)]$ is a nonlinearity with order of memory, $m$, that can be adequately modeled for hysteresis effects [9] using a p-th order Taylor series expansion

$$\underline{y}(n) = f[\underline{x}(n)] = \sum_{i=n-m}^{n} \sum_{j=1}^{p} \hat{a}_j \underline{x}^j(i) \qquad (8)$$

Combining equations (6) and (8) for the estimated sequence makes,

$$\hat{d}(n) = \sum_{i=0}^{N-1} \sum_{l=n-m}^{n} \sum_{j=1}^{p} w_i \hat{a}_j x^j(l-i) = \underline{w}^T \hat{\underline{y}}(n) \qquad (9)$$

In the Recursive Least Squares (RLS) algorithm the vector $\underline{w}(n)$ is sought which minimizes the cost function of the accumulated square of the residuals up to time n:

$$J(n) = \sum_{i=0}^{n} [d(i) - \underline{w}^T(n)\hat{\underline{y}}(i)]^2 \qquad (10)$$

Taking the derivative of equation (10) with respect to $\underline{w}(n)$ and setting it to zero we obtain an expression for the weights of the linear dispersive system $\underline{w}(n)$.

$$\underline{w}(n) = \left[\sum_{i=0}^{n} \underline{\hat{y}}(i)\underline{\hat{y}}^T(i)\right]^{-1} \sum_{i=0}^{n} d(i)\underline{\hat{y}}(i) \qquad (11)$$

This result leads to a recursive algorithm for calculating $\underline{w}(n)$. We form an estimate of $d(n)$ based on the current estimate of $\underline{w}(n-1)$,

$$\hat{d}(n) = \underline{\hat{y}}^T(n)\underline{w}(n-1) \qquad (12)$$

Next, calculate the prediction error,

$$e(n) = d(n) - \hat{d}(n) \qquad (13)$$

Finally, update $\underline{w}(n-1)$ based on the prediction error,

$$\underline{w}(n) = \underline{w}(n-1) + \underline{K}(n)e(n) \qquad (14)$$

where the Kalman gain is defined by,

$$\underline{K}(n) = \left[\sum_{i=0}^{n} \underline{\hat{y}}(i)\underline{\hat{y}}^T(i)\right]^{-1} \underline{\hat{y}}(n) \qquad (15)$$

The Kalman gain can be calculated recursively by efficient fast algorithms [10, 11], which can reduce the computation time to $O(N)$ mult/div per sample time.

Next, the RMS error is determined. This RMS error could be determined on the $N^{th}$ error sample alone, or on several of the next possible error samples:

$$H_j = \left[\frac{\sum_{m=N-1}^{n} e^2(m)}{n - N + 2}\right]^{1/2} \qquad (16)$$

The resulting error is a function of $\underline{y}(n)$ and therefore a function of $\hat{a}_i's$. This error is returned to QMFA for nonlinear minimization. Until the optimal nonlinear parameters are reached, the above procedure is repeated iteratively as the temperature is lowered in the QMFA algorithm.

# 3 Digital Subscriber Loop Example

A model of a full duplex digital subscriber loop with digital to analog converter (DAC) nonlinearity with hysteresis [12, 13] was implemented in the CAPSIM [14] simulation package. In the simulation three nonlinear coefficients and twenty filter taps were used to model a nonlinear system with memory of $\tau = 3$ time samples. Figure 3 shows hysteresis loops due to the effect of the nonlinearity with memory. As can be seen, the hysteresis loop widens as the order of memory is increased. Pseudorandom sequences of 100 samples with mean and standard deviation of 0.0 and 72.4 respectively were used for each QMFA optimization run. Figure 2 shows the RMS error as a function of the number of iterations for QMFA optimization during the training stage. As can be seen, convergence occurs after about 38 parallel iterations or 920 average independent runs which required approximately 2.3 cpu minutes on a DECstation 3100. In these analysis, the starting and ending temperatures and the rate of temperature decrease were $10^4$, $10^0$ and 0.4 respectively. Elimination of the nonlinear constraint in the training stage reduces the optimization to a Least Squares problem in the tracking stage.

A reasonable way to evaluate the performance of QMFA algorithm for this problem is to compare it to that of ordinary RLS/FTF and steepest-descent (DA) algorithms. In the case of the RLS/FTF, it is assumed that no other optimization is used for the determination of the nonlinear coefficients and the memory element. The performance of QMFA algorithm, DA, and RLS/FTF are determined in the tracking stage by monitoring the error signal $e(n)$ as a function of time samples. The input driving the system in all cases was white gaussian noise of 1000 samples. Figure 4 shows the error signal as a function of time samples for the three cases. As can be seen from the figure, the QMFA algorithm was able to improve the performance by about 40 dBs compared to the RLS/FTF algorithm and by about 37 dBs compared to the DA algorithm. Clearly the gradient descent algorithm did not perform well since it was only able to improve the system performance by about 3 dBs compared to the RLS/FTF algorithm. This demonstrates the power of QMFA in solving for the true MMSE estimate, as opposed to solving for an estimate which satisfies suboptimal equations. It is interesting to note that the run time for QMFA was less than the run time for DA.

# 4 Conclusions

This paper presents a technique for adaptive filtering of nonlinear systems with hysteresis. Unlike other methods which rely on Volterra and Wiener models, our technique can efficiently handle large order nonlinearities with or without hysteresis effects. Adaptive filtering of nonlinear systems can become viable and practical by using QMFA and RLS/FTF algorithms. QMFA can be used to determine the parameters of the nonlinear element in the training stage, when the dispersive linear element is stationary. RLS/FTF algorithms can be used to determine the weights of the time varying linear dispersive system in the tracking stage, while the nonlinear element remains unchanged. In simulations this method demonstrates the elimination of the nonlinear constraint during the training stage, hence reducing to a Least Squares problem.

# References

[1] M. J. Coker and D. N. Simkins. A nonlinear adaptive noise canceler. In *Proceedings of ICASSP 80*, pages 470–473, 1980.

[2] C. E. Davila, A. J. Welch, and H. G. Rylander III. A second-order adaptive volterra filter with rapid convergency. *IEEE Transactions on Acoustics, Speech, Signal Processing*, ASSP-35(9):1259–1263, September 1987.

[3] T. Koh and E. J. Powers. Second order volterra filtering and its application to nonlinear system identification. *IEEE Transactions on Acoustics, Speech, Signal Processing*, ASSP-33(6):1445–1455, December 1985.

[4] M. Schetzen. *The Volterra and Wiener Theory of the Nonlinear Systems*. Wiley, New York, 1980.

[5] S. Kirkpatrick, C. Gelatt, and M. Vecchi. Optimisation by simulated annealing. *Science*, 220(4598):671–680, May 13 1983.

[6] J. Hopfield and D. Tank. Neural computations of decisions in optimisation problems. *Biological Cybernetics*, 52:141–152, 1985.

[7] G. Bilbro, R. Mann, T. Miller III, W. Snyder, D. E. Vandenbout, and M. White. Mean field annealing and neural networks. In *Advances in Neural Information Processing Systems*, pages 91–98. Morgan-Kaufmann Publishers, Inc., 1989.

[8] C. F. N. Cowan and P. F. Adams. Nonlinear system modeling: Concept and application. In *Proceedings of ICASSP 84*, pages 456.1–4, 1984.

[9] M. A. Krasnosel'skii and A. V. Pokrovskii. *Systems With Hysteresis*. Springer-Verlag, New York, 1989.

[10] L. Ljung, M. Morf, and D. D. Falconer. Fast calculation of gain matrices for recursive estimation schemes. *Int. Journal of Control*, 27(1):1–19, 1978.

[11] J. M. Cioffi and T. Kailath. Fast, recursive-least-squares transversal filters for adaptive filtering. *IEEE Transactions on Acoustics, Speech, Signal Processing*, ASSP-32(2):304–337, April 1984.

[12] O. Agazzi, D. G. Messerschmitt, and D. A. Hodges. Nonlinear echo cancellation of data signals. *IEEE Transactions on Communications*, COM-30(11):2421–2433, Nov. 1982.

[13] The Engineering Staff of Analog Devices Inc. *Analog-Digital Conversion Handbook, Third Edition*. Prentice-Hall, Englewood Cliffs, NJ, 1986.

[14] R. A. Nobakht, P. W. Pate, and S. A. Ardalan. CAPSIM: A graphical simulation tool for communication systems. In *Proceedings of IEEE Global Telecommunications Conference (Globecom)*, pages 1692–1696, Hollywood, Florida, Nov. 28 - Dec. 1, 1988.

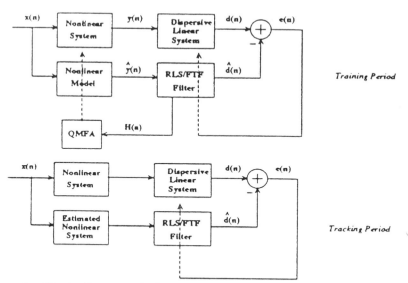

**Figure 1:** Block diagram of adaptive filtering for a nonlinear system during the training and tracking periods.

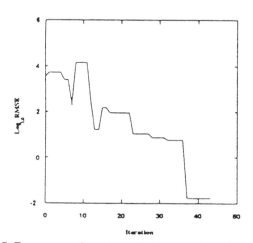

**Figure 2:** RMS Error as a function of the number of iterations for QMFA optimization during the training stage.

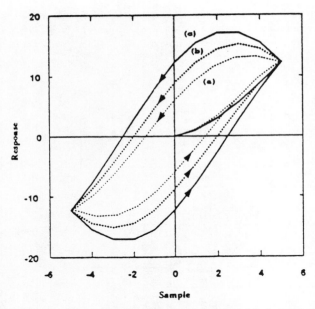

**Figure 3:** Hysteresis loops due to the effect of the third order nonlinearity and order of memory: **a.** 3, **b.** 4, **c.** 5

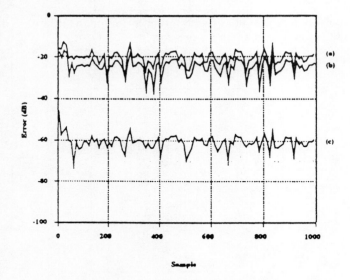

**Figure 4:** Error signal e(n) as a function of time samples using the following algorithms: **a.** RLS/FTF, **b.** DA, **c.** QMFA

# AN OUTER PRODUCT NEURAL NETWORK FOR EXTRACTING PRINCIPAL COMPONENTS FROM A TIME SERIES

L. E. Russo
Surveillance Systems Division
Lockheed Sanders
Hudson, NH 03051

Abstract-An outer product neural network architecture has been developed based on subspace concepts. The network is trained by auto-encoding the input exemplars, and will represent the input signal by k-principal components, k being the number of neurons or processing elements in the network. The network is essentially a single linear layer. The weight matrix columns orthonormalize during training. The output signal converges to the projection of the input onto a k-principal component subspace, while the residual signal represents the novelty of the input. An application to extracting sinusoids from a noisy time series is given.

## INTRODUCTION

Artificial neural networks draw from a variety of disciplines: neurobiology, optimization theory, pattern recognition, statistics, etc. Application of neural networks relies on abstracting from these disciplines an architecture which allows potential solution of a problem and on selection of a cost or objective function which, when minimized, enables the architecture to map the stimulus into the solution space. There may be several combinations of architecture and cost function; however, the effectiveness of the neural network will be based on how well these choices match the problem at hand.

In this paper, a simple architecture, suggested by subspace methods in pattern recognition, is used to extract principal components from a time series. The number of principal components extracted matches the number of neurons in the network. Although the particular application is relatively simple, principal component extraction has wide application to signal enhancement, noise reduction and signal discrimination. Since the network is trained by auto-encoding, the objective function is simple although it converges to an interesting result.

Oja first discussed using neural network to extract a single principal component from a time series[1]. His related work on subspace methods[2][3] has strongly influenced the approach in this paper. Oja's training rule in vector form is:

$$\delta C = \gamma \times (yx^T - diag(yy^T))C. \quad (1)$$

where $x$ is the input vector exemplar, $y$ is the output vector, and $\gamma$ is the training coefficient. This rule was used to update a single neuron architecture, hence, to extract a single principal component. Sanger[4] took Oja's rule and added the vector form of Gram-Schmidt orthogonalization:

$$\delta C = lower(yy^T)C \quad (2)$$

to get:

$$\delta C = \gamma \times (yx^T - LT(yy^T))C, \quad (3)$$

where *'diag'* indicates diagonal entries, *'LT'* means lower triangular and *'lower'* indicates entries on and above the diagonal are zero. Normalization of the orthogonal components was performed as a separate step. Sanger's approach allowed multiple neurons to be trained to extract more than one principal component. However, training required intra-layer connections among the neurons to enforce orthogonality. The most recent work of Kung[5] retains the flavor of Sanger's work in that there are excitory and inhibitory intra-layer connections combined in a coupled training rule. The intra-layer or anti-Hebbian connections orthogonalize the network as in Sanger's approach. Kung claims this rule may be extended to the *Constrained Principal Components* problem.

With the exception of Oja, most work has focussed on a modification of Oja's rule, i.e., on the training rule. The focus of this paper is somewhat different: a simple, yet suggestive, architecture is selected for which meaning may be attributed to network structure, objective function and network residual. The resulting network has no intra-layer connections between neurons.

## OUTER PRODUCT NEURAL NETWORK

Consider the outer product neural network shown in Figure 1. This network is linear because the output $y = CC^T x$, i. e., there are no sigmoids in the network. In general, it is desirable that the number of neurons, p, be less than the size of the exemplar vector, $x$. The number of neurons determines the rank of the non-negative definite matrix, $CC^T$, and the rank of $CC^T$ determines the subspace spanned by the principal components of $x$. Using fewer neurons than inputs is a well established principle for stimulating a neural network to extract features[6], and is a fundamental concept behind using neural networks to extract principal components. However, this was usually done by restricting neurons in the hidden layers. In the outer product neural network, there are no hidden layers as such, but reducing the number of neurons with respect to the input vector size has the same effect.

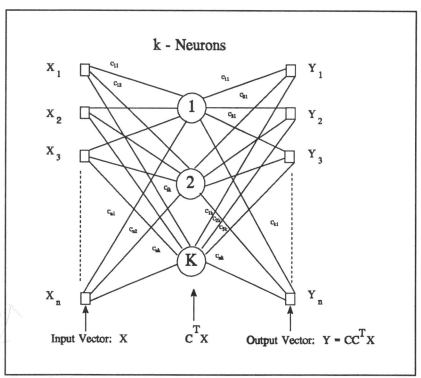

**Figure 1. Outer Product Neural Network.**

Extraction of the principal components of $x$ implies that the output, $y$, represents the best rank-p estimate of $x$. Therefore, $\|x - y\|^2$, which represents the mean-squared error in the auto-encoding of $x$, is the objective function to be minimized. Once the architecture and objective function have been fixed, the training rule may be determined by minimizing the objective function with respect to the coefficients of $C$. Given the training rule, the question remains: To what will the matrix $C$ converge? The training rule and convergence of $C$ will be addressed in the following paragraphs.

Note that the network of Figure 1 has two layers of weights which are strongly coupled, and that the training update rule will be symmetric because of the symmetry of $CC^T$. The residual of the network is $r = (x - y) = (I - CC^T)x$, which is the component of $x$ not 'explained' by the network. Minimizing the squared norm of this residual will yield the training rule.

## DERIVATION OF THE TRAINING RULE

To derive the training rule, the square of the residual of the network, $r^2 = \|(I-CC^T)x\|^2$ must be minimized with respect to the coefficients of $C$, $C_{mn}$. Therefore,

$$\delta C_{mn} = -\sum_{i=1}^{n} r_i \times (\frac{d}{dC_{mn}} CC^T x)_i \qquad (4a)$$

$$\delta C_{mn} = -\sum_{i=1}^{n} r_i \times (C_{in}x_m + \delta_{mi} \times \sum_{k=1}^{n} C_{kn}x_k) \qquad (4b)$$

$$\delta C_{mn} = -((r^T C)_n x_m + r_m (x^T C)_n) \qquad (4c)$$

$$\delta C_{mn} = \frac{d}{dC_{mn}} .5 \times \|(I - CC^T)x\|^2, \qquad (4d)$$

where $\delta_{ml}$ is the Kronecker-$\delta$, and the coefficient .5 has been included for convenience. This equation may be expressed in matrix form as:

$$\delta C = -(xr^T C + rx^T C). \qquad (5)$$

The update training rule for the network may be expressed as:

$$C = C + \gamma \times (xr^T C + rx^T C), \qquad (6)$$

where $\gamma$ is the training coefficient, typically a small positive constant and $r$ is the residual. Since the update rule has products of inputs and outputs (via the residual), the update rule is Hebbian. No intra-layer connections among the neurons are required for training.

## A BRIEF DISCUSSION OF SUBSPACE CONCEPTS

Before inquiring as to the convergence of the weight matrix $C$ once the network is trained, a brief review of subspace concepts is appropriate. Any $pxn$ matrix, $C$, may be decomposed via the Singular Value Decomposition (SVD) so that:

$$C = U \times \Lambda \times V^T \qquad (7)$$

where $\Lambda$ is a diagonal matrix of the singular values of $C$, and $U$ and $V$ are orthonormal matrices whose columns are the singular vectors of $C$. Given an orthonormal matrix, $U$, derived from matrix $C$, one may form the matrix operator, $P = UU^T$, which projects onto the subspace spanned by the columns of $C$. The projection operator, $P$, defines the subspace; knowing the projection operator is equivalent to knowing the subspace. The complement space of the subspace spanned by $P$ is $I - P$; therefore, a complement operator may be defined: $N = I - UU^T$, called the *novelty* of $P$ (or $C$). The novelty represents the 'unexplained' or 'novel' component of an input, $x$, not spanned by the columns of $C$. The novelty will typically contain the noise subspace and any signal not in the $P$-subspace. Both $P$ and $N$ are projection operators and are, therefore, idempotent, i. e., $N^2 = N$, $P^2 = P$. With these definitions, one may come to an understanding of the network.

## CONVERGENCE OF THE OUTER PRODUCT NETWORK

Consider again the objective function to be minimized for the auto-encoding problem:

$$\|r\|^2 = \|x - CC^Tx\|^2 \tag{8a}$$

$$= (x^T - x^TCC^T) \times (x - CC^Tx) \tag{8b}$$

$$= x^Tx - 2x^TCC^Tx + x^TCC^TCC^Tx. \tag{8c}$$

To simplify this expression, form the SVD of $C$ and consider $CC^T$:

$$CC^T = U\Lambda V^T \times V\Lambda U^T \tag{9a}$$

$$= U \times \Lambda^2 \times U^T, \tag{9b}$$

where the orthonormality of $V$ has been used. The expansion of $CC^T$ may be used to simplify the expression for the objective function:

$$\|r\|^2 = x^Tx - 2x^TU\Lambda^2U^Tx + x^TU\Lambda^2U^TU\Lambda^2U^Tx \tag{10a}$$

$$= x^T(I - 2U\Lambda^2U^T + U\Lambda^4U^T)x \tag{10b}$$

$$= x^T(I - UU^T + UU^T - 2U\Lambda^2U^T + U\Lambda^4U^T)x \tag{10c}$$

$$= x^T(I - UU^T)x + x^TU(I_p - 2\Lambda^2 + \Lambda^4)U^Tx \tag{10d}$$

$$= x^TN_cx + x^TUU^TU(I_p - \Lambda^2)^2U^TUU^Tx \tag{10e}$$

$$= \|N_cx\|^2 + \|(I_p - \Lambda^2)U^TP_cx\|^2, \tag{10f}$$

where the fact that $U^TU = I_p$ has been used liberally as well as the fact that $N_c^2 = N_c = I - UU^T$. Equation (10f) shows that, at any given iteration, the residual may be expressed as a component based on the projection of $C$ and a component based on the novelty of $C$. Furthermore, the residual will be minimized only when the p-non-zero singular values of $C$ are unity! In this case the network itself becomes a projector onto the subspace, $C$ approaches $U$ and becomes orthonormal, and the residual approaches the novelty. To minimize $\|r\|^2$ at convergence, the novelty must be minimized. The novelty is minimized when the maximal part of $x$ which can be expressed in a p-subspace lies in the subspace spanned by $C$, i. e. when the network becomes a projector onto the subspace spanning the p-principal components of the space of the $x$-exemplars. Based on this somewhat heuristic argument, the conclusion is that the network will converge to the projection matrix spanning the space of the first p-principal components, and the residual will converge to the novelty of that space. Since $C$ orthonormalizes, the column vectors of $C$ should converge to the principal singular vectors themselves.

Note that the outer product is linear and has no intra-layer connections among the neurons; hence, rapid convergence is expected. A k-neuron network will extract k-principal components. Since the network residual depends purely on the novelty which is also required for $C$-matrix update, analyzing the residual with another outer product network of size m-neurons will result in the extraction of (k+m)-principal components. Hence, networks may be cascaded as in Figure 2 and trained in parallel, the computation required per network being modest if the number of principal components per network is kept small. Thus the network has

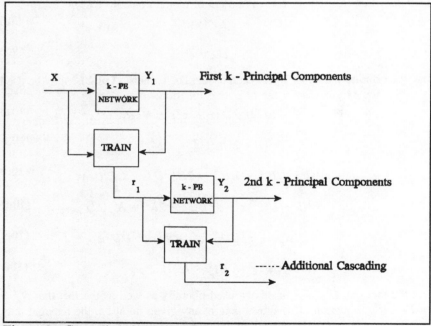

**Figure 2. Cascading Outer Product Neural Networks.**

an inherent building block structure, and a cascade of networks of this type is amenable to VLSI implementation.

# EXPERIMENTS

To test the validity of these ideas, several experiments were conducted simulating the outer product neural network using MATLAB. As an initial test case, two sinusoids of frequencies f1 = .25 and f2 = .26 were embedded in noise to form a time series. These frequencies are similar those used in other tests reported in the literature.[7] Amplitude was set at unity for both signals, and SNR (peak signal to broadband, computer simulated white Gaussian noise) was set to 0 dB. A cascade network consisting of two subnetworks of two neurons each was trained using 32 unique 16-point exemplars from a 512-point noisy time series. Sixteen experiments were conducted using unique noise samples. The results are shown in Figures 3 through 6. Figures 3 and 4 show the first and second column vectors of the network weight matrix, $C$, for all tests (dotted lines) superposed on the original signals at f1 and f2. The short length of the segments obscures the difference in periods, but the principal component are clearly coherent from run to run. In contrast to this, the third column vector of the weight matrix, plotted in the same way in Figure 5, appears random. Figure 6 shows the spectra of the column-wise sum of each of the four columns of the $C$-matrix over all 16 runs. A large FFT of the data for each column padded with zeroes reveals the separation of the frequencies of columns one and two of the weight matrix (dotted and dashed lines) while the randomness and incoherence of the remaining two columns (dot and dot-dash lines) is also apparent. The peaks for the spectra of the first two column vectors do indeed appear at frequencies .25 and .26 as expected. The fact that the second cascade network converged randomly while the first consistently extracted the sinusoids demonstrates the ability of the outer product network to separate subspaces. However, there is no constraint on the columns of the $C$-matrix, so columns within one subnetwork may have to be rearranged based on orthonormality. Experiments with a 4-neuron block network were conducted with somewhat less success: SNR had to be raised several dB to get similar discrimination of the peaks. It must be noted, however, that this test is quite severe. Identification of sinusoids which are not orthogonal on the interval, and which are separated by about one-sixth the Rayleigh frequency limit, is being achieved at 0 dB SNR!

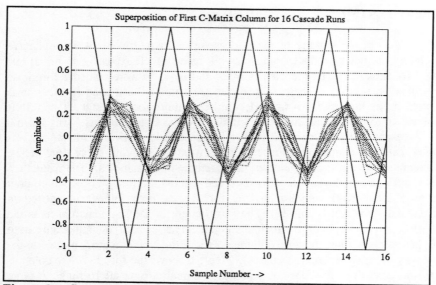

**Figure 3.** Superposition of First Converged C-Matrix Column for 16 Cascade Runs.

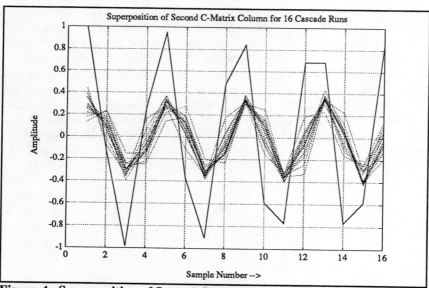

**Figure 4.** Superposition of Second C-Matrix Column for 16 Cascade Runs.

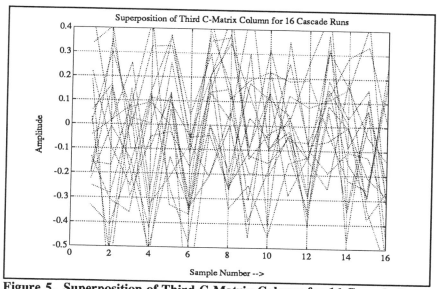

**Figure 5. Superposition of Third C-Matrix Column for 16 Cascade Runs.**

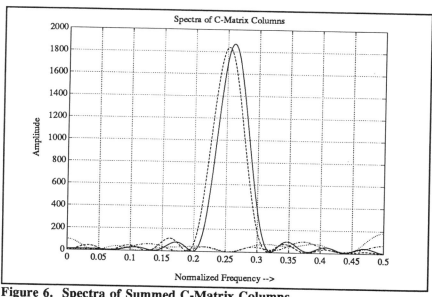

**Figure 6. Spectra of Summed C-Matrix Columns.**

## SUMMARY AND CONCLUSIONS

This paper has introduced the outer product neural network and demonstrated its ability to extract principal components from a time series under adverse conditions. Because the network is linear, rapid convergence is expected which admits the possibility of real-time tracking. The network is relatively simple to implement and may be realized in block or cascade forms; hence, VLSI implementation is possible. The results shown here suggest the network can extract signals in stages until only noise is left. The subspace derivation is the basis for the cascade implementation of the network; however, a corollary of the derivation is that the converged outer product network a projector for a reduced rank representation of the signal which implies the possibility of noise reduction. Finally, the network weight matrix orthonormalizes naturally in training, and there are no intra-layer neuron connections.

Further research will apply this network to other principal components problems, examine convergence properties of the outer product neural network more closely and seek to quantify some of the observations made in this paper.

1. E. Oja, "A simplified neuron model as a principal component analyzer," J. Math. Biology, vol. 15, pp. 267-273, 1982.

2. E. Oja and T. Kohonen, "A subspace learning algorithm as a formalism for pattern recognition and neural networks," in Proceedings IEEE Neural Network Conference, 1988, vol. I, pp. 270-284.

3. E. Oja, Subspace Methods for Pattern Recognition, Research Studies Press, Letchworth, England (also John Wiley, NY, NY) 1982.

4. T. Sanger, "Optimal unsupervised learning in a single layer feed forward neural network," presented at NIPS, Denver, CO, November 29-December 1, 1988, pp. 1-17.

5. S. Y. Kung and K. I. Diamantaras, "A neural network learning algorithm for Adaptive Principal Component Extraction (APEX)," in Proceedings ICASSP, 1990, vol. 2, pp. 861-864.

6. G. W. Cottrell and P. Munro, "Principal components analysis of images via back propagation," in SPIE Visual Communications and Image Processing, vol. 1001, 1988, pp. 1070-1077.

7. D. W. Tufts and R. Kumaresan, "Improved spectral resolution II," in Proceedings ICASSP, 1980, pp. 592-597.

# Part 2:

# Pattern Recognition

# PATTERN RECOGNITION PROPERTIES OF NEURAL NETWORKS

John Makhoul
BBN Systems and Technologies
Cambridge, MA 02138
makhoul@bbn.com

**Abstract** - Artificial neural networks have been applied largely to solving pattern recognition problems. We point out that a firm understanding of the statistical properties of neural nets is important for using them in an effective manner for pattern recognition problems. This paper gives an overview of pattern recognition properties for feedforward neural nets, with emphasis on two topics: partitioning of the input space into classes and the estimation of posterior probabilities for each of the classes.

## 1 INTRODUCTION

Rarely has there been an area of technical endeavor that has cut through many branches of scientific inquiry and been embraced quickly by numerous scientists as the area of artificial neural networks, or neural nets. Although much of the interest in neural nets has arisen as a result of certain exaggerated claims about their ability to mirror human neuronal, learning, and recognition capabilities, their popularity has been enhanced by the fact that they are relatively easy to use, in principle, and that they have been rather successful in pattern recognition problems. For some, neural nets offered the possibility of performing pattern recognition tasks without the need to learn or understand the statistical concepts usually associated with pattern recognition methods.

In this paper we point out that a full understanding and effective application of neural nets in solving pattern recognition problems cannot be achieved without a real understanding of their classification and statistical pattern recognition properties. Accordingly, we present in this paper two aspects of the pattern recognition properties of neural nets. First, we offer a simple description of how neural nets partition the input space into regions associated with the different classes to be classified or recognized. Second, we present several of the known probability estimation properties of neural nets, especially the ability of neural nets to estimate *a posteriori* or posterior probabilities. Throughout the paper, we shall deal exclusively with feedforward neural nets, which have been the main architecture employed in pattern recognition problems.

## 2 PARTITIONING PROPERTIES

One of the objectives in pattern recognition is to partition the input feature space into regions, with one region (possibly noncontiguous) associated with each of the classes. Given a feature vector that lies in a particular region, we classify the feature vector as belonging to the class corresponding to that region. We now explore in what manner a neural net partitions the input space into regions.

## 2.1 Neural Nets with Threshold Units

Although most neural nets used in pattern recognition problems have employed sigmoidal nonlinearities, it is instructive to examine threshold units (i.e., units whose output is binary) because their partitioning properties are simpler to analyze and the conclusions can be considered to be applicable to neural nets with sharp sigmoidal nonlinearities.

In neural nets with threshold units, the first layer, often called the first hidden layer, uses hyperplanes to partition the input space into a number of cells. The function of all additional layers then is simply to group these cells into regions for classification purposes. Let the input to the net be a $d$-dimensional real-valued feature vector $\mathbf{u} = [u_1 u_2 \cdots u_d]$ and the output a binary-valued scalar $y$. Having a binary output $y$ corresponds to a partitioning of the input space $\mathbf{u}$ into two decision regions. All inputs $\mathbf{u}$ to the network which result in an output $y = 1$ form one decision region and the inputs for which $y = 0$ then form the complement decision region in the input space.

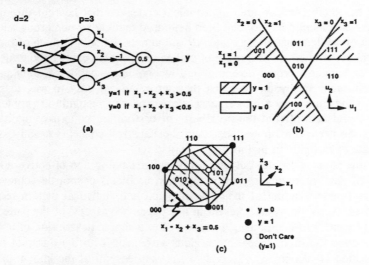

Figure 1: A two-layer neural net with 2 inputs ($d = 2$) and 3 hidden units ($p = 3$). Each unit has a threshold nonlinearity with binary output. In (c), the corner labeled 101 corresponds to a *virtual* cell.

Let us now consider a two-layer net of the type shown in Fig. 1a. The first layer has $p$ nodes, with the feature vector $\mathbf{u}$ as input and $x_i, 1 \leq i \leq p$, as the binary outputs:

$$x_i = g\left[\sum_{j=0}^{d} v_{ij} u_j\right], 1 \leq i \leq p, u_o = 1, \qquad (1)$$

$$g(z) = 1, \; z \geq 0$$
$$= 0, \; z < 0. \qquad (2)$$

The terms $\{v_{ij}\}$ are weights that determine the positions of the $p$ planes in the first layer. The output of the second layer is then given by

$$y = g\left[\sum_{i=0}^{p} w_i x_i\right], x_0 = 1, \qquad (3)$$

where $\{w_i\}$ are weights that determine the position of the single plane in the second layer of a two-layer net. (We shall assume a single output $y$; generalization to many outputs is straightforward and will not be pursued here.) Fig. 1a shows an example of a two-layer net with two inputs ($d = 2$), three nodes in the hidden layer ($p = 3$), and one output. The 3-node hidden layer specifies three planes which divide the input space into seven cells, as shown in Fig. 1b. Each cell is labeled with a binary number $x_3 x_2 x_1$ that specifies on which side of each plane the points in that cell lie. In the 3-dimensional x space, the points $x_3 x_2 x_1$ correspond to the corners of a unit cube as shown in Fig. 1c. The plane $x_1 - x_2 + x_3 = 0.5$ determined by the weights in the output layer then divides the unit cube into two regions, with $y = 0$ on one side of the plane and $y = 1$ on the other side. The input space is then divided into two regions as shown in Fig. 1b: the hatched region corresponds to $y = 1$ and the remaining region corresponds to $y = 0$.

In Fig. 1c, note that if two corners of the cube are adjacent, then the cells corresponding to those corners will be adjacent (connected) to each other in the input space. Since the plane in the second layer divides the cube into two regions, each consisting of corners that are connected to each other, the corresponding cells for each region in the input space must also be connected. It has been observed previously that two-layer nets are capable of forming convex decision regions [1] or non-convex but connected decision regions [2] in the input space. We note, however, that in the example of Fig. 1b, the hatched region is in fact disconnected. Given our analysis above, how is the formation of disconnected regions possible with a two-layer net? To answer this question, we first need to find the number of cells formed in the input space.

It is important to note that the first layer serves a special function that is quite different in nature from other layers. Once the first layer partitions the input space into cells, the other layers can only group these cells together to form decision regions. Therefore, the resolution with which decision regions can be specified, as determined by the size and density of the cells, is completely determined by the number and placement of the first-layer planes in the input space. Since portions of these planes will serve to define the boundaries of decision regions, one can think of the first layer as providing the planes that will result in a piece-wise linear (planar) approximation to the boundaries between decision regions.

## 2.2 Number of Cells

Let $C(p, d)$ be the number of cells formed by $p$ planes in *general position* in $d$ space. (Planes are in general position if every set of $d$ planes intersect at a point and no more than $d$ planes intersect at any single point. These conditions are almost always met in practice.) $C(p, d)$ obeys the important Schläfli recursion [3]:

$$C(p, d) = C(p-1, d) + C(p-1, d-1). \qquad (4)$$

By starting with the initial conditions $C(0, d) = 1, d \geq 0$, and $C(p, 0) = 1, p \geq 1$, one can show that [3]

$$C(p, d) = \begin{cases} 2^p & , p \leq d \\ \sum_{i=0}^{d} \binom{p}{i} & , p > d. \end{cases} \qquad (5)$$

It is clear from (5) that the number of cells $C(p, d)$ grows rapidly as $p$ and $d$ increase. For $p >> d$, the expression for $C(p, d)$ can be approximated by [4]

$$C(p, d) \approx \left(\frac{\alpha p}{d}\right)^d, p >> d \qquad (6)$$

where $\alpha = e/(2\pi d)^{1/2d}$ is in the range $1 < \alpha < e$. It is clear from (6) that $C^{1/d}$ is linear in $p/d$ for $p >> d$. One can think of $C^{1/d}$ as roughly the number of cells per dimension, i.e., the average number of divisions per dimension. For example, to obtain roughly 10 divisions per dimension for $d = 10$, or a total of $10^{10}$ cells, we have from (6) that $p/d \approx 5$ or $p = 50$ nodes in the first layer. Therefore, depending on the resolution needed in representing the classification regions of interest, one can use (6) to help estimate the needed number of nodes in the first layer.

## 2.3 Virtual Cells and Disconnected Regions

From (5) we note that, for $p > d, C(p, d) < 2^p$. This means that, there are $p$-digit binary numbers at the output of the first layer which simply cannot occur, irrespective of the feature values at the input. We give here two explanations for this non-intuitive phenomenon: geometric and algebraic.

Let us call any point at which planes intersect a *vertex*. Let us now assume that we have $p - 1$ intersecting planes and we add a $p$th plane that intersects all the others. This plane can be located physically only on one side of each of the vertices. This means that, for every vertex, there is a cell that will not be divided by the $p$th plane. Since $p - 1$ planes in $d$ space intersect in $\binom{p-1}{d}$ vertices, it means that there are that many cells that will not be divided by the new plane.

Therefore, the value of $C(p, d)$ is double that of $C(p-1, d)$ minus the number of vertices. Indeed, one can show from (4) that

$$C(p, d) = 2C(p-1, d) - \binom{p-1}{d}. \tag{7}$$

Therefore, $C(p, d) < 2^p$ for $p > d$.

Algebraically, we try to find inputs **u** that correspond to each of the $2^p$ possible binary vectors **x**. For each choice of **x**, (1) and (2) constitute $p$ inequalities in the $d$ unknowns $\{u_i\}$. When the number of inequalities is greater than the number of unknowns (i.e., $p > d$), it is not always possible to find solutions to the inequalities. In fact, a maximum of only $C(p, d)$ out of the $2^p$ sets of inequalities will have solutions.

In Fig. 1, for example, $C(3, 2) = 7 < 2^3$. In Fig. 1b, we note that there is no cell that corresponds to the binary number 101. This means that, no matter what values the inputs $u_1$ and $u_2$ take, the output vector **x** from the first layer can never equal 101. Since a cell labeled 101 cannot exist in this example, we shall call such a cell a *virtual* cell [4]. In general, the number of virtual cells is equal to

$$C_v(p, d) = 2^p - C(p, d) = \sum_{i=d+1}^{p} \binom{p}{i}. \tag{8}$$

Since for each cell the output of the first layer corresponds to one corner of the unit $p$-dimensional hypercube, $C_v(p, d)$ is the number of corners of the hypercube that will *never* occur.

We mentioned above that the region corresponding to one side of the plane in the second layer consists of cells that are connected to each other. However, because some of those cells are virtual cells, the region in the input space may be connected together through a set of virtual cells, which may lead to a decision region being disconnected. The reason that the three disconnected hatched cells in Fig. 1b can form a single decision region is that the three cells can be thought of as being connected through the virtual cell corresponding to the corner of the cube labeled 101.

The number of virtual cells increases as $p$ increases. For $p > 2d + 1$ the number of virtual cells becomes larger than the number of real cells [4]. Given the relatively large number of virtual cells, even for modest values of $p$, one would expect the number of possible disconnected decision regions that can be formed by a two-layer net to be large indeed.

Even though it is possible to form connected as well as disconnected decision regions with a two-layer network, it is not possible to form arbitrary decision regions. It is well known that, to be able to realize all $2^{C(p,d)}$ possible decision regions for a specific partitioning of the input space, requires a 3-layer network with *two* hidden layers. However, recent theoretical results have shown that one can *approximate* any decision region in a finite portion of the space arbitrarily closely with only a two-layer net [5]. The ability to form such decision regions would not be possible without the appropriate use of virtual cells.

# 3 POSTERIOR PROBABILITY ESTIMATION

## 3.1 Classification

Let us assume that the input vector **u** belongs to one of $M$ classes, $c_i$, $1 \leq i \leq M$. The main objective in pattern recognition is to decide to which of the $M$ classes the vector **u** belongs. Typically, we use a decision rule that takes **u** as input and gives the class as output. The decision rule can be based on some form of discriminant function that is a function of **u**. A more general decision rule that guarantees the minimum classification error is what is known as the *Bayes decision rule* [6, 7], namely, choose the class $c^*$ for which the posterior probability $P(c_i|\mathbf{u})$ is maximum for a given **u**. All one requires, therefore, is the ability to estimate $P(c_i|\mathbf{u})$ for each of the classes.

Instead of estimating $P(c_i|\mathbf{u})$ directly, most methods rely on Bayes' theorem

$$P(c_i|\mathbf{u}) = \frac{p(\mathbf{u}|c_i)P(c_i)}{p(\mathbf{u})} \qquad (9)$$

where $p(\mathbf{u}|c_i)$ is the conditional density (also known as the likelihood) and $P(c_i)$ is the *a priori* probability of the class $c_i$. Now, since $p(\mathbf{u})$ is the same for all classes, a decision rule based on maximizing the joint density $p(\mathbf{u}, c_i) = p(\mathbf{u}|c_i)P(c_i)$ is equivalent to maximizing $P(c_i|\mathbf{u})$. $P(c_i)$ is simple to compute and the literature abounds with methods to estimate likelihoods $p(\mathbf{u}|c_i)$ [6, 7].

Neural nets serve as a point of departure from traditional pattern recognition methods in that they can be made to estimate posterior probabilities *directly*, as we show below.

## 3.2 Least Squares Training of Neural Nets

For this part of the discussion we shall assume that we have a 2-class problem (classes $c$ and $\bar{c}$) and only a single output $y(\mathbf{u}, \theta)$, where the dependence of $y$ on the parameters $\theta$ of the network is made explicit. A popular method of training the neural net is to determine the parameters $\theta$ that minimize the mean (expected) squared error between the output $y$ and a target function $t(\mathbf{u})$:

$$E_{LS} = E[y(\mathbf{u}, \theta) - t(\mathbf{u})]^2 \qquad (10)$$

$$t(\mathbf{u}) = \begin{cases} a, & \mathbf{u} \in c \\ b, & \mathbf{u} \in \bar{c} \end{cases} \qquad (11)$$

where typically $a = 1$ and $b = 0$. One can show that (10) can be rewritten as [8]:

$$E_{LS} = E[y - E(t|\mathbf{u})]^2 + E[E(t|\mathbf{u}) - t]^2. \qquad (12)$$

Since in (12) the second term does not depend on the parameters $\theta$, minimizing (12) is equivalent to minimizing

$$E_{LS} = E[y - E(t|\mathbf{u})]^2. \qquad (13)$$

But,
$$\begin{aligned} E(t|\mathbf{u}) &= aP(c|\mathbf{u}) + bP(\bar{c}|\mathbf{u}) \\ &= (a-b)P(c|\mathbf{u}) + b \end{aligned} \quad (14)$$

since $P(\bar{c}|\mathbf{u}) = 1 - P(c|\mathbf{u})$. For $a = 1$ and $b = 0$, $E(t|\mathbf{u}) = P(c|\mathbf{u})$ and (14) reduces to

$$E_{LS} = E[y - P(c|\mathbf{u})]^2. \quad (15)$$

Therefore, by minimizing the mse between the output and the target function $t$ with $a = 1$ and $b = 0$, *the neural net gives as output a least squares estimate of the posterior probability* $P(c|\mathbf{u})$. This result gives us an explicit expression for the manner in which the neural net provides an estimate of the posterior probability. It also explains the apparent success of neural nets in solving various pattern recognition problems. The result has been derived in various ways by different authors [8 - 11].

A number of researchers have advocated using values of $a$ and $b$ different from 1 and 0; for example, $a = 0.9$ and $b = 0.1$. We see from (14) that setting $a$ and $b$ to other than 1 and 0 results in a biased estimate of $P(c|\mathbf{u})$. The bias is both multiplicative and additive. However, if the neural net is being used only for classification, then a bias of this kind will not affect the results. But, there are applications for which the probability values are important, in which case $a = 1$ and $b = 0$ should be used. The reason that researchers have used other values of $a$ and $b$ is because of training difficulties when the classes are perfectly separable, i.e., in cases where $P(c|\mathbf{u})$ is identically 0 or 1 in different regions. Clearly, for such problems, a nonlinearity that goes to 0 and 1 for *finite* values is needed. For general problems where the classes are not separable, values of $a = 1$ and $b = 0$ do not cause any training difficulties.

## 3.3 Kullback-Leibler Criterion

Since the neural net is providing an estimate of $P(c|\mathbf{u})$, it is worth asking if there are distance measures that might result in better probability estimates than the mse. After all, it is well known that the mse tends to give better estimates at large values than at small values. In a problem where many probabilities are combined to produce a final decision rule, it is important to estimate accurately the low probability values as well as the large ones.

One criterion that has been used extensively in estimating probabilities is the Kullback-Leibler (K-L) directed divergence [12]. In the K-L criterion, the distance between the true probability $P(c|\mathbf{u})$ and the estimated probability $y$ is given by:

$$E_{KL} = \sum_{i=1}^{M} \int_{\mathbf{u}} P(c_i|\mathbf{u}) \log \frac{P(c_i|\mathbf{u})}{y_i(\mathbf{u}, \theta)} p(\mathbf{u}) \, d\mathbf{u} \quad (16)$$

where $y_i$ is the network output corresponding to class $c_i$. $E_{KL}$ is always nonnegative; it is equal to zero if and only if $y_i(\mathbf{u}, \theta) = P(c_i|\mathbf{u})$ for all $\mathbf{u}$. Minimizing (16) over $\theta$ is equivalent to minimizing

$$E_{KL} = -\sum_{i=1}^{M} \int_{\mathbf{u}} p(c_i, \mathbf{u}) \log y_i(\mathbf{u}, \theta) \, d\mathbf{u}. \tag{17}$$

Given $N$ training vectors $\mathbf{u}(n)$, (17) can be estimated from [9]:

$$E_{KL} = -\frac{1}{N} \sum_{i=1}^{M} \sum_{\mathbf{u}(n) \in c_i} \log y_i(n). \tag{18}$$

Note that the contribution of each output to the overall error is activated only when the input belongs to the corresponding class.

Another expression for $E_{KL}$, in which the constraint $\mathbf{u}(n) \in c_i$ on the summation does not appear, can be obtained by noting that each of the outputs $y_i$ actually acts as a discriminator between the class $c_i$ and its complement $\bar{c}_i$ (i.e., not $c_i$). If $y_i(n)$ is an estimate of $P(c_i|\mathbf{u})$, then $1 - y_i(n)$ will be the estimate for $P(\bar{c}_i|\mathbf{u}) = 1 - P(c_i|\mathbf{u})$. By using the K-L criterion to estimate $y_i(n)$ and $1 - y_i(n)$ simultaneously, one can show that

$$E'_{KL} = -\frac{1}{N} \sum_{i=1}^{M} \left[ \sum_{\mathbf{u}(n) \in c_i} \log y_i(n) + \sum_{\mathbf{u}(n) \in \bar{c}_i} \log(1 - y_i(n)) \right] \tag{19}$$

which can be shown to reduce to [13]

$$E'_{KL} = -\frac{1}{N} \sum_{n=1}^{N} \sum_{i=1}^{M} \frac{1}{2} \log[y_i(n) - \bar{t}_i(n)]^2 \tag{20}$$

where $\bar{t}_i(n)$ is the complement target function to $t_i(n)$, i.e., $t_i(n) + \bar{t}_i(n) = 1$. In minimizing $E'_{KL}$, one is in effect maximizing the distance between the output and the complement of the target function. One can show that for output values where $|y_i(n) - t_i(n)| \ll 1$, $E'_{KL}$ is approximately equal to

$$E'_{KL} \approx \frac{1}{N} \sum_{n=1}^{N} \sum_{i=1}^{M} |e_i(n)| \tag{21}$$

where $e_i(n) = y_i(n) - t_i(n)$. Thus, $E'_{KL}$ depends on the *absolute* value of the error between the output and the target, while $E_{LS}$ depends on the square of the error [13]. In general, $E'_{KL}$ is likely to do a better job in estimating low probability values than $E_{LS}$.

## 3.4 Outputs as Probability Estimates

Minimizing the mse or the K-L criterion does not necessarily ensure that the outputs are in fact probability estimates, i.e., that they obey the rules of probability:

$$0 \leq y_i(n) \leq 1, \text{ for all } c_i \text{ and } \mathbf{u}(n) \tag{22}$$

$$\sum_{i=1}^{M} y_i(n) = 1, \text{ for each } \mathbf{u}(n). \tag{23}$$

The first condition can be met easily by having the output node nonlinearity be a sigmoid of the type

$$y_i = f(z_i) = \frac{1}{1 + e^{-z_i}} \tag{24}$$

where $z_i$ is the input to the nonlinearity (see Fig. 2). The second condition states

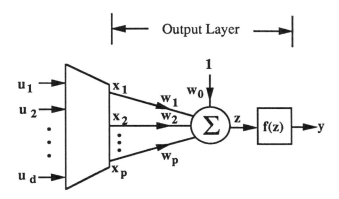

Figure 2: A general feedforward neural network.

that the sum of the outputs for each input vector must sum to unity; otherwise, the outputs are not probabilities, even if they obey (22). Minimizing $E_{LS}$ or $E_{KL}$ does not guarantee that condition (23) will be met. A constraint on the structure of the network must be placed to ensure (23). Instead of using a sigmoid as the nonlinearity at the output of each node, it is possible to implement a generalized sigmoid function [14]:

$$y_i = \frac{e^{z_i}}{\sum_{j=1}^{M} e^{z_j}}. \tag{25}$$

This function reduces to (24) for a 2-class problem. The use of (25) in the output layer complicates the training, but it is necessary if true probability values are desired. We note, however, that in many classification problems where true probabilities are not necessary, it may not be important to use the generalized sigmoid to obtain good results.

Note that if $E_{KL}$ in (18) is used as the error measure of choice, then output normalization must be used; otherwise, the resulting network will be degenerate, since each output $y_i$ is trained using inputs belonging to $c_i$ only. In contrast, $E'_{KL}$ in (20) can be used without normalization since each output is trained on all the data. Each output $y_i$ is then an estimate of $P(c_i|\mathbf{u})$, but the outputs will usually not add up to one for each input $\mathbf{u}$.

## 3.5 Likelihoods vs. Posterior Probabilities

We mentioned above that traditional methods have emphasized the estimation of likelihoods $p(u|c)$ from which the posterior probabilities are computed, while neural nets estimate $P(c|u)$ directly. Which of the two approaches is better will depend on the models used for each, the methods to estimate the model parameters, and the specific problem to be solved. The fact that with neural nets one is estimating $P(c|u)$ directly does not give it any specific advantage.

One advantage that is often cited for neural nets is that they are trained as discriminators, i.e., the output is trained to discriminate between the class and its complement. However, this advantage is achieved at a cost: one must train each output on inputs from that class as well as the other classes. In contrast, estimating likelihoods $p(u|c_i)$ can be done for each class by looking at data from that class only. Also, $p(u|c)$ can be estimated by a simple one-pass counting procedure if the input u is vector quantized first.

There are applications, such as in continuous speech recognition, in which the likelihood is actually desired, because it is then combined with the grammar score (i.e., the *a priori* probability of word sequences) to find the total score that is used for making the recognition decision. From Bayes' theorem, $p(u|c)$ can be estimated from $P(c|u)$

$$p(u|c_i) = \frac{P(c_i|u)p(u)}{P(c_i)}. \tag{26}$$

Since $p(u)$ is the same for all classes and does not affect the final decision, it is sufficient to divide $P(c_i|u)$ by $P(c_i)$ and use that quantity for classification.

## 3.6 Neural Net as a Discriminant

Fig. 2 shows a representation of a neural net where the mapping from the input u to x contains the effects of all hidden layers, and the output layer is shown explicity (here with a single output). The hidden layers can be thought of as implementing a set of feature extractors $x_i(u)$, $1 \leq i \leq p$, mapping the $d$-dimensional input space u into a $p$-dimensional space x. The last stage then implements a hyperplane $(x \cdot w)$ which separates class $c$ from its complement $\bar{c}$. In other words, the hidden layers map the boundary between $c$ and $\bar{c}$ in u space to a linear boundary in x space. Therefore, we can view neural nets as implementing a mapping from one feature space into another where a class can be separated from the rest of the world by a linear discriminant. Note that this interpretation is independent of what $f(z)$ is in Fig. 2, as long as it has a monotonic relation to $z$. One method that enhances the ability of a neural net to perform the requisite boundary mapping to a plane is to transform the input into a much higher dimensional space where the classes are separable by planes.

We note in passing that while a linear discriminant is the best discriminant for two Gaussians with equal variance, a single-node network with a sigmoid nonlinearity gives the exact posterior probability as its output.

# 4 NEURAL NET TRAINING

The ability of a neural net to have its parameters estimated from examples (often called "learning") has been considered one of its major strengths. Such an ability is of course very important, but because of the highly nonlinear nature of neural nets, the large amount of computation required for training must be considered the weakest point of neural nets. Great care, therefore, must be taken to achieve the required performance level at the lowest possible cost.

## 4.1 Error Minimization

The least squares error measure $E_{LS}$ in (10) has been the most popular thus far. It is simple and has given good results. In particular, back propagation [15] has been used to implement an on-line (or sample by sample) gradient descent algorithm. Off-line (or batch) methods, such as conjugate gradient and quasi-Newton methods [16], have also been used to good advantage. Some have started with back propagation to obtain an initial estimate, followed by one of the batch methods. The general experience has been that multiple local minima makes it almost impossible to find the global minimum in a reasonable amount of time. What is not known, however, is whether or not the global minimum results in significantly better recognition performance for real-life problems.

### A Very Special Case

In the very special case where the K-L criterion is used *and* the nonlinearity $f(z)$ in the output layer is given by the sigmoid (24), the output layer will have a *single minimum*, which is also the global minimum [13]. For this special case, the first gradient of (20) with respect to the weights $w_j$ is given by [13]:

$$\frac{\partial E'_{KL}}{\partial w_j} = \frac{1}{N} \sum_{n=1}^{N} x_j(n) e(n), \, 1 \leq j \leq p, \qquad (27)$$

where x is the input to the last layer and $e(n)$ is the error $y(n) - t(n)$. The second derivative can then be shown to be always positive, thus resulting in a single minimum. The single-minimum property for a single layer is true only if the probability normalization in (25) is not used.

The gradient in (27) should be compared with the familiar gradient for least squares:

$$\frac{\partial E_{LS}}{\partial w_j} = \frac{1}{N} \sum_{n=1}^{N} x_j(n) e(n) y(n) [1 - y(n)]. \qquad (28)$$

Clearly, (27) is simpler than (28). Also, the second derivative in (28) is not always positive, and therefore results in multiple minima, even for a single layer. For the multi-layer case, both criteria would be expected to have multiple minima, but we would expect the K-L criterion with the sigmoid nonlinearity at the output to result in fewer minima and to have faster convergence. Another advantage of the special case is that the gradient is affected by the error for all values of $y(n)$, while in (28) the gradient is not affected when the output is close to 0 or 1, thus contributing to the slower convergence of the mse criterion.

We note finally that the gradient (27) happens to be identical to the gradient when linear least squares is used with no nonlinearity at the output. However, the error $e(n)$ in (27) assumes that the signal has gone through the sigmoid nonlinearity.

In practice we have found that the K-L criterion converges faster than least squares and gives better recognition results. Although the differences have not been dramatic, they are almost always there.

## 4.2 Network Structure

Irrespective of the structure of the hidden layers, it appears that the output layer with a sigmoid nonlinearity has some very special properties that make it the structure of choice for posterior probability estimation. This, however, does not imply that the hidden layers should have the same nonlinearity. Other structures that use, for example, radial basis functions in a hidden layer can be quite effective. It is important to note that even a one-layer network should have a sigmoid at its nodes. In general, a one-layer net with a sigmoid nonlinearity gives much better recognition performance than a single layer with no nonlinearity.

## 4.3 Data Sampling

One problem in training a network to classify among a large number M of classes, where each output corresponds to one of the classes, is that the training data from outside the·class can dominate training time. If $M = 50$, for example, then roughly 2% of the data will belong to any one class, but the remaining 98% of the data will dominate the training time. One way to reduce training time is to reduce the number of nonclass samples used in training a particular network output. Improper sampling of the training data, however, can lead to biased estimates of the posterior probability. One sampling method that has been shown to reduce computation significantly, without any deleterious effects on probability estimation, is to choose a small random sample of the nonclass data but weight the error criterion for those samples proportional to the amount of reduction in the data [17]. For example, if the training data was reduced by a factor of 10 then the error criterion should be multiplied by 10 for the nonclass samples.

Note that the sigmoid, because of its sharp transition, already plays an important role in that it focusses the attention during training on data samples that are confusable. Basically, it helps specify the boundary between the samples within and outside the class. Thus, in general, much of the training data does not participate in determining the network parameter values. The issue of data sampling given above is concerned with how to reduce training time, especially in light of the fact just mentioned.

## 4.4 Network Optimality

One of the main concerns in training a particular network is whether the estimated parameters lead to the best recognition performance for that network. If the available training data is large relative to the number of free parameters in the network (at least 50 training samples per parameter), then the only issue really is whether the training process itself leads to the optimal parameter values. Usually, ensuring that the error criterion is actually minimized is sufficient (even though it does not guarantee optimum recognition performance).

The real problem in network training, however, concerns the prevalent case where the amount of training data is small relative to the number of parameters. The problem manifests itself in high recognition accuracy on the training data and poor performance on an independent test set.

There is a host of *regularization* methods that are used to deal with the sparse data problem. One such method that is often used is *cross validation*. The idea here is to use a separate development test set during training to test the performance of the network after every pass through the data. The iterations are stopped if the performance on that test set reaches a peak and then begins to fall.

The basic idea of regularization is to limit the range (or the freedom) that the network parameters can have, thus reducing the "effective" number of parameters and, therefore, rendering the network more robust. One such method is to add to the error criterion a term that is a function of the square of the parameters. For example, a criterion of the form

$$J = E_{KL} + \lambda \sum_j w_j^2 \tag{29}$$

where $\lambda$ is a Lagrange multiplier and the summation is over all the weights of the neural net, will clearly limit the size of the weights since $J$ is being minimized. Many other functions of $w_j^2$ can be used instead [18].

## Probability Estimation

Another issue regarding network optimality is whether the output is in fact a good estimate of the posterior probability. There are simple necessary tests that can be performed which are indicative of how well a probability estimate we are getting. We give three such tests below.

If a network is optimal in the sense that $y = P(c|\mathbf{u})$ for all $\mathbf{u}$, then one can show that

$$P(c|y) = y \tag{30}$$

where $P(c|y)$ is the probability of class $c$ given $y$. Since $y$ is a scalar, $P(c|y)$ can be estimated quite easily as the relative number of times that the input was in class $c$ and the output value was equal to $y$. (Usually, we measure these values within a small range around $y$.) Thus, a plot of $P(c|y)$ versus $y$ should be linear. Although the test in (30) is necessary but not sufficient for optimality, it provides a good indicator that the outputs are probabilities.

Another simple necessary condition is to check if

$$\frac{1}{N}\sum_{n=1}^{N} y(n) = P(c) \tag{31}$$

where $P(c)$ is obtained by counting the frequency of class $c$ in the training data. Yet another simple necessary condition for optimality, taken from the K-L criterion, is to test if

$$\sum_{n=1}^{N} y(n)\log y(n) = \sum_{u(n)\in c} \log y(n). \tag{32}$$

All the above tests can be performed separately for each of the classes.

## 5 CONCLUSION

In this paper we reviewed several pattern recognition properties of neural nets, especially class partitioning and posterior probability estimation properties. Experience with neural nets will no doubt teach us more about how to use them. However, a firm understanding of their statistical properties will be needed for using them optimally.

**Acknowledgement** - The author would like to thank the following colleagues for many useful discussions and joint work: S. Austin, A. El-Jaroudi, H. Gish, and R. Schwartz.

## References

[1] R.P. Lippman, "An introduction to computing with neural nets," *IEEE ASSP Magazine*, pp. 4-22, April 1987.

[2] A. Wieland and R. Leighton, "Geometric analysis of network capabilities," *IEEE Int. Conf. Neural Networks*, San Diego, CA, June 1987, pp. III-385-392.

[3] L. Schläfli (1814-1895), *Gesammelte Mathematische Abhandlungen*, Vol. 1, Birkhäuser, Basel, 1950, pp. 209-212, (in German).

[4] J. Makhoul, A. El-Jaroudi, and R. Schwartz, "Partitioning capabilities of two-layer neural networks," *IEEE Trans. Signal Processing*, Vol. 39, No. 6, pp. 1435-1440, June 1991.

[5] G. Cybenko, "Approximation by superpositions of a sigmoidal function," *Mathematics of Control, Signals and Systems*, Aug. 1989.

[6] K. Fukunaga, *Introduction to Statistical Pattern Recognition*, Academic Press, New York, 1972.

[7] R. Duda and P. Hart, *Pattern Classification and Scene Analysis*, Wiley, New York, 1973.

[8] H. White, "Learning in artificial neural networks: A statistical perspective," *Neural Computation*, pp. 425-464, 1989.

[9] J. Patterson and B. Womack, "An adaptive pattern classification system," *IEEE Trans. Systems Science and Cybernetics*, pp. 62-67, Aug. 1966.

[10] A. Barron, "Statistical properties of artificial neural networks," *IEEE Conf. Decision and Control*, Tampa, FL, pp. 280-285, 1989.

[11] H. Gish, "A probabilistic approach to the understanding and training of neural network classifiers," *IEEE Int. Conf. Acoust., Speech, Signal Processing*, April 1990.

[12] S. Kullback, *Information Theory and Statistics*, Wiley, New York, 1959.

[13] A. El-Jaroudi and J. Makhoul, "A new error cirterion for posterior probability estimation with neural nets," *Int. Joint Conf. Neural Networks*, San Diego, CA, Vol. III, pp. 185-192, June 1990.

[14] J. Bridle, "Training stochastic model recognition algorithms as networks can lead to maximum mutual information estimation of parameters," in *Advances in Neural Information Processing Systems 2*, D. Touretzky (ed.), Morgan Kaufmann, 1990.

[15] D. Rumelhart, G. Hinton, and R. Williams, "Learning representations by error propagation," in *Parallel Distributed Processing: Explorations in the Microstructure of Cognition*, D. Rumelhart and J. McClelland (eds.), MIT Press, Cambridge, MA, Vol. 1, pp. 318-362, 1986.

[16] D. Luenberger, *Linear and Nonlinear Programming*, Addison-Wesley, Massachusetts, 1984.

[17] S. Austin, G. Zavaliagkos, J. Makhoul, and R. Schwartz, "A hybrid continous speech recognition system using segmental neural nets with hidden Markov model," *IEEE-SP Workshop on Neural Networks for Signal Processing*, Princeton, NJ, 1991.

[18] L. Bottou, "Minimizing approximations of the empirical risk," *Neural Networks for Computing Conference*, Snowbird, Utah, 1991.

# EDGE DETECTION FOR OPTICAL IMAGE METROLOGY USING UNSUPERVISED NEURAL NETWORK LEARNING[1]

Hamid K. Aghajan    Charles D. Schaper    Thomas Kailath

Information Systems Laboratory
Department of Electrical Engineering
Stanford University
Stanford, CA 94305

## Abstract

Several unsupervised neural network learning methods are explored and applied to edge detection of microlithography optical images. Lack of a priori knowledge about correct state assignments for learning procedure in optical microlithography environment makes the metrology problem a suitable area for applying unsupervised learning strategies. The methods studied in this paper include a self-organizing competitive learner, a bootstrapped linear threshold classifier, and a constrained maximization algorithm. The results of the neural network classifiers were compared to the results obtained by a standard straight edge detector based on the Radon transform and good consistency was observed in the results together with superiority in speed for the neural network classifiers. Experimental results are presented and compared with measurements obtained via Scanning Electron Microscopy.

## INTRODUCTION

Edge detection and enhancement is one of the most demanding tasks in optical image processing for artificial vision and image matching works.

The edge detection process serves to simplify the analysis of images by drastically reducing the amount of data to be processed, while still preserving useful information about the image. Several approaches have been developed for edge detection. Among them, one may mention the Gaussian filters [1], which are the basis for a series of algorithms for detecting *sharp* edges. Several other methods for the detection of straight edges are based on producing a set of likely edge points by first applying some edge detection schemes [2, 3, 4], and then combining the resulting data to estimate the line coordinates of an edge; this step is usually done either by least squares fitting or by the Hough transform [5, 6]. There is also a projection-based detection method for straight line edges that analyzes the peaks in projection space to estimate the parameters of a line representing an edge [7].

In several applications, however, the assumption that the edges can be represented by sharp discontinuities is a poor one. Microlithography and wafer pattern analyzing and matching in IC-processing is one of these cases, where the edge profiles are smoothed out and blurred and the corners are rounded off by both process-introduced defects such as imperfect etching,

---

[1] This work was supported by the Advanced Research Projects Agency of the Department of Defense and was monitored by the Air Force Office of Scientific Research under contract F49620-90-C-0014 (the United States Government is authorized to reproduce and distribute reprints for governmental purposes notwithstanding any copyright notation hereon). The views and conclusions contained in this document are those of the authors and should not be interpreted as necessarily representing the official policies or endorsements, either expressed or implied, of the Air Force Office of Scientific Research or the U.S. Government.

and by filtering effects and aberrations introduced by the optical imaging system. Furthermore, the optical images taken from the wafer contain noise due to several sources such as random local changes in reflectivity of the wafer surface and the noise introduced by the imaging system. In this environment, some useful results have been obtained by S. Douglas and T. Meng [8], who proposed to use a neural classifier element to recognize the position of an edge by classifying the pixels into edge/non-edge categories. They use a modified sigmoid-LMS algorithm and in order to teach the filter (or adapt the weights), they artificially generate a raster-scan image in which the edges arrive as the result of a Markov finite state process, and then they add Gaussian noise to the resulting image. In this manner, their filter weights are adapted through comparing the filter output and the assumed desired response.

In the present work, neural elements are used in a different way to learn to classify the pixels of an image into two classes of different height levels without requiring any *a priori* assumption about the distribution of pixels heights or the characteristics of the noise on the image. In other words, since the correct state assignments are not available, unsupervised learning schemes are more suited for this application. The cost of this lack of information is that the classification itself has to define the edge and therefore a calibration may be needed to derive the edge locations defined by other techniques than optical imaging. This calibration is a general necessity also in using other image processing methods, because the intensity profile obtained by optical imaging naturally needs this kind of interpretation.

In this paper we study the application of several unsupervised neural network learning methods to edge detection in lithography images. These methods include a self-organizing competitive learner, a bootstrapped linear threshold classifier, and a constrained maximization algorithm. Due to the superiority in speed of these techniques over standard methods, the neural network techniques can be used to implement on-line linewidth measurement for lithography process control.

The rest of the paper is organized as follows. First, we study several unsupervised learning methods for edge detection. Then, methods of line fitting to a set of detected edge points are studied. Experimental results of applying the studied methods to real optical images from lithography environment are then presented and compared with the results obtained via standard methods. Finally, we summarize our results and provide a brief conclusion.

## EDGE DETECTION USING NEURAL NETWORKS

In this section, we explore the application of three unsupervised learning techniques to the problem of detecting edges in optical images from microlithography environment.

### Bootstrapped Linear Threshold Classifier

A schematic of the bootstrapped linear threshold classifier is shown in Fig. 1(a). In this approach the output of the linear threshold classifier is used as a state assignment for the learning input pattern. The error is then defined as the difference between this assignment and the output of the adaptive linear filter. By minimizing this error, the neural network adaptively classifies pixels on the image into high and low heights, and hence provides an edge. Before proceeding, a comparison is made between the adaptation properties of the bootstrapped threshold classifier and a standard supervised LMS algorithm. We first note that in the supervised LMS technique, the weights are changed in a manner to minimize the mean-square distance between a given response and the filter output. Both approaches use the assumption that the patterns in the observation space exist in two compact unimodal clusters and that the separation between the two clusters is large compared with the variation within each cluster. A supervised classifier adapts the weights such that the linear projection of all patterns onto the set of all possible outputs of the form $\mathbf{w}^T\mathbf{x}$ minimizes the mean-square distance between the projection of a pattern and the corresponding given response. Here $\mathbf{w}$ is the filter weight vector. However, in the unsupervised learning strategy, the binary output of the linear threshold element is used as the correct state assignment and thus also used in the definition of the bootstrap mean-square-error function. Consequently, the weight vector is chosen such that the linear projection of all patterns onto the set of all possible outputs of the form $\mathbf{w}^T\mathbf{x}$ minimizes the mean-square distance between the projection of a pattern and the *closer* of the two scalar values, $\pm 1$.

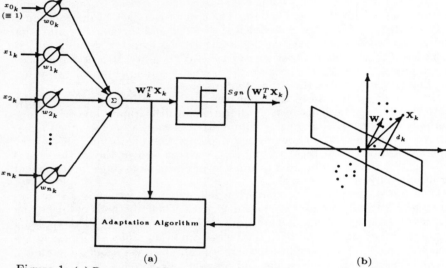

Figure 1: (a) Bootstrapped linear threshold classifier. (b) Classifying hyperplane.

The bootstrap mean-square-error is defined by the relation:

$$\overline{\epsilon^2} = E\left\{\left[\text{Sgn}\left(\mathbf{w}^T\mathbf{x}\right) - \mathbf{w}^T\mathbf{x}\right]^2\right\} \quad (1)$$

where

$$\text{Sgn}(a) = \begin{cases} +1 & \text{if } a \geq 0 \\ -1 & \text{if } a < 0 \end{cases}$$

In practice, sample averaging is used instead of the expectation opertaion and the optimization criterion becomes to find:

$$\mathbf{w}_{\text{opt}} = \arg\min_{\mathbf{w}} \Sigma_{k=1}^{N} \left[\text{Sgn}\left(\mathbf{w}^T\mathbf{x}_k\right) - \mathbf{w}^T\mathbf{x}_k\right]^2 \quad (2)$$

where $N$ is the number of pattern vectors. Based on this criterion, the LMS-bootstrap adaption algorithm [9, 10] which is a steepest descent algorithm is

given by:
$$\epsilon_k = \text{Sgn}\left(\mathbf{w}_k^T \mathbf{x}_k\right) - \left(\mathbf{w}_k^T \mathbf{x}_k\right) \quad (3)$$
$$\mathbf{w}_{k+1} = \mathbf{w}_k + 2\mu_k \epsilon_k \mathbf{x}_k \quad (4)$$

where $\mu_k$ is the step size of the iteration and can be chosen a constant or a function of the iteration number $k$. This algorithm provides an approximate gradient descent on the bootstrap mean-square-error surface. Each incremental change in the weight vector is made in the direction of an unbiased estimate of the negative gradient of the bootstrap mean-square-error surface, evaluated at $\mathbf{w} = \mathbf{w}_k$.

The convergence of the weight vector using a finite data set in the learning phase, to the value that achieves a local minimum of the bootstrap mean-square-error function is assured if the following conditions are true [9]:

i) The $k^{th}$ pattern vector $\mathbf{x}_k$ is chosen according to the probability density function $f(\mathbf{x}) = \frac{1}{N}\sum_{j=1}^{N} \delta(\mathbf{x} - \mathbf{x}_j)$, where $\delta(.)$ is the delta fuction, for all $k$.

ii) The sequence of proportionality constants $\mu_k$ satisfies $\mu_k > 0$ and we have $\sum_{k=1}^{\infty} \mu_k = \infty$ and $\sum_{k=1}^{\infty} \mu_k^2 < \infty$.

iii) The matrix $\mathbf{R}_N \doteq \sum_{k=1}^{N} \mathbf{x}_k \mathbf{x}_k^T$ is positive definite.

The choice of $\mu_k$ is selected to tradeoff between rate of convergence and residual error. We chose the sequence $\mu_k = \frac{1}{k}$ for the step size factor. With this sequence, the weights converge towards their optimum values with relatively large steps, while as adaptation proceeds, $\mu_k$ becomes gradually smaller allowing the weights to settle at locally optimum values.

To apply the bootstrap learning method to edge detection problem, we need to define the input patterns in terms of image pixels. For our case of a one-dimensional classifier, a specific number of adjacent pixels are introduced to the classifier each time as an input pattern. The neural network is trained on a relatively small portion of the image and when the weights have converged to a certain tolerance, they are applied to the entire image.

## Classification by Constrained Maximization

Assume that two clusters of points in the $r$-dimensional space are randomly generated such that a point is equally likely to belong to each cluster. Then a classifying hyperplane can be defined as follows.

A classifying hyperplane divides the data point space into two half-spaces in a manner that the sum of squared distances of all the points from that plane be a maximum between all choices of hyperplanes for which this sum is a minimum. This definition for the classifying hyperplane is in a way a dual to the definition for a line (or plane) that best fits a set of data points in the $r$-dimensional space. In that case, the fitting line (or plane) is defined such that the sum of squared normal distances of all the points from it be a minimum between all choices of lines for which this sum is a minimum. It is well-known that for a given and fixed direction, the line (or plane) parallel to this direction with the property of minimizing squared sum-of-distances of all the points will pass through the center of mass of the sample space. In the case of line (or plane) fitting, we are interested in the minimum between all these directions, but in the case of classification, we seek the maximum between them. The problem of line fitting to a set of data points is explored in a later section. In summary, the classifying hyperplane passes through the mean of the points and has such an orientation such that the sum of squared distances of all points to it is maximum. By subtracting the mean coordinates from all points, we translate the coordinate system origin to the mean point, but still will use $\mathbf{x}$ to denote the transformed data vector.

Consider the situation in Fig. 1(b), where $\mathbf{x}_k$ is a data point and $\mathbf{w}$ is the unit vector normal to the classifying hyperplane originating at the origin. The distance $d_k$ of each data point from the hyperplane can be written in terms of the inner product of $\mathbf{w}$ and $\mathbf{x}_k$:

$$d_k = |\mathbf{w}^T \mathbf{x}_k| \tag{5}$$

The sum of squared distances can then be expressed by:

$$\sum_{k=1}^{N} d_k^2 = \sum_{k=1}^{N} \mathbf{w}^T \mathbf{x}_k \mathbf{x}_k^T \mathbf{w} = \mathbf{w}^T \left( \sum_{k=1}^{N} \mathbf{x}_k \mathbf{x}_k^T \right) \mathbf{w} = \mathbf{w}^T \mathbf{R} \mathbf{w} \tag{6}$$

where $\mathbf{R} = \sum_{k=1}^{N} \mathbf{x}_k \mathbf{x}_k^T$ is an $r \times r$ symmetric positive semidefinite matrix. The objective is to maximize $\sum_{k=1}^{N} d_k^2$ with appropriate choice of $\mathbf{w}$. It is easy to show that $\sum_{k=1}^{N} d_k^2$ will be maximized if the vector $\mathbf{w}$ is chosen along the direction of the eigenvector corresponding to the largest eigenvalue of $\mathbf{R}$. This scheme has been implemented and applied to the classification of $r$-dimensional pattern vectors. This approach however is computationally intensive due to large number of possible input patterns.

An iterative steepest ascent method can be also used to find the vector $\mathbf{w}_{\text{opt}}$ that maximizes $\sum_{k=1}^{N} |\mathbf{w}^T \mathbf{x}_k|^2$ (with the constraint of $\|\mathbf{w}\| = 1$). The goal is then to determine the orientation of the weight vector that satisfies the following:

$$\mathbf{w}_{\text{opt}} = \text{argmax}_{\|\mathbf{w}\|=1} \Sigma_{k=1}^{N} |\mathbf{w}^T \mathbf{x}_k|^2 \tag{7}$$

The constraint $\|\mathbf{w}\| = 1$ is imposed to avoid the trivial answer of an unbounded $\mathbf{w}$. In fact, we are only interested in the direction of the weight vector and its magnitude does not have any effect on the classification of patterns. At each iteration, the sharpest ascent direction on the function surface is determined and the weight update is chosen along its projection on the surface $\mathbf{w}^T \mathbf{w} = 1$ by the projection matrix $(\mathbf{I} - \mathbf{w}\mathbf{w}^T)$. The result is:

$$\Delta \mathbf{w}_k = 2\mu_k (\mathbf{I} - \mathbf{w}_k \mathbf{w}_k^T) \mathbf{w}_k^T \mathbf{x}_k \mathbf{x}_k^T \tag{8}$$

The choice of projected gradient guarantees that $\Delta \mathbf{w}_k$, the change in the weight vector, is perpendicular to $\mathbf{w}_k$ at each step. Since $\Delta \mathbf{w}_k$ is not infinitesimal, it will correspond to a a finite motion along the tangent to the surface at the point $\mathbf{w}_k$. Thus, the new point $\mathbf{w}_{k+1} = \mathbf{w}_k + \Delta \mathbf{w}_k$ will be slightly off the surface and must be renormalized to have unit norm:

$$\mathbf{w}_{k+1} = \frac{\mathbf{w}_k + 2\mu_k (\mathbf{I} - \mathbf{w}_k \mathbf{w}_k^T) \mathbf{w}_k^T \mathbf{x}_k \mathbf{x}_k^T}{\|\mathbf{w}_k + 2\mu_k (\mathbf{I} - \mathbf{w}_k \mathbf{w}_k^T) \mathbf{w}_k^T \mathbf{x}_k \mathbf{x}_k^T\|} \tag{9}$$

## Self-Organizing Competitive Learning

Another neural network architecture that is suitable for unsupervised learning and classification is the self-organizing network introduced by T. Kohonen[11, 12]. A simple diagram of this network is shown in Fig. 2. In the clustering problem, the units *competitively* become sensitive to specific features of the input patterns. The original idea behind *competitive learning* is the following:

Consider a set of patterns $\mathbf{x}(k) \in \mathcal{R}^r$ sequentially introduced as the result of a statistical event. Let $\{\mathbf{w}_i(k), \mathbf{w}_i \in \mathcal{R}^r, i = 1, \ldots, p\}$ denote a set of variable reference vectors. Assume that $\mathbf{w}_i(0)$ is initialized in some manner. At each time instant $k$, a certain measure of distance between the pattern $\mathbf{x}(k)$ and each reference vector $\mathbf{w}_i(k)$ is calculated. The reference vector closest to $\mathbf{x}(k)$ is then updated in a manner to better match $\mathbf{x}(k)$. Over time, the competing reference vectors will tend to become specifically sensitive to different features of the input pattern space.

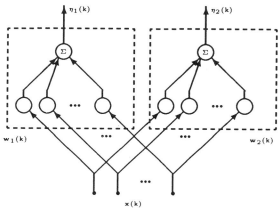

Figure 2: Self-organizing competitive classifier used for edge detection. Each of the two units competitively becomes sensitive to one cluster of input patterns.

In edge detection, a network with two units is trained to identify the two clusters of patterns. As before, the set of input patterns is generated by forming a set of a group of adjacent pixels chosen at random locations on the image. Denoting the input pattern at iteration $k$ as $\mathbf{x}(k)$, and the reference weight vectors of the units as $\mathbf{w}_i(k)$, $i = 1, 2$, the discriminant measure is chosen as:

$$\eta_i(k) = \mathbf{w}_i^T(k)\mathbf{x}(k), \quad i = 1, 2 \tag{10}$$

Because $\eta_i$ is a measure of matching between $\mathbf{x}(k)$ and $\mathbf{w}_i(k)$, the unit with the maximum value of $\eta$ is selected. We denote this unit by $l$. Then for unit $l$ the following adaptive process can be employed to update $\mathbf{w}_l(k)$:

$$\mathbf{w}_l(k+1) = \frac{\mathbf{w}_l(k) + \alpha_k \mathbf{x}(k)}{\|\mathbf{w}_l(k) + \alpha_k \mathbf{x}(k)\|} \tag{11}$$

where $\alpha_k$ is a gain parameter in adaptation which has a role similar to that of $\mu_k$ in bootstrap learning procedure in controlling the convergence rate and stability of adaptation.

## EIGENVECTOR LINE FITTING

In order to define a straight edge for the classified image, a method of line fitting is needed. several methods of line fitting are investigated. One approach applies the Hough transform [5, 6, 13] to the edge points and then uses a two dimensional search to find the parameters that determine a line. These parameters are $s$, the normal distance of each line from the origin, and $\theta$, the angle that it makes with positive $x$−axis. The Hough transform method is especially good for cases where noise contamination is fairly high. However, the required exhaustive two dimensional search makes it time consuming.

Least squares methods can also be used for line fitting. The criterion here can be minimizing the sum of squared distances from the desired line in $x$ or $y$ directions. In total least squares (TLS) method [14, 15], the sum of normal distances of the points from the fitted line is minimized and thus this method is independent of the coordinate system definition.

Here, we use an eigenvector approach [16] to find the line parameters that minimize the sum of normal distances. First we note that this line passes through the mean point of all the points. Therefore, by subtracting the mean coordinates from all points, we translate the origin to the mean point. The distance $d_i$ of each data point from the line can be written now as the inner product of $\mathbf{n}$, the unit normal vector to the line and the vector $\mathbf{v}_i$, including new coordinates of the $i^{\text{th}}$ data point:

$$d_i = |\mathbf{n}^T \mathbf{v}_i| \qquad (12)$$

The goal is now to minimize $\sum_{i=1}^{n} d_i^2$ with appropriate choice of $\mathbf{n}$. It can be shown that $\sum_{i=1}^{n} d_i^2$ is minimized if the vector $\mathbf{n}$ is chosen along the eigenvector that corresponds to the smaller eigenvalue of $\mathbf{V}$. Straightforward geometry is then used to extract the parameters $s$ and $\theta$.

## APPLICATION AND COMPARISON

The neural network structures introduced earlier are applied to edge detection for the purposes of critical dimension (linewidth) measurement in integrated circuit fabricaion. Since most features on IC chips are rectangular, the edges that are to be found are straight lines. As was mentioned earlier, several methods exist for detecting edges in optical images. The neural network edge detectors are thus compared to an alternative method of detecting straight lines. The method that the neural network approach is compared with is a projection-based approach using the Radon transformation [7], in which the image is projected along different directions and then the angle of projection with the sharpest transition profile is chosen as the estimated edge angle. Another searching phase for finding peaks in the gradient of this projection estiamtes the edge location.

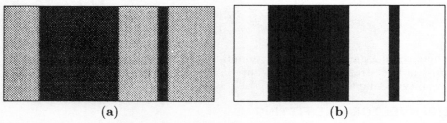

Figure 3: An example of applying neural network classifier to edge detection. (a) Optical image taken from a part of a chip. The darker parts show two etched trenches. The image size is 150 × 400 pixels. (b) Output of the self-organizing competitive classifier. Similar results were obtained by the other two classification schemes.

The neural network classifiers and the projection-based edge detector are applied to images of patterned (or layered) silicon wafers. The images were digitized in grey scale over 256 levels and quantized into matrix form. For the neural networks, an array of vectors, $\mathbf{x}_k$ were constructed to contain the grey levels of a set of columnwise adjacent pixels. These vectors are used in both adaptation and classification phases.

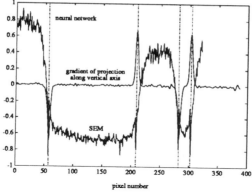

Figure 4: Comparison of the detected edges by the explored methods and the SEM measurement. Accurate observation and comparison with the SEM measurement shows that the relative error in locating edges for different schemes studied here is within 2.5%.

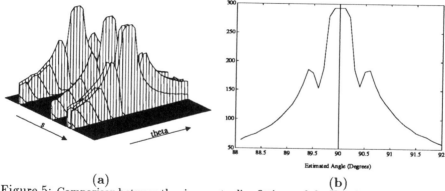

(a) (b)

Figure 5: Comparison between the eigenvector line fitting and the Hough transform result for estimating the direction of the detected edges in Fig. 3(b). (a) Hough transform output plane. (b) The curve shows the average of profiles in $\theta$ in (a) that are located at places where the $s$ parameter is maximum. The line is the result of applying the eigenvector line fitting.

The image in Fig. 3(a) shows two trenches etched on a chip. The difference in brightness is due to both the focal plane location and the differences in reflectivity. Fig. 3(b) shows the result of applying the neural network approach to this image, where the image pixels are separated into two classes. It is noted that clustering results are virtually the same for different approaches. Fig. 4 gives a comparison between the explored methods and an SEM (Scanning Electron Microscopy) imaging of the same wafer. The SEM measurement is known to have superior accuracy and resolution to other techniques, but due to its invasive nature, it is restricted to off-line measurements. In Fig. 5 the results of applying Hough transform and the eigenvector line fitting technique to the edge pixel set obtained from Fig. 3(b) are presented.

Another experiment was conducted using the neural network classifier on an image that was rotated before digitizing. Fig. 6 shows this image together with the result of applying neural network classification to it. The estimated edge orientation and distances from the origin by both Hough transform and the eigenvector line fitting are presented in Figs. 7(a) and 7(b), respectively.

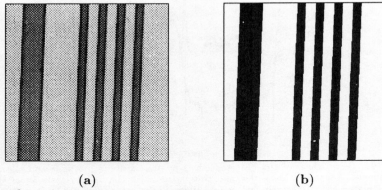

Figure 6: Another example of applying neural network classifiers to edge detection in lithography images. (a) Optical image. (b) Output of the neural network classifier.

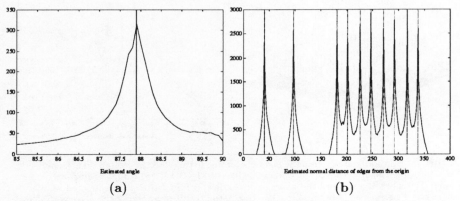

Figure 7: (a) Estimated rotation angle and (b) estimated distances from the origin for the detected edges in Fig. 6(b) by the eigenvector line fitting method and the Hough transform. The curves are obtained by the Hough transform. The straight lines are the result of the eigenvector line fitting method.

## CONCLUSION

Several unsupervised neural network learning methods have been explored and applied to detecting edges on real wafer images. The methods included a self-organizing competitive learner, a bootstrapped linear threshold classifier, and a constrained maximization algorithm. Experiments showed good robustness and repeatability for the self-organizing classifier while the other two methods, although producing excellent results in several cases, showed sensitivity to the average brightness of the image. The results of the neural network classifiers were compared to the results obtained by a standard straight edge detector based on the Radon transform [7]. Also a line fitting method was explored to extract the parameters of edges detected by the neural networks. This scheme was based on fitting a line along the dominant eigenvector of the covariance matrix of the detected edge pixels [16]and was equivalent to a total-least-squares line fitting method. Comparison was made between the result of this method and the lines fitted by the standard method of Hough transform. All the comparisons showed good consistency between the proposed methods and the standard schemes. However, the neural network classifiers proved superior in speed over the standard methods and this advantage can be exploited to implement on-line linewidth measurement for lithography process control.

## REFERENCES

[1] J. F. Canny, Finding edges and lines in images, Report AI-TR-720, MIT Artificial Intell. Lab., Cambridge, MA, 1983.

[2] D. C. Marr and E. C. Hildreth, Theory of edge detection, In *Proc. of Royal Society of London*, pages 187–217, 1980.

[3] J. F. Canny, A computational approach to edge detection, *IEEE Pattern Analysis and Machine Intelligence*, 8(6):679–698, 1986.

[4] V. Torre and T. Poggio, On edge detection, Report AIM-768, MIT Artificial Intell. Lab., Cambridge, MA, 1984.

[5] P. Hough, Method and means for recognizing complex patterns, U.S. Patent 3069654, 1962.

[6] R. O. Duda and P. E. Hart, Use of the Hough transform to detect lines and curves in pictures, *Communications of the ACM*, 15:11–15, 1972.

[7] D. Petkovic, W. Niblack, and M. Flickner, Projection-based high accuracy measurement of straight line edges, *Machine Vision and Applications*, 1(3):183–199, 1988, Springer Int.

[8] S. Douglas and Th. Meng, An adaptive edge detection method using a modified sigmoid-LMS algorithm, In *23rd Annual Asilomar Conf. on Signals, systems, and computers*, pages 252–256, 1989.

[9] W. Miller, *A Modified Mean-Square-Error Criterion for Use in Unsupervised Learning*, PhD thesis, Stanford university, Stanford, CA, 1967.

[10] B. Widrow, Bootstrap learning in threshold logic systems, a paper presented to theory committee, American Automatic Control Council, for IFAC meeting, London, June 1966.

[11] T. Kohonen, Self-organized formation of topologically correct feature maps, In *Biological Cybernetics*, pages 59–69. Springer-Verlag, 1982.

[12] T. Kohonen, *Self-Organization and Associative Memory*, Springer-Verlag, 3rd edition, 1989.

[13] A. K. Jain, *Fundamentals of Digital Image Processing*, Prentice-Hall, Englewood Cliffs, NJ, 1989.

[14] G. H. Golub and C. F. Van Loan, *Matrix Computations*, John Hopkins University Press, Baltimore, MD, 1984.

[15] S. van Huffel and J. Vandewalle, The total least squares technique: Computation, properties and applications, In E. F. Deprettere, editor, *SVD and Signal processing: Algorithms, Applications and Architectures*, pages 189–207. Elsevier Science Publishers, 1988.

[16] R. O. Duda and P. E. Hart, *Pattern Classification and Scene Analysis*, John Wiley and Sons, New York, 1973.

# Improving Generalization Performance in Character Recognition

Harris Drucker
AT&T Bell Laboratories and Monmouth College
Monmouth College, West Long Branch, NJ 07764

Yann Le Cun
AT&T Bell Laboratories
Holmdel, NJ 07733

**Abstract:** One test of a new training algorithm is how well the algorithm generalizes from the training data to the test data. A new training algorithm termed double backpropagation improves generalization by minimizing the change in the output due to small changes in the input. This is accomplished by minimizing the normal energy term found in backpropagation and an additional energy term that is a function of the Jacobian.

### Introduction

Backpropagation [1] has been a popular supervised training algorithm for a number of years. The general procedure is to learn on a training set and see how well the generalization is on a test set. In an attempt to improve generalization performance, one can create different architectures, or for a specific architecture one can impose additional constraints. Two of the latter techniques are weights decay [2-4] and training with noise [3,5,6].

The normal energy term in backpropagation, here denoted as $E_f$ (f for forward) is of the form $E_f = (D - X)(D - X)^t/2$ where t indicates transpose, and $D$ and $X$ are the desired and actual output row matrices (of size m corresponding to the m output components), respectively. In weight decay we add to $E_f$ a term of the form $\alpha ||W||^2$, i.e., a constant times the norm squared of the weight vector. The idea here is to force the weights to be small therefore keeping the output of the nonlinear elements out of the saturated regions. Although weight decay does work, it is not always obvious in advance what the value of $\alpha$ should be. The rationale of adding noise to the input patterns is to move the search out of a local minimum of the weight space in addition to providing variations of the input. The problem with this approach is determining how much noise should be added thereby requiring multiple runs to determine the noise level. Generalization may be increased by picking appropriate architectures or starting with an architecture and then pruning or adding units [7-9].

## Double Backpropagation

In double backpropagation, in addition to the normal energy term $E_f$ we try to minimize a term of the form:

$$E_b = \frac{1}{2}\left[\frac{\partial E_f}{\partial i_1}\right]^2 + \frac{1}{2}\left[\frac{\partial E_f}{\partial i_2}\right]^2 + ... + \frac{1}{2}\left[\frac{\partial E_f}{\partial i_n}\right]^2$$

where $i_j$ is one of n input components. The rationale for this approach is that if the input changes slightly the energy function $E_f$ should not change. One measure of this change is just the derivative of $E_f$ with respect to all the inputs. Therefore, by forcing $E_b$ to be small we force the appropriate derivatives to be small.

We can show that the above equation is equivalent to $E_b = (D-X)JJ^t(D-X)^t$ where $J$ is the m by n Jacobian matrix (corresponding to the m output components and n input components. The elements of $J$ are composed of the derivatives of each output component to every input component.

If we now add the normal energy term to this last term we obtain for the total energy term $E_t$:

$$E_t = (D-X)\left[\frac{1}{2}I_m + \alpha JJ^t\right](D-X)^t$$

where $I_m$ is the identity matrix of square size m and $\alpha$ is a multiplicative constant which will be related to the learning rate of the neural network. The constant $\alpha$ is greater than one, the rationale being that near the minimum, $(D-X)(D-X)^t$ is close to zero and hence $JJ^t$ will have small effect near this minimum unless $\alpha$ is large. In practice the optimum solution is fairly insensitive to the choice of $\alpha$ and no time is spent searching for the optimum $\alpha$. $\alpha$ is related to the choice of learning constant in the neural network.

One of the key ideas in double backpropagation is that the additional energy term $E_b$ can be calculated in one backward pass through a neural network or one forward pass through what we will call an appended network. Let us examine a simple architecture (Fig 1 - below the dashed horizontal line). The layers are fully connected (to the layer immediately above) but not all weights are shown (nor is the bias). The input layer has linear neurons while the other layers have the nonlinear transfer function. Therefore $a_j^{(r)}$, represents the summed input of neuron number j at layer r and $x_j^{(r)} = f\left[a_j^{(r)}\right]$ where f is the hyperbolic tangent.

In this case the forward energy function is:

$$E_f = \frac{1}{2}\left[d_1 - x_1^{(2)}\right]^2 + \frac{1}{2}\left[d_2 - x_2^{(2)}\right]^2. \qquad (1)$$

Now, let us look at some of the terms obtained by backpropagating through the network. First, the derivative of the forward energy function with respect to the input of the first neuron of the output layer:

$$-\frac{\partial E_f}{\partial a_1^{(2)}} = \left[d_1 - x_1^{(2)}\right] f'\left[a_1^{(2)}\right]. \qquad (2)$$

where f' is the derivative.

The change in weight A is now due to the term $-\frac{\partial E_f}{\partial a_1^{(2)}}$ multiplied by both $x_1^{(1)}$ and the learning constant. Similarly we can update the other weights. In normal backpropagation, these are all the gradients needed. However, we can proceed further and calculate the derivative of the forward energy function with respect to the input (recalling that the input neurons are linear):

$$-\frac{\partial E_f}{\partial i_1} = \left[E\frac{\partial E_f}{\partial a_1^{(1)}} + F\frac{\partial E_f}{\partial a_2^{(1)}}\right]$$

Terms of the form $\frac{\partial E_f}{\partial i_j}$, where j ranges over the n input components, are called the *input gradients*. It is important to note that the calculation of the input gradients is a linear operation. We now show that the input gradients can be calculated by appending a network (the top part of Figure 1) to the original network.

The appended network, which is a "mirror image" about the dashed line, uses linear neurons:

$$y_j^{(r)} = k_j^{(r)} b_j^{(r)}$$

$$k_j^{(r)} = \begin{cases} f'(a_j^{(r)}) & r > 0 \\ 1 & r = 0 \end{cases}$$

where k is the multiplicative constant whose value is related to the derivative of the input state of a neuron in the lower network. Note that the superscripts decrease as one approaches the top of the appended layer.

Although not all weights are shown consider the input state of the first neuron of the appended layer:

$$b_1^{(2)} = \left[x_1^{(2)} - d_1\right]$$

$$y_1^{(2)} = k_1^{(2)} b_1^{(2)}$$

$$= f'\left[a_1^{(2)}\right] \left[x_1^{(2)} - d_1\right]$$

which is equation (2). Proceeding in this fashion, we see that the output of the appended network is just the input gradient of the forward energy function.

Therefore, three steps are involved in calculating the input gradient: (1) forward propagate through the lower network (2) copy the derivatives of the input states from the lower network to the appended network as the multiplicative constants of the linear neurons and (3) forward propagate through the appended network. Once training is complete, the appended network is no longer needed and is removed.

Now we can form the backward energy function, so named because it could be obtained by backpropagating through the lower network:

$$E_b = \frac{1}{2}\left[y_1^{(0)}\right]^2 + \frac{1}{2}\left[y_2^{(0)}\right]^2 + \frac{1}{2}\left[y_3^{(0)}\right]^2$$

$$= \frac{1}{2}\left[\frac{\partial Ef}{\partial i_1}\right]^2 + \frac{1}{2}\left[\frac{\partial Ef}{\partial i_2}\right]^2 + \frac{1}{2}\left[\frac{\partial Ef}{\partial i_3}\right]^2 \quad (3)$$

We will now minimize the sum of a constant times the backward energy function (3) and the forward energy function (1). The general idea is to backpropagate through the lower network to minimize the forward energy function and do another backpropagation starting at the top of the appended network to minimize the backward energy function (hence the description of the training algorithm as double backpropagation).

Backpropagation through the upper network has some subtleties because the weights are shared between the upper network and the lower network. First let us find the derivative of the backward energy function with respect to the state of the first neuron in the top layer recalling both that the neuron is linear and that for the top layer, the multiplicative constant is 1.

$$\frac{\partial Eb}{\partial b_1^{(0)}} = y_1^{(0)}.$$

Now, the gradient with respect to the the weight F:

$$\frac{\partial E_b}{\partial F} = \frac{\partial E_b}{\partial b_1^{(0)}}\frac{\partial b_1^{(0)}}{\partial F} + \frac{\partial E_b}{\partial a_2^{(1)}}\frac{\partial a_2^{(1)}}{\partial F} = \frac{\partial E_b}{\partial b_1^{(0)}} y_2^{(1)} + \frac{\partial E_b}{\partial a_2^{(1)}} i_1 .$$

The first term of the sum is found in normal backpropagation, the second is not. However, we will get the equivalent result by backpropagating through the whole network. Therefore, the algorithm proceeds as follows:

1. Present the input pattern and propagate it to the output (the top of the lower network).

2. Backpropagate the gradient of the forward energy function through the lower network. Compute the change in weights but do not change the weights yet.

3. Copy the appropriate derivatives from the lower network to the appended network.

4. Propagate forward through the upper network

5. Now backpropagate the backward energy function from the top of the appended network down through the original network, calculating the weights changes but do not change the weights yet.

6. Finally, change the weights using the weight changes calculated in steps 2 and 5.

**Experimental Results**

Each results presented below is the average error rate on the test set for ten runs for a particular architecture and particular training algorithm. Each training cycle was followed by one test cycle and the best test results were retained. Five learning algorithms were used:

1. Backpropagation

2. Full double backpropagation which was described above.

3. Partial double backpropagation is a modified form of the computationally expensive full double backpropagation. In this case we form a backward energy term that is a function of the gradients at the input to the hidden layer: $\frac{\partial E_f}{\partial a_j^{(1)}}$. The rationale is that small changes in the input to the hidden layer should not affect the output. In this case, the appended network required to calculate these derivatives is smaller (minus the uppermost layer in Figure 1).

4. Normal backpropagation followed by full double backpropagation. The reasoning is that backpropagation is faster and full double backpropagation following normal backpropagation should require less training cycles. In this case, full double backpropagation starts with the results of the network with the best test score found in (1) above.

5. Normal backpropagation followed by partial double backpropagation.

A database consisting of 320 training examples and 180 test samples were used on four locally constrained architectures that had been previously shown to give good results using backpropagation [10]. Twelve examples of each of the ten digits were hand drawn by a single person on a 16 by 13 bitmap using a mouse. Each image was then used to generate four examples by putting the original image in four consecutive horizontal positions on a 16 by 16 bitmap. Thus the architectures and training algorithms will be specifically tested against a database consisting strictly of translations.

The four architectures :

1. local-net: A locally connected architecture with two hidden layers (figure 2). The output of the second hidden layer is fully connected to the output.
2. local2-net: A network with two hidden layers and weight sharing (figure 2).. All units in the first hidden layer share the same weights.
3. local21-net: Same as local2-net except that the weights in going from the first to second hidden layer are also shared.
4. local4-net: Two hidden layers, the first of which consists of four 8x8 feature maps and the second four 4x4 feature maps. All the units in a feature map share the same weights, the receptive fields being of size 3x3 in going from the input to first hidden layer and of size 5x5 going from the first to second hidden layer.

In these cases of multiple hidden layers, partial double backpropagation means that the appended network consists of two layers: the mirror image (around the dashed line of Figure 1) of the output layer and the hidden layer closest to the output. The resultant error rates for the four architectures and five learning algorithms are shown in Table 1. Except for two cases of the local21 architecture double backpropagation improves performance. Our local21 architecture was never able to learn the training set in those cases where double backpropagation gives worse results. Of special interest is that fact that the local4 architecture which gives the best results using normal backpropagation has the most significant increase in performance (to 2.2%) using double backpropagation.

Our next trial was on a very large database consisting of 9709 training samples and 2007 test samples taken from handsegmented zip code data obtained from the U.S. Postal Service. The architecture is fully explained in [11,12] and consists of 4645 neurons, 2578 weights (some shared), and 98442 connections. We therefore did one run starting from the best network configuration to date. That best result using a Newton version of regular backpropagation has previously been reported as 5.03%. Using partial double backpropagation starting from these best results, the error rate was reduced to 4.68% after twenty-three iterations (approximately twelve days).

Analysis of the distribution of weights shows the reason why double backpropagation improves performance. In double backpropagation, the distribution of weights to the first hidden layer have a unimodal distribution with significantly smaller variance than that of a network trained using backpropagation. By inputting all the test patterns we can also obtain a distribution of the summed input (essentially corresponding to the $a_j^{(1)}$ of figure 1.) This distribution is bimodal with smaller variance when trained using double backpropagation. The transfer function of the sigmoid is linear over a small region centered around zero. For networks trained using double backpropagation, many more of the signals at the input to the first hidden layer are in the linear region (typically 20%) than when trained used backpropagation (where the typical number is 5%). This type of behavior is not exhibited at the higher layers, therefore the improved performance is due to the different distribution of weights at the input to the first hidden layer.

**Conclusions**

Double backpropagation has been shown to be a technique that improves performance by forcing the output to be insensitive to incremental changes in the input. The improvements are especially significant for those architectures which show very good performance when trained using backpropagation. The penalty paid is an increased running time which is not too large a penalty if partial double propagation is used. Double backpropagation can be used following normal backpropagation but generally does not give as good results as double backpropagation alone. It was furthermore shown that double backpropagation creates a weight distribution at the input to the first hidden layer that has a smaller variance than that generated using backpropagation.

| ARCHITECTURE | local | local21 | local2 | local4 |
|---|---|---|---|---|
| BACKPROPAGATION | 8.6 | 8.2 | 4.5 | 3.8 |
| FULL DOUBLE BACKPROPAGATION | 5.0 | 6.2 | 3.3 | 3.1 |
| PARTIAL DOUBLE PROPAGATION | 6.9 | 8.5 | 2.8 | 2.2 |
| FULL DOUBLE BACKPROPAGATION FOLLOWS BACKPROP | 5.8 | 6.1 | 3.2 | 2.9 |
| PARTIAL DOUBLE BACKPROPAGATION FOLLOWS BACKPROP | 6.5 | 9.5 | 3.7 | 2.6 |

Table 1.  Error rate in percent for four architectures and five training algorithms. 320 items in training set and 180 in test set."

**References**

[1] D.E.Rumelhart, et. al., "Learning internal representations by error propagation," **Parallel Distributed Processing: Explorations in the**

Microstructure of Cognition, 1 Rumelhart and McClelland (eds.), MIT Press, Cambridge MA, 1986, pp. 318-362.

[2] G.E. Hinton, "Learning Distributed Representations of Concepts", **Proceeding of the Eight Annual Conference of the Cognitive Science Society**, pp. 1-12 (Amherst 1986), Hillsdale: Erlbaum, pp. 1- 12.

[3] R. Scaletter and A. Zee, "Emergence of Grandmother Memory in Feed Forward Networks: Learning with Noise and Forgetfulness. **Connectionist Models and Their Implications: Readings from Cognitive Science**, D. Waltz and J.A. Feldman (eds.), pp. 309-332. Norwood: Ableex, 1988.

[4] A. H. Kramer and A. Sangiovanni-Vincentelli, "Efficient Parallel Learning Algorithms for Neural Networks, **Advances in Neural Information Processing Systems I**, (Denver 1988), D.S. Touretzky (ed.), San Mateo: Morgan Kaufmann, 1989.

[5] A. von Lehman, et. al., "Factors Influencing Learning by Back Propagation, **IEEE International Conference on Neural Networks**, (San Diego 1988), vol I, pp. 335-341, New York, IEEE,

[6] J. Sietsma and R.J.F. Dow, "Neural Net Pruning--Why and How", **IEEE International Conference on Neural Networks** (San Diego 1988), vol I, pp. 325 - 333, New York: IEEE.

[7] S.E. Fahlman and C. Lebiere, "The Cascade-Correlation Learning Architecture, **Advances in Neural Information Processing Systems II**, D.S. Touretzky (ed.), 1990, pp. 524-532, San Mateo: Morgan Kaufmann.

[8] M. Mezard and J.-P. Nadal, "Learning in Feedforward Layered Networks: The Tiling Algorithm, **Journal of Physics A 22**,(1989) pp. 2191-2204.

[9] S.I. Gallant, "Optimal Linear Discriminants, **Eighth International Conference on Pattern Recognition**, (Paris 1986) pp. 849-852 New York: IEEE.

[10] Y. LeCun, "Generalization and Network Design Strategies, **Connectionism in Perspective**, Pfeifer, et. al. (eds.), 19 Zurich, Switzerland: Elsevier

[11] Y. Le Cun, et. al., "Backpropagation Applied to Handwritten Zip Code Recognition, **Neural Computation 1**, (1989), pp. 541-551.

[12] Y. Le Cun, et. al., "Handwritten Digit Recognition with a Back-Propagation Network", **Advances in Neural Information Processing Systems**, (Denver - 1989) D.S. Touretzky (ed.), pp. 396-404, San Mateo: Morgan Kaufmann.

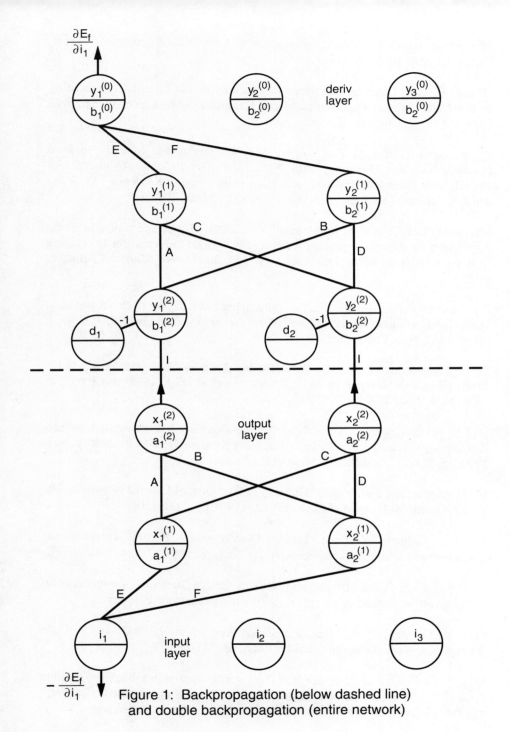

Figure 1: Backpropagation (below dashed line) and double backpropagation (entire network)

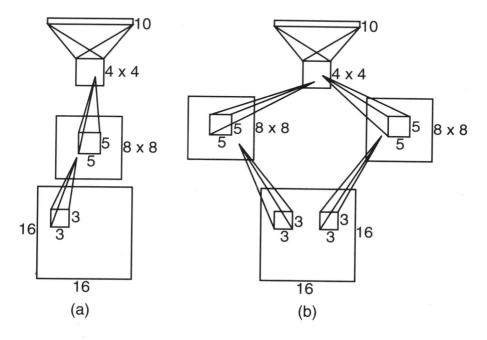

Fig. 2. (a) local architecture
(b) local21 and local2 architecture

# Neural Networks for Sidescan Sonar Automatic Target Detection

Michael J. LeBlanc[*]
Fault-Tolerant Systems Division
The Charles Stark Draper Laboratory
MS 6F
555 Technology Square
Cambridge, MA 02139

mjl@draper.com

Elias Manolakos
Communications and Digital
Signal Processing Center for Research
and Graduate Studies
Electrical & Computer Engineering
Dept.
Northeastern University
Boston, MA 02115
elias@northeastern.edu

**Abstract:**
The goal of this research is to develop a multi-layer feedforward neural network architecture which can distinguish targets (in this case, mines) from background clutter in sidescan sonar images. The network is to be implemented on a hardware neurocomputer currently in development at CSDL, with the goal of eventual real-time performance in the field. A variety of neural network architectures are developed, simulated, and evaluated in an attempt to find the best approach for this particular application. It has been found that classical statistical feature extraction is outperformed by a much less computationally expensive approach that simultaneously compresses and filters the raw data by taking a simple mean.

THE PROBLEM: SIDESCAN SONAR

A stylized sidescan sonar image is depicted in Figure 1. It is collected as follows: A ship tows a sonar sensor called a *towfish* behind it at a speed of 5-10 knots (1 knot = 1.15 mph = 0.515 m/s). The towfish periodically emits 50KHz - 500KHz sound pulses, called "pings", at a rate of 1 to 30 pings per second. Objects in the water column and on the sea floor itself reflect some portion of the emitted sound energy, depending upon their size, distance from the transducer, and

---

[*] This work was funded by the CSDL Internal Research and Development program through a Draper Labs Fellowship, in cooperation with the CDSP Center at Northeastern University during the author's term there as a graduate student.

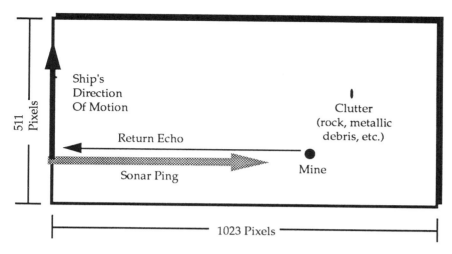

Figure 1: Sidescan Sonar Image

the materials of which they are composed. These return echoes are detected and recorded as they are generated by each ping, and used to create a single pixel-row of the image. When a number of rows are placed side by side, a two-dimensional (511x1023) image of the ocean floor is made, in a fashion analogous to a raster scan. These sonar images contain *only* echo-intensity information, which is directly translated into grey-scale levels in the final 2D image: No frequency-domain information is contained in the images. This distinguishes the data from that used in previous research on neural networks in sonar, in which frequency domain characteristics were used to identify the object returning the echo [1]. In our case, we are merely trying to determine if the object returning the echo is a mine or not, independent of the type of mine, based only upon the amplitude of the return echo at each pixel. Once a sidescan sonar image is produced, the problem becomes essentially an image-processing one, and the sonar nature of the image does not really influence our choice of analysis technique.

Analysis of these images is made difficult by the fact that the exact operating conditions under which they were collected are unknown, and are most likely not the same from image to image. Due to the methods used in processing the image, these conditions are not important to the target detection application.

THE INCA/1 NEUROCOMPUTER

Design constraints are imposed by the Integrated Neurocomputing Architecture (INCA), a multi-purpose and reconfigurable hardware neurocomputer currently in development at CSDL. This neurocomputer uses analog neural network building block chips developed by the Jet Propulsion Laboratory's Advanced Microelectronics group in Pasadena, CA [2]. Candidate architectures must satisfy the size and connectivity constraints of the INCA, and it must be verified that the developed architecture can still operate given the actual transfer function of the hardware neuron (rather than the perfectly smooth and symmetrical transfer functions such as the logistic sigmoid or hyperbolic arctangent used in simulation). We must also verify that the reduced weight precision of analog circuitry (11 bits) will not adversely affect the network's performance.

INCA limits us to a maximum of four layers, with a maximum of 64 neurons in each layer (including bias units). Interconnectivity between layers can be full, but only feedforward connections are allowed: We cannot have lateral or feedback connections. Since future versions of the INCA will most likely relax some of these constraints, the investigated architectures are not necessarily limited to those that will fit within these specifications, however. An additional constraint is the goal of real-time performance: While this does not directly influence the network architecture (since the INCA will be an analog implementation and not a digital simulation-- the anticipated feedforward processing time of the INCA is 300 µs), it *does* influence the amount and types of preprocessing we can perform to convert each token window into a form digestable by the particular network.

## GENERAL APPROACH

Since it is impractical to scan the entire image at once, the image is subdivided into small, overlapping subwindows (NxN, $10 \leq N \leq 30$) called *tokens* which are presented to the detector. The detector must determine if the token presented contains a mine, or only clutter. The most obvious distinguishing characteristic of a mine in these images is its apparent brightness, corresponding to the relatively loud sonar echo it returns. However, the problem of finding a mine is more difficult than simply finding the brightest spot in the image. There are many clutter objects such as large rocks on the sea floor or metallic debris which could be mistaken for a mine. The level of background noise in the image is also relatively high, and could easily obscure a mine echo.

Our particular dataset contains 167 images, with one and only one target in each image. We have also been provided with ground truth information that indicates the true location of the target in each image. This information is used to generate two sets of data: A training set (750 images, half mine tokens, half clutter tokens) and a test set (250 images). All neural networks were trained on a SPARCStation with a commercial neural simulation package, NeuralWorks Professional II [3], using a minor variant of backpropagation. Networks generally took between 75 000 and 100 000 presentations to train (100 to 133 randomly-ordered iterations through the training set).

Once trained, each network was tested on both the initial training data, and the novel test data. Evaluation on the training data lets us determine the innate ability of a network to learn the detection function, while evaluation on the testing data give a reasonable measure of how a network could perform in the field. Three quantitative performance measurements are made on each candidate network. The "Correct Detection/Rejection rate" indicates the percentage of correctly classified token windows: i.e. mine tokens correctly detected as mines, plus clutter tokens correctly rejected as clutter. The "Missed Target rate" indicates the percentage of targets present that were classified as clutter. The "False Alarm Rate" indicates the percentage of clutter tokens that were mistaken for mines. The latter two metrics are standard in ATR parlance: The first metric is not, but serves as a simple, single-figure metric for comparing architectures.

FOUR-FEATURE METHOD

The "classical" approach to this type of pattern recognition problem is to extract some set of features from the raw data, and to present these features to the classifier. In our case, the extracted features are the sample estimates of the first four statistical moments (mean, standard deviation, skew and kurtosis, depicted in Table 1), but many other sets could have been selected. The advantage of this approach is that our network can be very small, since the set of features is so small, and that much of the noise in the image will be filtered out by the feature extraction process. However, the primary risk of this approach is that if the set of features we choose to extract is not optimal, we may actually be *discarding* important information present in the image, information which could otherwise prove useful to the classifier. There is also a large amount of preprocessing necessary before the neural network can be used, a potential bottleneck with which we must be concerned if we are to keep our goal of real-time performance.

**Table 1. The Four Statistical Features**
(for an NxN token window containing $N^2$ points, $x_{ij}$)

| $m_p$ | $\frac{1}{N^2} \sum_{i=1}^{N} \sum_{j=1}^{N} x_{ij}^p$ |
|---|---|
| mean | $m_1$ |
| standard deviation ($\sigma$) | $\sqrt{m_2 - m_1^2}$ |
| skew | $\dfrac{m_3 - 3m_2 m_1 + 2m_1^3}{\sigma^3}$ |
| kurtosis | $\dfrac{m_4 - 4m_3 m_1 + 6m_2 m_1^2 - 3m_1^4}{\sigma^4}$ |

Table 2 reports the results and topology of this and all other networks mentioned here. Note that, while not exceptional, the four-feature network *is* the most consistent performer between the training and test data. This in fact complies with what we expect from feature extraction: Good generalization. The statistical feature set eliminates much of the unnecessary information, which allows the network to apply what it has learned from the training set to novel data without much performance degradation. However, the relatively low detect/reject rate indicates that this feature set is most likely not optimal. One possible approach to improve the performance of this network would be to embark upon a rigorous investigation of various feature sets, to determine the "optimal" one. Due to the large amount of preprocessing which feature extraction requires, however, we would ideally like to escape from having to extract features at all. This saves us from having to accept a slowdown in performance, or from investing in expensive and bulky floating-point hardware.

RAW DATA

Since backpropagation itself is capable of feature extraction, another possible approach would be to simply present raw token window data to the network and let the learning algorithm determine which features deserve attention. The advantage to this is that *all* information in the image is now available to the net. The disadvantages are that we now have to worry about image noise (since it is no longer being filtered out by feature extraction), and the network size will also be much larger than before, so that even a fairly small token window size of 10 x 10, with 100 pixels, is much too large to fit on the INCA/1. To see how capable backpropagation is in extracting features on its own, raw data was presented to a very large neural network. Both 10x10 and 8x8 (which can fit on the INCA/1) token window sizes were used.

The results show that, of all the networks tested, the 10x10 raw data network is the one most capable of learning the training set. Presumably, access to all of the pixels in each token window has enabled the network to "memorize" the training set quite well. Also note that the lesser performance of the 8x8 token window can best be explained when we consider target sizes: These smaller token window is *too* small, and actually misses some of the larger targets.

However, these networks exhibit poor generalization, as indicated by their poor performance on the novel test data. This is most likely due to image noise: The nets can easily memorize the training set by simply memorizing the noise as well, but they have not learned what *distinguishes* features from noise, and so are not capable of carrying generalizations over to the novel test data. Again, another possible approach would be to investigate pre-processing algorithms, especially low-pass filtering, to alleviate the effects of noise, but for reasons stated above this was not the path chosen. Instead, a very simple approach which simultaneously addresses the problems of high-frequency noise in the image, as well as the large size of network required by raw data, was taken.

THE MEAN-COMPRESSION FILTER (MCF)

This simple algorithm compresses the raw data and filters it in a single stroke. Basically, small MxM pixel chunks ($2 \leq M \leq 4$) are averaged, and the result taken as a "virtual pixel" in a "virtual image", as depicted in Figure 2. This virtual image is much smaller than the original token window (allowing a smaller network) but retains *most* of the character of the raw data, so the network can still "decide for itself" what features are worthwhile, the primary advantage of the raw-data approach. The method itself is also far less computationally expensive than statistical feature extraction, allowing the calculations to be performed in real-time with an inexpensive microprocessor. We also get zero-mean noise filtering, which makes the net more robust to novel data sets (see results). This approach will only work in applications in which the noise has a mean of zero, but fortunately this seems to be the case with this particular data set.

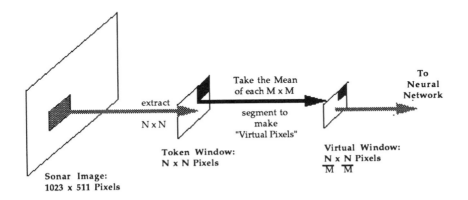

Figure 2: The Mean-Compression Filter

This net, while still larger than the four-feature network, is considerably smaller than that necessary for raw image processing, and gives very good results. Token window sizes of 10x10, 15x15, and 20x20 were tried, with the virtual window size held constant at 5x5. As table 2 indicates, the MCF networks perform nearly as well as the raw data networks on the training data. While performance still degrades somewhat for the novel data set, this degradation is not as marked as for the raw data net, allowing the MCF net to consistently give correct detection/rejection rates over 90 percent (see table). Note that as the size of the token window increases, so does performance, at least for $10 \leq N \leq 20$: A more in-depth investigation of the effects of token window size over a greater range is taken in [4]. This is the architecture which will most likely be demonstrated on the INCA/1 in September 1991.

"RECEPTIVE FIELD " NETWORKS

The last approach in this investigation is a so-called receptive field architecture. These networks are still feedforward, and still trained with backpropagation. However, instead of full interconnectivity between the input layer and the first hidden layer, the input layer is parsed into a number of overlapping 2x2 portions, each connected to a single hidden unit, as shown in Figure 3. Since it (slightly) resembles the way in which the visual receptors in the retina are organized to scan the visual field, it was hoped that it would be interesting to observe how backpropagation deals with this arrangement. Also, some previous work indicates that sparse interconnections can reduce the propagation of errors and/or noise from the input layer, thereby increasing the fault tolerance of a neural network [5].

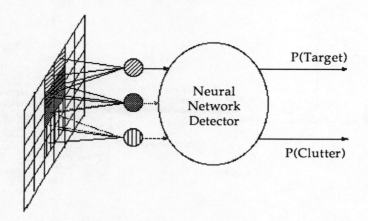

Figure 3: The "Receptive Field" Network

Both 10x10 raw data and 5x5 MCF data generated from 10x10 data were tried as inputs to the receptive field network. Note that these networks exhibited the lowest correct detection/rejection rates of any of those investigated, and that the miss rates were particularly high. This may be due to the localization of the RF approach: Whereas in previous networks, each hidden unit in the first hidden layer had access to the entire input window, here we are severely limiting the amount of information available to each of these units. However, this is only a hypothesis. Also note that, for unexplained reasons, the RF-MCF net performance is better for the test set than for the training set: This was a consistent result.

CONCLUSION

The primary conclusion drawn from this work is that standard feature extraction may not always be the best approach. For this particular application, the conceptually and computationally simple mean-compression filter outperforms

Table 2: Results

| Input Data<br>Network Topology* | Training Data Performance<br>750 examples | | | Test Data Performance<br>250 examples | | |
|---|---|---|---|---|---|---|
| | Correct Detect/Reject | Missed Target Rate | False Alarm Rate | Correct Detect/Reject | Missed Target Rate | False Alarm Rate |
| Four-Feature Data<br>4-5-5-2 | 89.7% | 10.0% | 10.6% | 87.9% | 11.6% | 12.4% |
| Raw Data (10x10)<br>100-200-100-2 | 97.9% | 3.2% | 1.0% | 82.6% | 18.0% | 16.8% |
| Raw Data (8x8)<br>64-64-64-2 | 94.1% | 10.2% | 1.6% | 86.8% | 17.6% | 8.8% |
| 5x5 MCF from 10x10 window<br>25-50-50-2 | 96.9% | 4.2% | 1.8% | 88.4% | 8.8% | 14.4% |
| 5x5 MCF from 15x15 window<br>25-50-50-2 | 96.4% | 5.4% | 1.8% | 91.6% | 9.6% | 7.2% |
| 5x5 MCF from 20x20 window<br>25-50-50-2 | 97.5% | 2.4% | 2.6% | 92.8% | 4.0% | 10.4% |
| Receptive Field on 10x10 Raw Data<br>100-81-25-2 | 80.3% | 30.6% | 8.8% | 78.4% | 32.0% | 11.2% |
| Receptive Field on 5x5 MCF from 10x10<br>25-16-9-2 | 76.9% | 25.0% | 21.0% | 82.0% | 31.2% | 4.8% |

statistical feature extraction, at least for one set of features. Of course, if the images were subject to non-zero-mean noise, the MCF approach would most likely fail if unmodified, but the important conclusion here is that we should not rule out the use of raw or near-raw data when choosing input sets for neural networks, despite the drastic increase in the size of network required.

---

* The network topology field is read as follows: "4-5-5-2" indicates a "zeroth" input layer of 4 nodes (linear), a first layer of 5 neurons, a hidden layer of 5 neurons, and 2 output neurons. The zeroth layer is not a true neural network layer, and so all of these networks are three-layer networks.

Several interesting questions were also raised as to the use of raw data. Assuming that we can ignore the practical problems imposed by the implementation of large networks, is there any way in which we could improve the performance of these networks, so that they perform as well on novel data as on the training set? A variety of pre-processing algorithms could be explored, as well as modifications to the network: Pre-processing could even be integrated into the network itself.

Despite the apparent failure of the receptive field approach, the RF investigation was very precursory, and we intend to further work in this direction, and particularly to find how lateral interconnections between the RF neurons affect the net's performance. If the RF network can be improved, we gain all the benefits of less weights and potential fault tolerance described above. Recent simulations on networks trained under slightly different conditions show much better performance for this type of net than reported here.

REFERENCES

[1] R.P. Gorman and T.J. Sejnowski, "Analysis of Hidden Units in a Layered Network Trained to Classify Sonar Targets", Neural Networks, vol. 1, p. 75 - 89, Jan. 1988.

[2] S. Eberhardt, T.Duong, A. Thakoor, "A VLSI Analog Synapse 'Building-Block' Chip for Hardware Neural Network Implementations", in Proceedings of the Third Annual Parallel Processing Symposium, Fullerton, CA, March 29-31, 1989.

[3] NeuralWorks Professional II Documentation, NeuralWare, Inc., Pittsburgh, PA, 1990.

[4] M.J. LeBlanc, "Neural Networks for Automatic Target Detection in Sidescan Sonar Images", Northeastern University M.S.E.E. Thesis, *in progress*.

[5] M.J. Dzwonczyk, "Quantitative Failure Models of Feed-Forward Neural Networks", MIT S.M. Thesis, CSDL-T-1068, February 1991.

[6] H.L. Roitblat and P.W.B. Moore, "Dolphin Echolocation: Identification of Returning Echoes Using a Counterpropagation Network", IJCNN 1989, page I-295.

[7] Sidescan Sonar Record Interpretation, Klein Associates, Inc., Salem, NH, 1985.

# AN EFFECTIVE METHOD FOR VISUAL PATTERN RECOGNITION

*Ioannis N. M. Papadakis*
Decision and Control Systems Laboratory
Department of Electrical and Computer Engineering
The Pennsylvania State University
University Park, PA 16802
Email: I1P@ECL.PSU.EDU

Abstract-An effective method for visual pattern recognition is presented. It is shown that it can be successfully used for visual recognition of deformed letters. The main advantages of the presented method are its intuitive appeal, simple implementation and analytical justification.

## INTRODUCTION

The problem of visual pattern recognition has been the topic of much recent research. The results of the work of professor Fukushima et al. [1], [2] and his Neocognitron architecture constitute an impressive evidence of the capabilities of Artificial Neural Networks (ANN). The impressive capabilities of the Neocognitron architecture demonstrated by the authors of [1] are justified if one observes how the main idea on which the Neocognitron is based is implemented : There are cell planes (S-planes) that perform feature extraction and planes (C-planes) that perform blurring of the presented pattern connected alternately to form the sequence: S > C > S > C .... so that feature extraction is gradual and takes place spatially and temporally "before" the next blurring occurs. One would expect that given the Neocognitron architecture and the demonstration of its capabilities a series of applications of the Neocognitron would follow since any successful visual pattern recognition method has a vast field of real-life applications to be tested on. To the best knowledge of the author, such flourishing of applications of the Neocognitron has not taken place and one can possibly see the reasons for that. Although the main idea appears to be simple and appealing, the general implementation of the Neocognitron is not a simple task. Furthermore, it involves training of the feature extracting cells with training patterns, the "best" and "simplest" set of which, is hard to be determined for a particular problem. Finally, and perhaps more importantly in view of the previous remark, is that, if one starts with a simple training set which (almost always) will not produce the desired results, one does not have a definitive clue on how to alter the initial training set in order to improve the previous performance of the ANN.

From this standpoint, a simpler architecture that revolves around the main ideas of the Neocognitron without possibly sacrificing performance is necessary. The presented method, although motivated by the Neocognitron, possesses several distinguishing characteristics to be described in detail in the following section. For reasons that become evident in the sequel, we call the presented method the Blur Method (BM).

## THE BLUR METHOD

The BM involves two key ideas. First, opposite to the Neocognitron, it is the stored patterns (to be called "Pure Patterns" (PPs) in the sequel) that are blurred and not the presented ("Deformed") Pattern (DP). The blurring is uniform and gradual (some alternatives are currently investigated) "around" the PP. At each blur-stage a suitably defined similarity index called Product Ratio ($R_p$) is computed. Finally, a decision rule combines all $R_p$ values to make a decision associating the DP with one of the stored PPs.

All patterns are discretized to a given resolution and are therefore assumed to be binary valued (two dimensional) arrays (0/1). Each element of this array can also be thought as a tile, active (=1) or inactive (=0), depending on the value of the corresponding pixel.

Let $S_d$, $S_p$ denote the sets of **active** tiles of the DP and one of the PPs. The set $S_d \cap S_p$ represents the intersection of $S_d$ and $S_p$. Let $S_b$ be the set of the active tiles of the blurred PP at a certain blur stage. For example, initially (no blur) we have that $S_b = S_p$ and at any subsequent stage $card(S_b) \geq card(S_p)$ since the blurred pattern (denoted by BP) has always more active tiles than the initial PP. Some PPs are shown in Figure 1. The implementation of the blurring process is shown in Figure 2. Figure 2 (i) shows how the units are interconnected. Note that all interconnection weights are set equal to 1. Each unit (circle) shown in Figure 2 (i), represents the three McCulloch-Pitts neurons interconnected as shown in Figure 2 (ii). Each unit accepts (at most) four inputs which correspond to the outputs of its neighbors as shown in Fig. 2 (i). The leftmost cell in (ii) performs an OR operation and the second an AND operation with a clock pulse, common to all units, which is needed so that the blurring can be performed uniformly and in stages. The rightmost cell in (ii) is a one-bit buffer. The input $P_i$ is 1 only if the corresponding unit is an element of $S_p$. This implies that in the first stage (no-blur), only the units that correspond to the active tiles of the PP have their outputs equal to 1. After the first clock pulse, each of the active units "infects" (activates) its four neighbors so that their outputs become also equal to 1. This process is repeated until a significant amount of units (say 90%) are active. Notice that for a given resolution, say 29x29, at most 29 stages will take place.

At each stage of blurring, the following two ratios are determined :

$$R_d := \frac{\mu(S_d \cap S_b)}{\mu(S_d)} \qquad (1.1)$$

$$R_b := \frac{\mu(S_d \cap S_b)}{\mu(S_b)} \qquad (1.2)$$

where $\mu(\cdot)$ is taken to be the cardinality function card($\cdot$) although this is not necessary, i.e. a different suitably defined function may be used. The following, are two important remarks concerning $R_d$ and $R_p$ :

1) As the blurring increases, $R_d$ saturates to 1 and $R_b$ decreases to its minimum value $R_{b(min)} = \mu(S_d)/c$, where c is the total number of pixels (i.e. c = 29x29 = 841).

2) The ratio $R_d$ is a satisfactory measure of similarity between the DP and the PP which is blurred. However, as the blurring increases, the $R_d$ conveys less information and does not differ significantly between PPs for the same blur level. Some confidence should be therefore attached to this similarity measure to account for the gradual feature loss due to blurring. An intuitive suggestion is to define as a similarity measure the product of the two ratios. The resulting quantity is called product ratio $R_p$ and is given by

$$R_p := \frac{[\mu(S_d \cap S_b)]^2}{\mu(S_d)\, \mu(S_b)} \tag{1.3}$$

where $R_p$ takes values in the interval [0,1].

The values of $R_p$ for the assumed PP are computed at each blur stage and the maximum $R_{p(max)}$ is compared with the corresponding maxima of all other PPs. Finally, the overall maximum is determined and the PP associated to the presented DP is the one which this overall maximum corresponds to.

An example of visual pattern recognition using the above procedure is presented below. The stored PPs are the capital letters E, F, H, I, L, N, T, Z as shown in Figure 1. The resolution is 29x29. For each of the PPs, two DPs are presented with the second DP always more deformed than the first one, so that information about the sensitivity of the recognition scheme may be obtained. The DPs are denoted with their corresponding PP along with their number separated by an underscore, i.e. E_2 represents the most deformed of the DPs corresponding to the PP E. The plots of the values or $R_p$ as a function of the blur level (equivalent to blur stage) ranging from 0 (no blur) to 29 (100% blur) for the DPs are not shown since the curves for each DP are extremely clustered. In general they have a bell shaped form with a clear overall maximum. Since the decision scheme described above involves only the maxima of these curves, the results may be presented more clearly as shown in Figure 3. In Figure 3, for each DP we only plot the maximum value of $R_p$ identified with a letter corresponding to the associated PP. The DP is shown in the box for each case. All DPs were succesfully recognized. The DPs shown in Figure 3 correspond to E_2, F_2, L_1, L_2, T_2, and Z_2. The recognition of a particular DP is successful if the maximum value of $R_p$ corresponds to the correct PP. For example, in the upper left case of Figure 3, i.e. the case for E_2, it is clear that the overall maximum corresponds to the PP E. It is "followed" by F, Z, L, N, M etc. Notice that although some of the letters overlap, one can deductively find the position of each letter on the plots. As it is clear from Figure 3, all presented DPs are successfully recognized. The "hardest" DP case seems to be Z_2 (last plot of Figure 3), however the maximum clearly corresponds to the PP Z. Note that the scale of $R_p$ is kept the same for all cases although the range of values of $R_p$ varies between them.

In the following section, an analytical justification of the BM is presented.

**ANALYSIS**

In this section, we justify the use of the overall maximum $R_M$ given by

$$R_M := \max_{PP}(R_{p(max)}) \qquad (2.1)$$

as a measure of the similarity between the presented DP and the PPs. Let $M_d$, $M_p$, $M_b$ denote the binary (0/1) matrices of a DP, a PP and a BP (at a certain blur stage) with elements $m_{ij}^d$, $m_{ij}^p$, $m_{ij}^b$ respectively. From (1.3), $R_p$ can be expressed as

$$R_p = \frac{\left[\sum_{i=1}^{r}\sum_{j=1}^{r} m_{ij}^d\, m_{ij}^b\right]^2}{\left[\sum_{i=1}^{r}\sum_{j=1}^{r} m_{ij}^d\right]\left[\sum_{i=1}^{r}\sum_{j=1}^{r} m_{ij}^b\right]} \qquad (2.2)$$

where r denotes the resolution used. Using more compact notation one may rewrite (2.2) as

$$R_p = \frac{<M_d,M_b>^2}{<M_d,M_d><M_b,M_b>} \qquad (2.3)$$

where $<X,Y>$ is defined by

$$<X,Y> := \sum_{i,j} x_{ij} y_{ij} \qquad (2.4)$$

for any two binary matrices X, Y of the same dimensions. In the following, it is shown that $<X,Y>$ is a valid inner product.

Let $\mathcal{X}_{mn}$ be the vector space of all complex mxn matrices over $\mathbb{C}$. For any $X, Y \in \mathcal{X}_{mn}$, we define a mapping $(X,Y) \longrightarrow <X,Y>$ from $\mathcal{X}_{mn} \times \mathcal{X}_{mn}$ to $\mathbb{C}$ such that

$$<X,Y> := \sum_{i,j} x_{ij} y_{ij}^* \qquad (2.5)$$

where * stands for complex conjugate. For any X, Y, Z in $\mathscr{X}_{mn}^*$ and a, b in $\mathbb{C}$, it is not difficult to establish the following:
1) $<aX + bY, Z> = a_*<X,Z> + b<Y,Z>$.
2) $<X,Y> = <Y,X>^*$.
3) $<X,X> \in (0,\infty)$ for all nonzero $X \in \mathscr{X}_{mn}^*$.

and therefore [3], $<X,Y>$ defined by (2.5) is a valid inner product in $\mathscr{X}_{mn}^*$. Furthermore, for real (and therefore binary as well) matrices, eq. (2.4) is an equivalent expression for $<X,Y>$ given by (2.5).

Let $\mathscr{B}_{mn} \subset \mathscr{X}_{mn}$ be the space of binary mxn matrices and assume that the resolution is taken equal to mxn. From (2.3) and the Schwartz inequality we readily conclude that

$$R_b \leq 1 \qquad (2.6)$$

with the equality satisfied if and only if

$$M_b = \lambda M_d \qquad (2.7)$$

which in $\mathscr{B}_{mn}$ accepts only one solution, i.e. $\lambda = 1$. This would imply that at some blur stage, the DP is identical to the PP something that is almost always not true. Therefore, the next best strategy appears to be the choice of the PP corresponding to the maximum value of $R_p$, since that would automatically make $R_p$ to be as close to 1 as possible, given the constrain that it is only a small subset of $\mathscr{B}_{mn}$ in which the scheme searches for the maximum $R_p$.

From the preceding analysis, we observe that the PP corresponding to the overall maximum value of $R_p$ is "more similar" to the presented DP in the sense explained above which justifies the effectiveness of the BM method.

## CONCLUSION

An effective method for visual pattern recognition was presented. It is shown that it can successfully be applied to visual recognition of letters with some degree of deformation. Its intuitive appeal was justified by a vector space approach, in which the Schwartz inequality plays a central role.

**REFERENCES**

[1] K. Fukushima, S. Miyake and T. Ito, "Neocognitron: A Neural Network Model for a Mechanism of Visual Pattern Recognition," IEEE Transactions on Systems, Man, and Cybernetics, Vol. 13 (5), September/October 1983, pp. 826-34.

[2] K. Fukushima, "Analysis of the Process of Visual Pattern Recognition," Neural Networks, Vol. 2, pp. 413-20, 1989.

[3] G. B. Folland, Real Analysis, J. Wiley & Sons, New York, 1984.

**Figure 1.** The 9 Pure Patterns (PPs) used, on a 29x29 resolution.

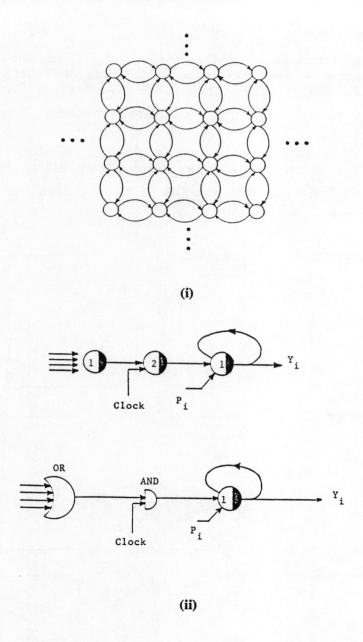

**Figure 2.** (i) Descriptive structure of interconnected units for implementation of gradual uniform blurring ; (ii) Detailed configuration of each unit.

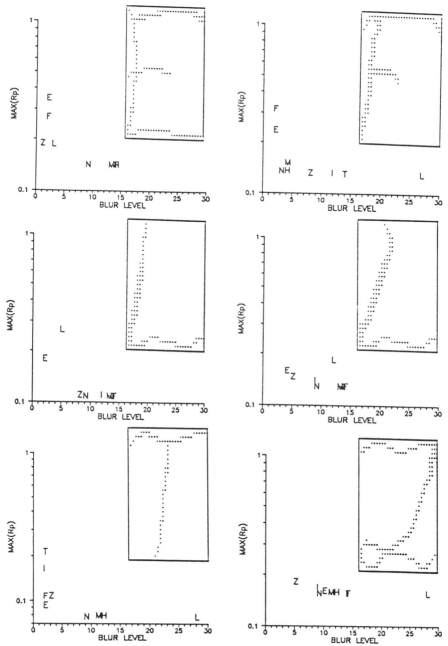

**Figure 3.** Examples of letter recognition. The Deformed Pattern (DP) is shown in the box and each of the small letters corresponds to one of the PPs. The recognition is successful if the maximum value of $R_p$ corresponds to the correct PP.

# FINGERPRINT RECOGNITION USING NEURAL NETWORK

W. F. Leung, S. H. Leung, W. H. Lau and Andrew Luk

Department of Electronic Engineering
City Polytechnic of Hong Kong
83 Tat Chee Avenue
Hong Kong

## ABSTRACT

This paper describes a neural network based approach for automated fingerprint recognition. Minutiae are extracted from the fingerprint image via a multilayer perceptron (MLP) classifier with one hidden layer. The backpropagation learning technique is used for its training. Selected features are represented in a special way such that they are simultaneously invariant under shift, rotation and scaling. Simulation results are obtained with good detection ratio and low failure rate. The proposed method is found to be reliable for system with a small set of fingerprint data.

## 1. INTRODUCTION

Fingerprints are imprints formed by friction ridges of the skin in fingers and thumbs. They have long been used for identification because of their immutability and individuality. Immutability refers to the permanent and unchanging character of the pattern on each finger, from before birth until decomposition after death. Individuality refers to the uniqueness of ridge details across individuals, the probability that two fingerprints are alike is about 1 in $1.9 \times 10^{15}$.

The use of computers in fingerprint recognition is highly desirable in many applications, such as building or area security and police work to identify criminals. Recently, automated fingerprint classification techniques have been investigated [1][2]. The most prevalent current model for automated fingerprint identification systems is the Minutiae-Coordinate model (also called the FBI representation of prints). Most commercial systems are based on the basic FBI model. Classification is usually performed by locating the positions of certain distinctive features, known as minutiae, which are the beginnings and endings of ridges or forks (bifurcations) in the ridge lines of the print. Current fingerprint recognition systems are sensitive to errors in the positioning of prints at acquisition, and hence require to position the prints correctly for coordinates calculation or reference line placement. Our proposed method is based on a data model for fingerprints that is structural rather than coordinate. This structural data model is robust with respect to translation, rotation, and distortions.

Most of the pattern recognition systems are composed of four building blocks (see Figure 1). The first step is image acquisition, i.e. converting a scene into an array of numbers that can be manipulated by the computer. The second part is preprocessing, which involves removing noise, enhancing the picture, and, if necessary, segmenting the image into meaningful regions to be analyzed separately. The third phase is feature extraction, whereby the image is represented by a set of numerical "features" to remove redundancy from the data and reduce its dimension. The fourth building block is classification. This is the last stage of an image recognition system, where a class label is assigned to the unknown image/object by examining its extracted features and comparing them with class representations that the classifier has learned during its training stage.

The main focus of this paper is on the feature extraction and classification stages. Neural networks enable solutions to be found to problems where algorithmic methods are too computationally intensive or do not exist. They also offer significant speed advantages over conventional techniques. The problems of feature extraction and classification therefore seem to be a suitable application for neural nets. In this paper we have concentrated on the multilayer perceptron network using backpropagation learning technique [3]. This combination is the simplest and most widely used neural network. In the next section, we will discuss the configuration of the preprocessing system. Sections 3 and 4 describe in more detail the feature extraction and recognition procedures involved and the difficulties encountered. Some results and discussions will be presented in section 5 and the paper will be summarized with a brief conclusion section.

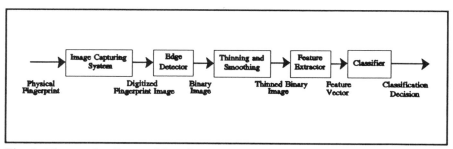

Fig. 1  Fingerprint Recognition System

## 2. PREPROCESSING SYSTEM

A 'clean' fingerprint image should be obtained to facilitate recognition, therefore methods of simple, standard image processing techniques [4] for edge enhancement and binarization or fingerprint filter design [5][6] are required. The first phase of our work is to capture the fingerprint image and convert it to a digital representation with a resolution of 512x512 by 256 gray levels. It is then subsampled and a valid region is selected based on visual inspection. Histogram equalization technique is sometimes used to increase the contrast if the illumination condition is poor. Although the original image is in gray-scale, we are only interested in binary information, i.e. the foreground fingerprint ridges and the background valleys. Therefore binarization is usually performed by using a Laplacian edge detection operator followed by a thresholding technique. The binary image is further enhanced by a thinning algorithm (modified from [7]) which reduces the image ridges to a skeletal structure with thickness of only one pixel and without changing the connectivity of the print. This algorithm is of the "banana peeling" type, where contour points are peeled off according to certain rules.

After obtaining the binary form of the fingerprint image, there may be some irregularities in the print. This is because of the imperfections of the fingerprint such as ridge gaps, which are usually caused by: (a) skinfolds and contiguous ridges, or (b) the spreading of ink due to finger pressure, or (c) in the worse cases by excessive inking or by smearing during rolling of the finger. Unfortunately, these irregularities cannot be avoided. To remedy this problem, smoothing of the binary image is necessary. The smoothing operations include: (a) filling holes, (b) deleting redundant points, (c) removing noisy points, and (d) filling potential missing points. For more details, see Ref. [8].

## 3. FEATURE EXTRACTION AND SELECTION

Extraction of appropriate features is one of the most important tasks for a recognition system. Because it is impractical to match a given input image or image representation with all the image templates stored in the system, it is necessary to find a compact set of features which can represent much of the useful information (in the sense of discriminability) present in the original data. Selection of "good" features is a crucial step in the process since the next stage sees only these features and acts upon them. "Good" features are those satisfying two requirements: (i) small intraclass invariance -- slightly different shapes with similar general characteristics should have numerically close values, and (ii) large interclass separation -- features from different classes should be quite different numerically. The next phase of our work involves the extraction of features from the thinned binary image. A multilayer perceptron network of three layers is trained to detect the minutiae in the thinned print image of size 128x128. The first layer of the network has nine units associated with the components of the input vector. The hidden layer has five units and the output layer has one unit corresponding to the number of the classes. The network is trained to output a '1' when the input window is centred on the feature to be located and it outputs a '0' if minutiae are not present. Figure 3 shows the initial training patterns which are composed of 16 samples of bifurcations in eight different orientations and 36 samples of non-bifurcations. The network is trained by using the backpropagation learning technique as described by Rumelhart et al [3] with the weight change is updated according to the equation,

$$\Delta w_{ij}(n) = -\eta \frac{\partial E}{\partial w_{ij}} + \alpha \Delta w_{ij}(n-1)$$

where  $E$  is the energy function which is defined as the sum of the square difference between the desired output response and the actual output response of each training example,

$w_{ij}$  is the connection weight between unit i and unit j of the network,

$\Delta w_{ij}(n)$  is the weight change in the $n^{th}$ cycle,

$\eta$  is the learning rate,

and  $\alpha$  is the momentum term.

In all experiments, a learning rate of 0.3 and a momentum of 0.9 are used. The training procedure is performed on a 80386-based PC run at 25MHz with a floating point coprocessor. It takes about 30 seconds for the network to achieve satisfactory convergence.

The trained network is then used to analyze the complete image by raster scanning the fingerprint via window of size 3x3. The network is proved to be very effective at identifying the positions of the minutiae and gives only few false responses. These false responses mainly come from the blurred areas in the original image where the signal level is low. In order to prevent the falsely reported features and select "significant" minutiae, two more rules are added to the system to guarantee perfect ridge forks are detected while excluding all other features. They are: (i) at those potential minutiae feature points, we re-examine them by increasing the window size to 5x5, and (ii) if two or more minutiae are too close together, we ignore all of them. Figure 2 shows the distribution of minutiae on two identical fingerprints (a) before and (b) after applying the rules. Note that on the left hand side of Fig. 2(a), there are a lot of feature points that are identified by the network but they are too closely linked together to be of significance in classification. When the above two rules are applied to the extracted features, these points are correctly removed. The network is then further tested on a number of fingerprint images and the important minutiae are also correctly selected.

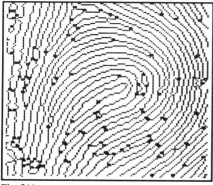

Fig. 2(a)

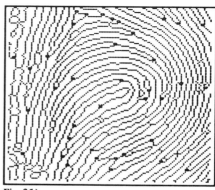

Fig. 2(b)

## 4. INVARIANT RECOGNITION

An important issue in applying neural networks to image recognition is the representation of feature data as input to the network. Additionally, a flexible recognition system must be able to recognize an object regardless of its orientation, size, and location in the field of view. This requirement is similar to the translation-, rotation-, and scale-invariancy properties for the extracted features. The location of a reference point (or a centre point) of the fingerprint is important for invariant recognition and has to be determined. To accomplish this, an approach of contour tracing [4] is used to find one or more turning points (i.e. points with maximum rate of change of tracing movement). These points are then used to find the reference point of the fingerprint.

The Euclidean distances $d(i)$ from each feature point i to the centre point are calculated. The referencing of the distance data to the centre point confers the property of positional invariance. The data is now sorted in ascending order from $d(0)$ to $d(N)$ and this operation gives the data the property of rotational invariance. In order to make the data becomes invariant to scale change, it is normalized to unity by the shortest distance $d(0)$, i.e. $d_{norm}(i) = d(0)/d(i)$, $i = 0..N$. This normalization will weight those feature points nearer to the centre more heavily because these points are usually more significant in classification. On the other hand, feature points at the borders are usually of poor quality or more noisy (due to excessive inking or smearing when obtaining the physical fingerprint) and, therefore, should be weighted less. Consequently the centroidal data patterns output in the form of amplitude spectrum should be shift, scale and rotation independent. Also the invariant feature vectors are in the range [0,1], they can be directly used as the training/stored vectors in the MLP classifier.

## 5. RESULTS AND DISCUSSIONS

Of the 30 fingerprints that we have collected so far, 10 were used to train the network. Using the method we proposed in sections 3 and 4, all the fingerprints can be recognized correctly. The invariant amplitude spectra of the first four fingerprints (shown in Figs. 5 and 6) are shown in Figure 4. It can be seen that the discrimination of prints is easy. Figures 5(a)-(f) show six digitized fingerprint images and Figures 6(a)-(f) show their corresponding thinned binary images with "significant" minutiae are marked by 'o' and the centre point is marked by an 'X'. Notice that Figures 5(a) and (b) are the same fingerprints with different positions while Figures 5(c) and (d) are of different types of fingerprints.

The recognition rate of fingerprints depends much on the quality of fingerprints and effectiveness of the preprocessing system, such as the threshold level used in edge detection which is a subjective issue. Also if there are too many broken lines or noisy points in the image, the preprocessing system of contour tracing may fail. Thus an intelligent connection algorithm to recover broken lines and suppress spurious irregularities is necessary.

## 6. CONCLUSIONS

We have presented a fingerprint recognition system that uses local features to identify fingerprints. Preprocessing techniques are first applied to produce a clean, thinned binary fingerprint ridge structure which is ready for feature extraction. Backpropagation-trained neural network are successfully used to extract the minutiae in the print. Good features are then selected and represented in a simple data model which allows correct recognition of prints even in the presence of positioning errors. The use of neural network supports fast classification rate and saves the user from spending enormous amount of time to derive rule-based databases for matching.

## 7. ACKNOWLEDGMENT

This research work was performed with support from the City Polytechnic of Hong Kong Strategic Indicated Grant #700-073.

## 8. REFERENCES

[1] A. K. Hrechak and J. A. McHugh, "Automated Fingerprint Recognition Using Structural Matching," Pattern Recognition, vol. 23, no. 8, pp. 893-904, 1990.

[2] K. Rao and K. Black, "Type Classification Of Fingerprints: A Syntactic Approach," IEEE Trans. Pattern Anal. Mach. Intell., PAMI-2, pp. 223-231, 1980.

[3] D. E. Rumelhart, G. E. Hinton and R. J. Williams, "Learning Internal Representations by Error Propagation," chapter 8 of vol. 1 in Parallel Distributed Processing, MIT Press, Cambridge, Mass., USA, 1986.

[4] A. Luk, S. H. Leung, C. K. Lee and W. H. Lau, "A Two-level Classifier For Fingerprint Recognition," in Proc. IEEE 1991 International Symposium on CAS, Singapore, June 11-14, 1991, pp. 2625-2628.

[5] L. O'Gorman and J. V. Nickerson, "An Approach To Fingerprint Filter Design," Pattern Recognition, vol. 22, no. 1, pp. 29-38, 1989.

[6] M. T. Leung, W. E. Engeler and P. Frank, "Finger Image Processing Using Neural Network," in Proc. of IEEE Region 10 Conference on Comp. and Commun. Sys., Hong Kong, 1990, pp. 582-586.

[7] G. Feigin and N. Ben-Yosef, "Line Thinning Algorithm," in Proc. of SPIE : Applications of Digital Image Processing, vol. 397, Switzerland, 19-22 April 1983, pp. 108-112.

[8] A. Luk, S. H. Leung, W. K. Suen and C. Y. Cheng, "A Low Cost Fingerprint Recognition Subsystem For Security Systems," in Proc. of International AMSE Conf. on Signals and Systems, Brighton, UK, AMSE Press, vol. 1, 1989, pp. 221-232.

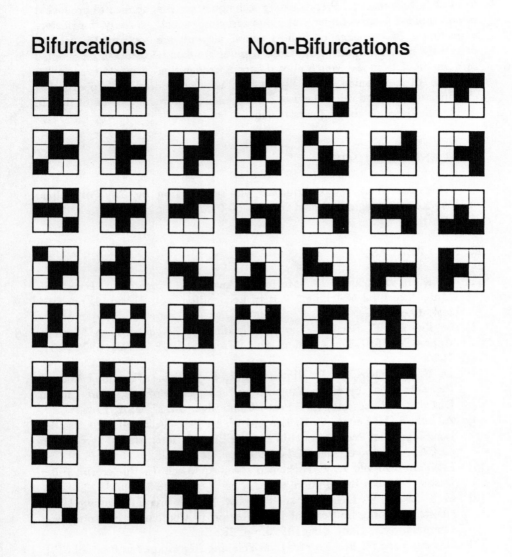

Fig. 3  16 bifurcated and 32 non-bifurcated training samples

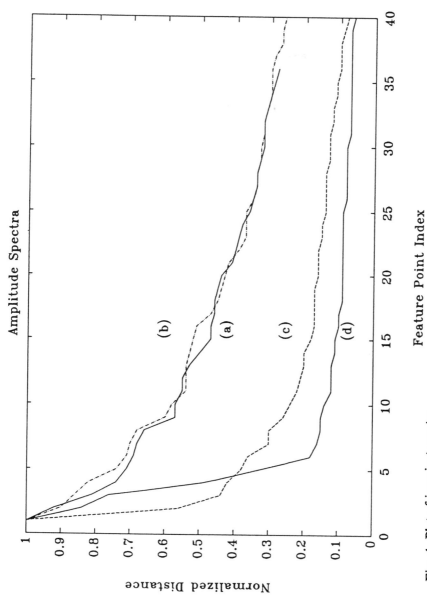

Fig. 4 Plot of invariant spectra

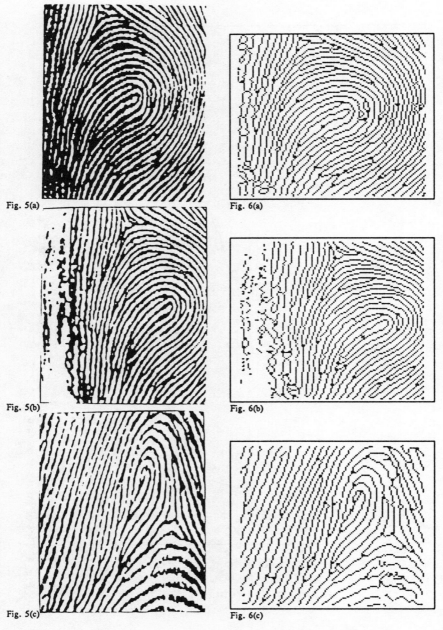

Fig. 5(a)

Fig. 6(a)

Fig. 5(b)

Fig. 6(b)

Fig. 5(c)

Fig. 6(c)

Fig. 5 Original gray-level fingerprints

Fig. 6 Feature points extracted from thinned prints

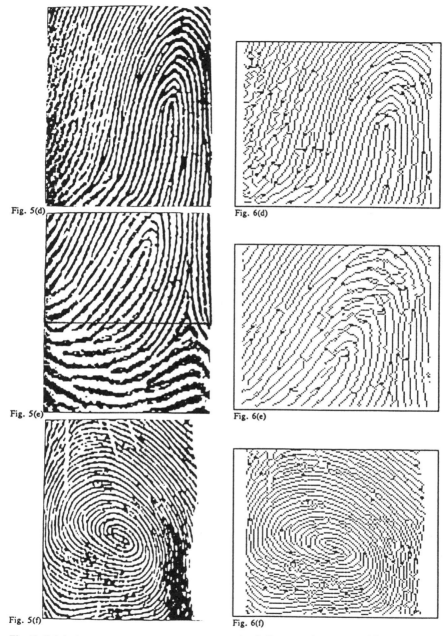

Fig. 5(d)  Fig. 6(d)

Fig. 5(e)  Fig. 6(e)

Fig. 5(f)  Fig. 6(f)

Fig. 5 Original gray-level fingerprints

Fig. 6 Feature points extracted from thinned prints

# A Comparison of Second-order Neural Networks to Transform-based Methods for Translation- and Orientation-Invariant Object Recognition

Russ Duren
General Dynamics Corporation
Fort Worth Division
Mail Zone 1527
P.O. Box 748
Fort Worth, TX 76101

Behrouz Peikari
Electrical Engineering Department
School of Engineering and Applied Science
Southern Methodist University
Dallas, TX 75275

Abstract - Neural networks can use second-order neurons to obtain invariance to translations in the input pattern. Alternatively transform methods can be used to obtain translation invariance before classification by a neural network. This paper compares the use of second-order neurons to various translation-invariant transforms. The mapping properties of second-order neurons are compared to those of the general class of fast translation-invariant transforms introduced by Wagh and Kanetkar and to the power spectra of the Walsh-Hadamard and discrete Fourier transforms. A fast transformation based on the use of higher-order correlations is introduced. Three theorems are proven concerning the ability of various methods to discriminate between similar patterns. Second-order neurons are shown to have several advantages over the transform methods. Experimental results are presented that corroborate the theory.

## INTRODUCTION

Pattern recognition problems frequently require the use of features that do not vary when the input pattern undergoes some set of transformations. Typical transformations include translation, rotation and changes in scale. This paper examines techniques that provide invariance to translation. Invariance to orientation is obtained by combining translation invariance with polar sampling.

Various methods can be used to obtain translation invariance. One set of methods is based on the use of transforms. One example of this group makes use of the discrete Fourier transform (DFT) [1]. In this method the magnitude of the DFT of a pattern is used as the input to a neural network. The magnitude of the DFT is invariant to cyclic shifts of the pattern. A similar technique is based on the power spectrum of the Hadamard-ordered Walsh-Hadamard transform [2]. Wagh and Kanetkar have defined a general class of fast translation-invariant transforms [3]. Members of this class of transforms yield values that are translation invariant without requiring a separate magnitude calculation. The

R-transform is one member of this class that has been used by previous researchers [4].

Within the field of neural networks several additional methods have been used to obtain translation invariance. One such method entails learning invariance by training on examples that have undergone a representative set of transformations [5]. Another method involves building invariance into the network. This can be accomplished using averaging techniques such as employed by Rumelhart, et al. [6]. Alternatively, Giles and Maxwell suggest using higher-order neurons to build invariance into neural networks [7].

This paper compares the use of higher-order neurons to various transform-based methods. It begins by examining the use of second-order neurons for translation-invariant feature extraction. This is followed by a brief review of the general class of fast translation-invariant transforms. Next, the discrimination power of the transform-based methods (i.e., their ability to distinguish between similar patterns) is compared to that of second-order neurons. Three theorems concerning discrimination power are included as part of the comparison. The comparison concludes with a discussion of computational issues. Finally, the results of several experiments are presented.

## SECOND-ORDER NEURONS

The most common neural network architectures use first-order neurons. A first-order neuron receives two or more inputs, multiplies each input by a weight, sums the weighted values, and then optionally passes the sum through some form of thresholding or squashing function known as an activation function. Mathematically, the operation of a first-order neuron is given by

$$o = f\left(\sum_{i=0}^{N-1} w_i x_i - \theta\right) \qquad (1)$$

where o is the output of the neuron, $f(\cdot)$ is the activation function for the neuron, N is the number of inputs to the neuron, $w_i$ is the weight connecting input $x_i$ to the neuron, and $\theta$ is a trainable threshold.

As defined by Giles and Maxwell, second-order neurons calculate a sum of weighted products of inputs taken two at a time [7]. The general equation for a second-order neuron is

$$o = f\left(\sum_{i=0}^{N-1} w_i x_i + \sum_{j=0}^{N-1} \sum_{k=0}^{N-1} w_{j,k} x_j x_k - \theta\right) \qquad (2)$$

where $w_{j,k}$ is the weight corresponding to the product of two inputs to the neuron, $x_j$ and $x_k$, and the other terms are similar to those of (1).

A second-order neuron can be made translation invariant by constraining the weights to be a function of the distance between the inputs. When this is done the weights can be factored out of the sums. The resulting operation is given by

$$o = f\left( w_0 \sum_{i=0}^{N-1} x_i + \sum_{d=1}^{N-1} w_d \sum_{j=0}^{N-1-d} x_j x_{j+d} - \theta \right) \quad (3)$$

where $x_j$ and $x_{j+d}$, are two inputs separated by distance d and the lower limit on the second summation has been set to eliminate those terms that involve the product of an input with itself.

For the comparisons presented in this paper it will be convenient to model the second-order neuron of (3) with a first-order neuron. This can be done by treating the first and last summations as inputs to a first-order neuron. The resulting first-order neuron is defined by

$$o = f\left( \sum_{d=0}^{N-1} w_d y_d - \theta \right) \quad (4)$$

where the $y_d$ terms represent the inputs to the first-order model. They are given by

$$y_d = \begin{cases} \sum_{j=0}^{N-1} x_j & ; d = 0 \\ \sum_{j=0}^{N-1-d} x_j x_{j+d} & ; d = 1, 2, \cdots, N-1. \end{cases} \quad (5)$$

Two types of translation-invariant second-order neurons can be implemented. They will be referred to as linearly-invariant neurons and cyclicly-invariant neurons. The distinction between these two types of neurons can be demonstrated with a binary sequence of five digits. Linearly-invariant neurons classify the binary sequences 00011, 00110, 01100 and 11000 into one group, but not in the same group as the sequence 10001. Cyclicly-invariant neurons classify all five sequences into the same group.

Equations (4) and (5) define a linearly-invariant neuron. To obtain a cyclicly-invariant neuron (5) must be replaced by one of two equations. The equation used depends upon whether the number of inputs to the second-order neuron is an even number or an odd number. Equation (6) is used if N is an even number. Equation (7) is used for odd N. The upper limit of the sum in (4) must also be modified to correspond to the number of $y_d$ terms defined by the appropriate equation.

The first-order neuron model provides a direct comparison to transform-based methods for obtaining translation invariance. The values defined by (5), (6) and (7) can be compared directly to the values of the power spectrum of the DFT or

Hadamard transform and to the values produced by any member of the general class of fast translation-invariant transforms.

$$y_d = \begin{cases} \sum_{j=0}^{N-1} x_j & ; d = 0 \\ \sum_{j=0}^{N-1} x_j \, x_{((j+d) \bmod N)} & ; d = 1, 2, \dots, \dfrac{N}{2} - 1 \\ \sum_{j=0}^{\frac{N}{2}-1} x_j \, x_{(j + \frac{N}{2})} & ; d = \dfrac{N}{2} \end{cases} \qquad (6)$$

$$y_d = \begin{cases} \sum_{j=0}^{N-1} x_j & ; d = 0 \\ \sum_{j=0}^{N-1} x_j \, x_{((j+d) \bmod N)} & ; d = 1, 2, \dots, \dfrac{N-1}{2} \end{cases} \qquad (7)$$

## THE GENERAL CLASS OF FAST TRANSLATION-INVARIANT TRANSFORMS

The algorithm for the general class of fast translation-invariant transforms is shown in Figure 1 [3]. The exact transform implemented by the algorithm of Figure 1 depends upon the functions chosen for $f_1$ and $f_2$. The R-transform is obtained when $f_1 = a_1 + a_2$ and $f_2 = |a_1 - a_2|$.

As will be discussed in a later section, use of the second-order neurons can be computationally expensive for large N. This is because the values defined by (5), (6) and (7) cannot be computed using a fast transformation technique. The mathematical operation of multiplying two input terms together is known as a second-order correlation. If combinations of higher-order correlations are allowed, a fast translation-invariant transform can be defined by choosing the functions $f_1 = a_1 + a_2$ and $f_2 = a_1 \times a_2$. The resulting transform will be called the *fast correlation transform*. This transform will be examined as a possible substitute for (5) through (7).

## MAPPING PROPERTIES OF TRANSLATION INVARIANT METHODS

As discussed previously, second-order neurons can be either linearly invariant or cyclicly invariant. A linearly-invariant neuron is more powerful in the sense that

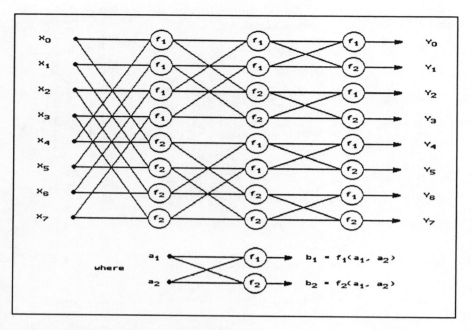

Figure 1. The general class of fast translation-invariant transforms, where $f_1$ and $f_2$ can be any commutative functions of two inputs.

it can distinguish more sequences than a cyclicly-invariant neuron.

The Fourier magnitude technique also can be used for either type of invariance. When an N-point Fourier transform is applied to an N-point sequence the resulting magnitude is cyclicly invariant. A linearly invariant system can be obtained by appending N zeros to the original input sequence. The desired features are the magnitude values of the resulting 2N-point Fourier transform.

Burkhardt and Müller have described the mapping properties of the general class of fast translation-invariant transforms in detail [8]. They demonstrated that all members of this class of transforms are invariant to wide variety of permutations of the input sequence. The permutations include cyclic shifts, dyadic shifts and the set of interchanges of the $i$th element of the first N/2 members of the input sequence with the corresponding element of the last N/2 members. The following theorems compare the mapping properties of second-order neurons to those of the various transform-based methods.

**Theorem 1:** No member of the general class of fast translation-invariant transforms can be made linearly invariant by appending a constant-valued sequence to the original input sequence.

**Proof:** The proof of Theorem 1 can be seen by examining the flow graph of Figure 2. Let the vector notation $x$ represent the sequence $[x_0, x_1, \ldots, x_{N-1}]^T$. Let $x$ be the original N-point input. Let $c$ be an N-point vector with every member

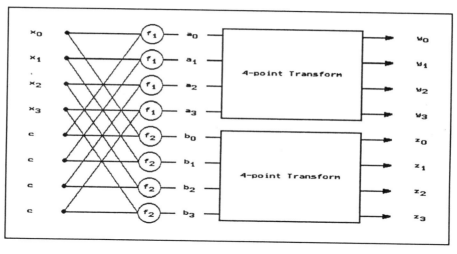

Figure 2. A general class transformation of the 8-point sequence generated by appending a 4-point constant valued sequence to the 4-point sequence $x$.

having the same value. The input vector to the 2N-point transform is the column concatenation of $x$ with $c$. The output of the first stage of the transform is seen to be the column concatenation of two vectors $a$ and $b$ where $a = f_1(x, c)$ and $b = f_2(x, c)$. Since $c$ is constant, any permutation in the vector $x$ will result in identical permutations in the vectors $a$ and $b$. The output of the last stage of the transform is seen to be the column concatenation of two vectors $y$ and $z$ that are N-point transforms of the vectors $a$ and $b$ respectively. Each of these N-point transforms are identical in form to the original N-point transform. Therefore, both $y$ and $z$ inherit the same invariances to permutations in $x$ as an ordinary N-point transform of $x$.

**Theorem 2:** The power spectrum of the Walsh-Hadamard transform cannot be made linearly invariant by appending a constant-valued sequence to the original input sequence.

**Proof:** Burkhardt and Müller demonstrated that the Walsh-Hadamard power spectrum can be mapped onto the flow graph of Figure 1 [8]. Therefore Theorem 2 follows as a result of Theorem 1.

Together these theorems yield the primary result of this paper: Both the magnitude of the discrete Fourier transform and second-order neurons can be used as linearly-invariant systems and therefore can discriminate between more patterns than both the power spectrum of the Walsh-Hadamard transform and all members of the general class of fast translation-invariant transforms, which are limited to cyclicly-invariant systems.

Within the general class of fast translation-invariant transforms the discrimination power of the transform depends upon the functions used for $f_1$ and $f_2$. Theorem 3 addresses the discrimination power of the fast correlation transform.

**Theorem 3:** The fast correlation transformation has a greater discrimination power than the R-transform for real-valued input sequences.

**Proof:** Referring to the butterfly calculation of Figure 1, the operation of the R transform can be defined as

$$b_1 = a_1 + a_2 \quad \text{and} \tag{8}$$
$$b_2 = |a_1 - a_2|.$$

Following the method of Burkhardt and Müller [8], these equations can be rearranged as

$$a_1 = \frac{1}{2}(b_1 \pm b_2) \quad \text{and} \tag{9}$$
$$a_2 = \frac{1}{2}(b_1 \mp b_2)$$

which result in real values for any real $b_1$ and $b_2$. In a similar manner, the operation of the fast correlation transform can be defined as

$$b_1 = a_1 + a_2 \quad \text{and} \tag{10}$$
$$b_2 = a_1 \times a_2.$$

These equations can be rearranged as

$$a_1 = \frac{b_1}{2} \pm \sqrt{\left(\frac{b_1}{2}\right)^2 - b_2} \quad \text{and} \tag{11}$$
$$a_2 = \frac{b_1}{2} \mp \sqrt{\left(\frac{b_1}{2}\right)^2 - b_2}$$

which result in real values if and only if

$$\left(\frac{b_1}{2}\right)^2 \geq b_2. \tag{12}$$

Clearly the R-transform always has multiple real sequences with the same transform. On the other hand, the fast correlation transform does not. Therefore the fast correlation transform has a greater discrimination power than the

R-transform for real input sequences. It should be noted, however, that Theorem 1 demonstrates the fast correlation transform is not as powerful as linearly-invariant second-order neurons.

Theorem 3 can be demonstrated by a simple example. The R-transform of the sequence $[3, 6, 7, 5]^T$ is $[21, 1, 5, 3]^T$. Using (9) it is found that the sequence $[7.5, 5.5, 3.5, 4.5]^T$ has the same transform. The fast correlation transform of $[3, 6, 7, 5]^T$ is $[21, 110, 51, 630]^T$. Using (11) it is found that no other real-valued sequence has the same transform. Instead, it is found that the complex sequence $[5 + j2.24, 8.54, 5 - j2.24, 2.46]^T$ has the same fast correlation transform as $[3, 6, 7, 5]^T$.

## COMPUTATIONAL CONSIDERATIONS

Given that both the Fourier magnitude used in combination with first-order neurons and second-order neurons are of similar discrimination power, the choice of which method to use for a particular application may depend on computational complexity. In this section the computational complexity of second-order neurons will be compared to that of the various transform-based methods.

By examining (5), (6) and (7) it can be seen that, neglecting weights, both linearly- and cyclicly-invariant second-order neurons require the same number of multiplications and additions. From (6) it can be seen that, for even values of N, calculation of the input product terms requires $N(N/2)$ real multiplications and additions. For odd values of N this becomes $N(N-1)/2$. A discrete Fourier transform requires $N^2$ complex multiplications and additions. Therefore second-order neurons are always more efficient than an ordinary discrete Fourier transform. The fast Fourier transform, or FFT, requires $Nlog_2N$ complex multiplications and additions. It is easy to verify that the FFT requires fewer operations than second-order neurons for $N \geq 128$ for a linearly-invariant system.

One other point should be made when comparing the FFT to second-order neurons. The FFT can only be applied when the number of inputs is an integer power of two. Second-order neurons can be used with an arbitrary number of inputs. Depending upon the application this may change which method is most desirable.

For two types of input data second-order neurons have additional advantages over the FFT. If the application involves binary valued inputs, all the input products for the second-order neurons can be realized as logical ANDs of the inputs. If the inputs are integer valued then the second-order neurons can use integer mathematics. The FFT requires complex mathematics regardless of the type of input data.

The fast correlation transform was defined in an attempt to reduce the computational requirements of second-order neurons. It accomplishes this but not without several drawbacks. One problem is that it loses the ability to implement a linearly-invariant system.

Another problem involves the dynamic range of the transform outputs. By examining Figure 1 it can be seen that the last term of the transform, $Y_{N-1}$, is the joint product of all N inputs. This can quickly lead to problems. The example at the end of the previous section demonstrates this. The input sequence has four terms, the largest of which is 7. The last term in the transformed sequence is $3*6*7*5 = 630$, which is much larger than any of the inputs. This effectively limits the use of the fast correlation transform to binary sequences.

## EXPERIMENTAL CORROBORATION

The difference in discrimination power of the various methods can be demonstrated by a simple experiment. Consider the four-bit sequences corresponding to the binary values for the numbers 0 through 15. Cyclicly-invariant neurons, the power spectrum of an N-point DFT or of the Hadamard transform, the R-transform and the fast correlation transform results in the classification into the same six groups. Linearly-invariant neurons and a 2N-point DFT spectrum classify these sequences into eight groups.

In previous work a handwritten digit recognition system was presented that used translation-invariant second-order neurons in combination with semicircular sample paths to obtain invariance to rotations of plus or minus 90 degrees [9], [10]. The system used linearly-invariant neurons and achieved a recognition rate of 94.8%. The experiments were repeated for this paper using cyclicly-invariant neurons. The number of neurons was increased over the original network so that the total number of weights in each network was approximately equal (cyclicly-invariant neurons have approximately half as many weights for the same number of inputs.) The recognition rate decreased to 92.8% indicating the loss of discrimination power due to the use of cyclicly-invariant neurons.

## CONCLUSION

This paper compared the use of the second-order neural networks to various transform-based techniques for translation-invariant pattern recognition. The most important result discussed in the paper is a proof that both second-order neural networks and first-order networks used in combination with the magnitude of the discrete Fourier transform are more powerful than first-order neural networks used in combination with any member of the general class of fast translation-invariant transforms. Superiority over the power spectrum of the Walsh-Hadamard transform was also demonstrated. The fast correlation transform was introduced proven to be more powerful than the R-transform.

The computational requirements of the second-order neural network method were compared to those of the Fourier magnitude technique. The relative computational requirements were shown to be application dependent. In some applications, particularly applications with binary inputs, the second-order method has a large advantage. In other applications the Fourier method requires less computation.

# REFERENCES

[1] D. E. Glover, "An optical Fourier / electronic neurocomputer automated inspection system," in <u>Proceedings of the International Conference on Neural Networks,</u> 1988, vol. I, pp. 569-576.

[2] N. Ahmed and K. R. Rao, <u>Orthogonal Transforms for Digital Signal Processing,</u> New York: Springer-Verlag, 1975.

[3] M. D. Wagh and S. V. Kanetkar, "A class of translation invariant transforms," <u>IEEE Transactions on Acoustics, Speech and Signal Processing,</u> vol. ASSP-25, pp. 203-205, April 1977.

[4] H. J. Reitboeck and J. Altmann, "A model for size- and rotation-invariant pattern processing in the visual system," <u>Biological Cybernetics,</u> vol. 51, pp. 113-121, 1984.

[5] H. Yang and C. C. Guest, "Performance of backpropagation for rotation invariant pattern recognition," in <u>Proceedings of the International Conference on Neural Networks,</u> 1987, vol. IV, pp. 365-370.

[6] D. E. Rumelhart and J. L. McClelland, <u>Parallel Distributed Processing: Explorations in the Microstructure of Cognition. Volume 1: Foundations,</u> Cambridge: The MIT Press, 1986, ch. 8, pp. 348-352.

[7] C. L. Giles and T. Maxwell, "Learning, invariance, and generalization in high-order neural networks," <u>Applied Optics,</u> vol. 26, pp. 4972-4978, December 1987.

[8] H. Burkhardt and X. Müller, "On invariant sets of a certain class of fast translation-invariant transforms," <u>IEEE Transactions on Acoustics, Speech, and Signal Processing,</u> vol. ASSP-28, pp. 517-523, October 1980.

[9] R. Duren and B. Peikari, "A new neural network architecture for rotationally invariant object recognition," to appear in <u>Proceedings of the 34th Midwest Symposium on Circuits and Systems,</u> 1991.

[10] R. W. Duren, <u>An Efficient Second-order Neural Network Architecture for Orientation Invariant Character Recognition,</u> Ph.D. dissertation, School of Engineering and Applied Science, Southern Methodist University, Dallas, TX, 1991.

# SHAPE RECOGNITION WITH NEAREST NEIGHBOR ISOMORPHIC NETWORK

Hung-Chun Yau    Michael T. Manry
Department of Electrical Engineering
University of Texas at Arlington
Arlington, Texas 76019

Abstract - The nearest neighbor isomorphic network paradigm is a combination of sigma-pi units in the hidden layer and product units in the output layer. Good initial weights can be found through clustering of the input training vectors, and the network can be successfully trained via back propagation learning. We show theoretical conditions under which the product operation can replace the Min operation. Advantages to the product operation are summarized. Under some sufficient conditions, the product operation yields the same classification result as the Min operation. We apply our algorithm to a geometric shape recognition problem and compare the performances with those of two other well-known algorithms.

## INTRODUCTION

As pointed out by Duda and Hart [1] and Fukunaga [2], the nearest neighbor classifier (NNC) approximates the minimum error Bayesian classifier in the limit as the number of reference vectors gets large. When the feature vector joint probability density is unknown, the NNC would be the preferred classifier, except for two problems. First, a prohibitive amount of computation is required for its use. Second, the NNC's performance is usually not optimized with respect to the training data.

As more hardware for parallel processing becomes available, the first problem will be solved. Several neural networks which are isomorphic to NNC's have been developed to attack the second problem. These include the learning vector quantization (LVQ) of Kohonen [3], the counter-propagation network of Hecht-Nielsen [4], the adaptive-clustering network of Barnard and Casasent [5], and the nearest neighbor isomorphic network (NNIN) of Yau and Manry [6]. In this paper we discuss properties of product units that allow them to substitute for Min units, in the NNIN. We compare its performance to that of LVQ2.1 for the geometric shape recognition problem.

# NETWORK TOPOLOGY AND TRAINING

The NNIN, which is a back propagation (BP) network isomorphic to a type of NNC, uses fairly conventional sigma-pi units [7] in the hidden layer and units similar to the product units [8] in the output layer. A set of good initial weights and thresholds can be directly determined from the reference feature vectors via appropriate mapping equations.

Each pattern is represented by a dimension-$Nf$ feature vector, $x = [x(1)\ x(2) \cdots x(Nf)]^T$. As shown in Fig. 1, the first processing layer consists of $Nc \cdot Ns$ sigma-pi units which are connected to the $Nf$ input features. Here $Nc$ is the number of classes and $Ns$ is the number of clusters per class. Let $Snet(i,j)$, $S\theta(i,j)$, and $Sout(i,j)$ denote the net input, threshold, and output of the $j$th unit of the $i$th class, respectively. $Sw1(i,j,k)$ and $Sw2(i,j,k)$ denote the connection weights from the $k$th input feature to the $j$th sigma-pi unit of the $i$th class. The sigma-pi unit net and activation are respectively

$$Snet(i,j) = S\theta(i,j) + \sum_{k=1}^{Nf} \left[ Sw1(i,j,k)\,x(k) + Sw2(i,j,k)\,x^2(k) \right]$$

$$Sout(i,j) = \frac{1}{1 + \exp[-Snet(i,j)]}$$

The second processing layer is composed of product units [8] with each feature raised to the first power. Let $P\theta(i)$, $Pnet(i)$ and $Pout(i)$ denote the threshold, net input and output of the $i$th unit in the second layer. The $Pw(i,\hat{\imath})$ are the connection weights for $Pin(\hat{\imath})$ and $Pnet(i)$.

$$Pin(i) = \prod_{j=1}^{Ns} Sout(i,j)\,, \qquad Pnet(i) = P\theta(i) + \sum_{i=1}^{Nc} Pw(i,\hat{\imath})\,Pin(i)\,,$$

$$Pout(i) = \frac{1}{1 + \exp[-Pnet(i)]}$$

The NNIN can be initialized and trained as follows. Let $r_{ij} = [r_{ij}(1),\ r_{ij}(2), \cdots,\ r_{ij}(Nf)]^T$ and $v_{ij} = [v_{ij}(1),\ v_{ij}(2),\ \cdots,\ v_{ij}(Nf)]^T$ respectively be the mean and variance vectors of the $j$th cluster of the $i$th class. Define the squared distance of the vector $x$ to $r_{ij}$ as

$$D_{ij}^2 = \sum_{k=1}^{Ng} \frac{[x(k) - r_{ij}(k)]^2}{v_{ij}(k)}$$

Comparing $Snet(i,j)$ with $D_{ij}^2$, we may assign the initial weights and threshold of the $j$th sigma-pi unit of the $i$th class as

$$S\theta(i,j) = \sum_{k=1}^{Nf} \frac{r_{ij}^2(k)}{v_{ij}(k)}\,, \quad Sw1(i,j,k) = \frac{-2r_{ij}(k)}{v_{ij}(k)}\,, \quad Sw2(i,j,k) = \frac{1}{v_{ij}(k)}$$

It is simple to initialize the second layer as $Pw(i,\hat{i}) = \delta(i-\hat{i})$ and $P\theta(i) = 0$.

Let $Tout(i)$ be the desired output. $Np$ denotes the total number of training patterns. Using the $q$-norm, which is the $p$-norm of [9], the system performance measure is defined as

$$E_T = \sum_{p=1}^{Np} \left[ \sum_{i=1}^{Nc} [Tout(i) - Pout(i)]^{2q} \right]^{1/q}$$

where $q$ is a positive integer. In practice we alter the error criterion from the least square approach ($q=1$) to the minimax approach ($q=$infinity) or vice versa when the learning process slows. In our experiments, this adaptive-$q$ technique results in an increase in learning speed.

## APPROPRIATENESS OF THE PRODUCT UNIT

In a classical NNC implementation, Min units would be used in the output layer to determine the classification result. A Min unit passes the minimum value of its inputs through to its output. In the NNIN, it is quite possible to use Min units in the output layer. This would result in a network very similar to those of Kohonen. In this section, we consider alternative units. In the NNC, the Min units have two important properties that allow the generation of complicated (disjoint for example) decision regions. These are as follows.

**Property 1**: Let $r_{\hat{i}j}$ be the vector among the $r_{ij}$ from which the random vector $x$ has the smallest distance $D_{\hat{i}j}$. If $x \to r_{\hat{i}j}$, then $D_{\hat{i}j} \to 0$, and $D_{\hat{i}j} = \min D_{ij}$ for all $i \in [1,Nc], j \in [1,Ns]$.

**Property 2**: Property 1 still holds if new reference vectors are added to any class.

Two possible replacements for the Min unit are the product unit and the sum unit, whose activations are respectively

$$Pin(i) = \prod_{j=1}^{Ns} D_{ij}^2, \qquad S(i) = \sum_{j=1}^{Ns} D_{ij}^2 \qquad (1)$$

The product unit has both of the properties given above. As $x \to r_{\hat{i}j}$, then $D_{\hat{i}j}$ and $Pin(\hat{i})$ both approach 0. Assuming all the $r_{ij}$ vectors are unequal, then $D_{ij} > 0$ and $Pin(i) > 0$ for all $i \neq \hat{i}$ and Property 1 is satisfied. When a new reference vector is added, Property 1 and therefore Property 2, are still satisfied. Therefore the product units have the required properties and can function as Min units.

For the sum unit activation of (1), as $x \to r_{\hat{i}j}$, $D_{\hat{i}j}$ approaches 0 but $S(i)$ does not. Property 1 is not generally satisfied for the sum unit. If a sum unit did have Property 1 and if a new example vector is added to the $\hat{i}$th class, $S(\hat{i})$ can no longer be driven to 0 as $x$ approaches $r_{ij}$. Thus the sum unit does not have properties 1 or 2, and is not a suitable replacement for the Min unit.

## COMPARISONS OF THE PRODUCT AND MIN UNITS

The two advantages to using product units rather than Min units are that (a) derivatives of products are much simpler to calculate than derivatives of a minimum operation, and (b) a back-propagation iteration for product units results in changes for all weights, unlike a similar iteration for min units.

Assume that $Pw(i,\hat{i}) = \delta(i-\hat{i})$. For each class, put the cluster outputs $Sout(i,j)$ in increasing order such that $Sout(i,1) \leq Sout(i,2) \leq \cdots \leq Sout(i,Ns)$. Assume that the correct class is class 1. The Min unit assigns the vector $x$ to class 1 if

$$Sout(1,1) < Sout(i,1) \quad for \quad i > 1 \qquad (2)$$

The product unit assigns $x$ to class 1 if $Pin(1)$ is smaller than $Pin(i)$ for every $i > 1$, which is the same condition as

$$Sout(1,1) < Sout(i,1) \left[ \prod_{j=2}^{Ns} Sout(i,j) \bigg/ \prod_{j=2}^{Ns} Sout(1,j) \right] \qquad (3)$$

$$= Sout(i,1) \cdot Rs(i)$$

for all $i > 1$. If (2) is true, (3) is true if either of the following sufficient conditions are true: (a) $Sout(1,j) < Sout(i,j)$ for all $i > 1$ and $j \geq 2$; and (b) $Rs(i) < Rs(1)$ for all $i > 1$. Also, if (8) is true, either of the above conditions is sufficient for (7) to be true. Thus under certain conditions, product units give the same answer as Min units.

## SHAPE RECOGNITION EXPERIMENT

In this section, we compare our algorithm's performance to that of LVQ2.1 and the NNC for a geometric shape recognition problem. The four primary geometric shapes used in this experiment are ellipse, triangle, quadrilateral, and pentagon. Each shape image is binary-valued with 64 × 64 pixels. There were 200 training patterns and 200 testing patterns, each with varying amounts of rotation, scaling, translation, and oblique distortions. Example shapes are shown in Fig. 2. The feature sets used include circular harmonic expansion (CHE) features, log-polar transform (LPT) features, and ring-wedge energy features (RWE). Detailed feature definitions are given in the appendices. The feature vectors were of dimension 16. All three classifiers were initialized with 5 reference clusters per class through the K-mean clustering algorithm [10].

In figures 3-5, we plot the recognition error percentage versus the iteration number for training. The classifier performances are summarized in Table 1. In this experiment, the NNIN generally has better training speed and classification performance than the NNC and LVQ2.1.

**TABLE 1.** ERROR PERCENTAGE IN TRAINING/TESTING

| Classifier | CHE | LPT | RWE |
|---|---|---|---|
| NNC | 3.38/6.88 | 4.38/5.13 | 5.38/8.25 |
| LVQ2.1 | 1.50/2.88 | 2.13/4.30 | 3.00/7.13 |
| NNIN | 0.13/2.75 | 0.00/1.75 | 0.38/6.13 |

## CONCLUSIONS

In this paper we have analyzed the properties of the product unit, and shown that it is a valid replacement for the Min unit in the NNIN. We have demonstrated that good initial weights enhance learning in the NNIN. It has been shown that the NNIN can be trained efficiently using the gradient descent backpropagation algorithm. The NNIN is found to perform better and learn faster than the NNC and LVQ2.1 for the particular feature sets and recognition problems tried. Since the classification performance of the network is improved for both training and testing data, it is apparent that the classifier is generalizing, and not just memorizing the training data.

Acknowledgement; This work was funded by the State of Texas through the Advanced Technology Program.

## REFERENCES

[1] R.O. Duda and P.E. Hart, Pattern Classification and Scene Analysis, Wiley, New York, 1973.

[2] K. Fukunaga, Introduction to Statistical Pattern Recognition, Academic Press, New York, 1972.

[3] T. Kohonen, "Improved versions of learning vector quantization", IJCNN, San Diego, Ca., Vol.1, 1990, pp.545-550.

[4] R. Hecht-Nielsen, "Applications of counterpropagation network", Neural Networks, Vol.1, 1988, pp.131-139.

[5] E. Barnard, and D. Casasent, "Image processing for image understanding with neural network", IJCNN, Washington, D.C., Vol.1, 1989, pp.111-115.

[6] H.C. Yau and M.T. Manry, "Sigma-pi implementation of a nearest neighbor classifier", IJCNN, San Diego, Ca., Vol.1, 1990, pp.667-672.

[7] D.E. Rumelhart and J.L. McClelland (Eds.), Parallel Distributed Processing, Vol.1, The MIT Press, Cambridge, Massachusetts, 1986.

[8] R. Durbin and D.E. Rumelhart, "Product units: a computationally powerful and biologically plausible extension to backpropagation networks", Neural Computation, Vol.1, 1989, pp.133-142.

[9] E.W. Cheney, Introduction to Approximation Theory, Chelsea Publishing, New York, 1982.

[10] M. Anderberg, Cluster Analysis for Applications, Academic Press, New York, 1973.

[11] Y-N Hsu and H.H. Arsenault, "Pattern discrimination by multiple circular harmonic components", Applied Optics, Vol.23, 1984, pp.841-844.

[12] D. Psaltis and D. Casasent, "Deformation invariant optical processors using coordinate transformations", Applied Optics, Vol.16, 1977.

[13] J.L. Smith and R.D. Douglas, "Hybrid optical/electronic pattern recognition with both coherent and noncoherent operations", SPIE Vol.938, 1988, pp.170-177.

[14] N. George, S. Wang and D.L. Venable, "Pattern recognition using the ring-wedge detector and neural-network software", SPIE Vol.1134, 1989, pp.96-106.

[15] R. Wu and H. Stark, "Rotation-invariant pattern recognition using a vector reference", Applied Optics, Vol. 23, 1984, pp.838-840.

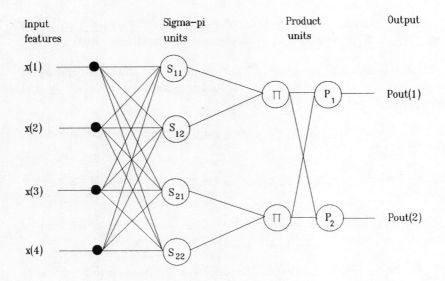

**Figure 1.** Topology of Nearest Neighbor Isomorphic Network

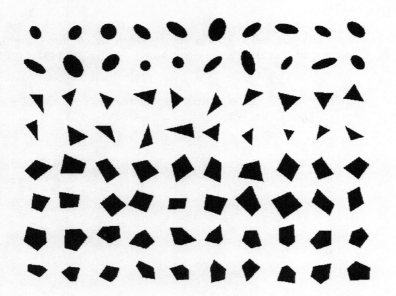

**Figure 2.** Typical geometric shapes

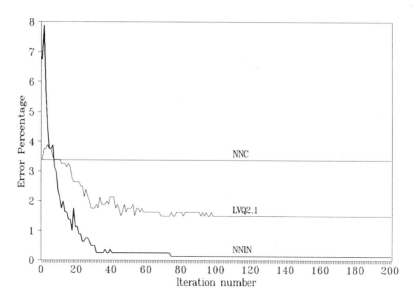

**Figure 3.** Neural networks in training for CHE feature set

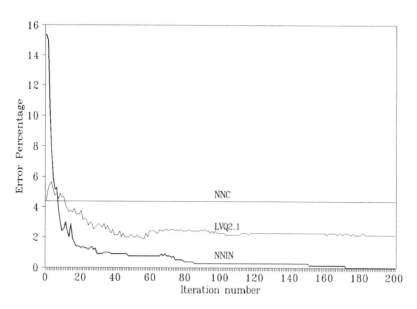

**Figure 4.** Neural networks in training for LPT feature set

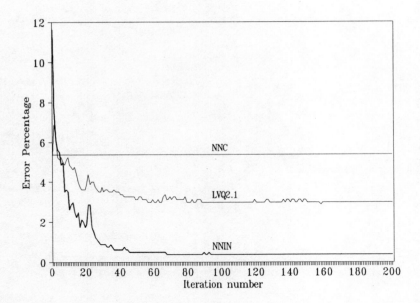

**Figure 5.** Neural networks in training for RWE feature set

## APPENDIX A. CIRCULAR HARMONIC EXPANSION FEATURES

Hsu and Arsenaul [11] have used the circular harmonic expansion (CHE) in the design of optical pattern recognition systems. After choosing the origin in polar coordinates, the circular harmonic expansion of the image, $f(r,\theta)$ can be represented as the summation of the circular harmonics,

$$f(r,\theta) = \sum_{m=-\infty}^{\infty} f_m(r) \, e^{jm\theta} , \qquad f_m(r) = \frac{1}{2\pi} \int_0^{2\pi} f(r,\theta) \, e^{-jm\theta} \, d\theta$$

We use the image's centroid as the origin in the expansion. The CHE features are defined as the energy of the circular harmonics over radius by

$$g(m) = \frac{\int_0^\infty |f_m(r)|^2 \, r \, dr}{\int_0^\infty |f_0(r)|^2 \, r \, dr}$$

## APPENDIX B. LOG-POLAR TRANSFORM FEATURES

Psaltis and Casasent [12] have proposed log-polar transform (LPT) techniques combining a geometric coordinate transformation with the conventional Fourier transform. Smith and Devoe [13] have applied the LPT to the magnitude-squared Fourier transform of the image.

The first step in taking the LPT is to Fourier transform the input image $f(x,y)$ to get $F(R,\psi)$, where $R$ and $\psi$ denote frequency domain radius and angle. Next, define $\rho = \ln(R)$ and $L(\rho,\psi) = e^{2\rho}|F(e^{\rho},\psi)|^2 = R^2|F(R,\psi)|^2$. The Fourier transformation of the LPT is taken as

$$M(\zeta,\xi) = \int_{-\infty}^{\infty} \int_{0}^{\pi} L(\rho,\psi)\, e^{-j(\zeta\rho+\xi\psi)}\, d\psi\, d\rho$$

The $(\zeta,\xi)$ domain is referred to here as the Mellin domain. We define the LPT features $g_1$ and $g_2$ as

$$g_1(m) = \frac{|M(m,0)|}{|M(0,0)|}, \qquad g_2(n) = \frac{|M(0,n)|}{|M(0,0)|}$$

## APPENDIX C. RING-WEDGE ENERGY FEATURES

George and his coworkers [14] developed a set of energy features which is very useful in image spectrum analysis. Wu and Stark [15] showed successful optical recognition process by using of these energy features. These features consist of energies in several ring and wedge-shaped areas in the Fourier transform domain $F(R,\psi)$. Let $\Delta_R$ and $\Delta_\psi$ denote the step sizes for $R$ and $\psi$, respectively. Define the surface area of the $m$th ring as $S_r(m) = \{(R,\psi) \mid (m-1)\Delta_R \leq R \leq m\Delta_R,\ 0 \leq \psi \leq \pi\}$ and the surface area of the $n$th wedge as $S_w(n) = \{(R,\psi) \mid 0 \leq R \leq \infty,\ (n-1)\Delta_\psi \leq \psi \leq n\Delta_\psi\}$. The energy in the $m$th ring and the energy in the $n$th wedge are respectively

$$E_r(m) = \iint_{S_r(m)} |F(R,\psi)|^2 R\, dR\, d\psi, \qquad E_w(n) = \iint_{S_w(n)} |F(R,\psi)|^2 R\, dR\, d\psi$$

To introduce rotation invariance, we circularly rotate $E_w(n)$ such that

$$E_w(1) = \max_{n} E_w(n)$$

Since the rotation angle $\phi$ is not always a multiple of $\Delta_\psi$, we normalize the ring-wedge features as

$$g_r(m) = \frac{E_r(m)}{\sum_k E_r(k)}, \qquad g_w(n) = \frac{E_w(n)}{\sum_k E_w(k)}$$

# DIMENSIONALITY REDUCTION OF DYNAMICAL PATTERNS USING A NEURAL NETWORK

S. Nakagawa    Y. Ono    Y. Hirata
Department of Information and Computer Sciences
Toyohashi University of Technology
Tempaku-cho, Toyohashi, 441 Japan

Abstract—To recognize speech with dynamical features, we should use feature parameters including dynamical changing patterns, that is, time sequential patterns. The K-L expansion has been used to reduce the dimensionality of time sequential patterns. This method changes axes of feature parameter space linearly as minimizing the error between original and reconstructed parameters. In this paper, the dimensionality of dynamical features is reduced by using a non-linear dimensional compressing ability of a neural network. We compared the proposed method on the speech recognition using a continuous HMM (hidden Markov model) with the reduction method using K-L expansion and the feature parameters of regression coefficients in addition to original static features.

## INTRODUCTION

Traditional HMMs have less ability to describe dynamical changes of time sequential feature parameters. Therefore, to use dynamical features, several approaches have been studied such as using many states [1,2], using regression coefficients of time series [3], using models with a linear or non-linear prediction [4,5].

In this paper, to use dynamical changes of feature parameters in speech, successive four frames are combined to one segment vector, where each frame corresponds to a feature vector representing a short time spectrum. However, a HMM consists of many parameters according to the number of frames in a segment (i.e., feature parameters, dimensions). Therefore the parameters of the model can not be estimated precisely if the training data is not enough. In practical application, the dimensionality of parameters should be reduced.

Bocchieri and others [6] and Brown [7] did principal component analysis of successive two frames using the K-L expansion. They applied the reduced parameters to DTW (dynamic time warping) and HMM for speech recognition. The K-L expansion linearly transforms the feature parameter space as minimizing the error between original and reconstructed parameters.

On the other hand, noting the non-linear transformation of a neural network, it has been used to reduce the dimensionality. Cottrell and others [8] did the image compression using a three layer neural network. Usui and others [9] compressed and analyzed the color information using a five layer neural network. It shows that the information compression with the five layer neural network has less reconstructed errors than one with K-L expansion.

In this study, we reduce the dimensionality of input segments with a similar five layer neural network, and regard the compressed vectors as observation vectors of continuous HMM.

## REDUCTION OF DIMENSIONALITY

### Dimensional reduction with neural network

A feed-forward neural network is composed of an input layer, several hidden layers, and an output layer. If the unit number in a hidden layer is less than the unit number in the input and output layer, this architecture can reduce the dimensionality by training it with the same target parameters for output units as input parameters [8,9] (identical mapping).

We used the five layer feed-forward neural network like Figure 1, because the ability of dimensional reduction by the three layer neural network is equal to or less than one of the K-L expansion [10]. The first three layers of five layers compress the dimension of input parameters and the latter three layers reconstruct the compressed parameters into original input parameters. In practice, since a four layer neural network is better than a three layer network in the pattern recognition [11,12], a seven layer neural network may be better than the five layer network for the reduction, but the training may become more difficult. So we used the five layer neural

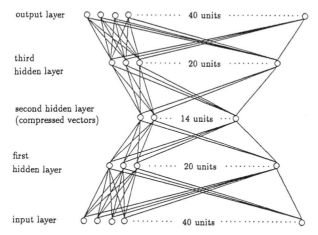

FIGURE 1. THE FIVE LAYER FEED-FORWARD NEURAL NETWORK
USED TO DIMENSIONAL REDUCTION
(CASE OF COMPRESSING 40 DIMENSIONS TO 14 DIMENSIONS)

network.

It is expected that the neural network having non-linear sigmoid function may compress the dimension more efficiently than the K-L expansion which transforms linearly a feature space. In this study, the output values of the first and third hidden layer are non-linearly transformed by a sigmoid function, but the output values of the input, output and second hidden layer are not transformed by it.

## Dimensional reduction with K-L expansion

There is the Karhunen-Loeve (K-L) expansion for the principal component analysis. This method compresses input parameters by the linear transformation as minimizing the error between input and reconstructed parameters. The following is the procedure of this method.

$N$ : dimension of input vectors
$x_i$ : sample vector $(i = 1, \ldots, I)$
$\bar{x}$ : mean of sample vectors
$z_i$ : $x_i - \bar{x}$ $(z_i = (z_i^1, z_i^2, \ldots, z_i^N))$
$p$ : dimension of compressed vectors
$y_i$ : compressed vector

1. estimate the covariance matrix $A = [a_{lm}]$ from sample vectors.

$$a_{lm} = \frac{1}{I} \sum_i z_i^l z_i^m$$

2. calculate eigenvalues $\lambda_j$ and eigenvectors $\phi_j$.

$$A\phi_j = \lambda_j \phi_j$$

3. sort eigenvalues and eigenvectors as $\lambda_1 \geq \lambda_2 \geq \lambda_3 \geq \cdots \geq \lambda_p$.
4. calculate the transformation matrix $B$.

$$B = [\phi_1 \; \phi_2 \; \phi_3 \; \cdots \; \phi_p]^T$$

5. calculate compressed vector.

$$y_i = B x_i$$

## SYLLABLE RECOGNITION IN CONTINUOUS SPEECH

### Structure of HMM

Figure 2 shows the structure of HMM for speech recognition. This model consists of five states, four transition probabilities and four continuous output probability density functions. A Baum-Welch estimation algorithm is used for the parameter estimation in training. A multidimensional Gaussian distribution is assumed as the output probability density functions.

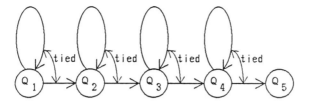

FIGURE 2. STRUCTURE OF HMM USED IN SYLLABLE RECOGNITION

## Speech data

1776 words (5216 syllables) in total were uttered by six male speakers as speech data. These are used for sample data of dimensional reduction and training data of the HMM. Test data for syllable recognition are 1885 syllables taken from 20 sentences per speaker.

The neural network was trained by using 220, 440, 880, and 1760 syllables selected at random from 5216 training syllables. All of 5216 syllables (176712 segments) were used for sample data of the K-L expansion.

The syllable based HMMs were trained and tested on the following 36 categories appearing the test data. The same HMM for each syllable was commonly used to six speakers, that is, the recognition mode was a multi-speaker style. The mean number of samples per category is 82.3 syllables for training and 8.7 syllables for test. Table 1 shows the analysis condition of speech data.

[kinds of syllables]

/a/   /i/    /e/    /o/
/ka/  /ki/   /ku/   /ke/   /ko/
/sa/  /si/   /su/   /se/   /so/
/ta/  /tu/   /te/   /to/   /ni/   /no/   /hi/
/ma/  /mi/   /me/   /ra/   /ri/   /ru/   /re/
/wa/  /N/
/zi/  /zu/   /da/   /de/   /do/   /fa/

TABLE 1. ANALYSIS CONDITION OF SPEECH DATA

| sampling frequency | 12kHz |
|---|---|
| window function | Hamming window 21.33ms (256 points) |
| frame period | 5ms (60 points) |
| speech analysis | 14-order Linear Predictive Coding (LPC) |
| feature parameters | LPC melcepstrum coefficients (10 dimensions) |

## Evaluation of dimensional reduction

A segment is composed of successive four frames or successive seven frames (four frames every two frames). As the dimension of a frame is 10, the dimension of segment is 40 in all. All segments of syllable are compressed to 10, 14 or 20 dimensions while shifting segment at every frame.

We used the five layer neural network, where the number of units in each layer are 40-15-10-15-40 for 10 dimensional reduction, 40-20-14-20-40 for 14 dimensional reduction and 40-30-20-30-40 for 20 dimensional reduction, respectively.

The network was trained by the backpropagation algorithm. Weights of the network were changed by every segment or training sample.

The activation values of the second hidden layer's units were regarded as the observation vector of HMMs while 40 dimensional parameters in each segment are inputted to the input layer. Training data of a neural network were increased from 220 to 1760 syllables every two times until the recognition rate of HMM saturated.

Figure 3 illustrates the reconstruction errors between the original input vector and the output vector that is compressed and reconstructed by the neural network, while training it. The reconstruction errors using the neural network trained for enough times are less than the errors using the K-L expansion. Figure 4 also illustrates the syllable recognition rates using HMM, where the compressed vectors are used as the input vectors.

## Evaluation experiment

Syllables taken from continuous speech were recognized by continuous HMMs on a multi-speaker mode. Each method was compared each other in terms of the average recognition rate.

Table 2 shows the recognition rate of the basic HMM not using dynamical features, that is, only frame by frame. Table 3 shows the recognition rate using the regression coefficients over 9 frames in addition to the basic parameters, that is, 20 dimensional parameters in a frame. Table 4 shows the recognition rate using vectors compressed with neural networks and the mean square error per frame between original and reconstructed vectors when the highest recognition rate was obtained. Table 5 shows the recognition rate using the K-L expansion.

Table 6 shows the mean recognition rate of three cases using the untrained initial neural network (in this case, weights of network are random values from $-0.5$ to $+0.5$). Also, to look the correlation among dimensions of compressed vectors, we performed the syllable recognition by HMM using only diagonal components of covariance matrix, assuming that there was no correlation between dimensions.

The K-L expansion selects orthogonal coordinates and compresses dimensions as minimizing reconstruction errors. We performed the recognition experiments three times using vectors transformed by orthogonal coordinates selected at random. That is, 40 vectors corresponding to eigenvectors were made at random, then they were orthogonalized by the Gram-

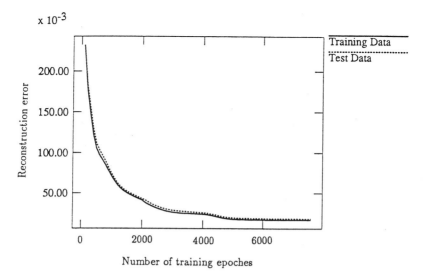

FIGURE 3. THE RECONSTRUCTION ERRORS THAT IS COMPRESSED AND RECONSTRUCTED BY THE NEURAL NETWORK
(CASE OF COMPRESSING 40 DIMENSIONS TO 14 DIMENSIONS)

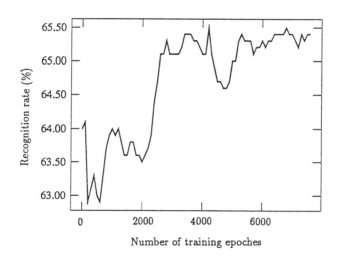

FIGURE 4. THE SYLLABLE RECOGNITION RATES USING HMM. THE VECTORS COMPRESSED BY THE NEURAL NETWORK WERE USED AS THE INPUT VECTORS OF HMM
(CASE OF COMPRESSING 40 DIMENSIONS TO 14 DIMENSIONS)

TABLE 2. RECOGNITION RATE BY BASIC HMM

| basic HMM | 58.1% |
|---|---|

TABLE 3. RECOGNITION RATE USING REGRESSION COEFFICIENTS IN ADDITION TO ORIGINAL VECTORS

| regression coefficients | 64.2% |
|---|---|

TABLE 4. RECOGNITION RATES AND RECONSTRUCTION ERRORS USING DIMENSIONAL COMPRESSION BY NEURAL NETWORK

| segment | dimensions | reconstruction errors | | recognition rates (%) |
|---|---|---|---|---|
| | | training data | test data | |
| 4 frames | 10 | 0.0395 | 0.0420 | 59.4 |
| | 14 | 0.0170 | 0.0184 | 65.4 |
| | 20 | 0.0066 | 0.0068 | 67.5 |
| 4 frames (no correlations, diagonal matrix) | 10 | 0.0395 | 0.0420 | 55.5 |
| | 14 | 0.0170 | 0.0184 | 56.8 |
| | 20 | 0.0066 | 0.0068 | 55.4 |
| 7 frames | 10 | 0.0806 | 0.0864 | 61.2 |
| | 14 | 0.0506 | 0.0547 | 63.6 |
| | 20 | 0.0267 | 0.0288 | 66.2 |

Schmidt orthogonalization. In this case, the reconstruction error is not minimized.

Since these tables, vectors compressed by neural network or K-L expansion improved the recognition rate from 58% to 64–67% in comparison with the basic HMM. Compressing to 14 and more dimensions can get higher recognition rate than the regression coefficients in addition to original vectors, which have been often used in speech recognition.

The expansion of the window of segment from four to seven frames didn't increase the recognition rate. We think the reason that the compression of a long segment is difficult why the dynamical feature changes hastily in a consonant part.

Compressing the dimension by the un-trained neural network and random orthogonal transformation gets higher recognition rate than the basic HMM. These methods may not be useful from the viewpoint of compressing and reconstructing of information, but it may be useful to speech recognition with segment data.

TABLE 5. RECOGNITION RATES AND RECONSTRUCTION ERRORS USING DIMENSIONAL COMPRESSION BY K-L EXPANSION

| segment | dimensions | reconstruction errors | | recognition rates (%) |
|---|---|---|---|---|
| | | training data | test data | |
| 4 frames | 10 | 0.0400 | 0.0428 | 59.7 |
| | 14 | 0.0170 | 0.0180 | 65.0 |
| | 20 | 0.0064 | 0.0065 | 67.4 |
| 4 frames (no correlations, diagonal matrix) | 10 | — | — | — |
| | 14 | 0.0170 | 0.0180 | 62.9 |
| | 20 | 0.0064 | 0.0065 | 61.5 |
| 7 frames | 10 | 0.0875 | 0.0936 | 60.4 |
| | 14 | 0.0406 | 0.0443 | 62.1 |
| | 20 | 0.0178 | 0.0195 | 64.7 |

TABLE 6. RECOGNITION RATES USING DIMENSIONAL COMPRESSION BY UN-TRAINED NEURAL NETWORK (THREE TRIALS)

| segment | dimensions | recognition rates (%) | | | |
|---|---|---|---|---|---|
| | | first | second | third | mean |
| 4 frames | 10 | 58.2 | 60.7 | 61.4 | 60.1 |
| | 14 | 64.1 | 64.5 | 64.3 | 64.3 |
| | 20 | 66.5 | 65.9 | 65.5 | 66.0 |
| 4 frames (no correlations, diagonal matrix) | 10 | 48.4 | 40.2 | 45.4 | 44.7 |
| | 14 | 44.5 | 47.2 | 49.0 | 46.9 |
| | 20 | 49.4 | 48.4 | 49.9 | 49.2 |

TABLE 7. RECOGNITION RATES USING DIMENSIONAL COMPRESSION BY RANDOM ORTHOGONALIZATION (THREE TRIALS)

| segment | dimensions | recognition rates (%) | | | |
|---|---|---|---|---|---|
| | | first | second | third | mean |
| 4 frames | 10 | 61.1 | 60.8 | 60.6 | 60.8 |
| | 14 | 65.1 | 64.2 | 65.4 | 64.9 |
| | 20 | 65.4 | 67.0 | 66.7 | 66.4 |

TABLE 8. RECOGNITION RATES USING HMM WITH 20 DIMENSIONAL PARAMETERS ACCORDING TO TWO COMBINED FRAMES

| segment | dimensions | recognition rates (%) |
|---|---|---|
| 2 successive frames | 20 | 64.3 |
| 2 out of 3 successive frames | 20 | 67.1 |

In final, we tried the recognition using HMM with 20 dimensional parameters according to two combined frames. There were two types. One was the two successive frames and the other was a current frame and one skipped frame (two frames out of three successive frames). Table 8 shows the recognition rusults. From this, we find that the dimensional reduction of four successive frames is effective. In particular, when we use the diagonal matrix to reduce the computation, the dimensional reduction method is effective.

## CONCLUSION

We performed the syllable recognition with compressed segments as input vectors to supplement the basic HMM lacking the ability to describe transitional changes. We compared five layer neural networks and K-L expansions as dimensional reduction methods.

Recognition rates using segmental vectors compressed by neural networks were 60.1% as 10 dimensional compressed vectors, 65.5% as 14 dimensions and 67.6% as 20 dimensions, respectively. In the case of K-L expansion, recognition rates are 59.7% as 10 dimensions, 65.0% as 14 dimensions and 67.4% as 20 dimensions, respectively. Above all recognition rates are higher than 58.1% as the basic HMM which used feature parameters frame by frame, and 64.2% as the regression coefficients in addition to the original feature parameters.

In this study, we ascertained that it is useful for speech recognition to use segmental statistics of the dimensional compressed feature vectors to extract dynamical feature. Also, the neural network having non-linear sigmoid function extracts dynamical features more or less efficiently than the K-L expansion. Besides, the dimensional compression by random weights is inferior to trained neural networks, but it may be useful for speech recognition considering training cost. This fact suggests that the randomization and orthogonalization of feature parameters may be "key" of pattern recognition.

## REFERENCES

[1] L. R. Bahl, P. F. Brown, P. V. de Souza, R. L. Mercer and M. A. Picheny : "Acoustic Markov models used in the TANGORA speech recognition system", Proc. ICASSP-88, pp. 497-500 (1988).

[2] S. Nakagawa, Y. Hirata and Y. Hashimoto : "Japanese phoneme recognition using continuous parameter hidden Markov models", The journal of the acoustical society of Japan, Vol. 46, No. 6, pp. 486-496 (1990, in Japanese).

[3] V. N. Gupta, M. Lennig and P. Mermelstein : "Integration of acoustic information in a large vocabulary word recognizer", Proc. ICASSP-87, pp. 697-700 (1987).

[4] P. Kenny et al. : "A linear predictive HMM for vector-valued observations with applications to speech recognition", IEEE Trans. Vol. ASSP-38, No. 2, pp. 220-225 (1990).

[5] E. Tsuboka, Y. Takada and H. Wakita : "Neural predictive hidden Markov model", Proc. ICSLP-90, Vol. 2, pp. 1341-1344 (1990).

[6] E. L. Bocchieri and G. R. Doddington : "Speaker-independent digit recognition with reference frame-specific distance measures", Proc. ICASSP-86, pp. 2699-2672 (1986).

[7] P. F. Brown : "The acoustic-modeling problem in automatic speech recognition", Ph. D thesis, Carnegie-Mellon University (1987).

[8] G. W. Cottrell, P. Munro and D. Zipser : "Image compression by back propagation: an example of extensional programming", Advances in Cognitive Science, Vol. 3 (1988).

[9] S. Usui, S. Nakauchi and M. Nakano : "Reconstruction of Munsell color space by a five-layered neural network", IJCNN-90, pp. 515-520 (1990).

[10] K. Funahashi : "On the approximate realization of identity mappings by three-layer neural networks", IEICE Trans., Vol. J73-A, No. 1, pp. 139-145 (1990, in Japanese).

[11] D. L. Chester : "Why two hidden layers are better than one", Proc. IJCNN, Vol. 1, pp. 265-268 (1990).

[12] A. Waibel et al. : "Phoneme recognition using time-delay neural networks", IEEE Trans. Vol. ASSP-37, No. 3, pp. 328-339 (1989).

# A CRITICAL OVERVIEW OF NEURAL NETWORK PATTERN CLASSIFIERS*

Richard P. Lippmann
Room B-349, MIT Lincoln Laboratory
Lexington, MA 02173-9108
rpl@sst.ll.mit.edu

ABSTRACT—A taxonomy of neural network pattern classifiers is presented which includes four major groupings. *Global* discriminant classifiers use sigmoid or polynomial computing elements that have "high" non-zero outputs over most of their input space. *Local* discriminant classifiers use Gaussian or other localized computing elements that have "high" non-zero outputs over only a small localized region of their input space. *Nearest Neighbor* classifiers compute the distance to stored exemplar patterns and *Rule Forming* classifiers use binary threshold-logic computing elements to produce binary outputs. Results of experiments are presented which demonstrate that neural network classifiers provide error rates which are equivalent to and sometimes lower than those of more conventional Gaussian, Gaussian mixture, and binary tree classifiers using the same amount of training data. Many neural network classifiers also provide outputs which estimate Bayesian *a posteriori* probabilities. Experiments used low-dimensional (2 to 55 input) phoneme classification tasks, a high-dimensional (360 pixel input) handwritten digit classification task, and machine learning data bases. They demonstrate that neural network classifiers provide new alternatives to more conventional approaches. They often provide reduced error rates and always allow other classifier characteristics to be traded off to best match the requirements of a particular problem. Characteristics which often differ dramatically across classifiers include classification time, training and adaptation time, ease of implementation, memory requirements, rejection accuracy, and usefulness of outputs as Bayes probability estimates.

## INTRODUCTION

The table in Figure 1 contains a taxonomy of five major types of neural network and conventional pattern classifiers that can be used to classify fixed-length patterns. The first row in this table represents conventional proba-

---

*This work was sponsored by the Defense Advanced Research Projects Agency and the Air Force Office of Scientific Research.

| GROUP | DECISION REGION | COMPUTING ELEMENT | REPRESENTATIVE CLASSIFIERS |
|---|---|---|---|
| PROBABILISTIC | 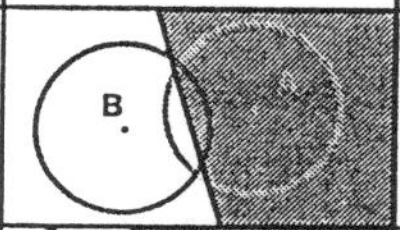 | DISTRIBUTION DEPENDENT | GAUSSIAN, GAUSSIAN MIXTURE |
| GLOBAL | 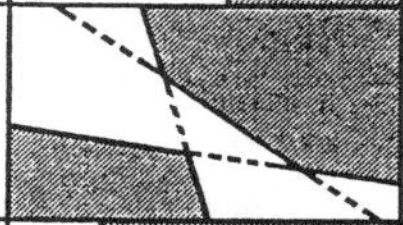 | SIGMOID 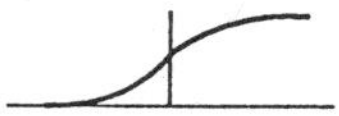 | MULTI-LAYER PERCEPTRON, HIGH-ORDER POLYNOMIAL NET |
| LOCAL | 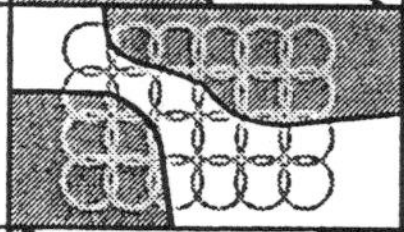 | KERNEL 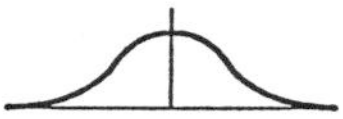 | RADIAL BASIS FUNCTION, KERNEL DISCRIMINANT |
| NEAREST NEIGHBOR | 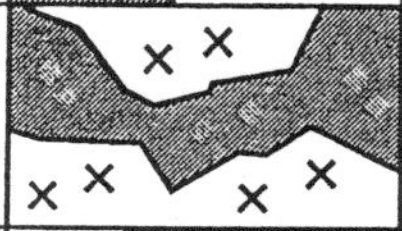 | EUCLIDEAN NORM  | K-NEAREST NEIGHBOR, LEARNING VECTOR QUANTIZER |
| RULE FORMING | 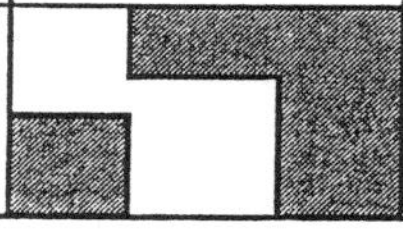 | THRESHOLD LOGIC 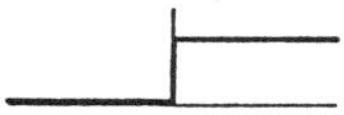 | BINARY DECISION TREE, HYPERSPHERE |

Figure 1: A Taxonomy Including Five Types of Conventional and Neural Network Pattern Classifiers.

bilistic classifiers which model likelihood distributions of pattern classes separately using parametric functions. Probabilistic classifiers used for speech recognition include Gaussian linear discriminant and Gaussian mixture classifiers. This most common approach to pattern classification provides good performance when the assumed functional form of class distributions matches real-world data distributions and when there is sufficient training data to estimate parameters. Performance can be poor when class distributions are not modeled well or when training data is limited.

The bottom four rows in Figure 1 include both neural network and conventional classifiers. Global classifiers form output discriminant functions from internal computing elements or nodes that use sigmoid or polynomial functions which have "high" non-zero outputs over a large region of the input space. These classifiers include multi-layer perceptrons trained with back-propagation (back-propagation classifiers), Boltzmann machines, and high-order polynomial networks. Local classifiers form output discriminant functions from internal computing elements that use Gaussian or other radially symmetric functions which have "high" non-zero outputs over only a localized region of the input space. These two types of classifiers make no strong assumptions concerning underlying distributions. They can form complex decision regions with only one or two hidden layers and are typically trained to minimize the mean-squared error between desired and actual network outputs.

Nearest neighbor classifiers perform classification based on the distance between a new unknown input and previously stored exemplars. Conventional k-nearest neighbor, neural network learning vector quantizer (LVQ) classifiers, and some forms of neural network adaptive resonance theory (ART) classifiers are examples of nearest-neighbor classifiers. These types of classifiers train extremely rapidly but can require large amounts of computation time on a serial processor for classification and also large amounts of memory.

Rule forming classifiers partition the input space into labeled regions using threshold-logic nodes or rules. Inputs are classified by the label of the region the input falls in. These classifiers have binary outputs and include binary decision trees such as the CART binary tree, the hypersphere classifier, perceptrons with hard-limiting nonlinearities trained using the perceptron convergence procedure, and many machine learning approaches that result in a small set of classification rules.

## NETWORKS ESTIMATE BAYES PROBABILITIES

Bayesian *a posteriori* probabilities, hereafter referred to as Bayes probabilities, are fundamental to statistical approaches to pattern classification. Selecting that pattern class with the highest Bayes probability for each input pattern minimizes the overall classification error rate [1]. Although rule-forming neural network classifiers only provide binary outputs, many other neural network classifiers provide continuous outputs which are estimates of Bayes probabilities. Continuous outputs are useful because they can provide confidence ratings for classification decisions and when input patterns can be rejected for human verification or to request further information. Such a strategy is common in medical diagnosis and when decoding monetary amounts on checks. Bayes probabilities are also required in problems such as speech recognition where outputs from many independent low-level classifiers must be integrated to make higher level decisions. They are also required to compensate for differences between pattern class probabilities in training and test data and to minimize alternative risk functions.

Many proofs have been presented, including those in [7], which demonstrate that outputs of local and global neural network classifiers minimize commonly used squared error and cross-entropy cost functions when they are Bayes probabilities. Figure 2 presents an example which illustrates that outputs of Radial Basis Function (RBF), Multi-Layer Perceptron (MLP) and high-order polynomial (GMDH) classifiers all estimate Bayes probabilities accurately. Estimation accuracy is best in the central region for this problem which has two input classes that have Gaussian mixture likelihood distributions. All networks in this example had one input, two outputs, and were trained with four thousand training samples per class. The sum of network outputs across all classes should be 1.0 if outputs accurately estimate Bayes probabilities. Figure 3 demonstrates that the sum of network outputs for the

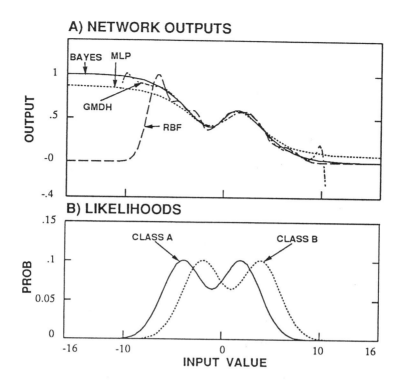

Figure 2: A) Actual Bayesian Probability For Input Class A and the Corresponding Outputs of Three Neural Networks and B) Gaussian Mixture Likelihood Functions For Two Input Classes Used to Generate Training Data and Calculate Actual Bayesian Probabilities.

above RBF, MLP, and GMDH classifiers is near 1.0 in the central region with sufficient training data. This occurred even though outputs were never explicitly constrained during training to sum to 1.0. Further simulations and a more thorough discussion of the relation between network outputs and Bayes probabilities are available in [7].

## CLASSIFIER COMPARISONS

A listing of many different neural network and conventional classifiers which groups classifiers using the above taxonomy but focuses on differences in network outputs is presented in Figure 4. Descriptions and references for many of these classifiers are available in [1, 4, 5, 6]. As noted above, probabilistic, global, and local classifiers can produce smoothly varying outputs which estimate either the likelihood of the input data or Bayes probabilities. Nearest neighbor and rule forming classifiers produce binary outputs that attempt to

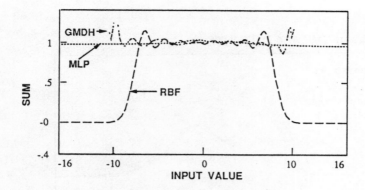

Figure 3: Sum of Two Outputs for Each of the Three Networks Whose Outputs are Plotted in Figure 2.

minimize the number of misclassifications. In addition to these differences, classifiers often differ dramatically in (1) Memory and computation requirements, (2) Training and classification times, (3) Complexity, (4) Rejection accuracy, (5) Ability to identify outliers, (5) Ease of adaptation, (6) Performance with noisy features and missing data, and (7) Performance with binary or continuous inputs.

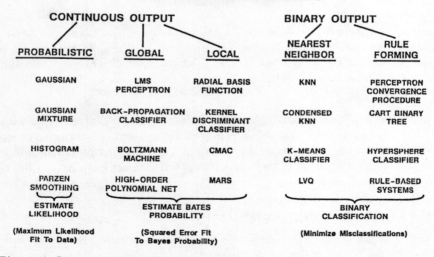

Figure 4: Important Neural Network and Conventional Pattern Classifiers.

Four studies described below compared many of the neural network and conventional classifiers in Figure 4 using real-world data from different application areas. In all studies, the complexity or size of classifiers was carefully adjusted to match the amount of training data provided, classification error rates were accurately estimated by partitioning the data base into training and test sets or by using cross-validation approaches, and classification parameters were carefully estimated from training data.

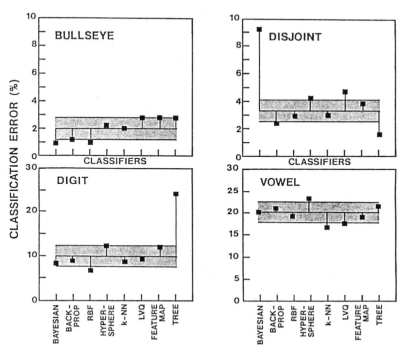

Figure 5: Classifier Error Rates on Two Speech and Two Artificial Data Bases.

One series of experiments summarized in [4, 6] compared eight neural network and six conventional classifiers using a talker-dependent recognition task (7 digits, 22 cepstral inputs per digit, 16 talkers, 70 training and 112 testing patterns per talker), a talker-independent vowel recognition task (10 vowels, 2 formant frequency inputs, 67 talkers, 338 training and 333 testing patterns), and two artificial tasks (Bullseye and Disjoint) with two input dimensions that require either a single convex or two disjoint decision regions. Error rates for Gaussian, Back-Propagation, Radial Basis Function, Hypersphere, k-nearest neighbor, learning vector quantizer, feature-map, and CART binary tree classifiers on these data bases are presented in Figure 5. The shaded area represents one binomial standard deviation above and below average classifier performance. Except for a few exceptions, classification error rates are statistically indistinguishable. Practical classifier characteristics on these problems including training time, classification time, and memory requirements, however, differed by more than an order of magnitude.

A second study used a more complex talker-dependent phoneme classification task with 4,600 training and 7,600 test frames extracted from the initial consonant segment of the letters "B", "D", and "G" from the TI-46 word data base [2]. Inputs to classifiers were adjoined cepstral vectors from one to five 10 msec speech frames (11 to 55 cepstral inputs). Classifiers were evaluated using the 10 training and 16 testing tokens available for each word and talker. Experiments comparing back-propagation, radial basis function, k-nearest neighbor, Gaussian, and Gaussian mixture classifiers demonstrated that radial basis function and Gaussian mixture classifiers provided error rates as low as the other classifiers. Error rates for these two classifiers obtained using different numbers of adjoined speech frames and different numbers of centers (Gaussian mixture components and radial basis function hidden nodes) are presented in Figure 6. As can be seen, neural network radial basis function classifiers provide error rates which are significantly lower than those of tied Gaussian mixture classifiers with similar numbers of parameters. This is presumably because the radial basis function classifier minimizes the overall error rate directly instead of approximating class distributions separately.

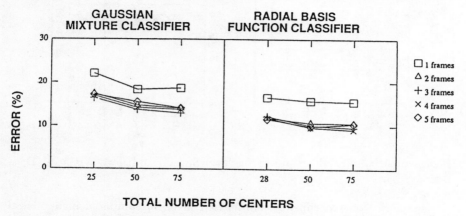

Figure 6: Classifier Error Rates for Gaussian Mixture and Radial Basis Function Classifiers on a Phoneme Classification Task Using 1 to 5 Adjoined Speech Frames as Inputs.

A recent study [3] demonstrated that global, local, and nearest neighbor classifiers can provide similar low error rates even with very high dimensional inputs (360 pixels) on a handwritten digit classification task. Classifiers were tested using a database with 30,600 training and 5,060 test patterns which were normalized to form 15x24 gray scale images with 360 pixels each. A multi-layer perceptron classifier was trained with back-propagation and had an architecture which was carefully tailored to this problem. It was compared to k-nearest neighbor and radial basis function classifiers. All classifiers were tuned to provide good performance and compared on this data base without rejections. Results are shown in Table 1.

|  | Back-Prop | KNN | RBF |
| --- | --- | --- | --- |
| Error Rate (No Rejections) | 5.15% | 5.14% | 4.77% |
| Free Parameters | 5,472 | 11,016,000 | 371,000 |
| Training Time (Hours) | 67.7 | 0.0 | 16.5 |
| Classification Time (Secs/Char) | 0.14 | 6.22 | 0.24 |

Table 1: Results of Handwritten Digit Recognition Experiments Using Multi-Layer Perceptron, k-Nearest Neighbor, and Radial Basis Function Classifiers.

Error rates across the three classifiers are statistically indistinguishable. This is somewhat surprising given the high dimensionality of the problem and the enormous differences (more than three orders of magnitude) in the number of parameters used by the different classifiers. Training time and classification time results in Table 1 clearly illustrate the differences in practical characteristics of the three classifiers. The back-propagation classifier requires days of training on a DECstation 3100 but provides the most rapid classification. The k-nearest neighbor classifier requires no training time but needs megabytes of storage and many seconds to classify each digit. The RBF classifier requires much less training time than the back-propagation classifier at the expense of more storage. Further experiments explored the number of patterns that must be rejected to reduce the error rate to a low 0.3% per digit. Patterns were rejected based on the outputs of back-propagation and RBF classifiers and the makeup and volume of space covered by the k-nearest neighbors for the k-nearest neighbor classifier. Outputs of the RBF classifier were most useful for accurately rejecting patterns that cause errors. A per-digit error rate of 0.3% was achievable with this classifier by rejecting only 19% of the digits. To achieve this same accuracy, the back-propagation classifier had to reject more than 30% of all patterns and the k-nearest neighbor classifier had to reject more than 66% of all patterns.

A number of studies have compared neural network and machine learning classifiers. Figure 7 plots error rates reported in [8] for four machine learning and statistical data bases using nearest neighbor, Gaussian, back-propagation, CART tree, and a machine learning rule-based classifier. The shaded area again represents one binomial standard deviation above and below average classifier performance. These four data bases, which are similar to those used to evaluate many machine learning algorithms, were surprisingly small. The maximum number of classes was 3, numbers of patterns were limited, and inputs were often binary and of low dimensionality. Similar low error rates were provided by back-propagation, CART binary tree, and rule classifiers. Higher error rates were provided by nearest neighbor and Gaussian classifiers for the cancer and hypothyroid problems where best performance was provided by classifiers which could ignore all but a few important input features.

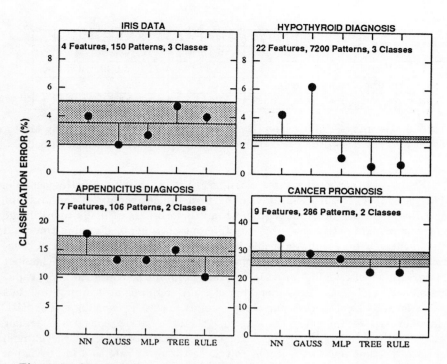

Figure 7: Classification Error Rates on Machine Learning Data Bases.

## SUMMARY

Many different neural network classifiers have been developed and compared to conventional classifiers over the past few years. Studies using speech, handwritten digit recognition, and machine learning data bases have demonstrated that error rates with these classifiers are as low as and sometimes lower than error rates of more conventional statistical and machine learning classifiers when using the same amount of training data.

Studies have demonstrated that neural network classifiers provide alternative tradeoffs in classifier characteristics including training and classification times, memory requirements, complexity, ease of implementation, and rejection accuracy. *Global* neural network classifiers include multi-layer perceptrons trained with back-propagation and Boltzmann machines. They generally have low memory requirements and provide rapid classification but have long training times. *Local* neural network classifiers include radial basis function classifiers and other kernel discriminant classifiers. They generally have higher memory requirements but shorter training times. *Nearest neighbor* neural network classifiers include learning vector quantizer, feature-map, and some versions of adaptive resonance theory classifiers. They generally have higher memory requirements but can provide rapid training and adaptation. *Rule Forming* neural network classifiers include binary perceptrons trained with the perceptron convergence procedure and the hypersphere classifier. They can provide rapid adaptation but may have high memory requirements. They are most appropriate when only a binary classification decision is required.

Simulations and theoretical proofs also demonstrate that many *Global* and *Local* neural network classifiers including back-propagation and radial basis function classifiers estimate Bayesian *a posteriori* probabilities. Such networks are useful when outputs of multiple networks must be used to make higher level classifications and when classification confidence must be estimated to reject patterns.

# References

[1] R. O. Duda and P. E. Hart. *Pattern Classification and Scene Analysis*. John Wiley and Sons, New York, 1973.

[2] William Y. Huang and Richard P. Lippmann. HMM speech recognition systems with neural net discrimination. In D. S. Touretzky, editor, *Advances in Neural Information Processing Systems 2*. Morgan Kaufman, San Mateo, CA, 1990.

[3] Yuchun Lee. Handwriten digit recognition using k nearest neighbor, radial basis function, and back propagation neural networks. *Neural Computation*, In Press.

[4] Yuchun Lee and Richard P. Lippmann. Practical characteristics of neural network and conventional pattern classifiers on artificial and speech problems. In D. S. Touretzky, editor, *Advances in Neural Information Processing Systems 2*. Morgan Kaufman, San Mateo, CA, 1990.

[5] Richard P. Lippmann. Pattern classification using neural networks. *IEEE Communications Magazine*, 27(11):47–54, November 1989.

[6] Kenney Ng and Richard P. Lippmann. A comparative study of the practical characteristics of neural network and conventional pattern classifiers. In R. P. Lippmann, J. Moody, and D. S. Touretzky, editors, *Advances in Neural Information Processing 3*. Morgan Kaufmann, San Mateo, CA, 1991.

[7] Mike D. Richard and Richard P. Lippmann. Neural network classifiers estimate Bayesian a posteriori probabilities. *Neural Computation*, In Press.

[8] Sholom M. Weiss and Casimir A. Kulikowski. *Computer Systems that Learn: Classification and Prediction Methods from Statistics, Neural Nets, Machine Learning, and Expert Systems*. Morgan Kaufmann, San Mateo, California, 1991.

# Part 3:

# Speech Processing

# WORKSTATION-BASED PHONETIC TYPEWRITER

Teuvo Kohonen
Helsinki University of Technology
Laboratory of Computer and Information Science
Rakentajanaukio 2 C, SF-02150 Espoo, Finland

Abstract – This presentation contains a general description of our "phonetic typewriter" system that transcribes unlimited speech into orthographically correct text. The purpose of this paper is to motivate certain choices made in the partitioning of the problem into tasks and describe their implementation. The combination of algorithms we have selected has proven effective for well-articulated dictation in a phonemic language such as Finnish and Japanese, whereas for English and many other languages that are organized differently in the phonological sense, an optimal solution may look completely different.

## 1. INTRODUCTION

It is possible to approach the problem of automatic speech recognition from different points of view. One is that of *phonetics*: the objective is thereby discovery and artificial implementation of those physiological, neural, and cognitive mechanisms that facilitate speech communication between humans, in spite of rather unstable biological components thereby applied. An almost opposite approach is that of *signal processing*: based on established methods of electro-acoustics, and making use of other good engineering solutions such as linear amplifiers, filters, and transforms that are computationally effective and numerically accurate, it is possible to achieve satisfactory results by straightforward means, without inclusion of the functions typical to biological systems. In the research of *artificial neural networks* one is aiming at a compromise between the above extreme views. Based on effective learning principles and massive computing capacity the models of which are in the biological nervous systems, one hopes to be able to automatically create some of the essential sensory and cognitive functions, while a greater degree of accuracy is achievable by virtue of the artificial computing principles thereby applied.

The ultimate goal in the research of automatic speech recognition is development of a system that can understand natural conversation, without the need of specific training for the present speaker. The state-of-the-art of commercial speech recognizers, however, is that they can recognize isolated standard words from a vocabulary that contains a few hundred words, and the equipment must be tuned to the voice of each speaker specifically. Under laboratory conditions there has been progress in the following directions: speaker independence, spotting and segmentation of words from connected speech, transcription of unlimited speech, and the recognition of emotional expressions.

The experimental speech transcription system being developed at Helsinki University of Technology is a genuine "phonetic typewriter" in the sense that it recognizes each phoneme separately and converts the utterances into text according to the orthography of the language. So far experimental setups have been implemented for Finnish and romanized Japanese. The principles I am reporting below have thus strictly been intended for the development of a "dictation machine", i.e., transcription of rather carefully articulated *unlimited* speech into acceptable text. It must also be pointed out that we are mainly working within the field of competence that is most natural to us, namely, the Finnish language. Definition of phonemes in Finnish is rather simple: for instance, the diphthongs are divisible into vowels. The problems in English are already very much different, since the number of phonemes is large (of the order of 40 to 60), and, for instance, the diphthongs are often regarded as indivisible phonological units.

At any rate the equipment we have implemented has been finished, and it deals with all the phonemes of speech, not just a subset of the easiest or phonetically most interesting cases.

## 2. FEATURE COMPUTATION

Every pattern recognition system needs a preprocessing stage by which a number of characteristic features from the primary signals is first extracted, before performing the statistical decisions that lead to the recognition of input patterns. For feature computation, in our experimental system the speech signal is first digitized at a rate of 12.8 kHz and transformed by a 256-point FFT every 10 ms using a 256-point (20 ms) Hamming window. The 128-component logarithmic power spectra are then computed from the output of the FFT.

Previously [1], the logarithmic power spectrum was smoothed and 15 components of it (i.e., single spectra) were picked as features: 12 equally spaced points in the range from 200 Hz to 3 kHz, and three points in the range from 3 kHz to 5 kHz. However, it has turned out to be a slight improvement to use *mel-scale* spectral features with linear frequency spacing below 1 kHz and exponential above that. These can be obtained as outputs, $X_k$, of a bank of triangular-shaped overlapping band-pass filters with mel-scale frequency spacing as described in [2].

Spectral analysis of the speech signal has been almost standard in speech processing, whereas there exist several choices for speech features derivable from it. For the present, the two main alternatives for feature sets based on speech spectra are the short-time Fourier amplitude spectra as such, and the nonlinear combination of two cascaded Fourier transformations, which results in another feature set called the *cepstrum*.

About seven years ago, the only preprocessing transformation we could compute in real time by available commercial signal processor chips (e.g., Texas Instruments TMS 32010) was the Discrete Fourier Transform. Frequency analysis has been traditional in acoustics, and speech analysis has almost exclusively been based on so-called *sonograms* (or *spectrograms*), i.e., sequences of short-time spectra plotted vs. time. In the automatic identification of these spectra and their division into phonemic classes there exists at least one principal problem: the speech voice is mainly generated by the glottal chords, and the pulsed flow of air causes characteristic acoustic resonances in the vocal tract. According to generally known system-theoretic principles, the vocal tract, approximately at least, can be described by a linear transfer function, and the operation of the glottal chords as a system input that consists of a periodic train of impulses. The latter has a broad spectrum of frequency components spaced periodically apart in the frequency scale at multiples of the fundamental (glottal) frequency. The output voice signal thus has a spectrum that is the product of the transforms of the glottal air flow and that of the vocal tract transfer function, respectively. It may then be clear that the voice spectrum is modulated by all the harmonics of the fundamental frequency. This modulation is not stable; its depth, and the locations of its maxima and minima in the frequency scale vary with articulation and the pitch of speech. It also depends on the speaker and his/her physical condition. In direct template comparison between two spectra, severe mismatch errors may therefore be caused due to this variable modulation.

In the *cepstral analysis*, the above periodic modulation can to a great extent be eliminated by a series of operations. From the logarithmized Fourier amplitude spectrum, a second Fourier amplitude spectrum is taken, whereby the periodicity of the modulation is converted into a single peak on the new transform (cepstrum) scale. This peak is easily filtered out by a band-rejection filter.

The *mel-scale cepstral coefficients, MFCC*, can be computed from the outputs $X_k$ by the cosine transformation

$$MFCC_i = \sum_{k=1}^{N} X_k \cos\left[i\left(k - \frac{1}{2}\right)\frac{\pi}{N}\right], i = 1, 2, \ldots, M, \quad (1)$$

where $X_k$ is the output of the $k$th filter, $N$ is the number of filters, and $M < N$ is the number of cepstral coefficients. The coefficients can then be weighted in various ways in order to reduce undesired variability in the speech spectrum (called liftering) [3].

The best parameter combination (the number of linearly spaced channels, the number of exponentially spaced channels, the division point between them, the number of cepstral coefficients, and the form of liftering) was determined experimentally.

Computation of the cepstrum in real time has now become possible on account of new signal processor chips, as well as special coprocessor boards using them (e.g., Loughborough Sound Images PCS/320C30, which uses the Texas Instruments TMS 320C30 chip). This is the new preprocessing method we have chosen for our newest design.

## 3. NEW STATISTICAL PATTERN RECOGNITION ALGORITHMS

Each of the feature vectors obtained above is classified into (identified with) one of the 21 phonemic classes of the Finnish language. There exist many alternative pattern recognition algorithms for this subtask. In order to achieve the highest possible speed without losing accuracy, we have used a special comparison method of our own called *Learning Vector Quantization (LVQ)* [4]. The basic idea is that each phonemic class is represented by a small number (say, one to a few dozen) *reference (weight, codebook) vectors*. The exact values of these reference vectors are determined in so-called supervised adaptive learning processes, of which we have developed three versions: LVQ1, LVQ2, and LVQ3. The central idea in such adaptive learning is the following. Assume that we have first collected a number of training samples, each one with its correct phonemic identity known. Some initial values for the reference vectors must first be set, for instance, from the vicinity of the respective class means. When the training samples are input, one at a time, the most similar reference vector is first identified. If it has the same phonemic class identity as the training sample, this reference vector is corrected towards the input sample. If the most similar reference vector has a different class identity, the correction is made in the opposite direction. When the corrections are small and the training samples are reiterated a sufficient number of times, all the reference vectors converge asymptotically into values that approximate to the class regions of the input patterns, even when the latter have very nonlinear borders or are made up of disconnected subregions. This approximation is comparable in accuracy to the best nonlinear statistical decision methods, but is orders of magnitude faster to compute.

Classification of a pattern vector is now a very fast, computationally light task: the unknown vector is simply compared with the list of reference vectors, and labeled according to the most similar one. With the fast signal-processor chips, comparison with even 1000 reference vectors can be made every 10 ms. The high classification accuracy results from careful prior optimization of the reference vectors.

Since speech is dynamic in nature, such straightforward classification of short-time feature vectors cannot be expected to yield the best accuracy in the labeling of the sequences. We made experiments in concatenating several consecutive cepstral feature vectors to longer "context vectors" to take time dependencies between adjacent speech segments into account, and then classified such vectors into phoneme classes. It turned out that while the recognition accuracy of manually picked phonemic samples from the region of best stationarity is better when wider context (time window) is used, in the recognition of continuous speech a sharper resolution is more favorable; thus we used concatenations of three consecutive vectors (30 ms time window).

## 4. DECODING OF QUASIPHONEME SEQUENCES INTO PHONEMES

The symbolic labels produced every 10 ms by the vector quantizer, called quasiphonemes, can be interpreted as output symbols of a discrete-observation hidden Markov model. A suitably structured model can then be used in phonemic decoding, i.e., correcting and grouping the quasiphonemes into real phonemic transcriptions.

The present system first models each phoneme as a separate Markov source [5, 6]. The simple topology shown on the right-hand side of Figure 1 was found to give good results. Three separate codebooks were used, specially in order to enhance the recognition of unvoiced plosives. The main LVQ1-codebook is trained using all phonemes except unvoiced plosives. An additional LVQ1-codebook is trained by unvoiced plosives only using three concatenated feature vectors. The third codebook represents the power of the speech signal. Thus each transition in the model has three distributions of discrete symbols. These models are then trained using the Baum-Welch algorithm [5]. Training data consists of up to two hundred phoneme segments per model. Phoneme models are then connected in parallel with empty transitions (i.e. transitions that do not produce output symbols). Initial and final states are also added to the combined model. The decoding algorithm is allowed to loop back from the final state of a phoneme model to the initial state of any other model. Thus the number of consecutive phonemes is not fixed and the model can be used to decode a sequence of any length. The resulting model appears as shown in Figure 1. This type of combined model has also been used in [6]. Probabilities for the transitions between phonemes were estimated separately from a large text corpus.

The Viterbi-search [5] is used in decoding quasiphoneme strings. This algorithm finds the most probable state path in the combined model. As a result, the phoneme models along this path give the most likely phoneme sequence. Durations of states in phoneme models are also recorded during the search. They are used in distinguishing long phonemes from the short ones by comparing the durations to individual threshold values. This is an important distinction because in Finnish about 10% of the phonemes appear as prolonged and in the orthography they are denoted by two identical successive symbols.

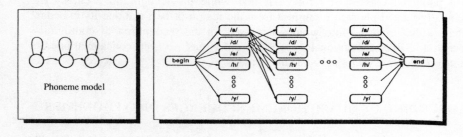

Fig. 1. The structure of the phoneme model and the combined model in HMM decoding.

## 5. TAKING LOCAL CONTEXT INTO ACCOUNT

At phoneme level the coarticulation effects and other systematic errors can be handled in symbolic form by a learning grammar called Dynamically Expanding Context [7]. The central idea behind it is to derive unique symbol to symbol(s) transformation rules from two streams of symbols: the source stream, and the desired target stream. In speech recognition the source stream is the phonemic transcription produced by the phonemic decoder, and the target stream is the correct transcription, or the orthographic form of the utterance.

The algorithm starts by creating initial context-free transformation rules from the aligned two streams. The initial rules are formed by segmenting the aligned streams by symbols. One segment of the source stream acts as the condition part of a rule, and the corresponding segment of the target string constitutes the production part of the rule. Now, these initial rules may conflict each other, which means that two rules with the same condition part may have different productions. These cases are specialized by dynamically expanding the discrete symbolic context around the initial condition parts by just a sufficient amount to make the condition part of the rules unique. The rules will then have condition parts of variable length. In this way the generality of the rules will be optimum, and still, all the conflicts within the training cases are solved. For details of the algorithm, see [7].

## 6. HARDWARE IMPLEMENTATIONS

Around 1984 we already had the prototype for the processor board, which was initially intended to accelerate the pattern recognition computations. Together with a smaller auxiliary board it had two TMS 32010 signal processors, one for to compute the 256-channel FFT every 10 ms, and the other mainly intended for the pattern recognition operations. It had 576 kilobytes of memory capacity, and could be programmed to perform a variety of tasks; the boards were under the control of an on-board Intel 80186 microprocessor chip. In principle we could have performed even the grammatical postprocessing computations on this board, to obtain the final text output. In practice it turned out that since many different modes of operation were provided, a dialogue language was indispensable in practical demonstrations and experiments; and since the board was then anyway hooked to a PC, we could more flexibly perform all the different postprocessing and output operations by the PC/AT type host computer. This system has been described in [1] and [8].

In order to implement the newer phase of our system, the coprocessor board of our own construction was replaced by the PCS/320C30 board mentioned earlier, which was interfaced with a PC. We wanted to use a workstation system for flexible experimentation and benchmarking with different combinations of algorithms. To this end, all the program modules were made transportable from one computer to another, and provided with fast intercommunication. The workstation demonstration was implemented for the ICANN-91 conference held at our university in June, 1991 [9]. Our goal in near future is to use the Silicon Graphics workstation solely, provided with its own signal processor, to perform all the computations; in the meantime, the external coprocessor board must be used, and it needs the PC that communicates with the workstation. The same program structure, however, can be used for both options.

# 7. EXPERIMENTS

Larger-scale experiments were performed for three male Finnish speakers in speaker-dependent mode. For each speaker, four repetitions of a set of 311 words were used. Each set contained 1737 phonemes. Three of the repetitions were used for training, and the remaining one for testing. Four independent runs were made by leaving one set at a time for testing. The main codebook contained 216 vectors, and the /k,p,t/ -codebook 100 vectors, respectively. For the LVQ training, only samples picked from the centers of the stationary regions of the phonemes were used.

Although we have performed quite an extensive benchmarking study of many alternative methods [9], it may suffice to report here figures from the configuration that we finally chose for our demonstration; only this option has been described in the present paper. The results have been reported in Table 1.

According to a widespread usage, *correctness* means the fraction of symbols correctly recognized from text, whereby insertion of new (wrong) symbols is not taken into account. A more realistic measure of performance is given by *error rate* that gives the percentage of all the different types of errors: insertion, deletion, and replacement.

*Table 1.* Overall recognition correctness and error rate (per cent)

|  | Correctness | Error rate |
|---|---|---|
| After the Hidden Markov Model | 93.1 | 9.4 |
| After the Dynamically Expanding Context | 96.0 | 5.8 |

# 8. DISCUSSION

Naturally there exist many alternative ways to implement the different subtasks, namely, feature selection, pattern recognition, decoding, and postprocessing. Furthermore, if the system had been designed to recognize isolated words from sentences, the word recognition accuracy would still be increased if some kind of language model (that predicts the sequences of words on the basis of the conditional probabilities of their occurrences) were used at the output stage. However, a "phonetic typewriter" is supposed to transcribe *unlimited* text, whereby, by definition, the use of any restrictive language models should be avoided.

It is often stipulated that a speech recognizer be speaker-independent, i.e., it would then not require any prior speech samples from a particular person. Especially if speech recognizers are used for making queries to data banks over telephone lines, it would be desirable that no voice samples need be given. Such speaker-independence has only been achieved in isolated-word recognizers where the vocabulary is not larger than a few dozen different words.

A compromise between "generality" and cost of the system can be reached if the recognizer is made "speaker-adaptive", as our system is. It means that the internal voice model (the reference vectors) of the pattern recognition stage has been preadjusted for an average speaker, and so the most time-consuming learning phases need not be repeated every time. When tuning the system to a new speaker, some 150 selected words have to be spoken first, and in ten minutes or so (including dictation of the samples) the system will be readjusted to the new speaker. Of course, such speaker-specific voice models can be kept in memory and recalled while logging in. The system later learns from errors committed during use, and then continually improves its performance.

In a search of fundamentally different approaches, however, one may find that improved preprocessing, if it could be implemented cost-effectively, might still improve the performance. The FFT and the cepstrum are not exactly what the biological hearing organs are computing; at least some kind of synchronization of the triggering of the neural cells to the pressure waves of speech seems to take place in the ear. The organs of the inner ear are adapting to variations in sound intensity with very small time constants. Notice also that there are two ears; the outputs from the many preprocessing stations in the auditory neural pathways then very probably correspond to some high-order correlation functions of the binaural signals that are not yet computed in the artificial speech recognizers. Some kind of attention mechanism that would adaptively control both preprocessing and recognition circuits could be used to make the system more selective to important features of the input and normalize the patterns. It would be very intriguing if all such functions, especially the peripheral ones, were formed adaptively and completely automatically in the more developed artificial neural networks and architectures. The research of artificial neural networks ought to take more steps in that direction.

**Acknowledgement.** This speech recognition system was implemented as a joint project of our laboratory [9]. It includes work by the following persons: K. Torkkola, J. Kangas, P. Utela, S. Kaski, M. Kokkonen, and M. Kurimo.

# REFERENCES

[1] T. Kohonen, "The Neural Phonetic Typewriter," *IEEE Computer*, vol. 21, pp. 11-22, March 1988.

[2] S.B. Davis and P. Mermelstein, "Comparisons of Parametric Representations for Monosyllabic Word Recognition in Continuously Spoken Sentences," *IEEE Transactions on Acoustics, Speech, and Signal Processing*, vol. 28. pp. 357-366, 1980.

[3] Y. Tohkura, "A Weighted Cepstral Distance Measure for Speech Recognition," *IEEE Transactions on Acoustics, Speech, and Signal Processing*, vol. 35, pp. 1414-1422, 1987.

[4] T. Kohonen, "The Self-Organizing Map", *Proceedings of the IEEE*, vol. 78, pp. 1464-1480, Sept. 1990.

[5] L.R. Rabiner, "A Tutorial on Hidden Markov Models and Selected Applications in Speech Recognition," *Proceedings of the IEEE*, vol. 77, pp. 257-286, 1989.

[6] A.-M. Deroualt, "Context-Dependent Phonetic Markov Models for Large Vocabulary Speech Recognition," in *Proceedings of IEEE International Conference on Acoustics, Speech, and Signal Processing (ICASSP87)*, Dallas, Tx., April 6-9, 1987, pp. 360-363.

[7] T. Kohonen, "Dynamically Expanding Context, with Application to the Correction of Symbol Strings in the Recognition of Continuous Speech," in *Proceedings of 8th International Conference on Pattern Recognition*, Paris, France, October 27-31, 1986, pp. 1148-1151.

[8] T. Kohonen, K. Torkkola, M. Shozakai, J. Kangas, and O. Ventä, "Phonetic Typewriter for Finnish and Japanese," *Proceedings of the IEEE International Conference on Acoustics, Speech, and Signal Processing (ICASSP88)*, New York City, USA, April 1988, pp. 607-610.

[9] K. Torkkola, J. Kangas, P. Utela, S. Kaski, M. Kokkonen, M. Kurimo, and T. Kohonen, "Status Report of the Finnish Phonetic Typewriter Project," in *Artificial Neural Networks*, T. Kohonen, K. Mäkisara, O. Simula, and J. Kangas, Eds., Proc. of the 1991 International Conference on Artificial Neural Networks (ICANN-91), Espoo, Finland, June 24-28, 1991, Amsterdam, New York, Oxford, Tokyo: Elsevier (North-Holland), 1991, vol. 1, pp. 771-776.

# WORD RECOGNITION WITH THE FEATURE FINDING NEURAL NETWORK (FFNN)

Tino Gramss

Drittes Physikalisches Institut der Universität Göttingen
Bürgerstraße 42–44, W-3400 Göttingen, Germany
Tel.: +49 551/397731, Fax: +49 551/397720
e-mail: gramss@up3spr1.gwdg.de

**Abstract** – An overview of the architecture and capabilities of the word recognizer FFNN ("Feature Finding Neural Network") is given. FFNN finds features in a self-organizing way which are relatively invariant in the presence of time distortions and changes in speaker characteristics. Fast and optimal feature selection rules have been developed to perform this task. With FFNN, essential problems of word recognition can be solved, among them a special case of the figure-ground-problem. FFNN is faster than the classical DTW and HMM recognizers and yields similar recognition rates.

## 1 PROBLEMS TO SOLVE

Five problems concerning the recognition of isolated words by machines will be discussed. FFNN provides solutions to each of these problems.

### 1.1 Invariance

Patterns used to train pattern recognition systems will not be exactly identical to those patterns presented during recognition mode. Therefore, one of the tasks an appropriate recognizer must solve is the search for invariant features and their extraction from the presented patterns. In speech recognition these invariances are not exactly known a priori. It is especially difficult to find invariances between different speakers. This is necessary for speaker-independent recognition. FFNN uses new fast algorithms to find invariant features. An important special case of the invariance problem is that of time normalization.

### 1.2 Time normalization

In general, the words used to train and test a recognizer differ in their temporal microstructure. Words may be spoken quickly or slowly. In addition, even if they had the same length, small distortions of the time axis might be necessary for an optimal projection from a training onto a test word. FFNN finds features in a self-organizing way that are relatively invariant against time distortions.

## 1.3 Generalization

This problem has a close relationship to the invariance problem. The projection from input to output patterns may be too coarse to achieve high recognition results or too fine to generate appropriate generalization. To select invariant features is essential to solve the generalization problem. Moreover, using a sufficient number of well selected, sparse coded features from sophisticated preprocessing allow to apply linear classifying networks. In our approach, linear classification shows better generalization capabilities than nonlinear classification.

## 1.4 Figure-ground-discrimination

This task involves separating the "figure" of interest from the insignificant background. In practice one has to deal with speech signals that are usually disturbed by noise and other speakers. Satisfactory technical solutions to this problem do not exist at present. FFNN is able to solve it in a special case: If two words from two speakers are presented simultaneously, the network is able to recognize both words.

## 1.5 Scaling of training times

While for recognition tasks neural nets are usually very fast compared to other methods, the training time is usually very long and scales badly with the size of vocabulary and the size of the network. This is especially a problem in cases where the user of the word recognizing system wants to train the net with his own vocabulary, or wants to extend the current one, i.e. if he wants to make use of the adaptivity in the neural net. The feature selector, as well as the linear net used to classify the features, both scale very moderately and predictable with the number of patterns to be learned and the size of the net.

# 2 PREPROCESSING

For the preprocessing of the acoustic signals, a simple model of the human ear and the first stages of the neural processing of the brain was simulated, based on the results of psychoacoustic, neurobiology and anatomical research. It has been shown that even with a simple and fast simulation of the peripheral acoustical processing of animals, it is possible to achieve a significant increase of the recognition rates in isolated word recognition experiments [10]. This preprocessing includes three stages.

1. Calculation of 19-channel short time spectra using a nonlinear psychoacoustic frequency (mel-)scale.
2. Transformation from power to loudness.
3. Contrasting the spectrograms by loudness adaptation.

In the simplest case, the contrasting is done by normalizing each bark spectrum (zero spectra excluded) and then differentating each bark channel over time. This has been done for the experiments described in this paper. See [10] for more sophisticated contrasting procedures and a detailed description of the preprocessing.

Our model for the loudness adaptation is beneficial in that the result is a sparse-coded spectrogram. It is well-known that a linear neural net has a high information capacity if the patterns used are sparsely coded. Moreover this kind of coding helps to solve the figure-ground-problem of speech (see section 5).

## 3 EXTRACTION OF FEATURES

With FFNN the extraction of features is done by special $\Sigma$-$\Pi$-cells subsequently to the preprocessing described above. Speaker independent experiments in particular have shown that some of the following features $z(\mathbf{p}, f, t)$ provide sufficient invariance.

$$z(\mathbf{p}, f, t) = z(f_0, t_0; \Delta f, \Delta t; f, t) = y_f(t) \sum_{f'=0}^{\Delta f - 1} \sum_{t'=0}^{\Delta t - 1} y_{f+f_0+f'}(t + t_0 + t') . \quad (1)$$

The idea of (1) is to extract second order terms which will detect, for example, the beginning, the end, or the direction and velocity of frequency changes. Invariance is improved by "blurring" the spectrogram within a rectangular window. For the classification, the values $z(\mathbf{p}, f, t)$ are summed over time.

The time delays $t_0 + t'$ in (1) are small (less than 200 ms). It has been found important to use small values for the time delays in order to extract features that are relatively invariant in presence of small distortions of the time axis. The invariance of short-time articulatory structures has already been described in [1]. This is the way FFNN by-passes time consuming dynamic programming.

In the first instance the free parameters collected in the parameter vector **p** have been chosen randomly and are only restricted by the number of bark channels and the maximum time delay of 200 ms, whereas for the parameters $f$ and $t$, all possible values have been taken. Parameters for features that are invariant against change of speaker and time distortions are then selected using fast algorithms (see [4] and section 4.2).

## 4 THE ADAPTIVE PARTS OF THE NET

There are two adaptive stages in the operation of FFNN: The selection of features and the classification of the features in order to recognize words.

### 4.1 The classification net

There already exist several approaches to isolated word recognition using neural networks, e.g. [5, 6, 7, 8, 9, 10, 11]. The probably most famous method (TDNN, [5]) uses a network with hidden layers. The advantage of hidden layers is their ability to classify input patterns in an arbitrary nonlinear manner. *Nevertheless experiments with FFNN have shown that there is considerable advantage to be gained using networks without hidden layers when applying our approach to isolated word recognition.* This should not be very surprising for the following three reasons:

1. Linear nets learn much faster. If a pseudoinverse method [2, 3] or the conjugate gradient descend is applied, the computation times scale only with $\min(PN^2, NP^2)$ where $P$ is the number of patterns and $N$ the number of features, i.e. input cells of the net. Moreover there are no problems associated with local minima.

2. It is simpler for a linear neural net to learn additional patterns. If a network has been previously trained with a pattern set, then in the case of most multilayer nets, learning an additional pattern involves presenting all the patterns again. Otherwize the old patterns may not be classified correctly any more. This implies that all the previous patterns have to be stored somehow. This is not possible for some applications. However this is not necessary in case of applying an iterative pseudoinverse method [3] for linear nets. Only one presentation of the new pattern is sufficient to learn this pattern in an optimal (least mean square) sense.

3. Linear nets show better generalization performance and thus yield higher recognition rates. The pruning algorithm described below selects relatively independent features which have to be classified. Even with a small number of these features it is not at all difficult to get 100% recognition rate on the

training data, even on a difficult fourty-word vocabulary. *In this case the classification problem is not to form complex decision regions but to solve the generalization task.* One does not require more general classification capabilities and in fact a nonlinear classifier gives poorer generalization because of its additional degrees of freedom [11].

For the experiments described below only linear classifiers were used. It should be emphasized that nevertheless our net works in a highly nonlinear manner. The nonlinearity (which is essential) is introduced by the preprocessing and feature extraction.

In matrix notation, the classification can be described by $\tilde{\mathbf{Y}} = \mathbf{W}\mathbf{X}$ . The columns of $\mathbf{X}$ (a $N \times P$ matrix) contain the input patterns. The output patterns are collected in $\tilde{\mathbf{Y}}$ ($M \times P$), and $\mathbf{W}$ ($M \times N$) is the weight matrix. $N$ is the number of features, which is a function of time, $M$ is the number of words in the vocabulary and $P$ the total number of pattern vectors to learn. The target patterns are $M$-dimensional unit vectors, each specifying one word of the vocabulary. A word is said to be recognized if the corresponding output cell shows highest activity, i.e. the actual output pattern is closest to the corresponding target pattern.

## 4.2 Selection of relevant features

It is not clear a priori which of the features described in section 3 are important to recognize words in our vocabularies, i.e. which features provide a relatively invariant description of certain words. Probably not all the input cells are necessary to achieve high recognition results. For generalization reasons it is even possible that recognition rates will increase if some feature detectors are skillfully removed. In any case it is advantageous to minimize the number of input cells, for computational reasons. Consequently algorithms have been developed to decrease the size of networks without reducing their performance [12, 13, 14].

In our case the "relevance" $\mathcal{R}_i$ of a feature detector $i$ is the difference between the total Euclidean error resulting from recognition experiments applying the network with and without input cell $i$:

$$\mathcal{R}_i = \Delta E_i = E(\text{with cell } i) - E(\text{without cell } i) \ . \tag{2}$$

$E$ is the total Euclidean squared error. In the case of linear nets

$$E = \text{Tr}\left[\left(\mathbf{Y} - \tilde{\mathbf{Y}}\right)\left(\mathbf{Y} - \tilde{\mathbf{Y}}\right)^{\text{T}}\right] \ , \tag{3}$$

where the rows of $\mathbf{Y}$ contain the target patterns.

Several fast methods to extract the more invariant features based on this relevance criterion have been developed, which will be published in [4]. One relatively simple algorithm which provides a good compromise between computational efficiency and programming simplicity will be described here. It is the following "pruning rule" which gives rise to astonishingly good results, especially in case of small sets of training data (i.e. $P < N$).

The pruning-rule reads as follows:

1. Remove features which are never active.
2. Add low-level-noise to the pattern vectors so that the pattern matrix $\mathbf{X}$ has maximal rank.
3. Set $\Theta = 0$, where $\Theta$ is the "pruning threshold".
4. Calculate the optimal weight matrix $\mathbf{W}$. We used the pseudoinverse method [2, 3].
5. Calculate the vector of relevances $\vec{\mathcal{R}} = (\mathcal{R}_1, \mathcal{R}_2, \ldots, \mathcal{R}_N)$. We used a fast algorithm, see below.
6. Normalize: $\vec{\mathcal{R}} = N(t)\vec{\mathcal{R}}/\|\vec{\mathcal{R}}\|$ .
7. Prune: Remove all input cells $i$ with $\mathcal{R}_i < \Theta$ .
8. Set $\Theta = \Theta + d\Theta$, $d\Theta \ll 1$ .
9. Go back to step 4 if the recognition rates on the test data are sufficiently high. Otherwize the pruning is finished.

An input cell $i$ is "pruned" away if its relevance $\mathcal{R}_i$ is smaller than $\Theta$.

For $P < N$ the direct computation of the relevance vector (step 5) applying (2) would be computationally most costly. The number of multiplications[1] then scales proportional to $N^2 PM$. However it is possible to use the identity

$$\mathbf{X}_i = \mathbf{X} - \mathbf{e}_i \mathbf{x}_i^T ,  \quad (4)$$

where $\mathbf{e}_i$ is the ith $N$-dimensional unit vector and $\mathbf{x}_i^T$ is the ith row of $\mathbf{X}$, i.e. $\mathbf{X}_i$ is the matrix for the case where the feature $i$ is pruned away. After some manipulation, the equations (2), (3) and (4) yield

$$\mathcal{R}_i = \sum_{k=1}^{M} \left( \mathbf{x}_i^T \mathbf{x}_i [\mathbf{w}_i]_k^2 + 2 [\mathbf{w}_i]_k [(\mathbf{Y} - \mathbf{WX}) \mathbf{x}_i]_k \right) . \quad (5)$$

Here $[\mathbf{a}]_k$ refers to the kth component of the vector $\mathbf{a}$, and $\mathbf{w}_i$ is the ith column of the optimal weight matrix $\mathbf{W}$. For $P < N$, the second term under the sum of (5) vanishes (the exact solution $\mathbf{Y} = \mathbf{WX}$ exists).

Applying (5) to calculate the vector of relevances involves only the order of $N(P+M)$ multiplications if $P \leq N$ and $N(MN+NP+PM)$ multiplications

---

[1] Here and in the following estimates for computational complexity, only the highest order terms are considered.

otherwise. Thus step 4 becomes the most computational stage if $P \leq N$. The order of $\min(PN^2, NP^2)$ multiplications are necessary in case of using the pseudoinverse method. Further reduction of the computation time is possible but less simple [4].

An important aspect of the pruning rule described is now noted: At the beginning of pruning when the network is large, a relatively large number of feature detectors are removed in each pruning cycle. For the large number of later cycles one only has to deal with a relatively small net.

The dependence of the recognition rates on the number of surviving cells is illustrated in figure 1.

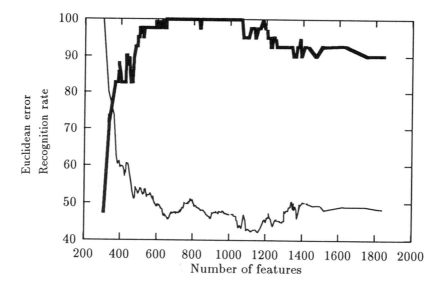

Figure 1: The dependence of recognition rate (thick line) and normalized mean Euclidean error (thin line) on the number of surviving cells. The network learned ten German digits from five speakers. Test data from a different speaker was used.

## 5 THE FIGURE-GROUND-PROBLEM

An important example of the figure-ground-problem in many applications is the recognition of speech with interference from a second speaker. This problem is very difficult to solve, because the signal (figure) has a similar long-time spectral structure as the noise (ground). This is why simple filtering techniques can not be applied.

With FFNN it is not necessary to separate the speech in order to recognize interferring words. To demonstrate this we added the time signals of words from two speakers. Recognition experiments on these combined signals were performed. The two words are said to be recognized if the activities of the two corresponding output cells were highest. FFNN is able to respond correctly to two words presented simultaneaously. No change of its architecture is necessary to solve this figure-ground-problem. The average recognition rate for a speaker-dependent experiment, with all pairs of the ten German digits is greater then 90%. This is probably due to the sparseness of the preprocessed spectral patterns: Features belonging to different words are extracted in such a way that they rarely overlap, so they can be recognized without confusion inspite of the highly nonlinear classification.

## 6 RECOGNITION EXPERIMENTS

We used a standard English database and a German vocabulary, specially recorded to fit our needs. The standard database was the RSRE-vocoder-database. The vocabulary we recorded (DPI-database) consisted of fourty German words. These words were choosen specifically to control the movements of a robot and included the German digits. The recognition rates and times were compared with a standard DTW-recognizer. Speaker-dependent as well as speaker-independent experiments with the RSRE- and the DPI-database were performed. The recognition rates and the comparison with other recognition methods are summarized in tables 1 and 2.

## 7 ACKNOLEDGEMENTS

We are grateful to Professor Dr.M.R.Schroeder for his stimulating interest in this work and to Dr.R.Moore at the RSRE for allowing the use of the RSRE digit database. This work was supported by the "Studienstiftung des deutschen Volkes" and by the Ministry for Research and Technology of Germany.

Table 1: SPEAKER INDEPENDENT RECOGNITION RATES

| model | DTW | HMM (estimated) | FFNN |
|---|---|---|---|
| Recognition rates | | | |
| RSRE-data | 98.1% | — | 97.7% |
| DPI-data | 84.0% | — | 97.2% |
| Comp. time for recognition in real-time units | 447 | 45 | 16 |

Speaker-independent experiments on ten English and German digits. 16 real-time units means that a SUN 3/60 is 16 times slower than would be necessary for a real-time operation. Although no processing was done in parallel (for which the neural nets are very predisposed) the neural nets had the shortest recognition times.

Table 2: SPEAKER DEPENDENT RECOGNITION RATES

| model | DTW | HMM (estimated) | FFNN |
|---|---|---|---|
| Recogniton rates | 98.0% | — | 98.7% |
| Comp. time for recognition in real-time units | 447 | 45 | 16 |

Speaker-dependent experiments on the 40 words from the DPI database.

# 8 REFERENCES

[1] O.Fujimura (1981): "Temporal organization of articulatory movements as a multidimensional phrasal structure", Phonetica, **38**, 66-83.
[2] R.Penrose (1956): Proc. Cambridge Philos. Soc. **52**, 17.
[3] T.Kohonen (1988): Self-Organization and Associative Memory, 2nd Ed., Springer Series in Information Sciences **8**, Chapters 5 and 7.
[4] T.Gramss (1991): "Fast algorithms to find invariant features for a word recognizing neural net.", to appear at the IEE Second International Conference on Artificial Neural Networks, Bournemouth, UK, 18-20 November 1991.
[5] K.J.Lang, A.H.Waibel, G.E. Hinton (1990): "A time-delay neural network architecture for isolated word recognition", Neural Networks **3**, No.1, 23-43.
[6] D.J.Burr (1988): "Neural net recognition of spoken and written text", IEEE Trans. Acoust., Speech and Signal Processing **36**, 1162-1168.
[7] B.Kämmerer, W.Küpper (1990): "Experiments for isolated-word recognition with single- and two-layer perceptrons", Neural Networks **3** No.6, 693-706.
[8] I.S.Howard, M.A.Huckvale (1989): "Two-level recognition of isolated word using neural nets." First IEE meeting on ANN, London.
[9] F.Kowalewski, H.W.Strube (1990): "Word Recognition with a Recurrent Neural Network", in: R.Eckmiller, G.Hartmann and G.Hauske (eds.), Parallel Processing in Neural Systems and Computers, Elsevier, 391-394.
[10] T.Gramß, H.Behme, H.W.Strube (1990): "Recognition and Reproduction of Isolated Words by Layered Neural Networks", in: J.Kindermann, A.Linden (eds.), GMD-Bericht Nr. 185: Distributed Adaptive Neural Information Processing, 23-35.
[11] T.Gramss, H.W.Strube (1990): "Recognition of isolated words based on psychoacoustics and neurobiology", Speech Commun. **9**, 35-40.
[12] M.C.Mozer, P.Smolensky (1989): "Using relevance to reduce network size automatically", Connection Science **1**, No.1, 3-16.
[13] E.D.Karnin (1990): "A simple procedure for pruning back-propagation trained neural networks", IEEE Transactions on Neural Networks **1**, No.2, 239-242.
[14] K.A.Fischer, H.W.Strube (1991): "A simple word-recognition network with the ability to choose its own decision criteria", Neural Networks for Signal Processing – Proceedings of the 1991 IEEE-SP Workshop, Princeton, New Jersey, September 29 - October 2, 1991.

# NEW DISCRIMINATIVE TRAINING ALGORITHMS BASED ON THE GENERALIZED PROBABILISTIC DESCENT METHOD

Shigeru Katagiri*, Chin-Hui Lee** and Biing-Hwang Juang**

* ATR Auditory and Visual Perception Research Laboratories,
Sanpeidani, Inuidani, Seika-cho, Soraku-gun, Kyoto 619-02, Japan

** Speech Research Department, AT&T Bell Laboratories
600 Mountain Avenue, Murray Hill, NJ 07974, USA

Abstract - We developed a *generalized probabilistic descent* (GPD) method [1] by extending the classical theory on adaptive training by Amari [2]. Our generalization makes it possible to treat *dynamic* patterns (of a variable duration or dimension) such as speech as well as *static* patterns (of a fixed duration or dimension), for pattern classification problems. The key ideas in GPD formulations include the embedding of time normalization and the incorporation of smooth classification error functions into the gradient search optimization objectives. As a result, a family of new discriminative training algorithms can be rigorously formulated for various kinds of classifier frameworks, including the popular dynamic time warping (DTW) and hidden Markov model (HMM). Experimental results are also provided to show the superiority of this new family of GPD-based, adaptive training algorithms for speech recognition.

## 1. INTRODUCTION

It has been shown that several discriminative training algorithms lead to performance improvements in speech recognition [3-6]. These algorithms, due to the lack of prior analytical results, mostly rely on heuristic rules and the optimality of the resultant classifiers and convergence properties of the algorithms were never rigorously addressed. In view of the increasing complexity in classifier designs for speech recognition, a theoretical framework for discriminative training could prove beneficial in furthering the advancement of the technology. Motivated by the above concern, we present in this paper a new theory on adaptive discriminative training based on a generalized probabilistic descent method [1], and introduce a new family of GPD-based discriminative training algorithms. The original probabilistic descent (PD) method was developed by Amari for classifying *static* patterns in a fixed-dimensional vector space [2]. For speech recognition where the observation patterns are *dynamic* (of variable duration and unspecified dimension), there is a serious need for extending the fundamental thoughts of Amari. The main results in this paper are the incorporation of smooth classification error functions, generalization for handling dynamic patterns, and several practically implementable algorithms. In particular, we develop new discriminative training algorithms for DTW and HMM based speech recognition systems. The effectiveness of this new family of GPD-based algorithms is demonstrated by the results obtained from a set of speech recognition experiments.

## 2. GENERALIZED PROBABILISTIC DESCENT METHOD

To extend the original PD-based formulations to the classification of dynamic patterns, we use a four-step approach: 1) define a new probability measure such that dynamic patterns can be properly handled, 2) give a general functional form of discrimination, including the use of likelihood as a *discriminant function*, 3) define a proper *cost function* that approximates the recognition error rate and that can be minimized by descent methods, and 4) prove the convergence of the optimization procedure.

We consider a dynamic observation pattern $\mathbf{x}_1^T = (\mathbf{x}_1, \ldots, \mathbf{x}_T)$, where $\mathbf{x}_t \in \Re^M$ is a static observation vector at time $t$ and $T$ is a variable but finite natural number, representing the duration of $\mathbf{x}_1^T$. We are given a set of training tokens $\vartheta = \left\{ \mathbf{x}_1^{T_1}, \mathbf{x}_1^{T_2}, \ldots, \mathbf{x}_1^{T_N} \right\}$, where each $\mathbf{x}_1^{T_n}$ is known to belong to one of $K$ classes $\left\{ C_j \right\}_{j=1}^K$ and $T_n$ is the duration of the $n$-th token $\mathbf{x}_1^{T_n}$. Based on $\vartheta$, our goal is to design a classifier such that the misclassification rate on $\vartheta$ is minimized. A classifier consists of a set of parameters $\Lambda = \left\{ \lambda_j \right\}_{j=1}^K$ and a decision rule. The classifier should also be able to classify patterns not seen in $\vartheta$. This is often referred to the generalization capability of a classifier design.

We first define a new probability space for these dynamic patterns (see the details in [1]). The discriminant function, denoted by $g_j(\mathbf{x}_1^T; \Lambda)$, is then introduced to indicate the degree to which $\mathbf{x}_1^T$ belongs to $C_j$. A probability measure defined on this new space allows us to use the various distance functions and probability distributions including the HMM probability in defining a general discriminant function. The simplest example may be that $g_j(\mathbf{x}_1^T; \Lambda)$ is a distance function $D_j(\mathbf{x}_1^T)$ or a likelihood function $\ln\left\{ p_\Lambda(\mathbf{x}_1^T | C_j) \right\}$.

Here, we simply use the following decision rule

$$C(\mathbf{x}_1^T) = C_\alpha, \quad \text{if } \alpha = \underset{j}{\operatorname{argmax}}\{g_j(\mathbf{x}_1^T;\Lambda)\} \tag{1}$$

where $C(\cdot)$ is a class recognized by the classifier. Next, assume that a token $\mathbf{x}_1^T \in C_k$, one of tokens in $\vartheta$, is given. The *misclassification measure*, denoted by $d_k(\mathbf{x}_1^T;\Lambda)$, is defined to measure the degree of confusion between the correct class and the other competing classes for a given input token. We use an $L_p$ norm for defining a general, differentiable misclassification measure, to overcome the discontinuity problem in the original formalization [2]. There are many possibilities for the definition, and the following is a fundamental example accompanied by Eq. (1),

$$d_k(\mathbf{x}_1^T;\Lambda) = -g_k(\mathbf{x}_1^T;\Lambda) + \ln\left[\frac{1}{K-1}\sum_{j,j\neq k}\exp\{\eta g_j(\mathbf{x}_1^T;\Lambda)\}\right]^{1/\eta}. \tag{2}$$

Similar to the original approach, a cost function $\ell_k(\mathbf{x}_1^T;\Lambda)$ to be evaluated in training is defined as a nondecreasing, differentiable function, such as the *sigmoid function*, of the misclassification measure; $\ell_k(\mathbf{x}_1^T;\Lambda) = \ell_k(d_k(\mathbf{x}_1^T;\Lambda))$. An overall expected cost function is then defined as

$$L(\Lambda) = \sum_k \int \ell_k(\mathbf{x}_1^T;\Lambda)1(\mathbf{x}_1^T \in C_k)dp(\mathbf{x}_1^T,C_k) \tag{3}$$

where $1(\wp)$ is an indicator function of a logic expression $\wp$ and gives 1 if the value of $\wp$ is true and 0 otherwise. However, since the distribution of training data is usually unknown, we use an empirical average cost to approximate Eq. (3) and we have

$$L(\Lambda) = \frac{1}{N}\sum_{n=1}^{N}\sum_{k=1}^{K}\ell_k(\mathbf{x}_1^{T_n};\Lambda)1(\mathbf{x}_1^{T_n} \in C_k). \tag{4}$$

This empirical cost is also necessary to make possible the use of the likelihood as a discriminant function. The cost function can be of various forms depending on the optimization objectives of the application. For pattern classification problems, we can use it to approximate the error count, i.e. to assign zero cost when an input is correctly classified and to assign a unit cost when an input is not properly classified. This idea leads to a new formalization of the *minimum error* (the minimum Bayes risk) classification. For clarity of presentation, the details on this idea will be separately presented [7].

In the GPD-based adaptive learning, $\Lambda$ is adjusted by a small amount $\delta\Lambda(\mathbf{x}_1^T,C_k,\Lambda)$ every time a training token $\mathbf{x}_1^T \in C_k$ is given; i.e.

$$\delta\Lambda(\mathbf{x}_1^T,C_k,\Lambda) = -\varepsilon\mathbf{U}\nabla\ell_k(\mathbf{x}_1^T;\Lambda) = -\varepsilon\mathbf{U}\nabla\ell'_k(\mathbf{x}_1^T;\Lambda)\frac{\partial d_k(\mathbf{x}_1^T;\Lambda)}{\partial\Lambda} \tag{5}$$

where **U** is a positive-definite matrix and $\varepsilon$ is a small positive real number representing the adjustment step size. It can be shown that the above adaptation in Eq. (5) locally minimizes the empirical cost in Eq. (4) in a probabilistic sense [1-2].

The above GPD-based formulation gives us a theoretical background to develop a new family of discriminative training algorithms for many classifier structures including the DTW system, a distance classifier to which LVQ has been applied, and recent feed-forward artificial neural networks [8]. The above results also allow us to train an observation probability of a continuous density HMM. However, to complete our generalization such that a PD-based algorithm can be realized on all the model parameters in both continuous and discrete HMM's, we further prove the following convergence characteristics,

$$\sum_k \sum_n \delta L(\Lambda) = \sum_k \sum_n \left\{ \left( \delta \Lambda \left( x_1^{T_n}, C_k, \Lambda \right) \right)_\Phi \cdot \nabla L(\Lambda) \right\}$$

$$= -\varepsilon \mathbf{U} \left\{ \sum_k \sum_n \nabla \ell_k \left( x_1^{T_n}; C_k \right) \right\} \cdot \left( \nabla L(\Lambda) \right)_\Phi$$

$$= -\varepsilon \mathbf{U} \nabla L(\Lambda) \cdot \left( \nabla L(\Lambda) \right)_\Phi = -\varepsilon \mathbf{U} \| \nabla L(\Lambda) \|^2 \cos(\gamma) \leq 0 \qquad (6)$$

where $(\cdot)_\Phi$ denotes an orthogonal projection mapping the parenthesized vector onto the subspace spanned with the probability constraint and $\gamma$ is the angle between $\nabla L(\Lambda)$ and $(\nabla L(\Lambda))_\Phi$. This proof promises that the constrained adjustment of the HMM-based parameters will converge in the same sense as that in the unconstrained case. Note that the discussion here can be used for any probability-based system as well as an HMM classifier.

## 3. IMPLEMENTATIONS

In the following, we demonstrate some implementation examples. For classification of static patterns, we present two algorithms, a multi-reference distance classifier and a likelihood-based classifier. For classification of dynamic patterns in which nonlinear time normalization is required, we present two algorithms, one based on the integration of DTW and a multi-reference distance classifier, and the other based on the integration of HMM and a likelihood-based classifier. Experimental results will be given in Section 4.

### 3.1 Static Patterns

Here we treat a static pattern $\mathbf{x} \in \Re^M$. Consider a multi-reference distance classifier where $n_j$ references are used to represent class $C_j$. First we define the discriminant function as

$$g_j(\mathbf{x};\Lambda) = \left[\frac{1}{n_j}\sum_{i=1}^{n_j}\{D_{ji}(\mathbf{x})\}^{-\zeta}\right]^{-1/\zeta} \tag{7}$$

where $D_{ji}(\mathbf{x})$ is a distance function for the $i$-th reference of $C_j$, and $\zeta$ is an arbitrary positive number. The classification rule is simply the following operation based on the discriminant functions, i.e.,

$$C(\mathbf{x}) = C_\alpha, \quad \text{if } \alpha = \underset{j}{\operatorname{argmin}}\{g_j(\mathbf{x};\Lambda)\}. \tag{8}$$

For a training token $\mathbf{x} \in C_k$, the misclassification measure $d_k(\mathbf{x};\Lambda)$ is defined as

$$d_k(\mathbf{x};\Lambda) = g_k(\mathbf{x};\Lambda) - \left[\frac{1}{K-1}\sum_{j,j\neq k}\{g_j(\mathbf{x};\Lambda)\}^{-\eta}\right]^{-1/\eta} \tag{9}$$

where $\eta$ is an arbitrary positive number. The GPD-based adjustment rule in Eq. (5) is then applied to obtain the classifier parameters. The selection of $D_{ji}(\mathbf{x})$, $\zeta$ and $\eta$ allows us to implement many variations of this algorithm.

For example, the distance function $D_{ji}(\mathbf{x})$ can be chosen as the Euclidean distance between the input vector $\mathbf{x}$ and the reference codeword $\mathbf{m}_{ji}$ in the class $C_j$, i.e.

$$D_{ji}(\mathbf{x}) = \sum_{r=1}^{M}(x_r - m_{jir})^2 \tag{10}$$

where the subscript $r$ represents the $r$-th element of each vector. Let $\Lambda = \{\mathbf{m}_{ji}\}$ be the set of parameters in designing the multi-reference distance classifier. Then in the GPD training for a training token $\mathbf{x} \in C_k$, the parameter $m_{jir}$ is adjusted according to Eq. (5) as follows

$$\frac{\partial d_k(\mathbf{x};\Lambda)}{\partial m_{jir}} = \frac{\partial g_k(\mathbf{x};\Lambda)}{\partial m_{jir}}$$

$$= -\frac{2}{n_j}\left[\frac{1}{n_j}\sum_{l=1}^{n_j}\left\{\frac{D_{jl}(\mathbf{x})}{D_{ji}(\mathbf{x})}\right\}^{-\zeta}\right]^{-\frac{\zeta+1}{\zeta}}(x_r - m_{jir}), \quad \text{for } j = k \tag{11}$$

$$\frac{\partial d_k(\mathbf{x};\Lambda)}{\partial m_{jir}} = -\frac{1}{K-1}\left[\frac{1}{K-1}\sum_{l,l\neq k}\left\{\frac{g_l(\mathbf{x};\Lambda)}{g_j(\mathbf{x};\Lambda)}\right\}^{-\eta}\right]^{-\frac{\eta+1}{\eta}} \times \frac{\partial g_j(\mathbf{x};\Lambda)}{\partial m_{jir}}, \quad \text{for } j \neq k. \tag{12}$$

In the limiting case, $\eta \to \infty$ results in a much simpler adjustment. Such a simplification leads to an interesting interpretation of the recent, heuristic but powerful, modified LVQ2 algorithm [5] as a special case of the GPD method [1,7]. One notes here that the Euclidean distance is the simplest likelihood-based distance. Using a more general likelihood-based distance allows us to have the modified LVQ2-L adjustment rule similar to Eq. (11-12).

The use of the likelihood-based distances reminds us of another important example of implementation, i.e., the likelihood-based classifier. Here we assume that each class is characterized by a multivariate mixture Gaussian density of the form

$$p_\Lambda(\mathbf{x}|C_j) = \sum_{l=1}^{L_j} \omega_{jl} N_{jl}(\mathbf{x}|m_{jl}, \sigma_{jl}^2) \qquad (14)$$

where $L_j$ is the number of mixture components in $C_j$, $\omega_{jl}$ is the mixture weight of the $l$-th mixture component with the constraint that $\sum_l \omega_{jl} = 1$ and $N_{jl}(\mathbf{x}|m_{jl}, \sigma_{jl}^2)$ is the multivariate Gaussian density of the $l$-th mixture component with $m_{jlr}$ and $\sigma_{jlr}^2$ being the mean and variance of the Gaussian density for the $r$-th element of the vector $\mathbf{x}$. The discriminant function can then be defined as the likelihood of the form

$$g_j(\mathbf{x}; \Lambda) = \ln\{p_\Lambda(\mathbf{x}|C_j)\} \qquad (15)$$

and the misclassification measure is expressed as in Eq. (2). Now using the same adjustment rule in Eq. (5) for a training token $\mathbf{x} \in C_k$, the parameters are adapted as

$$\frac{\partial d_k(\mathbf{x}; \Lambda)}{\partial \Lambda_j} = -\frac{\partial g_k(\mathbf{x}; \Lambda)}{\partial \Lambda_j}, \qquad \text{if } j = k \qquad (16)$$

$$\frac{\partial d_k(\mathbf{x}; \Lambda)}{\partial \Lambda_j} = \left[\sum_{l, l \neq k} \exp\{\eta g_l(\mathbf{x}; \Lambda) - \eta g_j(\mathbf{x}; \Lambda)\}\right]^{-1} \frac{\partial g_j(\mathbf{x}; \Lambda)}{\partial \Lambda_j}, \quad \text{if } j \neq k. \quad (17)$$

Instead of using the likelihood in Eq. (15) as the discriminant function, we can also use the mutual information $I(\mathbf{x}, C_j)$ between the input vector $\mathbf{x}$ and a class $C_j$. This will result in the same adjustment rule if the misclassification measure in Eq. (2) is used in both cases. We will discuss this interesting issue in more detail in Section 3.4.

### 3.2 DTW-based Implementation

Consider a DTW-based classifier consists of a set of variable-length reference patterns, each belonging to one of $K$ classes. Let

$\Lambda = \{\lambda_j\}_{j=1}^{K} = \{\mathbf{r}_{jb}, \mathbf{w}_{jb}; j=1,\ldots,K, \text{ and } b=1,\ldots,B_j\}$ be the set of all reference templates for the DTW-based recognizer, with $\mathbf{r}_{jb}$ and $\mathbf{w}_{jb}$ being the $b$-th reference template and associated weighting function for the class $C_j$, and $B_j$ being the number of reference templates for the class $C_j$. The template $\mathbf{r}_{jb}$ is composed of a variable but finite duration sequence of acoustic feature vectors. The idea of using an $L_p$ norm function again plays a key role in the implementation of discriminative training for the DTW framework. According to the GPD formulation, the following measurable distances are introduced. First we define a discriminant function, also called a *class (word) distance*, as follows,

$$g_j(\mathbf{x}_1^T; \Lambda) = \left[ \frac{1}{B_j} \sum_{b=1}^{B_j} \{D(\mathbf{x}_1^T; \mathbf{r}_{jb}, \mathbf{w}_{jb})\}^{-\zeta} \right]^{-1/\zeta}. \quad (18)$$

In the limiting case where $\zeta$ approaches infinity, only the reference template with the smallest overall distortion in class $C_j$ is used, i.e., a *KNN* rule with $k=1$. The function $D(\mathbf{x}_1^T; \mathbf{r}_{jb}, \mathbf{w}_{jb})$ in Eq. (18), called the *reference (template) distance* is defined as

$$D(\mathbf{x}_1^T; \mathbf{r}_{jb}, \mathbf{w}_{jb}) = \left[ \frac{1}{\Theta_{jb}} \sum_{\theta=1}^{\Theta_{jb}} \{D_\theta(\mathbf{x}_1^T; \mathbf{r}_{jb}, \mathbf{w}_{jb})\}^{-\xi} \right]^{-1/\xi} \quad (19)$$

with $D_\theta(\mathbf{x}_1^T; \mathbf{r}_{jb}, \mathbf{w}_{jb})$ being a *weighted path distance* accumulated along the $\theta$-th best (smallest path distance) warping path selected by the DP-matching between the test pattern $\mathbf{x}_1^T$ and the reference template $\mathbf{r}_{jb}$ along all $\Theta_{jb}$ possible paths. If we select $\xi \to \infty$, only the reference parameters on the path optimally selected by the DP search are adjusted. Here, the weighted path distance is expressed as

$$D_\theta(\mathbf{x}_1^T; \mathbf{r}_{jb}, \mathbf{w}_{jb}) = \frac{1}{T_{jb}} \sum_{t=1}^{T_{jb}} w_{jbt} h_{jbt} \quad (20)$$

with $T_{jb}$ being the duration of the template $\mathbf{r}_{jb}$, $w_{jbt}$ being a weighting factor corresponding to the $t$-th frame of the template $\mathbf{r}_{jb}$ and $h_{jbt}$ being the *frame distances* or local distances defined as the Euclidean distance between the reference frame at time $t$ and the corresponding test frame at time $t$ on the $\theta$-th best warping path. We can now define the misclassification measure having the same form as in Eq. (9). The set of adjustment equations are similar to the ones in Eq. (11-12) for multi-reference distance classifiers.

## 3.3 HMM-based Implementations

An HMM-based classifier consists of a set of variable-state HMM's for each of the $K$ classes. We here use a *generalized class (word) likelihood* $Q_\Lambda(\mathbf{x}_1^T|C_j)$ as the discriminant function

$$g_j(\mathbf{x}_1^T;\Lambda) = \ln\{Q_\Lambda(\mathbf{x}_1^T|C_j)\} \tag{21}$$

where

$$Q_\Lambda(\mathbf{x}_1^T|C_j) = \left[\sum_{S=1}^{\Theta} \{p_\Lambda(\mathbf{x}_1^T,S|C_j)\}^\xi\right]^{1/\xi} \tag{22}$$

with the *path likelihood* $p_\Lambda(\mathbf{x}_1^T,S|C_j)$ for the $S$-th best DP-matching path. For a training token $\mathbf{x}_1^T \in C_k$, we can now define the misclassification measure $d_k(\mathbf{x}_1^T;\Lambda)$ as the one shown in Eq. (2). Again, the choices of $\xi$ and $\eta$ lead to various HMM-based implementations. The reader is referred to [9] for an HMM implementation called *segmental GPD training* which uses only the most likely path (the Viterbi segmentation) as basis for GPD training. This is accomplished by letting $\xi \to \infty$ in Eq. (22). This implementation is similar to the so called *segmental k-means training algorithm*. Another example is to use $\xi = 1$ so that $Q_\Lambda(\mathbf{x}_1^T|C_j) = p_\Lambda(\mathbf{x}_1^T|C_j)$ which results in a formulation similar to the Baum algorithm for estimating HMM parameters. However the goal here is approximately minimizing the error rate instead of maximizing the likelihood of the training data.

The mutual information between the input pattern $\mathbf{x}_1^T$ and the class $C_j$ can also be used as a discriminant function, i.e.,

$$g_j(\mathbf{x}_1^T;\Lambda) = I(\mathbf{x}_1^T,C_j) = \ln\frac{p_\Lambda(\mathbf{x}_1^T|C_j)}{\sum_{l=1}^{K} p_\Lambda(\mathbf{x}_1^T|C_l)p(C_l)}. \tag{23}$$

It can be easily shown that using the two completely different discriminative functions in Eq. (21) ($\xi = 1$) and Eq. (23) results in the same misclassification measure if Eq. (2) is used to define the measure in both cases. Therefore the use of the likelihood formulation and the use of the mutual information formulation [3] give the same classifier design if GPD-based algorithms are used to find the classifier parameters. In both formulations, the same classification rule as defined in Eq. (1) is used.

## 4. EXPERIMENTS

To show the effectiveness of the GPD-based approaches, we perform speech recognition experiments using an E-set database, consisting of the nine confusable E-rhyme letters {b, c, d, e, g, p, t, v, z} spoken in an isolated-word mode. There are one hundred untrained 50 females and 50 males, in the database. Each speaker uttered each of the nine letters twice, one for training and one for testing. The recording was made over dial-up telephone lines. Two baseline systems for multi-speaker recognition were implemented and tested. For the baseline DTW system, the recognition rates were 59.8% with one reference template per word and 67.6% using 12 templates per word [10]. For the baseline HMM system, the recognition rates were 61.7% with one 5-state, 5-mixture HMM per word and 66.7% using one 10-state, 5-mixture HMM per word [6].

To test the GPD-based approaches, we started with an HMM/LVQ hybrid algorithm [6]. Each speech utterance is first converted into a fixed dimensional vector by HMM segmentation and state normalization. Then an LVQ training algorithm is used to design a multi-reference distance classifier for word recognition. The reader is referred to [6] for a detail of the implementation. Using a simplified training algorithm for the likelihood-based classifier, i.e., LVQ2-L, and weighting factors similar to $w_{jt}$ in Eq. (20), the hybrid system improved the recognition rate from 61.7% to 81.3% for a 5-state, 5-mixture HMM system. When each word is modeled by a 10-state, 5-mixture HMM, the recognition rate improved from 66.7% to 83.0%. This represents a 50% error reduction when discrimination factors were considered in the design of the classifier.

Some heuristic assumptions were made to simplify the GPD-based algorithm in the HMM/LVQ hybrid implementation. Three full GPD-based implementations are discussed in the following. The first was a DTW implementation [10] for adapting the weights $w_{jt}$ in Eq. (20). The GPD-based algorithm improved the recognition from 59.8% to 70.0% for a single template system. Furthermore, when 12 reference templates are used to represent each word, the recognition rate improved from 67.6% to 78.1%. The reader is referred to [10] for a detailed description of this DTW-based implementation.

The parameters in the reference templates $r_{jb}$ in Eq. (19) can also be adapted simultaneously with the weighting function $w_{jt}$. The adjustment equations used in this implementation [11-12] are similar to the set of equations in Eq. (11-12). For a single template system, the recognition rate improved from 58.0% to 78.4% when both reference templates and template weighting functions are adjusted. For the system based on four templates per word, the recognition rate improved from 63.8% to 84.4%. The reader is referred to [11-12] for a detailed description of this implementation.

A GPD-based HMM system was also implemented for comparative studies [9]. The result obtained so far showed that a GPD-based adjustment rule applied on the parameters of the HMM's gave the best performance on the E-set test data. Using the formulation similar to Eq. (14-17) and Eq. (21-22), the recognition rate of the HMM system improved from 69% to 85.7% for a 15-state, 5-mixture HMM system. This is the best performance reported on this database. Again, the reader is referred to [9] for a detailed description of this HMM-based implementation. It is here worth mentioning that in the above experiments (e.g., [11-12]), the error rate in testing the training data is approximately equal to the overall loss $L(\Lambda)$ defined in Eq. (4) and evaluated over all the training data. This implies that using the smoothed error function to approximate the recognition error gives a result which closely links the optimization objective to the recognition error rate [7,10,12].

## 5. SUMMARY

We presented a generalized probabilistic descent method for training classifiers for both static and dynamic patterns. The GPD-based algorithms provide a rigorous formulation for designing classifiers with a high discriminative power. A number of GPD-algorithms have been implemented and evaluated. When compared with the conventional maximum likelihood or minimum distortion training approaches, the GPD-based algorithms give much better results for isolated-word recognition. We also expect the same approach to be superior for connected-word recognition.

## REFERENCES

[1] S. Katagiri, C.-H. Lee, and B.-H. Juang, "A Generalized Probabilistic Descent Method," ASJ, *Fall Conf.*, 2-p-6, pp. 141-142, Nagoya, Japan, Sept. 1990.
[2] S. Amari, "A Theory of Adaptive Pattern Classifiers," *IEEE Trans. on Elec. Computers*, Vol. EC-16, No. 3, pp. 299-307, June 1967.
[3] L. R. Bahl, P. F. Brown, P. V. de Souza, and R. L. Mercer, "A New Algorithm for the Estimation of Hidden Markov Model Parameters," *ICASSP88*, pp. 493-496, New York, April 1988.
[4] A. Waibel, T. Hanazawa, G. Hinton, K. Shikano, and K. Lang, "Phoneme Recognition: Neural Networks vs. Hidden Markov Models," *ICASSP88*, pp. 107-110, New York, April 1988.
[5] E. McDermott; "LVQ3 for phoneme recognition," ASJ, *Spring Conf.*, pp. 151-152, March 1990.
[6] S. Katagiri and C.-H. Lee, "A New HMM/LVQ Hybrid Algorithm for Speech Recognition," *GLOBECOM90*, pp. 1032-1036, San Diego, Dec. 1990.
[7] B. H. Juang, and S. Katagiri, "Discriminative Learning for Minimum Error Classification," *in preparation*.
[8] S. Katagiri, C.-H. Lee, and B.-H. Juang, "Discriminative Multi-Layer Feed-Forward Networks," *IEEE-SP Workshop on NN for SP*, Princeton, Sept. 1991.
[9] W. Chou, B.-H. Juang and C.-H. Lee, "Segmental GPD Training of HMM Based Speech Recognizer," *submitted for publication*, July 1991.
[10] P.-C. Chang, S.-H. Chen and B.-H. Juang, "Discriminative Analysis of Distortion Sequences in Speech Recognition," *ICASSP91*, Toronto, May 1991.
[11] T. Komori, and S. Katagiri; "A New Discriminative Training Algorithm for Dynamic Time Warping Speech Recognition," *IEICE Technical Report*, SP91-10, pp. 33-40, June 1991.
[12] P.-C. Chang and B.-H. Juang, "Discriminative Training of Dynamic Programming Based Speech Recognizers," *submitted for publication*, March 1991.

# PROBABILITY ESTIMATION BY FEED-FORWARD NETWORKS IN CONTINUOUS SPEECH RECOGNITION

Steve Renals* Nelson Morgan* Hervé Bourlard[†]
* International Computer Science Institute, Berkeley CA 94704, USA
[†] L&H Speechproducts, 1780 Wemmel, Belgium

Abstract: We review the use of feed-forward networks as estimators of probability densities in hidden Markov modelling. In this paper we are mostly concerned with radial basis functions (RBF) networks. We note the isomorphism of RBF networks to tied mixture density estimators; additionally we note that RBF networks are trained to estimate posteriors rather than the likelihoods estimated by tied mixture density estimators. We show how the neural network training should be modified to resolve this mismatch. We also discuss problems with discriminative training, particularly the problem of dealing with unlabelled training data and the mismatch between model and data priors.

## INTRODUCTION

In continuous speech recognition we wish to estimate the posterior probability $P(\mathbf{W}_1^W | \mathbf{X}_1^T, \mathbf{M})$ of a word sequence $\mathbf{W}_1^W = \mathbf{w}_1, ..., \mathbf{w}_W$ given the acoustic evidence $\mathbf{X}_1^T = \mathbf{x}_1, ..., \mathbf{x}_T$ and the parameters of the models used $\Theta$. This probability cannot be estimated directly; however we may re-express it using Bayes' rule:

$$(1) \quad P(\mathbf{W}_1^W | \mathbf{X}_1^T, \Theta) = \frac{P(\mathbf{X}_1^T | \mathbf{W}_1^W, \Theta) P(\mathbf{W}_1^W | \Theta)}{P(\mathbf{X}_1^T | \Theta)}$$
$$= \frac{P(\mathbf{X}_1^T | \mathbf{W}_1^W, \Theta) P(\mathbf{W}_1^W | \Theta)}{\sum_{\mathbf{W}'} P(\mathbf{X}_1^T | \mathbf{W}', \Theta) P(\mathbf{W}' | \Theta)}.$$

Equation (1) separates the problem into two components: acoustic modelling and language modelling. The language model is used to estimate the prior probability of a word sequence $P(\mathbf{W}_1^W | \Theta)$. The acoustic model is used to estimate the likelihood of the acoustic evidence given the word sequence $P(\mathbf{X}_1^T | \mathbf{W}_1^W, \Theta)$. The normalising denominator of (1) is constant at recognition time; however, during training it is not constant, as the parameters of the models are changing.

Each unit of speech is modelled by a hidden Markov model. A typical unit is the phone; word models consist of concatenations of phone HMMs, according to a phone-structured lexicon. A HMM is defined by a set of states $q_l$, a topology specifying allowed transitions between states and a set of local probability density functions (PDFs) $P(\mathbf{x}_t, q(t)|q(t-1), \mathbf{X}_1^{t-1})$. Making the further assumptions that the output at time $t$ is independent of previous outputs and depends only on the current state, we may separate the local probabilities into state transition probabilities $p(q(t)|q(t-1))$ and output PDFs $P(\mathbf{x}_t|q(t))$. A set of initial state probabilities must also be specified.

The transition probabilities and the parameters of the output PDFs are frequently estimated using a maximum likelihood training procedure, the forward-backward algorithm (see e.g. [2]). This procedure is optimal if the true model is in the space of models being searched[1]. However, this is not the case for speech recognition. What is desired is not the best possible model of each class, but the best set of models for discrimination between classes. Thus, discriminative training would seem to be preferable to maximum likelihood training. In terms of (1), this means that the best acoustic model would be achieved by maximising the likelihood of the correct model, whilst simultaneously minimising the likelihoods of the competing models.

In practice, a full maximum likelihood procedure is rarely used for either recognition or training. Instead, the Viterbi criterion is used. Here, the maximisation of $P(\mathbf{X}_1^T|\mathbf{W}_1^W, \Theta)$ which should be computed over all allowable state sequences is replaced by an approximation that considers only the most probable state sequence. This computation may be efficiently performed using a dynamic programming algorithm. When used at recognition time this is referred to as Viterbi decoding.

We have used discriminatively trained classifiers to estimate the output PDFs [5, 14, 17]. It may be shown that a "1-from-n" classifier trained using a relative entropy (or a least mean squares) objective function outputs the posterior probabilities, $P(q_l|\mathbf{x})$, of each class given the input data [6]. However, the likelihoods $P(\mathbf{x}|q_l)$ are required; the prior probabilities, $p(q_l)$ are given by the allowable sentence models constructed from the basic HMMs using a phone-structured lexicon and the language model. Likelihood estimates may be obtained simply by dividing the output posteriors by the relative frequencies of each class[2].

The classifiers we have used are layered, feed-forward networks: multi-layer perceptrons (MLPs) and radial basis function (RBF) networks. MLPs consist of layers of units that define a hyperplane over the space of the previous layer, followed by a "soft" transfer function (typically a sigmoid). The outputs of such hidden units may be considered as the probabilities of certain "facts" about the previous layer. An RBF network generally has a single hidden

---
1. And if some other conditions are satisfied [11].
2. These are the estimates of $p(q_l)$ implicitly used during classifier training.

layer, whose units may be regarded as computing local (or approximately local) densities, rather than global decision surfaces. The resultant posteriors are obtained by output units that combine these local densities.

In this paper, we are mainly concerned with RBF networks. An isomorphism to tied mixture density modelling has been pointed out. We also remark on a mismatch between the posteriors estimated by discriminatively trained RBF networks and the likelihoods estimated in tied mixture density modelling. This mismatch is resolved by redefining the transfer function of the output units of the RBF network to implement Bayes' rule, relating the posterior to the likelihood. The issue of a mismatch between discriminative and maximum likelihood training is important and has implications regarding our current approach to HMM probability estimation. We survey this problem and discuss some possible solutions.

## TIED MIXTURE HMM

Tied mixture density (or semi-continuous) HMMs have proven to be powerful PDF estimators in continuous speech recognition [13, 3]. This method may be regarded as intermediate between discrete vector-quantised methods and separate continuous PDF estimates for each state. If a unified formalism for both discrete and continuous HMMs is adopted, then tied mixture density modelling may be regarded as an interpolation between discrete and continuous modelling [3]. Essentially, tied mixture modelling has a single "codebook" of Gaussians shared by all output PDFs. Each of these PDFs has its own set of mixture coefficients used to combine the individual Gaussians. If $f_k(\mathbf{x}|q_k)$ is the output PDF of state $q_k$, and $N_j(\mathbf{x}|\mu_j, \Sigma_j)$ are the component Gaussians, then:

(2)
$$f_k(\mathbf{x}|q_k, \Theta) = \sum_j a_{kj} N_j(\mathbf{x}|\mu_j, \Sigma_j)$$
$$\sum_j a_{kj} = 1 \quad 0 \leq a_{kj} \leq 1,$$

where $a_{kj}$ is an element of the matrix of mixture coefficients (which may be interpreted as the prior probability $p(\mu_j, \Sigma_j|q_k)$) defining how much component density $N_j(\mathbf{x}|\mu_j, \Sigma_j)$ contributes to output PDF $f_k(\mathbf{x}|q_k, \Theta)$.

## RADIAL BASIS FUNCTIONS

The radial basis functions (RBF) network was originally introduced as a means of function interpolation [16, 10]. A set of $K$ approximating functions,

$f_k(\mathbf{x})$ is constructed from a set of $J$ basis functions $\phi(\mathbf{x})$:

$$f_k(\mathbf{x}) = \sum_{j=1}^{J} a_{kj}\phi_j(\mathbf{x}) \qquad 1 \leq k \leq K \tag{3}$$

This equation defines a RBF network with $J$ RBFs (hidden units) and $K$ outputs. The output units here are linear, with weights $a_{kj}$. The RBFs are typically Gaussians, with means $\mu_j$ and covariance matrices $\Sigma_j$:

$$\phi_j(\mathbf{x}) = R \exp\left(-\frac{1}{2}(\mathbf{x}-\mu_j)^T \Sigma_j^{-1}(\mathbf{x}-\mu_j)\right), \tag{4}$$

where $R$ is a normalising constant. The covariance matrix is frequently assumed to be diagonal[3].

Such a network has been used for HMM output probability estimation in continuous speech recognition [17] and an isomorphism to tied-mixture HMMs was noted. However, there is a mismatch between the posterior probabilities estimated by the network and the likelihoods required for the HMM decoding. Previously this was resolved by dividing the outputs by the relative frequencies of each state. It would be desirable, though, to retain the isomorphism to tied mixtures: specifically we wish to interpret the hidden-to-output weights of an RBF network as the mixture coefficients of a tied mixture likelihood function. This can be achieved by defining the transfer units of the output units to implement Bayes' rule, which relates the posterior $g_k(\mathbf{x})$ to the likelihood $f_k(\mathbf{x})$:

$$g_k(\mathbf{x}) = \frac{f_k(\mathbf{x})p(q_k)}{\sum_{l=1}^{K} f_l(\mathbf{x})p(q_l)}. \tag{5}$$

Such a transfer function ensures the output units sum to 1; if $f_k(\mathbf{x})$ is guaranteed non-negative, then the outputs are formally probabilities. The output of such a network is a probability distribution and we are using '1-from-K' training: thus the relative entropy $E$ is simply:

$$E = -\log g_c(\mathbf{x}), \tag{6}$$

where $q_c$ is the desired output class (HMM distribution). Bridle has demonstrated that minimising this error function is equivalent to maximising the mutual information between the acoustic evidence and HMM state sequence [9].

If we wish to interpret the weights as mixture coefficients, then we must ensure that they are non-negative and sum to 1. This may be achieved using a normalised exponential (softmax) transformation:

$$a_{kj} = \frac{\exp(w_{kj})}{\sum_h \exp(w_{kh})}. \tag{7}$$

---

3. This is often reasonable for speech applications, since mel or PLP cepstral coefficients are orthogonal.

The mixture coefficients $a_{kj}$ are used to compute the likelihood estimates, but it is the derived variables $w_{kj}$ that are used in the unconstrained optimisation.

**Training**

Steepest descent training specifies that:

$$(8) \quad \frac{\partial w_{kj}}{\partial t} = -\frac{\partial E}{\partial w_{kj}}.$$

Here $E$ is the relative entropy objective function (6). We may decompose the right hand side of this by a careful application of the chain rule of differentiation:

$$(9) \quad \frac{\partial E}{\partial w_{kj}} = \sum_{l=1}^{K} \frac{\partial E}{\partial g_l(\mathbf{x})} \frac{\partial g_l(\mathbf{x})}{\partial f_k(\mathbf{x})} \sum_{h=1}^{J} \frac{\partial f_k(\mathbf{x})}{\partial a_{kh}} \frac{\partial a_{kh}}{\partial w_{kj}}.$$

We may write down expressions for each of these partials (where $\delta_{ab}$ is the Kronecker delta and $q_c$ is the desired state):

$$(10) \quad \frac{\partial E}{\partial g_l(\mathbf{x})} = -\frac{\delta_{cl}}{g_c}$$

$$(11) \quad \frac{\partial g_l(\mathbf{x})}{\partial f_k(\mathbf{x})} = \frac{\delta_{kl} p(q_k)}{\sum_{i=1}^{K} f_i(\mathbf{x}) p(q_i)} - \frac{p(q_k) f_l(\mathbf{x}) p(q_l)}{(\sum_{i=1}^{K} f_i(\mathbf{x}) p(q_i))^2}$$

$$= \frac{p(q_k)}{\sum_{i=1}^{K} f_i(\mathbf{x}) p(q_i)} (\delta_{kl} - g_l)$$

$$= \frac{g_k(\mathbf{x})}{f_k(\mathbf{x})} (\delta_{kl} - g_l)$$

$$(12) \quad \frac{\partial f_k(\mathbf{x})}{\partial a_{kh}} = \phi_h(\mathbf{x})$$

$$(13) \quad \frac{\partial a_{kh}}{\partial w_{kj}} = a_{kh}(\delta_{hj} - a_{kj}).$$

Substituting (10), (11), (12) and (13) into (9) we obtain:

$$(14) \quad \frac{\partial E}{\partial w_{kj}} = -\frac{1}{g_c(\mathbf{x})} \frac{g_k(\mathbf{x})}{f_k(\mathbf{x})} (\delta_{kc} - g_c(\mathbf{x})) a_{kj} \left( \phi_j(\mathbf{x}) - \sum_{h=1}^{J} \phi_h(\mathbf{x}) a_{kh} \right)$$

$$= \frac{1}{f_k(\mathbf{x})} (g_k(\mathbf{x}) - \delta_{kc}) a_{kj} \left( \phi_j(\mathbf{x}) - \sum_{h=1}^{J} \phi_h(\mathbf{x}) a_{kh} \right)$$

$$= \frac{1}{f_k(\mathbf{x})} (g_k(\mathbf{x}) - \delta_{kc}) a_{kj} \left( \phi_j(\mathbf{x}) - f_k(\mathbf{x}) \right).$$

The expression is simpler if we ignore the constraints on the weights (i.e. if $w_{kj} = a_{kj}$), although $f(\mathbf{x})$ is no longer guaranteed to be a PDF:

$$(15) \quad \frac{\partial E}{\partial w_{kj}} = \frac{1}{f_k(\mathbf{x})} (g_k(\mathbf{x}) - \delta_{kc}) \phi_j(\mathbf{x}).$$

The only difference between this gradient and the one obtained using a sigmoid output transfer function with a relative entropy objective function is the $1/f_k(\mathbf{x})$ factor, which may be regarded as a 'dimensional artifact'.

The required gradient is simpler if we construct the network to estimate log likelihoods, replacing $f_k(\mathbf{x})$ with $z_k(\mathbf{x}) = \log f_k(\mathbf{x})$:

(16) $$z_k(\mathbf{x}) = \sum_j w_{kj}\phi_j(\mathbf{x})$$

(17) $$g_k(\mathbf{x}) = \frac{p(q_k)\exp(z_k(\mathbf{x}))}{\sum_l p(q_l)\exp(z_l(\mathbf{x}))}.$$

Since this is in the log domain, no constraints on the weights are required. The new gradient we need is:

(18) $$\frac{\partial g_l(\mathbf{x})}{\partial f_k(\mathbf{x})} = g_l(\delta_{kl} - g_k).$$

Thus the gradient of the error is:

(19) $$\frac{\partial E}{\partial w_{kj}} = (g_k(\mathbf{x}) - \delta_{ck})\phi_j(\mathbf{x}).$$

Since we are in log domain, the "$1/f_k(\mathbf{x})$" factor is additive and thus disappears from the gradient. This network is similar to Bridle's softmax, except here uniform priors are not assumed; the gradient is of identical form, though. However in this case the weights do not have a simple relationship with the mixture coefficients obtained in tied mixture density modelling: thus we use the likelihood estimation of (3) and (5).

We may also train the means and variances of the RBFs by back-propagation of error; alternatively they can be trained by some self-organising process. The relevant partials for gradient descent training are (assuming a diagonal covariance matrix with diagonal elements $\sigma_{ji}$):

(20) $$\frac{\partial \phi_j(\mathbf{x})}{\partial \mu_{ji}} = \frac{\phi_j(\mathbf{x})(x_i - \mu_{ji})}{\sigma_{ji}}$$

(21) $$\frac{\partial \phi_j(\mathbf{x})}{\partial \sigma_{ji}} = \frac{-\phi_j(\mathbf{x})(x_i - \mu_{ji})^2}{2\sigma_{ji}^2}$$

If the determinant of the covariance matrix $\det(\Sigma_j)$ is used as a scale factor for $\phi_j(\mathbf{x})$, then (4) becomes:

(22) $$\phi_j(\mathbf{x}) = \frac{R}{\det(\Sigma_j)^{1/2}}\exp\left(-\frac{1}{2}(\mathbf{x}-\mu_j)^T\Sigma_j^{-1}(\mathbf{x}-\mu_j)\right),$$

and (21) becomes:

(23) $$\frac{\partial \phi_j(\mathbf{x})}{\partial \sigma_{ji}} = \frac{-\phi_j(\mathbf{x})}{2\sigma_{ji}}\left(\frac{(x_i-\mu_{ji})^2}{\sigma_{ji}}+1\right)$$

These expressions, used with the back-propagation algorithm, enable us to adapt the means and covariances in a discriminative fashion.

## GLOBAL OPTIMISATION

The above methods for HMM probability density estimation involve only a local optimisation of parameters. In speech recognition training we typically have a small amount of labelled training data (used for model bootstrapping) and a large amount of unlabelled training data. (Here labelled training data refers to speech labelled and time-aligned at a phone level; unlabelled training data refers to speech for which only the (non-time-aligned) word sequence is available.) The local optimisation we have used has involved an initial maximum likelihood (or Viterbi) training to generate a prototype segmentation of the unlabelled data. These labels are then used as the targets for neural network training (performed on a framewise basis). This is a local training, since only the most likely path given the initial parameter estimation is considered.

One approach to a global optimisation method is analogous to segmental k-means training. In this method after an initial network training on labelled data and Viterbi segmentation, the targets used in training the unlabelled data are updated by performing a Viterbi segmentation after each epoch of discriminative training. Such an approach has been referred to as embedded MLP [5] or connectionist Viterbi training [12]. It should be noted that the transition probabilities are still optimised by a maximum likelihood criterion (or the Viterbi approximation to it). It may be proved that performing a Viterbi segmentation using posterior local probabilities will also result in a global optimisation [6]: however, there is a mismatch between model and data priors here (see next section).

It is possible to attempt a global optimisation in which all the parameters of the HMM are optimised simultaneously according to some discriminative criterion. Such an approach was first proposed by Bahl et al. [1] who presented a training scheme for continuous HMMs in which the mutual information between the acoustic evidence and the word sequence was maximised using gradient descent. More recently, Bridle introduced the "alphanet" representation [8] of HMMs, in which the computation of the HMM "forward" probabilities $\alpha_{jt} = P(X_1^t, q(t) = j)$ is performed by the forward dynamics of a recurrent network. Alphanets may be discriminatively trained by minimising a relative entropy objective function. This function has similar form to (6) (i.e. the negative log of the posterior of the correct output): however here we are looking at the global posterior probability of the word sequence given the acoustic evidence $P(\mathbf{W}_1^W | \mathbf{X}_1^T, \boldsymbol{\Theta})$ (1), rather than the local posterior of a state given the one frame of acoustic evidence. From (1), this posterior is the ratio of the likelihood of the correct model to the sum of the likelihoods of all models. For continuous speech, a model here refers to a sentence model; thus the numerator is the quantity computed by the forward-backward algorithm in training mode (when the word sequence is constrained to be the correct word sequence, so only time-warping variations are considered). The denominator involves a sum over all possible models: this is equivalent to the sum computed if the forward-backward algorithm were to be run at recognition time (with the only constraints over the word sequence provided by the language model). Computation of this quantity would be prohibitive for both training and recognition. A simpler quantity to compute is just the sum over all possible phoneme sequences (unconstrained by language model). This is not desirable as it assumes uniform priors rather than those specified by the language model.

Initial work in using global optimisation methods for continuous speech recognition has been performed by Bridle [7] and Bengio [4]; both of these involved training the parameters of the HMM by a maximum likelihood process, using the "alphanets" method to optimise the input parameters via some (linear or non-linear) transform.

## PROBLEMS WITH DISCRIMINATIVE TRAINING

It has been shown, both theoretically and in practice, that the training and recognition procedures used with standard HMMs remain valid for posterior probabilities [6]. Why then do we replace these posterior probabilities with likelihoods?

The answer to this problem lies in a mismatch between the prior probabilities given by the training data and those imposed by the topology of the HMMs. Choosing the HMM topology also amounts to fixing the priors. For instance, if classes $q_k$ represent phones, prior probabilities $p(q_k)$ are fixed when word models are defined as particular sequences of phone models. This discussion can be extended to different levels of processing: if $q_k$ represents sub-phonemic states and recognition is constrained by a language model, prior probabilities $q_k$ are fixed by (and can be calculated from) the phone models, word models and the language model. Ideally, the topologies of these models would be inferred directly from the training data, by using a discriminative criterion which implicitly contains the priors. Here, at least in theory, it would be possible to start from fully-connected models and to determine their topology according to the priors observed on the training data. Unfortunately this results in a huge number of parameters that would require an unrealistic amount of training data to estimate them significantly. This problem has also been raised in the context of language modelling [15].

Since the ideal theoretical solution is not accessible in practice, it is usually better to dispose of the poor estimate of the priors obtained using the training data, replacing them with "prior" phonological or syntactic knowledge.

A second problem arises from a mismatch between the maximum likelihood and discriminant criteria. As is well known, if the models are correct, then maximum likelihood training is optimal. In speech recognition, we use discriminative training because it is known that the models being used are incorrect. The use of unlabelled data highlights a contradiction in our current training methodology. To give unlabelled data the labels that discriminative training requires, the current best model estimates are used. Thus discriminative training is employed because of a belief that the models are incorrect, yet the labels used by the discriminative training assume model correctness.

It maybe that this mismatch is responsible for the lack of robustness of discriminative training (compared with pure maximum likelihood training) in vocabulary independent speech recognition tasks [15]. The assumption of model correctness used to generate the labels may have the effect of further embedding specifics of the training data into the final models.


## CONCLUSION

We have a defined a feed-forward network that estimates Gaussian mixture densities using a discriminative training criterion. Additionally we have discussed a mismatch between maximum likelihood and discriminative training that is inherent in many discriminative training schemes. We are currently performing speech recognition experiments using the RBF networks and training procedure described above.


## ACKNOWLEDGEMENTS

Thanks to David MacKay and Richard Durbin for good discussions.


## REFERENCES

[1] Lalit R. Bahl, Peter F. Brown, Peter V. de Souza, and Robert L. Mercer. Maximum mutual information estimation of hidden Markov model parameters for speech recognition. In *Proceedings IEEE International Conference on Acoustics, Speech and Signal Processing*, pages 49–52, Tokyo, 1986.

[2] Lalit R. Bahl, Frederick Jelinek, and Robert L. Mercer. A maximum likelihood approach to continuous speech recognition. *IEEE Transactions on Pattern Analysis and Machine Intelligence*, PAMI-5:179–190, 1983.

[3] Jerome R. Bellegarda and David Nahamoo. Tied mixture continuous parameter modeling for continuous speech recognition. *IEEE Transactions on Acoustics, Speech and Signal Processing*, 38:2033–2045, 1990.

[4] Yoshua Bengio, Renato de Mori, Giovammi Flammia, and Ralf Kompe. Global optimization of a neural network - hidden Markov model hybrid. Technical Report TR-SOCS-90.22, McGill University School of Computer Science, 1990.

[5] H. Bourlard and N. Morgan. A continuous speech recognition system embedding MLP into HMM. In David S. Touretzky, editor, *Advances in Neural Information Processing Systems*, volume 2, pages 413–416. Morgan Kaufmann, San Mateo CA, 1990.

[6] H. Bourlard and C. J. Wellekens. Links between Markov models and multilayer perceptrons. *IEEE Transactions on Pattern Analysis and Machine Intelligence*, PAMI-12:1167–1178, 1990.

[7] J. S. Bridle and L. Dodd. An alphanet aproach to optimising input transformations for continuous speech recognition. In *Proceedings IEEE International Conference on Acoustics, Speech and Signal Processing*, pages 277–280, Toronto, 1991.

[8] John S. Bridle. Alpha-nets: a recurrent neural network architecture with a hidden Markov model interpretation. *Speech Communication*, 9:83–92, 1990.

[9] John S. Bridle. Training stochastic model recognition algorithms as networks can lead to maximum mutual information estimation of parameters. In David S. Touretzky, editor, *Advances in Neural Information Processing Systems*, volume 2, pages 211–217. Morgan Kaufmann, San Mateo CA, 1990.

[10] D. S. Broomhead and David Lowe. Multi-variable functional interpolation and adaptive networks. *Complex Systems*, 2:321–355, 1988.

[11] Peter F. Brown. *The Acoustic-Modelling Problem in Automatic Speech Recognition*. PhD thesis, School of Computer Science, Carnegie Mellon University, 1987.

[12] Michael A. Franzini, Kai-Fu Lee, and Alex Waibel. Connectionist Viterbi training: a new hybrid method for continuous speech recognition. In *Proceedings IEEE International Conference on Acoustics, Speech and Signal Processing*, pages 425–428, Albuquerque, 1990.

[13] X. D. Huang and M. A. Jack. Semi-continuous hidden Markov models for speech signals. *Computer Speech and Language*, 3:239–251, 1989.

[14] N. Morgan, H. Hermansky, H. Bourlard, C. Wooters, and P. Kohn. Continuous speech recognition using PLP analysis with multi-layer perceptrons. In *Proceedings IEEE International Conference on Acoustics, Speech and Signal Processing*, pages 49–52, Toronto, 1991.

[15] Douglas B. Paul, James K. Baker, and Janet M. Baker. On the interaction between true source, training and testing language models. In *Proceedings IEEE International Conference on Acoustics, Speech and Signal Processing*, pages 569–572, Toronto, 1991.

[16] M. J. D. Powell. Radial basis functions for multi-variable interpolation: a review. Technical Report DAMPT/NA12, Dept. of Applied Mathematics and Theoretical Physics, University of Cambridge, 1985.

[17] Steve Renals, David McKelvie, and Fergus McInnes. A comparative study of continuous speech recognition using neural networks and hidden Markov models. In *Proceedings IEEE International Conference on Acoustics, Speech and Signal Processing*, pages 369–372, Toronto, 1991.

# NONLINEAR RESAMPLING TRANSFORMATION FOR AUTOMATIC SPEECH RECOGNITION

Y. D. Liu, Y. C. Lee, H. H. Chen and G. Z. Sun

UMIACS/LPR/Department of Physics

University of Maryland, College Park, MD 20742

Abstract -A new technique for speech signal processing called Nonlinear Resampling Transformation (NRT) is proposed. The representation of speech pattern derived from this technique has two important features: first, it reduces redundancy; second, it effectively removes the nonlinear variations of speech signals in time. We have applied NRT to TI isolated-word database achieving a 99.66% recognition rate on a 10 digits multi-speaker task for a linear predictive neural net classifier. In our experiment, we have also found that discriminative training is superior to non-discriminative training for linear predictive neural network classifiers.

## INTRODUCTION

One of the major problems for speech recognition is to find a robust representation of speech signals. The representation must be able to preserve useful information, reduce redundancy and ignore nonlinear variations in time. Typically a speech signal is represented by a sequence of feature vectors which may be derived from LPC or FFT analysis such as cepstral coefficients. Most current speech recognition systems use this sequence directly as a representation of the speech signal. However, there are two disadvantages. First, this representation contains large amount of redundant information. Usually a feature vector is constructed from every 5 to 10 ms speech segment. The number of feature vectors for a typical word is about 50 to 100. Second, the representation is sensitive to the nonlinear variation of the speech patterns in time.

Since the beginning of speech recognition research, researchers are seeking an efficient invariant representation of speech signal. For example Yu (1978) and later Kuhn *et al* (1981) introduced the Trace Segmentation method (Hitchcock 1971) to isolated word recognition. Knowing the word boundaries, the trace of a word is segmented equally into a prescribed number (e.g. $N$) of pieces from which $N$ feature vectors are derived. Thus this $N$-vector representation is uniform for all words. With this representation, they simplified the template matching (Yu 1978) and reduced the computation time for the dynamic warping (Kuhn *et al* 1981). Krammerer and Kupper (1989,1990) used a quasilinear segmentation method (Aktas *et al* 1986) and explicitly introduced time variations in their training patterns to simulate the effect of small fluctuations. Instead of template matching they use neural networks as classifiers. These methods are able to reduce the redundancy and the temporal distortions of the signal in time. However, when applying these method, one has to know the word boundaries, therefore their applications are restricted to isolated word recognition. In this paper, we introduce a new technique called Nonlinear Resampling Transformation(NRT) which differs from the trace segmentation method in two respects. First, NRT use fixed resampling step instead of fixed number of segments. Second, NRT dose not depend on the use of the word boundary information.

## NONLINEAR RESAMPLING TRANSFREORMATION

After preprocessing (e.g. short time FFT), a speech signal (a word or a sentence for example) can be expressed as a sequence of feature vectors. We assume that the difference in feature vectors depends only on their orientations, thus feature vectors can be normalized to unit vectors. To reduce the effect of noise, we set a threshold to eliminate noise components before normalization. This threshold can be estimated from the signal. Therefore, a speech signal is taken to be a variable-length sequence of unit feature vectors:

$$X = \{x_1, x_2, ..., x_N\}$$

where $x_i$ is assumed to have $n$ components:

$$x_i = \{x_{i,1}, x_{i,2}, ..., x_{i,n}\}$$

Each feature vector $x_i$ is a point on the surface of a unit sphere in an $n$-dimensional feature space. The feature vector sequence in the feature space is called the trace of the signal. The length of this one-dimensional curve depends only on the feature contents of the speech signal, not the temporal duration of individual features. A monotonic signal of any duration is represented by a single point in the feature space, while a short utterance of a rapidly varying signal would be represented by a longer curve. Thus the distribution of points on the curve is not uniform. The points cluster together when the signal changes little but are sparsely spaced when the signal varies rapidly. This temporal variations can be effectively eliminated by resampling the feature curve with some fixed step $\delta L$. The resulting representation is invariant to temporal distortion.

The length of the resampling step $\delta L$ determines the amount of information discarded. If $\delta L$ is too long, the signal is poorly represented and important information is lost. On the other hand, too small a $\delta L$ leads to a over-redundant representation and inefficient computation. In our experiments, $\delta L$ is determined empirically. Thus the number of feature vectors in the new representation depends on the length of the curve $L$.

The length of the feature space curve is defined as

$$L = \sum_{k=1}^{N-1} d_k$$

where

$$d_k = \sum_{i=1}^{n} (x_{k+1,i} - x_{k,i})^2$$

We summarize the procedure for Nonlinear Resampling Transformation as the following:

1. cut the feature space curve into segments of equal length $\delta L$, the number of segment is approximately equal to $L/\delta L$,

2. extracts one feature vector from each segment.

## LINEAR PREDICTIVE NEURAL NETWORK CLASSIFIER

Recently much attention has been paid to the predictive neural model on speech recognition[7,8,9]. In this model a speech pattern (e.g. a word or a phoneme) is represented by linked predictive neural networks. The number of neural networks in each link depends on the speech pattern it represents, typically for a word model the number is about ten (Iso 1990). Recognition is based on the prediction error of each model on a given speech pattern. Dynamic programming technique is used to calculate the optimum prediction error. Because we use NRT, the number of feature vectors in the speech pattern representations is reduced and nonlinear variation in the time is eliminated. Therefore only one linear predictive neural network is needed for each word model.

Figure 2 shows the two-layer linear predictive neural network which we use as word models. There is one model for each individual word in the vocabulary. The input layer contains $2n$ units and the output layer contains $n$ units, where $n$ is the number of components of a feature vector. No hidden layers are used for the present study. At step $k$ we present the neural network with $s_k$ and $s_{k+1}$ at the input layer and the network is trained to predict $s_{k+2}$. The actual output of the $j$-th neural network at time $k$ is $o_k^j$.

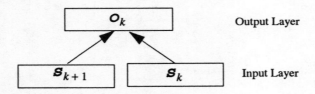

Figure 2. Linear Predictive Neyral Network

Given a speech pattern $S$, the prediction error of the $j$-th neural network model is defined as

$$\Delta^j = \sum_{k=1}^{N-2} (o_k^j - s_{k+2})^2$$

and the classification is done by choosing the model which gives minimum prediction error. We use a discriminative error-correction method to train the neural net model. The training algorithm is similar to Su(1991) and Chang(1991). Assume that the training pattern $S$ belongs to the $i$-th category and it is classified by the network as the $j$-th category. Then we have the following weight correction algorithm:

$$\delta W^i_{lm}(k) = \alpha W^i_{lm}(k) f(s_{k+2,\,l}, t_{k,\,m}, o_{k,\,l})$$

if $j$ is not equal to $i$, then we have

$$\delta W^j_{lm}(k) = (-\alpha) W^j_{lm}(k) f(s_{k+2,\,l}, t_{k,\,m}, o_{k,\,l})$$

where $\delta W^j_{lm}(k)$ is the change of the weight of the neural network model for the $j$-th category at step $k$, $\alpha$ is the learning rate, $t_{k,\,m}$ is equal to $s_{k,\,l}$ for $(m \leq n)$ and equal to $s_{k+1,\,m-n}$ for $(m > n)$, $o_{k,\,l}$ is the value of the $l$-th output neuron of the network at step $k$.

## EXPERIMENTS

In order to test the performance of NRT we have conducted a multi-speaker recognition experiments on a database of 10 isolated English digits ("ZERO", "ONE", "TWO", "THREE", "FOUR", "FIVE", "SIX", "SEVEN", "EIGHT", "NINE",) from 16 speakers(8 males and 8 females). Samples are from TI isolated word database. Each word is uttered 26 times by all speakers. Total number of samples is 4160. Data was sampled at 12.5KHz with 14-bit resolution. The speech data is preprocessed by the NRT front-end processor as shown in Figure 1. The input to the processor is speech signal $Y$ and the output is the resampled feature sequence $S$ which is used for training and recognition.

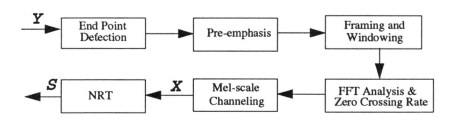

Figure 1. The NRT Front-end Feature Extractor

For a given task the experiments are repeated several times using different choice of the data as testing set. The network is trained using the rest of the data. There are no overlapping in the testing sets among different experiments of the task. Samples from all 16 speakers are used in training and testing.

Table 1 summarizes the result of the experiment. The average recognition rate is 99.66% for the experiments. We have also trained the classifier with non-discriminative error-correction method and the average recognition rate is about 98.5%. This demonstrates the advantage of using discriminative training over the non-discriminative training, which is different from the result of Tebelskis *et al* (1991) who used a different discriminative training algorithm.

TABLE 1

| Number of Testing Data | 800 | 800 | 640 | 640 | 640 | 640 |
|---|---|---|---|---|---|---|
| Number of Errors | 2 | 2 | 1 | 4 | 4 | 1 |

Result of Training of the Predictive Neural Net Classifier

## COMMENTS AND CONCLUSIONS

Nonlinear Resampling Transformation is different from trace segmentation method (Yu 1978, Kuhn *et al* 1981) in using a fixed segmentation step instead of fixed number of segments. The resulting number of feature vectors varies from sample to sample. The use of fixed segmentation step allows the NRT to be extended to the connected word and continuous speech recognition.

Nonlinear Resampling Transformation dose not depend on the feature extraction scheme. It can be applied to various feature analysis schemes such as LPC and FFT. We believe that conventional speech recognition systems based on HMM can also be improved by using NRT as a preprocessor.

## ACKNOWLODGEMENT

The authors are grateful to Dr. P. K. A. Wai for his valuable comment on the manuscript.

# REFERENCES

1. T. C. Yu, *A Real Time Speech Recognition System -- RTSRS(01)*, Chinese Journal of Physics (in Chinese), Vol. 27, p508, (1978)

2. M. H. Kuhn, H. Tomasschewski, H. Ney. *Fast Nonlinear Time alignment for Isolated Word Recognition*. In Proceedings of IEEE International Conference on Acoustics, Speech, and Signal Processing, Atlanta, (p. 736), (1981)

3. M. H. Hitchcock et al., *Method and Apparatus for Interpretation of Time -Varying Signals*, US patent No. 3,582,559, June 1st, 1971

4. A. Aktas., et al,. *Large-vocabulary isolated word recognition with fast coarse time alignment*. In Proceedings of IEEE International Conference on Acoustics, Speech, and Signal Processing, Tokyo, p709, (1986)

5. B. Kammmerer, & W. Kupper. *Perceptrons and multi-layer perceptrons in speech recognition: Improvements from temporal warping of training material*. In L. Personnaz, G, Dreyfus (EDs.), Neural networks from models to applications (p. 531). Paris: I.D.S.E.T., (1989)

6. B. Krammerer, & W. Kupper. *Isolated-Word Recognition with Perceptrons*, Neural Networks, Vol. 3 Number 6, 1990.

7. K. Iso and T. Watanabe. *Speaker-Independent Word Recognition Using a Neural Prediction Model*. In Proceedings of IEEE International Conference on Acoustics, Speech, and Signal Processing, p441, April 1990, Albuquerque, New Mexico.

8. J. Tebelskis and A. Waibel. *Large Vocabulary Recognition Using Linked Predictive Neural Networks*, In Proceedings of IEEE International Conference on Acoustics, Speech, and Signal Processing, p437, April 1990, Albuquerque, New Mexico.

9. E. Levin. *Speech Recognition Using Hidden Control Neural Network Architecture*. In Proceedings of IEEE International Conference on Acoustics, Speech, and Signal Processing, p 433, April 1990, Albuquerque, New Mexico.

10. J. Tebelskis, A. Waibel, B. Petek, and O Schmidbauer. *Continuous Speech Recognition Using Linked Predictive Neural Networks*, In Proceedings of IEEE International Conference on Acoustics, Speech, and Signal Processing, Vol. 1, p61, May 1991, Toronto

11. K. Y. Su and C. H. Lee, *Robustness and Discrimination Oriented Speech Recognition Using Weighted HMM and Subspace Projection Approaches*, In Proceedings of IEEE International Conference on Acoustics, Speech, and Signal Processing,

Vol. 1, p541, May 1991, Toronto

12. P. C. Chang, S. H. Chen and B. H. Juang, *Discriminative Analysis of Distortion Sequences in Speech Recognition*, In Proceedings of IEEE International Conference on Acoustics, Speech, and Signal Processing, Vol. 1, p549, May 1991, Toronto

# SPEECH RECOGNITION BY COMBINING PAIRWISE DISCRIMINANT TIME-DELAY NEURAL NETWORKS AND PREDICTIVE LR-PARSER

Jun-ichi TAKAMI[†]   Atsuhiko KAI[††]   Shigeki SAGAYAMA[†]

[†]ATR Interpreting Telephony Research Laboratories,
Sanpeidani, Inuidani, Seika-cho, Souraku-gun, Kyoto, 619-02 Japan

[††]Department of Information and Computer Sciences, Toyohashi University of Technology, Tempaku-cho, Toyohashi, Aichi, 441 Japan

Abstract—In this paper, a phoneme recognition method using Pairwise Discriminant Time-Delay Neural Networks (PD-TDNNs) and a continuous speech recognition method using the PD-TDNNs are proposed. It is known that classification-type neural networks have poor robustness against the difference in speaking rates between training data and testing data. To improve the robustness, we developed a phoneme recognition method using PD-TDNNs. This method has high performance owing to its particular mechanism, that is a majority decision by multiple less sharp discrimination boundaries. We tested these methods on both consonant recognition and phrase recognition, and obtained higher recognition performance compared with a conventional method using a Single TDNN.

## INTRODUCTION

In recent years, the investigation of speech recognition methods using artificial neural networks[1] has received considerable attention. It is well known that classification-type neural networks, such as the Time-Delay Neural Networks (TDNNs)[2], have high performance for phoneme discrimination. However, in the case of continuous speech recognition, using phoneme verifiers consisting of classification type neural networks trained with data extracted from isolated word utterances, there arise the following serious problems[3]:
- The difference in speaking rates between training data and testing data degrades the recognition performance.
- Neural networks trained with only the desired output values of 1 and 0 tend to produce outputs of almost 1 for the first candidate whether it is correct or not and almost 0 for the other candidates.

This is a problem of robustness against various speaking rates, and is most likely because an input data is discriminated into one phoneme category with little ambiguity by very sharp discrimination boundaries formed between each phoneme category even if the input data is ambiguous.

To overcome these problems, we believe that the following two solutions are effective:
- Discriminate each phoneme category using less sharp discrimination boundaries.
- Discriminate each phoneme category using multiple discrimination boundaries.

To realize these two solutions simultaneously, we developed a phoneme recognition method using Pairwise Discriminant Time-Delay Neural Networks (PD-TDNNs)[4][5]. In this method, it has been shown that high recognition accuracy and robustness can be achieved.

In this paper, we describe both a phoneme recognition method using PD-TDNNs and a continuous speech recognition method using phoneme verifiers consisting of the PD-TDNNs, and show the experimental results for both Japanese consonant recognition and Japanese phrase recognition.

## PAIRWISE DISCRIMINANT TIME-DELAY NEURAL NETWORKS

### Architecture of A PD-TDNN

Figure 1 shows the architecture of a PD-TDNN. A PD-TDNN is a kind of TDNN consisting of four layers in order to discriminate between two phoneme categories. The input layer consists of 112 units corresponding to input data (7 frames of 16 mel-scaled spectrum) and the output layer consists of only 1 unit for an output lying between 1 and 0.

A PD-TDNN for discrimination between two phoneme categories $p_i$ and $p_j$, $\text{Net}(p_i/p_j)$, is trained using the following particular algorithm: if input data belongs to category $p_i$, category $p_j$, or neither category, 1, 0, or 0.5 are given as the respective desired outputs. In this case, an output value of $\text{Net}(p_i/p_j)$ and the value calculated by subtracting the output value from 1 can be regarded as the pair discrimination score for categories $p_i$ and $p_j$, respectively.

In connection with this training algorithm, we improved the training efficiency by introducing a new non-linear function. If a PD-TDNN is trained using a common sigmoid function in the output unit, it is difficult to decrease the error for the data which should output the value 0.5, because the common sigmoid function has a maximum differential coefficient at the point corresponding to the output value of 0.5. Accordingly, the following non-linear function is used in the output unit:

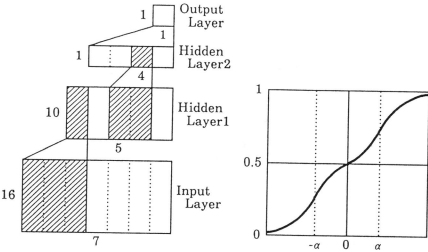

Figure 1 The architecture of a PD-TDNN.

Figure 2 A non-linear function in the output unit.

$$f(x) = \begin{cases} \dfrac{g(x+\alpha)}{2g(\alpha)} & (x<0), \\ 1 - \dfrac{g(-x+\alpha)}{2g(\alpha)} & (x \geq 0), \end{cases} \quad (1)$$

where $g$ is: $g(x) = \dfrac{1}{1+e^{-x}}$,

where $\alpha$ is a value to control the differential coefficient around the point corresponding to the output value of 0.5. We used $\alpha = 3.0$ for the results presented in this paper. Figure 2 shows the form of this function.

It has been confirmed that less sharp discrimination boundaries can be formed between each category by this training algorithm[5].

## Phoneme Recognition Method

Since $N$ categories produce ${}_NC_2$ pairs, ${}_NC_2$ PD-TDNNs are necessary for a phoneme recognition problem having $N$ categories.

First, a couple of pair discrimination scores are calculated from an output value of one PD-TDNN. Respective pair discrimination scores of $p_i$ and $p_j$ are calculated as follows:

$$\begin{aligned} S(p_i|p_i:p_j) &= \text{Out}(p_i/p_j), \\ S(p_j|p_i:p_j) &= 1 - \text{Out}(p_i/p_j), \end{aligned} \quad (2)$$

where $S(p_i|p_i:p_j)$ represents the pair discrimination score for category $p_i$, $\text{Out}(p_i/p_j)$ represents an output value of $\text{Net}(p_i/p_j)$. At this point, there are

$N$-1 pair discrimination scores per category.

Second, total scores are calculated by averaging the corresponding $N$-1 pair discrimination scores category by category. A total score for category $p_i$, $T(p_i)$, is obtained by the following equation:

$$T(p_i) = \frac{1}{N-1} \sum_{j \neq i} S(p_i | p_i : p_j) \ . \tag{3}$$

Finally, multiple phoneme candidates are selected in descending order of these total scores.

In this method, phoneme candidates are decided by majority decision of all PD-TDNNs, each of which has an individual less sharp discriminating boundary between the corresponding phoneme pair. Accordingly, even if wrong outputs are obtained from some PD-TDNNs, the total score $T(p_i)$ is smoothed by averaging the outputs of the individual PD-TDNNs. Because of this smoothing, improvement of the recognition performance can be expected.

## PHONEME RECOGNITION EXPERIMENTS

Phoneme recognition experiments were performed using the following two tasks:
- 6 Japanese consonants, /b,d,g,m,n,N/, which are confused easily with each other.
- 18 Japanese consonants, /b,d,g,p,t,k,m,n,N,s,sh,h,z,ch,ts,r,w,y/.

### Utterances and Analysis

As input data for each network, we used 7 frames of 16-channel mel-scaled spectral coefficients calculated from the phoneme data in the Japanese large vocabulary database uttered by a native male speaker[6]. All utterances in this database were digitized at 12kHz. The input data were calculated by the following process: a speech utterance was Hamming windowed and a 256-point FFT computed every 5ms, 16-channel mel-scaled spectrum coefficients were computed from the power spectrum, adjacent coefficients in time were averaged resulting in an overall 10ms frame rate and these coefficients (7 frames of 16 channel mel-scaled spectrum coefficients) were normalized to lie between -1 and +1, with the average at 0. For both training and testing, every end point of the hand-segmented data was aligned at the center frame of the input layer.

### Training Data and Testing Data

For network training and testing, the data extracted from the following utterances were used:

[Training]

- Even-numbered isolated words of the 5,240 common Japanese words. (5.68 mora/s)

[Testing]

- Odd-numbered isolated words of the 5,240 common Japanese words. ("word" : 5.68 mora/s)
- Conversational sentences from a task called "The International Conference Secretarial Service" uttered phrase by phrase. ("phrase" : 7.14 mora/s)
- Conversational sentences from a task called "The International Conference Secretarial Service" uttered continuously. ("continuous" : 9.56 mora/s)

## Network Training

For 6 Japanese consonant recognition, 3,000 training data (500 training data for each of 6 categories) were used, and for 18 Japanese consonant recognition, 2,600 training data (500 training data for each of corresponding categories $p_i$ and $p_j$, and 100 data for each of the remaining 16 categories) were used.

In both cases, the desired output for Net($p_i/p_j$) were given as 1 for phoneme $p_i$, 0 for phoneme $p_j$ and 0.5 for the other phonemes. During the training, the fast back-propagation learning method[7] was used.

## A Conventional Single TDNN for Comparison

On 6 Japanese consonant recognition, experiments using a Single TDNN were also performed to compare the performance. Figure 3 shows the architecture of this network. This network was trained with 3,000 training data using the following conventional training method: the desired output of 1.0 was given for the output unit corresponding to the correct phoneme category and that of 0.0 was given for the other output units.

## Experimental Results

TABLE 1 shows the experimental results of both 6 consonant recognition and 18 consonant recognition. For 6 consonant recognition, the results obtained using PD-TDNNs are shown in the upper line and the results obtained using a Single TDNN for comparison are shown in the lower line. From these, it was found that the method using PD-TDNNs has high recognition accuracy and robustness against differences in the speaking rates.

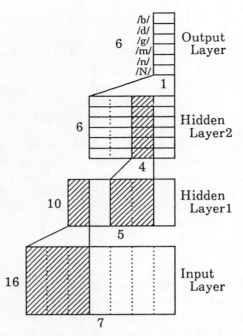

Figure 3  The Architecture of a Single TDNN.

TABLE 1  PHONEME RECOGNITION RATES FOR THE TOP 1 (TOP 3)

| testing data | for 6 consonants | for 18 consonants |
|---|---|---|
| "word" | 97.3 % (99.8 %)<br>95.2 % (99.8 %) | 93.9 % (99.8 %)<br>--------- |
| "phrase" | 86.2 % (98.1 %)<br>84.5 % (95.3 %) | 86.1 % (97.1 %)<br>--------- |
| "continuous" | 81.6 % (96.7 %)<br>77.8 % (92.2 %) | 76.8 % (92.9 %)<br>--------- |

upper line : using PD-TDNNs    lower line : using a Single TDNN

## INVESTIGATION OF NETWORK OUTPUT VALUES

In order to construct continuous speech recognition systems by combining phoneme recognizors with natural language processors, it is important to develop robust phoneme recognition methods. Nevertheless, it seems that a conventional TDNN does not have sufficient robustness so it is easy to cause unrecoverable errors for non-training data. This is because, owing to its sharp discrimination boundaries, the output values do not represent the likelihood of the candidate. However, in the method using PD-TDNNs, robust phoneme recognition can be performed because the output values represent the

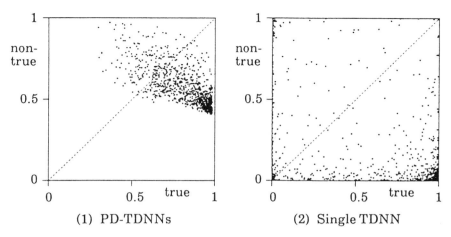

Figure 4  Scatter plots for "continuous" testing data.

likelihood of the candidates.

To support this claim, we observed the recognition scores using scatter plots for the "continuous" testing data. The scatter plots obtained from the PD-TDNNs and the Single TDNN, each of which was used for 6 Japanese consonant recognition, is shown in Figure 4(1) and Figure 4(2), respectively. In each figure, the horizontal axis shows the recognition score for the true phoneme category and the vertical axis shows the maximum recognition score for the non-true phoneme category. In these figures, the data located below the dotted line have been recognized correctly as the first candidates.

It was found that there is a great difference between these two approaches. That is to say, the Single TDNN causes a lot of errors each of which has a very small recognition score (i.e. unrecoverable errors) for the true phoneme category, and the recognition scores obtained from the PD-TDNNs do not include such errors.

From these results, we confirmed that the method using PD-TDNNs is more robust than one using the Single TDNN.

## CONTINUOUS SPEECH RECOGNITION EXPERIMENTS

### Combining PD-TDNNS and The Predictive LR-Parser

LR-parsers are well known in the field of programming languages, and can be applied to a large class of context-free grammars. The generalized LR-parser[8] is a kind of LR-parser, and has been extended to handle arbitrary context-free grammars. The predictive LR-parser for continuous speech recognition is based on the generalized LR-parser and can predict the next phonemes in input speech based on the currently processed phonemes. Until now, an HMM-LR method[9], which uses a

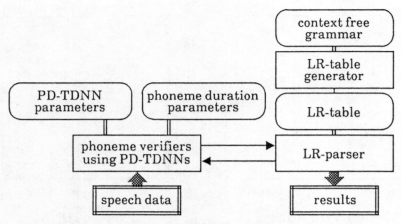

Figure 5 The block diagram of this speech recognition system

HMM phoneme recognition method as phoneme verifiers, and a TDNN-LR method[10], which uses the conventional TDNN phoneme recognition method as a phoneme verifier, were proposed. We tried to achieve accurate continuous speech recognition by combining phoneme verifiers consisting of PD-TDNNs, with the predictive LR-parser. The block diagram of this system is shown in Figure 5.

## Network Architecture and Training Samples

For these experiments, since 25 categories consisting of 24 phonemes of /b,d,g,p,t,k,m,n,N,s,sh,h,z,ch,ts,zh,r,w,y,a,i,u,e,o/ and a silence were used, 300 ($_{25}C_2$) PD-TDNNs were constructed by training. Each PD-TDNN has the same architecture shown in Figure 1 except that the first hidden layer consists of 16×5 units rather than 10×5 units.

All PD-TDNNs were formed by training with training data extracted from even-numbered isolated word utterances. At this time, training data was extracted from not only one point of each phoneme period but also multiple points of the phoneme period with a 10ms frame shift over the phoneme period because the phoneme verifiers should produce correct phoneme verification scores for the enter phoneme period. For each Net($p_i/p_j$) training, 6,900 training data (2,300 training data for each of the corresponding categories $p_i$ and $p_j$, and 100 data for each of the remaining 23 categories) were used.

## A Single TDNN for Comparison

We also performed recognition experiments using a phoneme verifier consisting of a conventional Single TDNN for comparison. The architecture of this network, which is similar to that shown in Figure 3, consists of an input layer having 16×7 units, a first hidden layer having

250×5 units, a second hidden layer having 25×4 units and an output layer having 25 units. For training this network, 50,000 data (2,000 data for each of 25 categories) were used.

## Conversion of Phoneme Verification Scores

As shown in Figure 4, the range of phoneme verification scores obtained by PD-TDNNs is, on average, narrower than that obtained by a Single TDNN, because the expectations of the phoneme verification scores, $T(p_i)$, lie not between 0 and 1 but between $0.5 \times (N-2)/(N-1)$ ($N$ is the number of categories) and 1. Hence, in this system, the phoneme verification scores obtained by PD-TDNNs are used after being converted to $T_{conv}(p_i)$ using equation (4) in order to expand the range.

$$T_{conv}(p_i) = \frac{T(p_i)-a}{1-a}(1-c)+c ,$$

where $a$ is: $a = 0.5 \times \frac{N-2}{N-1}$ . (4)

In this equation, $c$ is a flooring constant to prevent the phoneme verification scores from becoming too small. We used $c=0.1$ in this paper.

## Experimental Results

We tested this method on phrase recognition using both a small grammar (having 607 rules) and a large grammar (having 1,672 rules). TABLE 2 shows the experimental results. With the small grammar or the large grammar, we obtained phrase recognition rates of 81.7% or 74.1%, respectively, for the top 1 choice (9.8% or 9.4% are the respective improvements compared with a method using a Single TDNN). In addition, the top 5 choice recognition rates obtained using the PD-TDNN phoneme verifiers are also higher than those obtained using the Single TDNN phoneme verifier. From these results, we confirmed that the phoneme recognition (verification) method using PD-TDNNs has high recognition performance not only for phoneme recognition but also for phrase recognition.

## CONCLUSIONS

In this paper, we described a phoneme recognition method using Pairwise Discriminant Time-Delay Neural Networks (PD-TDNNs) and a

TABLE 2   PHRASE RECOGNITION RATES FOR THE TOP 1 (TOP 5).

|  | small grammar | large grammar |
|---|---|---|
| using the PD-TDNN | 81.7 %   (97.5 %) | 74.1 %   (92.4 %) |
| using a Single TDNN | 70.9 %   (91.7 %) | 64.7 %   (85.3 %) |

continuous speech recognition method using the PD-TDNNs as phoneme verifiers. Through the results of both 18 Japanese consonant recognition and phrase speech recognition, we confirmed that our phoneme recognition method has high accuracy and robustness for various speaking rates.

## ACKNOWLEDGMENTS

We would like to thank Dr.A.Kurematsu, President, ATR Interpreting Telephony Research Laboratories, for his continuous support of this work. We also acknowledge Dr.H.Sawai and Mr.Y.Komori for their help in carrying out the experiments and the other members of the Speech Processing Department for their discussion and encouragement.

## REFERENCES

[1] D.E.Rumelhart and J.L.McClelland, "Parallel Distributed Processing; Explorations in the Micro Structure of Cognition", MIT Press (1986).
[2] A.Waibel, T.Hanazawa, G.Hinton, K.Shikano and K.Lang, "Phoneme Recognition Using Time-Delay Neural Networks", IEEE Trans Acoust., Speech, Signal Processing, vol.37 (1989-3).
[3] Y.Minami, T.Hanazawa, H.Iwamida, E.McDermott, K.Shikano and M.Nakagawa, "On Sensitivity and Robustness of HMM and Neural Network Speech Recognition Algorithms", ICSLP'90, S31.3 (1990-11).
[4] J.Takami and S.Sagayama, "Phoneme Recognition by Pairwise Discriminant TDNNs ", ICSLP'90, S16.5 (1990-11).
[5] J.Takami and S.Sagayama, "A Pairwise Discriminant Approach to Robust Phoneme Recognition by Time-Delay Neural Networks", ICASSP'91, S8.12 (1991-5).
[6] K.Takeda, Y.Sagisaka and S.Katagiri, "Acoustic-Phonetic Labels in a Japanese Speech Database", European Conference on Speech Technology (1987-8).
[7] P.Haffner, A.Waibel, K.Shikano, "Fast Back-Propagation Learning Methods for Neural Networks in Speech", ASJ fall-meeting, 2-P-1 (1988-10).
[8] M.Tomita, "Efficient Parsing for Natural Language - A Fast Algorithm for Practical Systems", Kluwer Academic Publishers (1986)
[9] K.Kita, T.Kawabata and H.Shikano, "HMM Continuous Speech Recognition using Predictive LR Parsing", ICASSP'89, S13.3 (1989-5)
[10] H.Sawai, "The TDNN-LR Large-Vocabulary and Continuous Speech Recognition System," ICSLP'90, S31.4 (1990-11).

# SPEECH RECOGNITION USING TIME-WARPING NEURAL NETWORKS

Kiyoaki Aikawa
NTT Human Interface Laboratories
3-9-11 Midoricho, Musashino-shi, Tokyo 180 Japan
Email: aik@speech-sun.ntt.jp@relay.cs.net

Abstract- This paper proposes a time-warping neural network (TWNN) for phoneme-based speech recognition. The TWNN is designed to accept phonemes with arbitrary duration, whereas conventional phoneme-recognition networks have a fixed-length input window. The purpose of this network is to cope with not only variability of phoneme duration but also time warping in a phoneme. The proposed network is composed of several time-warping units which each have a time-warping function. The TWNN is characterized by time-warping functions embedded between the input layer and the first hidden layer in the network. The proposed network demonstrates higher phoneme recognition accuracy than a baseline recognizer based on conventional feed-forward neural networks and linear time alignment. The recognition accuracy is even higher than that achieved with discrete hidden Markov models.

## INTRODUCTION

One of the most important problems in speech recognition is time warping, which is observed not only at the word level but also at the phoneme level. The deviation of phoneme duration varies from phoneme to phoneme. Despite its small size, each phoneme shows not only linear time-expansion or compression but also nonlinear time warping. Conventional neural-network-based speech recognition methods handle time warping by using external time alignment mechanisms like dynamic time-warping algorithms or hidden Markov models (HMMs) [1, 2].

Feed-forward neural network architecture has proved to be a powerful tool for phoneme recognition. Moreover, the back-propagation algorithm can effectively train the feed-forward network [3]. The time-delay neural network (TDNN) is a feed-forward network that has been successfully applied to speech recognition [4], but it is difficult to recognize phonemes that vary in duration because the number of input cells is fixed. Overcoming this problem requires strategies for converting an arbitrary number of input frames into a fixed number to be fed into the feed-forward network. It is desirable that the strategy has a facility for handling nonlinear time warping. Several approaches have been proposed. An approach to embedded time alignment is the dynamic neural network, which has dynamic programming integration inside the network [5]. The temporally divided tied structure is another approach to achieving the robustness against time-shifted local features [6]. This paper investigates a feed-forward neural network which can accept phonemes of arbitrary duration coping with nonlinear time warping.

## BASELINE RECOGNIZER

A baseline recognizer that can accept phonemes of arbitrary duration is constructed by combining the TDNN with a linear time alignment mechanism. The input spectrum sequence is linearly converted into a fixed number of frames for feeding into the input layer of the TDNN. A nine-frame-input, four-layered architecture is used here. A first-hidden-layer cell receives input from three frames of the mapped spectrum sequence, and a second-hidden-layer cell receives signals from five columns of the first-hidden-layer cell matrix. The connections are tied along a time-correspondence axis. There seven tied steps between the input and the first hidden layers, and three between the first and the second hidden layers. The shift-invariant characteristics of the TDNN enable the baseline recognizer to cope with a time shift in the phonetic features. This phoneme recognizer can cope with linear time warping and time-shifting of acoustic-phonetic features, but it cannot cope with nonlinear time-warping.

## TIME-WARPING NEURAL NETWORK
### (A) Time Warping

The first-hidden-layer cell of the TDNN is designed to extract a phonetic feature from three adjacent frames of the original speech spectrum sequence, so, it cannot extract the features as well from the linearly time-aligned spectrum sequence. This is because the time alignment mechanism of the baseline recognizer is outside the fixed neural network structure. Moreover, the TDNN is not capable of nonlinear time warping but of only time shifting.

The input layer should access the original speech spectrum directly so that it can extract fine acoustical features without corruption by the time axis conversion. Therefore, the time-warping facility needs to be implemented within the network. This paper introduces a strategy in which a cell in the first hidden layer accesses the particular time of the original spectrum sequence by using its subordinate input cells like feelers. The time to be accessed is determined by a time-warping function. There is no need to convert an input spectrum sequence into a fixed number of input cells. The proposed phoneme recognizer is called a time-warping neural network (TWNN). Figure 1 shows the proposed network architecture. This network comprises several time-warping units that each have a time-warping function. The proposed approach is characterized by the time alignment mechanism implemented between the input layer and the first hidden layer.

The time-warping function $h(u)$ maps a normalized time axis to the actual time axis. This function must monotonically increase and must satisfy the boundary conditions $h(0) = 0$ and $h(1) = T$. This fixed-end-point condition is important when recognizing an utterance with concatenated phoneme recognizers. Here $T$ is the actual duration of the phoneme token. $T$ differs from token to token. This paper uses five time-warping functions:

$$t = Tu \tag{1}$$
$$t = T(u \pm \alpha \sin(\pi u)) \tag{2}$$
$$t = T(u \pm 0.5\alpha \sin(2\pi u)) \tag{3}$$

where $u$ is the normalized time and $t$ is the time elapsed since the beginning of a phoneme. The term $\alpha$ is the time-warping factor and is fixed at 0.3 on the basis of preliminary experimental results. Equation (2) represents time-warping toward the starting point of the phoneme when the sign is positive and toward the ending point when the sign is negative. Equation (3) represents time-warping toward the center of the phoneme when the sign is positive, and toward the sides when the sign is negative. This function set contains time compression, time expansion, and nonlinear time shift. This set of functions is called "complex time warping" (CTW). These functions are shown in Figure 2 (a).

Another function set, Equations (1), (2), and

$$t = T(u \pm 0.5\alpha \sin(\pi u)) \tag{4}$$

is also tested. Equation (4) represents less time warping than Equation (2). This function set is composed of only nonlinear time shifting, and is called "simple time warping" (STW). These functions are shown in Figure 2 (b).

This paper also compares function sets CTW and STW, with the time-shifting functions:

$$t = Tu + \Delta t n, \qquad n = -2, -1, 0, 1, 2 \tag{5}$$

where $\Delta t$ is the frame interval of the input spectrum sequence. This function set is called "linear time shifting" (LTS). It does not satisfy the fixed-end-point condition.

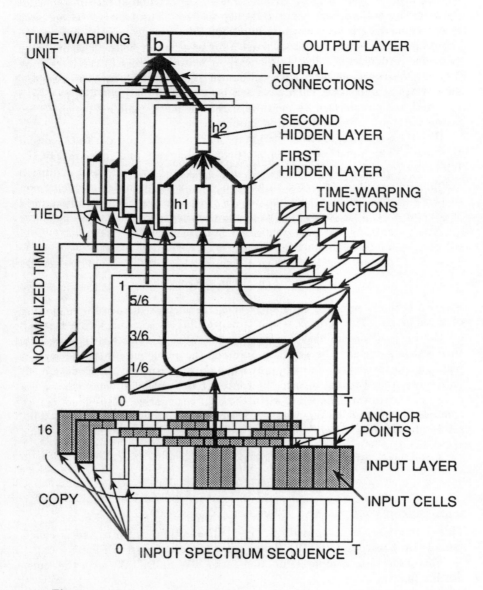

Figure 1: Time-warping neural network architecture for phoneme recognition.

T: Phoneme duration
h1: Number of cells in a first-hidden-layer cell block
h2: Number of cells in a second-hidden-layer cell block
⟶: A set of neural connections
INPUT SPECTRUM: 16-channel Mel-scale filterbank

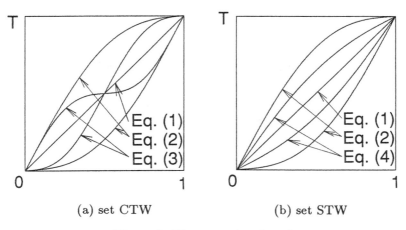

(a) set CTW  (b) set STW

Figure 2: Time-warping functions.

## (B) Network Architecture

The proposed TWNN recognizes a phoneme as a sequence of three segments, because three-state hidden Markov modeling is successful for phoneme representation. The first hidden layer in a time-warping unit is composed of three cell blocks, each corresponding to a segment. Three anchor points,

$$1/6, \ 3/6, \ 5/6, \qquad (6)$$

are set on the normalized time. These are the centers of the segments. Each time-warping function maps the three anchor points onto the actual time axis. A segment corresponds to a state in terms of HMM. The mapped segment boundary differs among time-warping units. The input frames in a mapped segment belong to a state. Three contiguous frames of the input spectrum centered on each mapped anchor point are fed into a set of input cells. The input cell set is fully connected to a cell block in the first hidden layer. A cell block in the first hidden layer has several cells so that it can detect several different phonetic features. Therefore, the cell block works as a multiple-template recognizer.

The corresponding cell blocks in each of the first hidden layers are designed to detect common phonetic features over five time-warping units. This copes with the difference in time warping among tokens. Tied connections are successfully applied to achieve this design. The connections from input cells to a first-hidden-layer cell are tied; that is, they have the same weight over these five time-warping units. The cell block sequence in the first hidden layer is capable of accepting several different feature sequences. Here, also, the cell block works as a multiple-template sequence recognizer.

A second hidden layer integrates the outputs of the first hidden layer independently of the other time-warping units. The first-hidden-layer cells are fully connected to the second-hidden-layer cells unit by unit. A block in the second hidden layer also has several cells for accepting different phonetic-feature sequences. The sequence of accessing time (mapped anchor point) of a time-warping unit preserves the temporal order, because the time-warping functions increase monotonically. Integration which preserves the temporal order of phonetic features is an important characteristic of the proposed approach.

The output cells integrate the responses of five time-warping units. The recognition mechanism is like hidden Markov modeling that integrates several probabilities of hidden state sequences. The TWNN is a discriminative network, so, the number of output cells is equal to the number of phonemes to be discriminated. A DC bias from a bias cell is supplied to every cell. These connections are tied over five time-warping units. The optimal bias for a cell is trained together with other connection weights.

The proposed neural network is trained by the error back-propagation algorithm. The supervisor signals for training tokens are 1 or 0. The error at an output cell is measured by the square distance between the supervisor signal and the output signal. The output function of the cell is the symmetrical sigmoid defined by

$$f(y_j) = \frac{1}{1+e^{-y_j}} - 0.5 \qquad (7)$$

$$y_j = bias_j + \sum_i w_{ij} x_i, \qquad (8)$$

where $x_i$ is the $i$-th input to the $j$-th cell, and $w_{ij}$ is the connection weight from the $i$-th cell to the $j$-th cell. At the output cell, a standard sigmoid is used.

The tied-connected feature-extracting mechanism over time-warping units is trained with tokens that include various kinds of time warping. The weights for integrating time warpings are also adjusted during the training process. The magnitude of the weight from an unimportant time-warping function is expected to decrease during training.

## EXPERIMENTS

The proposed phoneme recognizer was evaluated focusing on robustness against different utterance styles, particularly where the utterance length and speed are different from the training utterances. The speech database used for the experiment was spoken by a male speaker and sampled at 12 kHz. Mel-scale 16-channel filterbank outputs were obtained every 10 ms from a 256-point Fast Fourier Transform. Training and testing tokens were extracted according to the phoneme labels given by hand. The TWNNs were trained using phoneme tokens extracted from isolated-word utterances and tested on both isolated-word utterances and isolated-phrase utterances spoken faster than the isolated words.

The phoneme recognition experiments were carried out on 6-consonant recognition and 18-consonant recognition. The proposed TWNN performed much better than the baseline recognizer. Table 1 compares the proposed TWNNs and the baseline recognizers using linear mapping and TDNN architecture on 6-consonant recognition. TDNN3-4 specifies that the connections between the second hidden layer and the output layer are not tied. These tables show the superiority of the TWNN over the baseline recognizer for both the same utterance style and a different utterance style. For the different utterance style, the phoneme recognition rate was highest when the number of cells is 12x3x5 in the first hidden layer and 12x5 in the second hidden layer. For the same utterance style, the recognition rate was highest when the number of cells is 24x3x5 in the first hidden layer and 4x5 in the second hidden layer. The TWNN also achieved better cumulative recognition rates than TDNNs. Of the time-warping functions, the CTW set performed best. This is because the CTW function set can accept various types of time warpings. The best first-hidden-layer size for the baseline recognizer was 16x7. TDNN3-4 obtained a higher recognition rate than TDNN. Table 1 also compares the TWNN and the basic discrete hidden Markov modeling. Feature parameters, cepstrum, $\Delta$-cepstrum, and $\Delta$-power, are vector-quantized separately. The whole phoneme period is used for each experiment. This table shows that the TWNN also performs better than the discrete hidden Markov models both for the isolated-word utterance style and for the isolated-phrase utterance.

Table 2 shows that the superiority of the TWNN with time-warping function set CTW over the TDNN3-4 on 18-consonant recognition. Table 2 also shows the TWNN achieved better performance on 18-consonant recognition for the same utterance style.

Table 3 shows the relation between the recognition rate and number of iterations. One iteration is a cycle of training over all training tokens. Only a small decrease of the recognition rate due to the over-learning effect is seen through the long training span. This indicates that the training of the TWNN is stable.

Table 4 shows cumulative recognition rates of individual phonemes using the TWNN-CTW in the case of 18-consonant recognition for the different utterance style. Some phonemes show poor recognition rates for the top candidate, however, they are improved for the top-three cumulative recognition rate.

## CONCLUSIONS

The excellent performance of the time-warping neural network for phoneme recognition has been demonstrated. The time-warping neural network has

Table 1: Comparison among TWNN, conventional method, and HMM for /b,d,g,m,n,N/ recognition.

| Method | Number of Hidden Cells | | Testing Database | | | |
|---|---|---|---|---|---|---|
| | | | Word | | Phrase | |
| | 1st | 2nd | Top 1 | Top 3 | Top 1 | Top 3 |
| TWNN-CTW | 12x3x5 | 12x5 | 96.3% | 99.7 | 84.5 | 97.9 |
| | 12x3x5 | 8x5 | 96.1 | 99.7 | 80.8 | 97.0 |
| | 12x3x5 | 16x5 | 96.0 | 99.6 | 82.1 | 97.2 |
| | 8x3x5 | 12x5 | 96.4 | 99.8 | 79.2 | 96.9 |
| | 16x3x5 | 12x5 | 96.6 | 99.7 | 81.3 | 96.8 |
| | 24x3x5 | 4x5 | 96.8 | 99.9 | 83.2 | 97.3 |
| TWNN-STW | 12x3x5 | 12x5 | 96.5 | 99.7 | 81.8 | 98.3 |
| TWNN-LTS | 12x3x5 | 12x5 | 96.2 | 99.8 | 79.2 | 97.2 |
| TDNN | 16x7 | 6x3 | 95.3 | 99.4 | 80.3 | 96.8 |
| TDNN3-4 | 16x7 | 6x3 | 95.3 | 99.4 | 81.7 | 95.5 |
| HMM | Discrete | | 94.0 | 99.7 | 71.5 | 95.3 |

Training: Isolated-word utterance (5.7 mora/sec)
Testing: Word: Isolated-word utterance (5.7 mora/sec)
Phrase: Isolated-phrase utterance (7.7 mora/sec)

Table 2: Comparison among TWNN, conventional method, and HMM for 18-consonant recognition.

| Method | Number of Hidden Cells | | Testing Database | | | |
|---|---|---|---|---|---|---|
| | | | Word | | Phrase | |
| | 1st | 2nd | Top 1 | Top 3 | Top 1 | Top 3 |
| TWNN-CTW | 24x3x5 | 24x5 | 97.2% | 99.5 | 80.0 | 93.2 |
| TDNN3-4 | 32x7 | 18x3 | 96.4 | 99.3 | 74.1 | 88.9 |
| HMM | Discrete | | 96.3 | 99.5 | 81.9 | 96.6 |

Table 3: Relation between the number of iterations and recognition rate.

(a) 6-consonant recognition

| Number of Iterations | Testing Database | |
|---|---|---|
| | Word | Phrase |
| 500 | 95.5% | 81.6 |
| 600 | 95.8 | 83.2 |
| 700 | 95.8 | 83.9 |
| 800 | 95.7 | 84.4 |
| 900 | 95.8 | 84.5 |
| 1000 | 96.0 | 84.0 |
| 1100 | 95.9 | 84.3 |
| 1200 | 95.9 | 84.4 |
| 1300 | 96.1 | 84.3 |
| 1400 | 96.2 | 84.3 |
| 1500 | 96.3 | 84.2 |
| 1600 | 96.1 | 83.6 |

(b) 18-consonant recognition

| Number of Iterations | Testing Database | |
|---|---|---|
| | Word | Phrase |
| 1600 | 97.2% | 79.9 |
| 1700 | 97.2 | 79.3 |
| 1800 | 97.2 | 79.7 |
| 1900 | 97.2 | 79.8 |
| 2000 | 97.2 | 79.9 |
| 2100-2900 | 97.2 | 80.0 |

Training: Isolated-word utterance
Network: TWNN-CTW, first hidden layer: (a) 12x3x5, (b)24x3x5, second hidden layer: (a) 12x5, (b) 24x5

Table 4: Individual consonant recognition rates.

| Phoneme | Top 1 | Top 2 | Top 3 | Phoneme | Top 1 | Top 2 | Top 3 |
|---|---|---|---|---|---|---|---|
| p | 100.0% | 100.0 | 100.0 | t | 66.9 | 87.3 | 93.2 |
| k | 78.8 | 87.1 | 91.1 | ch | 83.3 | 86.7 | 86.7 |
| ts | 100.0 | 100.0 | 100.0 | s | 95.2 | 100.0 | 100.0 |
| sh | 100.0 | 100.0 | 100.0 | h | 47.7 | 67.7 | 76.9 |
| b | 89.4 | 95.7 | 97.9 | d | 86.8 | 94.5 | 97.0 |
| g | 90.0 | 98.3 | 99.2 | m | 59.5 | 76.8 | 85.8 |
| n | 65.9 | 80.8 | 86.6 | N | 95.7 | 100.0 | 100.0 |
| r | 85.5 | 92.8 | 94.0 | z | 100.0 | 100.0 | 100.0 |
| y | 93.5 | 98.7 | 98.7 | w | 95.1 | 98.8 | 98.8 |
| Average | | | | | 80.0 | 89.9 | 93.2 |

Training: Isolated-word utterance
Testing: Isolated-phrase utterance
Network: TWNN-CTW, first hidden layer: 24x3x5, second hidden layer: 24x5

proved to be robust against differences in utterance style, and has shown better consonant recognition rates than the baseline recognizer based on the TDNN and linear time alignment. The nonlinear function set CTW comprising various types of time-warping functions achieved best performance. The recognition rate is even higher than that achieved with discrete hidden Markov Models. The TWNN is stably trained and the over-learning effect is small.

## ACKNOWLEDGMENT

The author would like to thank Sadaoki Furui, Kiyohiro Shikano, and all members of the speech recognition research group of the NTT Human Interface Laboratories for their valuable discussions and suggestions.

## REFERENCES

[1] J. Tebelskis A. Waibel, "Large vocabulary recognition using linked predictive neural networks," ICASSP90, Vol. 1, pp. 437-440, 1990.

[2] K. Aikawa and A. H. Waibel, "Speech recognition using sub-phoneme recognition neural network," Inter National Conference on Spoken Language Processing, Vol. 1, pp. 685-688, 1990.

[3] D. E. Rumelhart, G. E. Hinton, and R. J. Williams, "Learning Internal Representations by Error Propagation, Parallel Distributed Processing: Explorations in the Microstructure of Cognition," Vol. 1: Foundations. MIT Press (1986).

[4] A.H. Waibel, et al., "Phoneme Recognition Using Time-Delay Neural Networks," IEEE Trans. Vol. ASSP-37, No. 3, pp. 328-339, 1989.

[5] H. Sakoe, R. Isotani, K. Yoshida, K. Iso, T. Watanabe, "Speaker-independent word recognition using dynamic programming neural networks," ICASSP89, Vol. 1 pp. 29-32., 1989.

[6] Y. Komori, Y. Minami, and K. Shikano, "Phoneme identification neural networks concerning phonetic temporal structure," Acoustical Society Japan spring meeting, Vol. 1, pp. 157-158, (1990) (in Japanese).

# A Hybrid Continuous Speech Recognition System Using Segmental Neural Nets with Hidden Markov Models

S. Austin, G. Zavaliagkos†, J. Makhoul and R. Schwartz

BBN Systems and Technologies, Cambridge, MA 02138
†Northeastern University, Boston, MA 02115

## Abstract

We present the concept of a "Segmental Neural Net" (SNN) for phonetic modeling in continuous speech recognition (CSR) and demonstrate how this can be used with a multiple hypothesis (or N-Best) paradigm to combine different CSR systems. In particular, we have developed a system that combines the SNN with a hidden Markov model (HMM) system. We believe that this is the first system incorporating a neural network for which the performance has exceeded the state of the art in large-vocabulary, continuous speech recognition.

By taking into account all the frames of a phonetic segment simultaneously, the SNN overcomes the well-known conditional-independence limitation of HMMs. However, the problem of automatic segmentation with neural nets is a formidable computing task compared to HMMs. Therefore, to take advantage of the training and decoding speed of HMMs, we have developed a novel hybrid SNN/HMM system that combines the advantages of both types of approaches. In this hybrid system, use is made of the N-best paradigm to generate likely phonetic segmentations, which are then scored by the SNN. The HMM and SNN scores are then combined to optimize performance.

## 1 INTRODUCTION

The current state of the art in CSR is based on the use of HMMs to model phonemes in context. Two main reasons for the popularity of HMMs are their high performance, in terms of recognition accuracy, and their computational efficiency (e.g. after initial signal processing, real-time recognition is possible on a Sun 4 [1]). However, the limitations of HMMs in modeling the speech signal have been known for some time. Two such limitations are (a) the conditional-independence assumption, which prevents a HMM from taking full advantage of the correlation that exists among the frames of a phonetic segment, and (b) the awkwardness with which segmental features (such as duration) can be incorporated into HMM systems. We have developed the concept of Segmental Neural Nets (SNN) to overcome the two HMM limitations just mentioned for phonetic modeling in speech. However, neural nets are known to require a large amount of computation, especially for training. Also, there is no known efficient search technique for finding the best scoring segmentation with neural nets in continuous speech. Therefore, we have developed a hybrid SNN/HMM system that is designed to take full advantage of the good properties of both methods: the phonetic modeling properties of SNNs and the good computational properties of HMMs. The two methods are integrated through a novel use of the N-best paradigm developed in conjunction with the BYBLOS system at BBN.

## 2 SEGMENTAL NEURAL NET STRUCTURE

There have been several recent approaches to the use of neural nets in CSR. The SNN differs from these approaches in that it attempts to recognize each phoneme by using all the frames in a phonetic segment simultaneously to perform the recognition. In fact, we define a SNN as a neural network that takes the frames of a phonetic segment as input and produces as output an estimate of the probability of a phoneme given the input segment. But the SNN requires the availability of some form of phonetic segmentation of the speech. To consider all possible segmentations of the input speech would be computationally prohibitive. We describe in Section 4 how we use the HMM to obtain likely candidate segmentations. Here, we shall assume that a phonetic segmentation has been made available.

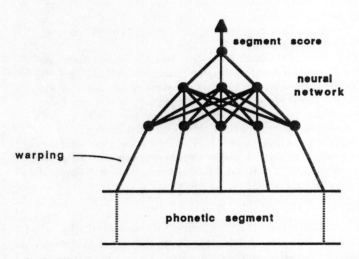

Figure 1: The Segmental Neural Network model samples the frames in a segment and produces a single segment score.

The structure of a typical SNN is shown in Figure 1. The input to the ne is a fixed number of frames of speech features (5 frames in our system). The features in each 10-ms frame consist of 16 scalar values: 14 mel-warped cepstral coefficients, power, and power difference. But the actual number of such frames in a phonetic segment is variable. Therefore, we convert the variable number of frames in each segment to a fixed number of frames (in this case, five frames). In this way, the SNN is able to deal effectively with variable-length segments in continuous speech. The requisite *time warping* is performed by a quasi-linear sampling of the feature vectors comprising the segment. For example, in a 17-frame phonetic segment, we would use frames 1, 5, 9, 13, and 17 as input to the SNN. In a 3-frame segment, the five frames used are 1, 1, 2, 3, 3, with a repetition of the first and third frames. In this sampling, we are using a result from stochastic segment models (SSM) in which it was found that sampling of naturally-occurring frames gives better results than strict linear interpolation [6].

Therefore, by looking at a whole phonetic segment at once, we are able to take advantage of the correlation that exists among frames of a phonetic segment. Also, by making explicit use of duration in the model, we are able to fully utilize duration information, thus ameliorating both limitations of HMMs. These properties of the SNN are also shared by the SSM [6]. The main difference between the two is in how the probability of a segment is computed. In the SSM, an explicit multi-dimensional probability model is used (usually Gaussian), with many simplifying assumptions so as to reduce the large amount of computation for training and recognition that would be needed in a model that has a complete covariance matrix. In contrast, the SNN has been shown to be capable of implicitly generating an estimate of the posterior probability without the need for an explicit model [2, 4]. In this way, we believe that the neural net will use as much correlation among frames as is needed to enhance performance.

In our experiments, we used SNNs with 53 outputs, each representing one of the phonemes in our system. The SNN outputs are trained to output 1 for the correct phoneme and a 0 for all the others. The input to the SNN was the 80 segment features (16 features per frame × 5 frames) derived from the segments via the time-warping procedure.

## 3 THE N-BEST RESCORING PARADIGM

In continuous speech recognition, many systems produce as output a single transcription that best matches the input speech, given some grammar. Because of imperfections in the recognition, the output may not be the correct sentence that was uttered and anything using this output (such as the natural language component of a speech understanding system) may be in error. One way to ameliorate this problem is to use a search that produces not only the single best-matching sentence, but also the N-best matching sentences [8], where N is taken to be large enough to include the correct sentence most of the time. The list of N sentences is ordered by overall score in matching the input utterance. For integration with natural language, we send the list of N sentences to the natural language component, which processes the sentences in the order given and chooses the first sentence that can be understood by the system.

In the hybrid SNN/HMM system, we use this N-best list differently. A spoken utterance is processed by the HMM recognizer to produce a list of the N best-scoring sentence hypotheses. The length of this list is chosen to be long enough to include the correct answer almost always. Thereafter, the recognition task is reduced to selecting the best hypothesis from the N-best list. The length of this list is usually less than 100, which means that the search space of possible word theories is reduced from a huge number (for a 1000 word vocabulary, even a two-word utterance has a million possible word hypotheses) to a much smaller and more manageable number. This means that each of the N hypotheses can be examined and scored using algorithms which would have been computationally impossible with a combinatorially large set of hypotheses. In addition, it is possible to generate several types of scoring for each hypothesis. This not only provides a very effective way of comparing the effectiveness of different speech models (e.g., SNN versus HMM), but it also provides an easy way to combine several radically different models.

The most obvious way in which the SNN could use the N-best list would be to derive a SNN score for each hypothesis in the N-best list and then reorder this list on the basis of these scores. The chosen answer would be the hypothesis with the best SNN score. However, it is possible to generate several scores for each hypothesis, such as SNN score, HMM score, grammar score, and the hypothesized number of words. We can then generate a composite score by, for example, taking a linear combination of the individual scores. After we have rescored the N-Best list, we can reorder it according to the hypotheses' new scores. If the CSR system is required to output just a single hypothesis, the highest scoring hypothesis is chosen. This is called *the N-best rescoring paradigm*.

The linear combination that comprises the composite score is determined by selecting the weights that give the best performance over a development test set. These weights can be chosen automatically [5].

Two factors influence the choice of N. While it is true that larger values of N will include the correct hypotheses in the list more often, we have found that when rescoring with a knowledge source that, on its own, performs worse than the HMM, a large list can produce worse results than a shorter one. The reason for this is that every system has a certain chance of giving an optimistic score to an incorrect hypothesis – the worse the system, the greater the chance. Longer N-best lists mean more incorrect hypotheses as well as more correct hypotheses, so that at some value of N, the advantage obtained from having more correct hypotheses in the list is overtaken by the disadvantage of the weaker system scoring incorrect hypotheses too highly. When using the SNN, we found that the best value of N lay somewhere between 2 and 20.

## 4  HYBRID SNN/HMM SYSTEM

As mentioned above, recognition in the hybrid SNN/HMM system is performed by using the SNN scores together with HMM and other scores to reorder the N-best list of likely hypotheses for the utterance. The process is shown schematically in Figure 2. First the N-best list is generated by using a HMM-based recognition system, such as BYBLOS. Then, scores are generated for the HMM and other knowledge sources for each hypothesis in turn. The generation of the HMM score also produces a segmentation for each hypothesis by finding the most likely state sequence according to that hypothesis. Of course, only one of these hypotheses can be correct, but this is not a problem since a bad segmen-

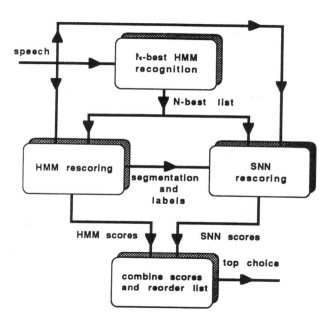

Figure 2: Schematic diagram of the hybrid SNN/HMM system using the N-best rescoring paradigm.

tation for the incorrect hypothesis will usually lead to a correspondingly poor SNN score. This means that the incorrect hypothesis will not only be penalized because of a bad acoustic match, but also because of a malformed segmentation.

The SNN uses the segmentation and phonetic sequence produced by the HMM under each hypothesis to construct feature vectors from each segment in the same way as in the training procedure. The logarithm of all these segment scores are computed and added together to produce a SNN score for the particular hypothesis. For each hypothesis, a total score is then computed by taking a linear combination of the SNN score, HMM score, the number of words in the hypothesis and the number of phonetic segments. The number of words and segments are included because they serve the same purpose as word and phoneme insertion penalties in a HMM CSR system. The weights for the linear combination are found by training on a development corpus that is different from the training corpus used to train the HMM and SNN. A different corpus is used since the acoustic scores generated from training data will be unrealistically optimistic. It is important to note that, because of the use of weighting to optimize performance in this hybrid system, overall recognition accuracy can never be worse than with the HMM system alone.

# 5 SNN REFINEMENTS

A number of refinements have been added to the SNN to improve performance and expedite training.

## 5.1 Durational Term

The SNN score for a segment is independent of the duration of the segment. In order to provide information about the duration to the SNN, we constructed a simple durational model. For each phoneme, a histogram was made of segment durations in the training data. This histogram was then smoothed by convolving with a triangular window, and probabilities falling below a floor level were reset to that level. The duration score was multiplied by the neural net score to give an overall segment score.

## 5.2 Log-Error Criterion

The SNN was originally trained using a mean-square error (MSE) criterion – i.e., the SNN was trained to minimize

$$E = \frac{1}{N} \sum_{n=1}^{N} (y_c(n) - d_c(n))^2$$

where $y_c(n)$ is the network output for phoneme class $c$ for the $n^{th}$ training vector and $d_c(n)$ is the desired output for that vector (1 if the segment belongs to class $c$ and 0 otherwise). This measure can lead to gross errors at low values of $y_c(n)$ when segment scores are multiplied together. Accordingly, we adopted the log-error training criterion [2], which is of the form

$$E = -\frac{1}{N} \sum_{n=1}^{N} \log \left(y_c(n) - [1 - d_c(n)]\right)^2.$$

This can be shown to have several advantages over the MSE criterion. When the non-linearity is the usual sigmoid function, this error measure has only one minimum for single layer nets. In addition, the gradient is simple and avoids the problem of "weight locking" (where large errors do not change because of small gradients in the sigmoid).

## 5.3 Reduction of Negative Training

For a 53-class problem, on average, only about 2% of the examples will be used to train positively. Therefore, most of the training examples for each output node will be used to train the node negatively. Since much computational effort is expended in training each node with every negative example, we trained each output node separately, and randomly discarded negative training examples to reduce their number by a factor of 20. To compensate for this, we multiplied the error for the negative samples we selected by 20, so that the proportion of negative and positive training vectors appeared unchanged. We discovered that this technique reduced the training time by about a factor of 6, but that the performance of the SNN remained the same.

## 5.4 Two-Layer SNN

Most of our experiments were performed with 1-layer neural networks. We also examined the performance of a SNN with a 2-layer network. The input consisted of the 80 segment features described in Section 2. This 2-layer network had 53 output nodes, 53 hidden nodes and the input was applied to both the first and second layers.

## 5.5 Left-Context SNNs

Some of the largest gains in accuracy for HMM CSR systems have been obtained with the use of context. Consequently, we implemented a version of the SNN that provided a simple model of left-context. In addition to the SNN previously described, which only models a segment's phonetic identity and makes no reference to context (i.e., phonetic identity of the segments before and after the scored segment), we trained 53 additional networks. Each of these 53 networks were identical in structure to the non-context SNN. Each left-context network was trained to recognize all 53 phonemes in a single left-context. For example, one network might correspond to the left-context of the phoneme 'AX', so its 53 outputs would be trained to recognize all 53 phonemes when preceded by 'AX'.

In training the context SNNs, we separate the input segments into 53 classes determined by the identity of the preceding data. Each left-context SNN can then be trained in a similar manner to the non-context SNN, but only on the subset of the training data that corresponds to the SNN's context.

In the recognition process, the $n^{th}$ segment score is obtained from combining the value of phoneme output of the non-context SNN, $y_c(n)$, where $c$ is the phonetic label of the segment, with the corresponding output of the SNN that models the left-context of the segment. This combination is a weighted average of the two network values, where the weights are determined by the number of occurrences of (a) the phoneme in the training data, and (b) the number of times the phoneme has its present context in the training data. This is similar to the way in which context HMMs are combined with non-context HMMs in the BYBLOS system [7] to yield robust estimates of HMM parameters.

## 6 EXPERIMENTAL CONDITIONS AND RESULTS

Experiments to test the performance of the hybrid SNN/HMM system were performed on the Speaker Independent (SI) portion of the Resource Management speech corpus, using the word-pair grammar (perplexity 60). The training set consisted of utterances from 109 speakers, 2830 from male and 1160 from female speakers, and the test set for the February 1989 DARPA Workshop was used for development of the system. The test set for the October 1989 DARPA Workshop was used for the final independent test.

In our initial experiments, we used BYBLOS, our HMM CSR system, to produce the 20-best hypothesis lists for the utterances in the the February '89 development set. Table 1 shows the word error rates for rescoring these 20-best lists under the different conditions described in Section 4. It should be noted that the figures do not reflect the unaided performance of the SNN in recognition,

Table 1: SNN development on February '89 test set

|   |   | Word Error (%) |
|---|---|---|
|   | Original SSN (MSE) | 13.7 |
| + | Duration | 12.7 |
| + | Log-Error Criterion | 11.6 |
| + | Two-Layer SNN | 11.0 |
| + | Left-Context (1-layer SNN) | 8.5 |

since the N-best list was generated by a HMM system, but instead illustrate the effectiveness of the respective improvements.

The original 1-layer, non-context SNN was trained with the MSE criterion giving an error rate of 13.7%. The incorporation of the duration term and the adoption of the log-error training criterion both resulted in some improvement. The addition of a second layer to the SNN resulted in only a slight improvement at the expense of a much greater amount of computation during training. Because of this, it was decided that the implementation of the modeling of left context would revert to 1-layer networks. Context modeling clearly produced the greatest improvement in error rate. Adding the duration modeling and log-error training criterion to the context gave a net reduction in error of almost 40%. This final condition was then used to generate the SNN score to examine the behavior of the hybrid SNN/HMM system.

Table 2 shows the results from combining the HMM and SNN scores in the re-ordering of the N-Best list. As mentioned in Section 3, the best value for N when combining the HMM and SNN scores lay between 2 and 20. Therefore, we used the combined SNN/HMM score to rescore N-best lists of 2, 4 and 20. Taking the top answer of the N-best list (as produced by the HMM system) gave an error rate of 3.5% on the February '89 development test set, but when re-ordering the list on the basis of the SNN score, the error rose to 8.5%. Given the large difference between these two error rates, we did not expect the error rate

Table 2: Hybrid SNN/HMM system: test results.

| System* | N | Feb '89 | Oct '89 |
|---|---|---|---|
| HMM | 1 | 3.5 | 4.1 |
| SNN | 20 | 8.5 | — |
| SNN+HMM | 2 | 3.5 | 4.1 |
| SNN+HMM | 4 | 3.0 | 3.2 |
| SNN+HMM | 20 | 3.2 | 3.5 |

* All systems include word and segment scores.

for the hybrid SNN/HMM system to be as low as 3.0%. Furthermore, it appears that the hybrid system yielded an even larger improvement on the independent test set, reducing the error rate from 4.1% in the HMM-based system to 3.5% in the SNN/HMM.

Based upon the results of tests on the February '89 test set, we picked the results of rescoring the 4-best utterances in the October '89 test set (which gave 3.2% error) and performed a significance test against the HMM system (4.1% error) using the Matched-Pairs test according to the advice of Gillick & Cox[3]. There proved to be a significant difference at a significance level of 95%. This indicates that the hybrid SNN/HMM system performs better than the system that incorporates only the HMM score.

# 7 CONCLUSIONS

We have presented the Segmental Neural Net as a method for phonetic modeling in large vocabulary CSR systems and have demonstrated that, when combined with a conventional HMM, the SNN gives a significant improvement over the performance of a state-of-the-art HMM CSR system.

We have used the N-best rescoring paradigm to achieve this improvement in two ways. Firstly, the N-best rescoring paradigm has allowed us to design and test the SNN with little regard to the usual problem of searching when dealing with a large vocabulary speech recognition system. Secondly, the paradigm provides a simple way of combining the best aspects of two systems, leading to a combined system which exceeds the performance of either one alone.

Future work will concentrate on improvements to the structure and context modeling abilities of the SNN and also the addition of other features as inputs to the networks.

## Acknowledgments

The authors would like to thank Amro El-Jaroudi of the University of Pittsburgh for his help in several aspects of this work. This work was sponsored by DARPA.

## References

[1] Austin, S., Peterson, P., Placeway, P., Schwartz, R., Vandegrift, J., "Towards a Real-Time Spoken Language System Using Commercial Hardware," *Proc. DARPA Speech and Natural Language Workshop*, Hidden Valley, PA, June 1990.

[2] El-Jaroudi, A. and Makhoul, J., "A New Error Criterion for Posterior Probability Estimation with Neural Nets," *International Joint Conference on Neural Networks*, San Diego, CA, June 1990, Vol III, pp. 185-192.

[3] Gillick, L. and Cox, S.J., "Some Statistical Issues in the Comparison of Speech Recognition Algorithms," ICASSP-89, Glasgow, UK, May 1989, pp. 532–535.

[4] Gish, H., "A Probabilistic Approach to the Understanding and Training of Neural Network Classifiers," ICASSP-90, Albuquerque, NM, April 1990, pp. 1361–1368.

[5] Ostendorf, M., Kannan, A., Austin, S., Kimball, O., Schwartz, R., Rohlicek, J.R., "Integration of Diverse Recognition Methodologies Through Reevaluation of N-Best Sentence Hypotheses," *Proceedings of the DARPA Speech and Natural Language Workshop*, Pacific Grove, CA, February 1991.

[6] Ostendorf, M. and Roukos S., "A Stochastic Segment Model for Phoneme-based Continuous Speech Recognition," *IEEE Trans. Acoustic Speech and Signal Processing*, Vol. ASSP-37(12), December 1989, pp. 1857–1869.

[7] Schwartz, R., Chow, Y.L., Roucos, S., Krasner, M., Makhoul, J., "Improved Hidden Markov Modeling of Phonemes for Continuous Speech Recognition," ICASSP-84, San Diego, CA, March 1984, pp. 35.6.1–35.6.4.

[8] Schwartz, R. and Chow, Y.L., "The N-Best Algorithm: An Efficient and Exact Procedure for Finding the N Most Likely Sentence Hypotheses," ICASSP-90, Albuquerque, NM, April 1989, pp. 81-84.

# CONNECTIONIST SPEAKER NORMALIZATION AND ITS APPLICATIONS TO SPEECH RECOGNITION

X.D. Huang, K.F. Lee, and A. Waibel
School of Computer Science
Carnegie Mellon University
Pittsburgh, PA 15213

## Abstract

Speaker normalization may have a significant impact on both speaker-adaptive and speaker-independent speech recognition. In this paper, a codeword-dependent neural network (CDNN) is presented for speaker normalization. The network is used as a nonlinear mapping function to transform speech data between two speakers. The mapping function is characterized by two important properties. First, the assembly of mapping functions enhances overall mapping quality. Second, multiple input vectors are used simultaneously in the transformation. This not only makes full use of dynamic information but also alleviates possible errors in the supervision data. Large-vocabulary continuous speech recognition is chosen to study the effect of speaker normalization. Using speaker-dependent semi-continuous hidden Markov models, performance evaluation over 360 testing sentences from new speakers showed that speaker normalization significantly reduced the error rate from 41.9% to 5.0% when only 40 speaker-dependent sentences were used to estimate CDNN parameters.

## 1. INTRODUCTION

Nonlinear mapping of two different observation spaces is of great interest for both theoretical and practical purposes. In the area of speech processing, nonlinear mapping has been applied to speaker normalization [3, 14, 4, 15, 6], noise enhancement [2, 21], articulatory motion estimation [18, 10], and speech recognition [9]. Recently, nonlinear mapping based on neural networks has attracted considerable attention because of the ability of these networks to optimally adjust the parameters from the training data to approximate the nonlinear relationship between two observed spaces. In speech recognition, speaker variability is one of the major error sources. For example, the error rate of speaker-dependent speech recognition is typically two to three times less than that of speaker-independent speech recognition [7]. To model speaker variability, use of either speaker clustered models [7], or speaker normalization may improve the performance of speaker-independent speech recognition. However, the latter can provide a more compact representation than the former. In addition, speaker normalization can be used to rapidly adapt speaker-dependent models for the new speaker as well as voice conversion for text-to-speech systems [1].

In this paper, a codeword-dependent neural network (CDNN) is presented for speaker normalization. The network is used as a nonlinear mapping function to transform speech data between two speakers. The mapping function is characterized by two important properties. First, the assembly of mapping functions enhances overall mapping quality. Second, multiple input vectors are used simultaneously in the transformation. This not only makes full use of dynamic information but also alleviates possible errors in the supervision data.

Based on the DARPA Resource Management task [17], large-vocabulary (1000 words) continuous speech recognition was chosen to study the effect of speaker normalization. Speaker-dependent semi-continuous hidden Markov models (SCHMM) [7] were estimated from 2400 sentences (the RM2 training set) [16]. Based on the speaker-dependent SCHMM of one speaker, performance evaluation over 360 testing sentences from the rest of speakers (three speakers) in the RM2 corpus was carried out. Without speaker normalization, the error rate was 41.9% for cross speaker speech recognition. When 40 speaker-dependent adaptation sentences were used, the error rate was reduced to 6.8% based on a single neural network for each new speaker. The CDNN further reduced the error rate from 6.8% to 5.0%. Overall, the error rate was comparable to that of speaker-independent speech recognition on the same testing data.

## 2. NEURAL NETWORK ARCHITECTURE

### 2.1. General Principal

Speaker normalization involves acoustic data transformation from one speaker to another. In general, let $\mathcal{X}^a = \mathbf{x}_1^a, \mathbf{x}_2^a, ...\mathbf{x}_t^a$ be a sequence of observations at time 1, 2, .. $t$ of speaker $a$. Here, each observation at time $k$ (a frame), $\mathbf{x}_k^a$, is a multidimensional vector, which usually characterizes some short-time spectral features. For speech observations $\mathcal{X}^a$ of speaker $a$, our goal is to find a mapping function $\mathcal{F}(\mathcal{X}^a)$ such that $\mathcal{F}(\mathcal{X}^a)$ resembles the observation sequences produced by the reference speaker.

Speaker variations include many factors such as sex, vocal tract, pitch, speaking speed, intensity, and cultural differences. Unfortunately, given two different speakers, there is no simple mapping function that can account for all these variations. At a given time $t$, $\mathbf{x}_t^a$ usually represents some spectral features for the speaker $a$. In this study, we are mainly concerned with spectral normalization, i.e, to find out a mapping function to transform $\mathbf{x}^a$ so that the normalized observation sequence of speaker $a$ resembles that of the corresponding phonetic realization of speaker $b$, $\mathbf{x}^b$. Thus, one of the objective functions is to minimize:

$$\sum_{corresponding\, pairs} (\mathcal{F}(\mathbf{x}^a) - \mathbf{x}^b)^2. \qquad (1)$$

Neural networks can be used to approximate any nonlinear mapping function [13]. To be useful for speaker normalization, a layered feedforward neural networks should have a number of features. First, we should have sufficient interconnections between multiple layers and sufficient connections between units in each of these layers so that the mapping network will have the ability to learn complex nonlinear mapping functions between different speakers. Second, as the neural network is suitable only to a small or medium task, the original acoustic space should be partitioned into different prototypes such that each network only performs its own work within the corresponding region.

## 2.2. Neural Network Topology

It has been found that dynamic information plays an important role in speech recognition [11, 8] As frame to frame normalization lacks use of dynamic information, the architecture of normalization network is thus chosen to incorporate multiple neighboring frames. One such architecture is shown in Figure 1. Here, the current

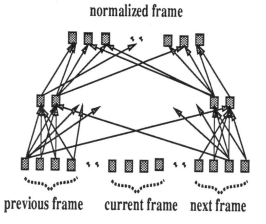

Figure 1: A normalization neural network.

frame and its left and right neighboring frames are fed to the multi-layer neural network as inputs. The network output is a normalized frame corresponding to the current input frame. By using multiple input frames for the network, the important dynamic information can be effectively used in estimating network parameters and in normalization.

If the dimension of observation space is $d$ and the number of input frames is $m$, we will have $d \times m$ input units in the normalization network. This will definitely increase the number of free parameters in the network. Although the increase in the number of free parameters lead to quick convergence during training, this nevertheless may not lead to improved generalization capability. Since the network is designed to normalize new data from a given speaker to the reference speaker, good generalization capability will be the most important concern.

### 2.3. Codeword-Dependent Neural Network

When presented with a large amount of training data, a single network is often unable to produce satisfactory results during training as each network is only suitable to a relatively small task. To improve the mapping performance, breaking up a large task and modular construction are usually required [22, 5]. As the nonlinear relationship between two speakers is very complicated, a simple network may not be powerful enough. One solution is to partition the mapping spaces into smaller regions, and to construct a neural network for each region as shown in Figure 2. As each neural

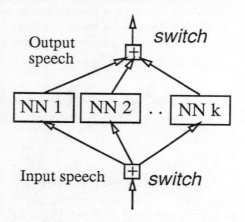

Figure 2: An assembly of neural networks.

network is trained on a separate region in the acoustic space, the complexity of the mapping required of each network is thus reduced. In Figure 2, the switch can be used to select the most likely network or top $N$ networks based on some probability measures of acoustic similarity. Functionally, the assembly of networks is similar to a huge neural network. However, each network in the assembly is learned independently with training data for the corresponding regions. This reduces the complexity of finding a good solution in a huge space of possible network configurations since strong constraints are introduced in performing complex constraint satisfaction in a massively interconnected network.

Vector quantization (VQ) has been widely used for data compression in speech and image processing. Here, it can be used to to partition original acoustic space into different prototypes (codewords). This partition can be regarded as a procedure to perform broad-acoustic pattern classification. The broad-acoustic patterns are automatically generated via a self-organization procedure based on the LBG algorithm [12].

When the codeword-dependent neural network (CDNN) was constructed from the data in the corresponding cell, it was found that learning for the CDNN converges very quickly in comparison with a huge neural network. The larger the codebook, the quicker it converges. However, the size of codebook relies on the number of available training data since codeword-dependent structure fragments training data. The size of codebook should be determined experimentally.

### 2.4. Modifications to Sigmoid Functions

The basic unit used in many neural networks computes the weighted sum of its inputs and passes this sum through a nonlinear function such as a threshold or sigmoid function [19] as:

$$sigmoid(x) = \frac{1}{1 + e^{-x}} \quad (2)$$

In the standard sigmoid function, the output dynamic range is between 0 and 1. If the mapping function input and output are out of this range, feature conditioning is generally required. However, if speaker normalization is applied to speaker adaptation, the original acoustic data can not be conditioned. Because of this, all the output units in the network are not associated with any sigmoid function. Instead, the linear function is used. In addition, the sigmoid function is generalized as:

$$SIGMOID(x) = \frac{\alpha}{1 + e^{-\beta * x}} - \gamma \quad (3)$$

Using $SIGMOID$ function, the dynamic range and the shape can be easily controlled according to the observation structure. In fact, these parameters can also be learned automatically during backpropagation.

## 3. SPEAKER NORMALIZATION EXPERIMENTS

We want to see if speaker normalization can minimize speaker variations with only a limited amount of training sentences. Consequently, we used 40 speaker-dependent sentences (about 5 minutes) to estimate the network parameters to map data of new speakers to the reference speaker.

The DARPA's resource management task [17] is used for the development of speaker-dependent models, which consists of 2400 training sentences from one male speakers (LPN) and 120 testing sentences (RM1, used in June 1990 evaluations) from 3 new speakers. The speaker-dependent training set includes 99% of the words in the vocabulary. The testing set includes 73% of the words in the vocabulary. For neural network training, 40 sentences are randomly extracted from the speaker-dependent training set. The word coverage is less than 19% in the normalization training set. Both testing and training have the same recording conditions. A number of experiments have been conducted on these data sets. The reader is referred to [7] for a summary of some recognition performance benchmarks.

Through this study, the feature extraction performed in both training and testing is a LPC-based cepstral coefficients with a 20-ms Hamming window and a 10-ms frame shift. Bilinear transformation of cepstral coefficients is employed to approximate mel-scale representation. Multiple-string features including first-order and second-order time derivatives are used to construct four independent codebooks [7]. Each codeword is modeled by a Gaussian density function. The top-4 codewords are used for the semi-continuous output probability density function. Only the cepstral vectors are considered for normalization. Once we have the normalized cepstral vector, the first-order and second-order time derivatives can be computed.

### 3.1. Benchmark experiments

As benchmark experiments, speaker-dependent speech recognition was first evaluated. The training set consists of 2400 sentences from each speaker. We used generalized triphone models derived from speaker-independent training set [11]. Based on speaker-dependent data, the SCHMM parameters and VQ codebook were estimated jointly starting with sex-dependent models and codebooks. The average error rate for four speakers was 1.4% as shown in Table 1. Here, the error rate of the

| Speaker | 2400 Training Sentences Word Error Rate |
|---|---|
| BJW | 1.0% |
| JLS | 2.7% |
| JRM | 1.5% |
| LPN | 0.4% |
| Average | 1.4% |

Table 1: Speaker-dependent results.

speaker-dependent system is about three times less that of the speaker-independent system [7]. When we used one male speaker (LPN) as the reference speaker, and designated the other three (one male and two female) as testing speakers, the word recognition error rate was 41.9% as shown in Table 2. We can see that the error rate of the female speakers (BJW and JRM) increased substantially.

| Speaker | Cross speaker recognition Word Error Rate |
|---------|-------------------------------------------|
| BJW     | 55.3%                                     |
| JLS     | 8.5%                                      |
| JRM     | 62.1%                                     |
| Average | 41.9%                                     |

Table 2: Cross speaker recognition results.

## 3.2. Normalization Results

To provide learning examples for the network, a DTW algorithm [20] was used to warp the target data to the reference data. For the given input frames, the desired output frame for network learning is the one paired by the middle input frame in DTW alignment. One important caution in applying the DTW alignment is that silence should be excluded.

The input of the network consists of three frames from the target speaker. Here, 12 cepstral coefficients and energy are used together. Thus, there are 93 input units in the network. The output of the network has 13 units corresponding the normalized frame, which is made to approximate the frame of the desired reference speaker. The energy output is discarded as it is relative unstable. The objective function for network learning is to minimize the distortion (mean squared error) between the network output and the desired reference speaker frame. The network has one hidden layer with 20 hidden units. Each hidden unit is associated with the $SIGMOID$ function, where $\alpha$, $\beta$ and $\gamma$ are predefined to be 4.0, 1.8, 2.0 respectively. They are fixed for all the experiments conducted here. Experimental experience indicates that 300 to 600 epochs are required to achieve acceptable distortion. We averaged results of three runs in following experiments.

When a single network was used for each speaker, the average word error rate was reduced from 41.9% to 6.8% as shown in Table 3 (VQ size = 1). Although neither the codebook nor the HMM parameter was adapted in this experiment, the error rate was already reduced by a factor of 6. It is also interesting to note that for female speakers (JRM and BJW), speaker normalization dramatically reduces the error rate.

To improve the generalization capability, one can increase the number of training data or reduce the number of free parameters. However, the complexity or dimensionality of the mapping network usually has to be increased to maintain accurate mapping between two speakers, which leads to the increased effective number of degrees of freedom in the networks. Therefore, it is important to smooth the less-well trained parameters. The nonlinear network output can be interpolated with the original network inputs [6]. A linear input-output feedforward path can added to the network. All the input units can be either fully or partially connected to the output units without passing through any nonlinear function. The rational to add an interpolation path is that the nonlinear network can not be well constructed. Such a network architecture interpolates the nonlinear network output with the original network input with the interpolation weights automatically determined by the error back propagation algorithm [19]. However, this topology only works well for some speakers. There is no significant overall difference in comparison with the basic topology.

When the CDNN was used, we observed additional 25% error reduction. The error rate was further reduced from 6.8% to 5.0% as shown in Figure 3. This error rate is comparable to that of the best speaker-independent performance on the same test set [7]. This indicates the assembly of mapping functions indeed enhances the overall mapping quality. The best performance was attained when the codebook size was between 4 to 8. Further increase in the codebook size, as shown in the figure, led to degraded performance because of too many free parameters.

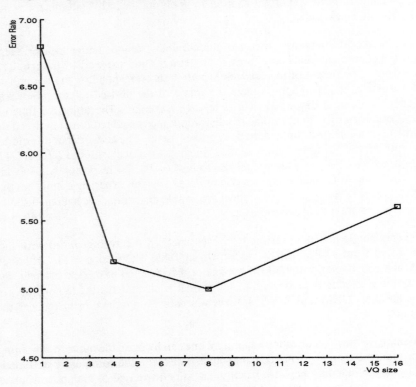

Figure 3: Codeword-dependent network performance.

## 4. SUMMARY

In this paper, the codeword-dependent neural network (CDNN) is presented for speaker normalization. The network is used as a nonlinear mapping function to transform speech data between speakers. Large-vocabulary continuous speech recognition is chosen to study the effect of speaker normalization. Using speaker-dependent SCHMMs on the DARPA RM task, performance evaluation from three new speakers showed that speaker normalization reduced the error rate from 41.9% to 5.0% when only 40 speaker-dependent sentences are used to estimate CDNN parameters. The error rate is comparable to that of the best speaker-independent performance. Our mapping function is characterized by two important properties. First, the assembly of mapping functions enhances the overall mapping quality. Second, multiple input vectors are used simultaneously in normalization.

Speaker-independent network can also be used as part of the front-end of the speaker-independent speech recognition system. The network can be built to reduce the difference among speakers before speaker-independent HMM training is involved such that speaker-independent models will have sharper distributions (better discrimination capability) in comparison with the conventional training procedure. Use of such normalization networks for speaker-independent speech recognition as well as unification of the SCHMM and speaker normalization could provide a new computational architecture for speech recognition.

### Acknowledgments

This research was sponsored by the Defense Advanced Research Projects Agency (DOD), Arpa Order No. 5167, under contract number N00039-85-C-0163. The authors would like to express their gratitude to Professor Raj Reddy for his encouragement and support, and other members of CMU speech group for their help.

### References

[1] Abe, M., Nakamura, S., and Shikano, K. *Voice conversion through vector quantization.* in: IEEE International Conference on Acoustics, Speech, and Signal Processing. 1988.

[2] Acero, A. and Stern, R. *Environmental Robustness in Automatic Speech Recognition.* in: IEEE International Conference on Acoustics, Speech, and Signal Processing. 1990, pp. 849–852.

[3] Choukri, K., Chollet, G., and Grenier, Y. *Spectral transformations through cannonical correlation analysis for speaker adapataion in ASR.* in: IEEE International Conference on Acoustics, Speech, and Signal Processing. 1986, pp. 2659–2552.

[4] Class, F., Kaltenmeier, A., Regel, P., and Trottler, K. *Fast speaker adaptation for speech recognition.* in: IEEE International Conference on Acoustics, Speech, and Signal Processing. 1990, pp. 133–136.

[5] Hampshire, J. and Waibel, A. *The Meta-Pi Network: Connectionist rapid adapatation for high-performance multi-speaker phoneme recognition.* in: IEEE International Conference on Acoustics, Speech, and Signal Processing. 1990, pp. 165–168.

[6] Huang, X. *A Study on Speaker-Adaptive Speech Recognition.* in: **DARPA Speech and Language Workshop**. Morgan Kaufmann Publishers, San Mateo, CA, 1991.

[7] Huang, X. and Lee, K. *On Speaker-Independent, Speaker-Dependent, and Speaker-Adaptive Speech Recognition.* in: **IEEE International Conference on Acoustics, Speech, and Signal Processing**. 1991, pp. 877–880.

[8] Huang, X., Lee, K., Hon, H., and Hwang, M. *Improved Acoustic Modeling for the SPHINX Speech Recognition System.* in: **IEEE International Conference on Acoustics, Speech, and Signal Processing**. 1991, pp. 345–348.

[9] Iso, K. and Watanabe, T. *Speaker-independnet word recognition using a neural prediction model.* in: **IEEE International Conference on Acoustics, Speech, and Signal Processing**. 1990, pp. 441–444.

[10] Kobayashi, T., Yagyu, M., and Shirai, K. *Applications of neural networks to articulatory motion estimation.* in: **IEEE International Conference on Acoustics, Speech, and Signal Processing**. 1991, pp. 489–4920.

[11] Lee, K., Hon, H., and Reddy, R. *An Overview of the SPHINX Speech Recognition System.* **IEEE Transactions on Acoustics, Speech, and Signal Processing**, January 1990, pp. 35–45.

[12] Linde, Y., Buzo, A., and Gray, R. *An Algorithm for Vector Quantizer Design.* **IEEE Transactions on Communication**, vol. COM-28 (1980), pp. 84–95.

[13] Lippmann, R. *Neural Nets for Computing.* in: **IEEE International Conference on Acoustics, Speech, and Signal Processing**. 1988, pp. 1–6.

[14] Montacie, C., Choukri, K., and Chollet, G. *Speech recognition using temporal decomposition and multi-layer feed-forward automata.* in: **IEEE International Conference on Acoustics, Speech, and Signal Processing**. 1989, pp. 409–412.

[15] Nakamura, S. and Shikano, K. *A comparative study of spectral mapping for speaker adaptation.* ICASSP, 1990, pp. 157–160.

[16] Pallett, D., Fiscus, J., and Garofolo, J. *DARPA Resource Management Benchmark Test Results June 1990.* in: **DARPA Speech and Language Workshop**. Morgan Kaufmann Publishers, San Mateo, CA, 1990, pp. 298–305.

[17] Price, P., Fisher, W., Bernstein, J., and Pallett, D. *A Database for Continuous Speech Recognition in a 1000-Word Domain.* in: **IEEE International Conference on Acoustics, Speech, and Signal Processing**. 1988, pp. 651–654.

[18] Rahim, M., Kleijn, W., Schroeter, J., and Goodyear, C. *Acoustic to articulatory parameter mapping using an assembly of neural networks.* in: **IEEE International Conference on Acoustics, Speech, and Signal Processing**. 1991, pp. 485–488.

[19] Rumelhart, D., Hinton, G., and Williams, R. *Learning Internal Representation by Error Propagation.* in: **Learning Internal Representation by Error Propagation**, by D. Rumelhart, G. Hinton, and R. Williams, edited by D. Rumelhart and J. McClelland. MIT Press, Cambridge, MA, 1986.

[20] Sakoe, H. and Chiba, S. *Dynamic Programming Algorithm Optimization for Spoken Word Recognition.* **IEEE Transactions on Acoustics, Speech, and Signal Processing**, vol. ASSP-26 (1978), pp. 43–49.

[21] Tamura, S. and Waibel, A. *Noise reduction using connectionist modelsnce Measure for Speech Recognition.* in: **IEEE International Conference on Acoustics, Speech, and Signal Processing**. 1988, pp. 553–556.

[22] Waibel, A., Sawai, H., and Shikano, K. *Consonant Recognition by Modular Construction of Large Phonemic Time Delay Neural Networks.* in: **IEEE International Conference on Acoustics, Speech, and Signal Processing**. 1989.

# A TIME-DERIVATIVE NEURAL NET ARCHITECTURE — AN ALTERNATIVE TO THE TIME-DELAY NEURAL NET ARCHITECTURE

K.K. Paliwal[1]

Speech Research Department
AT&T Bell Laboratories
Murray Hill, NJ 07974

**ABSTRACT** — Though the time-delay neural net architecture has been recently used in a number of speech recognition applications, it has the problem that it can not use longer temporal contexts because this increases the number of connection weights in the network. This is a serious bottleneck because the use of larger temporal contexts can improve the recognition performance. In this paper, a time-derivative neural net architecture is proposed. This architecture has the advantage that it can utilize information about longer temporal contexts without increasing the number of connection weights in the network. This architecture is studied here for speaker-independent isolated-word recognition and its performance is compared with that of the time-delay neural net architecture. It is shown that the time-derivative neural net architecture, in spite of using less number of connection weights, outperforms the time-delay neural net architecture for speech recognition.

## 1. INTRODUCTION

Hidden Markov modeling is a popular, and perhaps the most successful, technique today for speech recognition. Its main advantage lies in its ability to model the time variability of the speech signals. However, it has a drawback that it does not provide enough discrimination between classes (such as phonemes and words) as their models are usually obtained using the maximum likelihood algorithm. On the other hand, neural networks, and particularly multilayer perceptrons, provide good discrimination between classes and, hence, are being investigated by many researchers for speech recognition. The major problem with the multilayer perceptrons is that they are restricted to static patterns and it is difficult to extend them to the classification of time-varying speech signals. However, a number of neural net architectures have been recently proposed in the literature (see [1] for references) to overcome this problem. Notable among these is the time-delay neural net architecture proposed by Waibel et al. [2]. Time variability is incorporated in this architecture by utilizing temporal context in the form of time delays. In this

---

[1] On leave from Computer Systems and Communications Group, Tata Institute of Fundamental Research, Homi Bhabha Road, Bombay-400005, India.

architecture, the input to a neuron at a given time is computed as a weighted sum of not only the present outputs of the lower layer's neurons, but the outputs from the past as well (i.e., their time-delayed versions). Though the time-delay neural net architecture has been used in a number of applications, such as phoneme recognition [2], phoneme spotting [3, 4] and isolated word recognition [5], it has the problem that the number of connection weights become very large when longer temporal contexts have to be incorporated in the network. Therefore, this architecture can not be used for larger temporal contexts. This is a serious bottleneck because the use of larger temporal contexts can improve the recognition performance [6].

In the present paper, we propose an alternative to the time-delay neural net architecture. We call it the time-derivative neural net architecture. This architecture incorporates the time variability present in the speech signals by utilizing the temporal context in the form of time derivatives, instead of time delays. Here, the input to a neuron at a given time is obtained by taking the weighted sum of the outputs of the lower layer's neurons and their derivatives at that time. This architecture has the advantage that it can utilize information about longer temporal contexts without increasing the number of connection weights in the network. This is not possible with the time-delay neural net architecture, as mentioned earlier.

Some speech recognition experiments are performed in this paper to study the performance of the time-derivative neural net architecture and compare it with that of the time-delay neural net architecture. The recognizer is used here in a speaker-independent mode for recognizing an utterance from a 9-word vocabulary consisting of English e-set alphabets. It is shown in this paper that the time-derivative neural net architecture, in spite of using less number of connection weights, outperforms the time-delay neural net architecture for speech recognition.

The organization of this paper is as follows. The time-derivative neural net architecture is described in Section 2. In Section 3, this architecture is studied for speaker-independent isolated-word speech recognition and performance of this architecture is compared with that of the time-delay neural net architecture. Conclusions are reported in Section 4.

## 2. THE TIME-DERIVATIVE NEURAL NET ARCHITECTURE

In this section, the time-derivative neural net architecture is briefly described. This architecture uses a feed-forward fully-connected multi-layer perceptron type of neural network. It has one input unit, one output unit and a number of hidden units. However, in the architecture described below, only one hidden unit is used.

We use here the letter j for the index of a neuron in the output layer, i for the index of a neuron in the hidden layer, and k for the index of a neuron in the input layer. We denote the input to a neuron by the letter $x$ and its output by the letter $y$. Let $n_j$, $n_i$ and $n_k$ be the number of neurons in the output, hidden and input layers, respectively. The number of neurons in the input layer is equal to the number of features in the feature vector. The number of neurons in the output unit is equal to the number of classes (or, words in the vocabulary). The number of neurons in the hidden layer is selected depending on the complexity of the recognition problem.

Let the speech utterance to be recognized be represented by a sequence of feature vectors,

$$\mathbf{C} = \{\mathbf{c}(1), \mathbf{c}(2), \ldots, \mathbf{c}(T)\}, \quad (1)$$

where $\mathbf{c}(t)$ is the feature vector at time frame t and $T$ is the number of frames in the speech utterance. In order to recognize this speech utterance, the outputs of the neurons in the output layer have to be computed. This is done as follows.

For each time frame t, the components of the feature vector are applied to the inputs of the neurons in the input layer and outputs of these neurons are computed by assuming identity transfer function for each of these neurons. That is,

$$y_k(t) = c_k(t), \quad (2)$$

for $k = 1, 2, \ldots, n_k$ and $t = 1, 2, \ldots, T$.

The input to the i-th neuron in the hidden layer at time t is computed as the weighted sum of the outputs of the input layer's neurons and their derivatives at time t; i.e.,

$$x_i(t) = \sum_{k=1}^{n_k} \sum_{d=0}^{D_k} w_{ikd} y_k^{(d)}(t) + \theta_i, \quad (3)$$

where $D_k$ is the number of derivatives of the outputs from the input layer used for the computation of the inputs for the hidden layer, $\theta_i$ is the threshold for the i-th neuron in the hidden layer, and $w_{ikd}$ is the connection weight associated with the i-th hidden neuron, the k-th input neuron and the d-th derivative. $y_k^{(d)}(t)$ is the d-th derivative at time t of the output of the k-th neuron in the input unit and is computed as the d-th orthogonal polynomial coefficient over a finite length window as follows [7, 8, 9, 10]:

$$y_k^{(d)}(t) = K^{(d)}(L) \sum_{l=-L}^{L} a_l^{(d)}(L) y_k(t+l), \quad (4)$$

where $(2L+1)$ is the window length and $K^{(d)}(L)$ is a scaling constant which depends mainly on window length and order of derivative[2]. $a_l^{(d)}(L)$ is the d-th orthogonal polynomial defined over a window length of $(2L+1)$ frames. The first few orthogonal polynomials are [7]:

$$a_l^{(0)}(L) = 1, \quad (5)$$

$$a_l^{(1)}(L) = l, \quad (6)$$

---

[2] For the computation of the 0-th order derivatives, the window length is set to 1 frame; i.e., L=0, and the scaling constant K used in Eq. (4) is set to 1.

$$a_l^{(2)}(L) = l^2 - (L^2 - 1)/12, \tag{7}$$

$$a_l^{(3)}(L) = l^3 - (3L^2 - 7)l/20. \tag{8}$$

The output of the i-th neuron in the hidden layer at time t is computed from its input using the sigmoid nonlinear function as follows:

$$y_i(t) = 1/(1 + \exp(-x_i(t))). \tag{9}$$

The outputs of all the neurons in the hidden unit for all the time frames are computed by using Eqs. (3-9) for $i = 1, 2, \ldots, n_i$ and $t = 1, 2, \ldots, T$.

In a similar fashion, the outputs of the neurons in the output unit are computed as follows:

$$y_j(t) = 1/(1 + \exp(-x_j(t))), \tag{10}$$

for $j = 1, 2, \ldots, n_j$ and $t = 1, 2, \ldots, T$. Here, $x_j(t)$ is the input to the j-th neuron in the output layer and is computed as the weighted sum of the outputs of the hidden layer's neurons and their derivatives as follows:

$$x_j(t) = \sum_{i=1}^{n_i} \sum_{d=0}^{D_i} w_{jid} y_i^{(d)}(t) + \theta_j, \tag{11}$$

where the derivative $y_i^d(t)$ is computed as follows:

$$y_i^{(d)}(t) = K^{(d)}(L) \sum_{l=-L}^{L} a_l^{(d)}(L) y_i(t+l). \tag{12}$$

The outputs of the individual neurons in the output layer obtained from Eq. (10) for $t = 1, 2, \ldots, T$ define the a posteriori probabilities of the words associated with these neurons at different time frames [11, 12]. These can be normalized by their a priori probabilities to get the emission probabilities for time $t = 1, 2, \ldots, T$ [13]. The likelihood of the speech utterance (defined by the sequence C of the feature vectors) coming from the word associated with the i-th neuron in the output layer is obtained by multiplying these emission probabilities as follows:

$$L_i(\mathbf{C}) = \prod_{t=1}^{T} y_i(t)/p_i, \tag{13}$$

where $p_i$ is the a priori probability of the word associated with the i-th neuron in the output layer. The speech utterance is recognized by maximizing the likelihood; i.e., by using the maximum likelihood decision rule [14].

## 3. SPEECH RECOGNITION EXPERIMENTS AND RESULTS

In this section, the time-derivative neural net architecture is studied for speech recognition and its performance is compared with that of the time-delay neural net architecture. Speech recognition experiments reported in this section are conducted in a speaker-independent mode. Isolated-word speech recognizer is used in these experiments. The vocabulary of the recognizer consists of 9 English e-set alphabets (B, C, D, E, G, P, T, V and Z).

Speech data base used in the recognition experiments consists of 24 utterances per word from 4 speakers (2 male and 2 female) for training, and 40 utterances per word from the same 4 speakers for testing. These utterances are digitized at a sampling rate of 6.67 kHz. An 8-th order linear prediction analysis is performed frame-wise every 15 ms using a Hamming window of 45 ms, and each frame is represented by a feature vector of 12 liftered cepstral coefficients [15].

The time-derivative neural net architecture used in the recognition experiments has 3 layers: the input layer, the hidden layer and the output layer. The numbers of neurons in the three layers are 12, 16 and 9, respectively; i.e., $n_k = 12$, $n_i = 16$ and $n_j = 9$. In these experiments, we set $D_i = 0$. This means that the inputs to the output units are obtained as a weighted sum of the outputs of the hidden neurons and their derivatives are not used. Outputs and their derivatives are used only in the computation of the inputs to the hidden neurons; i.e., $D_k > 0$. For training the neural net, the error-back propagation algorithm is used with the total-squared error criterion [16].

As mentioned earlier, the time-derivative neural net architecture has the advantage that it can utilize longer temporal context without increasing the number of connection weights. This is done here by increasing the window length. In order to see how the longer temporal context improves the recognition performance, we use the time-derivative neural net with $D_k = 1$. This means that the total number of connection weights used in the neural net (including the thresholds) is 553. Performance of this net as a function of the duration of temporal context is shown in Table 1. It can be seen from this table that the time-derivative neural net architecture

Table 1: Recognition performance of the time-derivative neural net architecture as a function of the duration of the temporal context.

| Duration of the temporal context (in ms) | Number of connection weights | Recognition accuracy (in %) |
|---|---|---|
| 75 | 553 | 67.2 |
| 135 | 553 | 72.2 |
| 195 | 553 | 76.1 |
| 255 | 553 | 75.0 |

results in better recognition performance when the longer temporal context is used. Recognition performance is best for the temporal context of 195 ms duration. Note that Hanson and Applebaum [10] have made similar observations for the hidden Markov model based speech recognizer.

Next, we study the recognition performance of the time-derivative neural net architecture using higher order derivatives. For this, we fix the duration of the temporal context to 195 ms, and study the recognition performance as a function of the number of derivatives. When we say that the number of derivatives is 2 (for example), it means that we are computing the input to a neuron as a weighted sum of the outputs from neurons in the lower layer and their first and second derivatives. Results are shown in Table 2. It can be seen from this table that use of second

Table 2: Recognition performance of the time-derivative neural net architecture as a function of number of derivatives.

| Number of derivatives | Duration of the temporal context (in ms) | Number of connection weights | Recognition accuracy (in %) |
|---|---|---|---|
| 1 | 195 | 553 | 76.1 |
| 2 | 195 | 745 | 78.6 |
| 3 | 195 | 937 | 77.2 |

derivative improves the performance, but at the cost of more number of connection weights. However, when the third derivative is used the recognition performance goes down, in spite of the fact that we are using more number of connection weights. This happens because the amount of data available for training the speech recognizer is limited, a common problem in pattern recognition [17].

Thus, we have seen that the time-derivative neural net based speech recognizer performs better with the longer temporal context. Note that use of longer temporal context does not increase the number of connection weights in the neural net. Also, the inclusion of higher derivatives in the neural net improves the recognition performance, if sufficient amount of training data is available for training.

Now, we compare the recognition performance of the time-derivative neural net architecture with that of the time-delay neural net architecture. For this, we study the performance of the time-delay neural net based speech recognizer for different temporal delays. Here also, the inputs to the output neurons are computed from the outputs of the hidden neurons without using any temporal delays. The temporal delays are used only at the outputs of the input neurons to compute the inputs to the hidden neurons. Recognition results as a function of number of temporal delays are shown in Table 3. It can be seen from this table that we can get better recognition performance by using more number delays in the neural net. However, the performance does not improve after 3 delays, in spite of the fact that we are using more number of connection weights. This happens, as mentioned earlier, due to the limited amount of training data [17]. Note that if we want to use a longer temporal context, we have to use more delays. This means that we have to increase the number of connection weights. This increases the computational cost and memory requirements. Also, use of more connection weights does not always

Table 3: Recognition performance of the time-delay neural net architecture as a function of number of delays.

| Number of delays | Duration of the temporal context (in ms) | Number of connection weights | Recognition accuracy (in %) |
|---|---|---|---|
| 1 | 30 | 553 | 62.8 |
| 2 | 45 | 745 | 63.3 |
| 3 | 60 | 937 | 67.3 |
| 4 | 75 | 1129 | 67.5 |
| 5 | 90 | 1321 | 67.5 |

improve the recognition performance due to the limited amount of training data.

By comparing Table 3 with Table 2, we can see that for the same number of connection weights, the time-derivative neural net architecture performs much better than the time-delay neural net architecture. Also, comparison of Table 3 with Table 1 shows that the time-derivative neural net architecture can utilize information about longer temporal contexts without increasing the number of connection weights in the network, while this is not possible with the time-delay neural net architecture. The reason for this is that the time-derivative neural net architecture models the temporal context in terms of time derivatives and, thus, does not increase the number of connection weights for the longer temporal contexts. The time-delay neural architecture uses explicit time delays to incorporate the temporal context. Therefore, it has to increase the number of delays (and connection weights) to utilize longer temporal contexts.

It may be noted that we have not used in our recognition experiments the derivatives of the outputs of the hidden neurons to compute the inputs to the output neurons. This will be done in our future research work. Also, note that the present system can be thought of as a single-state hidden Markov model based speech recognizer where outputs of the output neurons are used as emission probabilities. Recognition performance can be improved by using these emission probabilities with the multi-state hidden Markov models, as done by Haffner et al. [18].

## 4. CONCLUSIONS

In this paper, a time-derivative neural net architecture is proposed for speech recognition. This architecture has the advantage that it can utilize information about longer temporal contexts without increasing the number of connection weights in the network. This is not possible with the time-delay neural net architecture. The time-derivative neural net architecture is studied here for speaker-independent isolated-word recognition and its performance is compared with that of the time-delay neural net architecture. It is shown that the time-derivative neural net architecture, in spite of using less number of connection weights, outperforms the time-delay neural net architecture for speech recognition.

# References

[1] H. Bourlard and C.J. Wellekens, "Speech dynamics and recurrent neural networks", in *Proc. IEEE Int. Conf. Acoust., Speech, Signal Processing* (Glasgow, Scotland), May 1989, pp. 33-36.

[2] A. Waibel, T. Hanazawa, G. Hinton, K. Shikano and K.J. Lang, "Phoneme recognition using time-delay neural networks", *IEEE Trans. Acoust., Speech, Signal Processing*, vol. ASSP-37, pp. 328-339, Mar. 1989.

[3] M. Miyatake, H. Sawai, Y. Minami and K. Shikano, "Integrated training for spotting Japanese phonemes using large phonemic time-delay neural networks", in *Proc. IEEE Int. Conf. Acoust., Speech, Signal Processing* (Albuquerque, NM), Apr. 1990, pp. 449-452.

[4] K.J. Lang, A.H. Waibel and G.E. Hinton, "A time-delay neural network architecture for isolated word recognition", *Neural Networks*, vol. 3, pp. 23-43, 1990.

[5] L. Bottou, F.F. Soulie, P. Blanchet and J.S. Lienard, "Experiments with time-delay networks and dynamic time warping for speaker independent isolated digits recognition", in *Proc. Eurospeech*, Sept. 1989.

[6] A.J. Robinson and F. Fallside, "A dynamic connectionist model for phoneme recognition", in *Proc. Eurospeech*, Sept. 1989.

[7] N.R. Draper and H. Smith, *Applied Regression Analysis*. New York: Wiley, 1981.

[8] S. Furui, "Cepstral analysis techniques for automatic speaker verification", *IEEE Trans. Acoust., Speech, Signal Processing*, vol. ASSP-29, pp. 254-272, Apr. 1981.

[9] F.K. Soong and A.E. Rosenberg, "On the use instantaneous and transitional spectral information in speaker recognition", *IEEE Trans. Acoust., Speech, Signal Processing*, vol. 36, pp.871-879, June 1988.

[10] B.A. Hanson and T.H. Applebaum, "Robust speaker-independent word recognition using static, dynamic and acceleration features: Experiments with lombard and noisy speech", in *Proc. IEEE Int. Conf. Acoust., Speech, Signal Processing* (Albuquerque, NM), Apr. 1990, pp. 857-860.

[11] H. Bourlard and C.J. Wellekens,"Links between Markov models and multilayer perceptrons", *IEEE Trans. Pattern Analysis and Machine Intelligence*, vol. 12, pp. 1167-1178, Dec. 1990.

[12] E.A. Wan, "Neural network classification: A Bayesian interpretation", *IEEE Trans. Neural Networks*, vol. 1, pp. 303-305, Dec. 1990.

[13] N. Morgan and H. Bourlard, "Continuous speech recognition using multilayer perceptrons with hidden Markov models", in *Proc. IEEE Int. Conf. Acoust., Speech, Signal Processing* (Albuquerque, NM), Apr. 1990, pp. 413-416.

[14] R.O. Duda and P.E. Hart, *Pattern Classification and Scene Analysis*. New York: Wiley, 1973.

[15] B.H. Juang, L.R. Rabiner and J.G. Wilpon, "On the use of bandpass liftering in speech recognition", *IEEE Trans. Acoust., Speech, Signal Processing*, vol. ASSP35, pp. 947-954, 1987.

[16] D.E Rumelhart, G.E. Hinton and R.J. Williams, "Learning representations by back-propagating errors", *Nature*, vol. 323, pp. 533-536, Oct. 1986.

[17] L.N. Kanal and B. Chandrasekaran, "On dimensionality and sample size in statistical pattern classification", *Pattern Recognition*, vol. 3, pp. 225-234, Oct. 1971.

[18] P. Haffner, M. Franzini and A. Waibel, "Integrating time alignment and neural networks for high performance continuous speech recognition", in *Proc. IEEE Int. Conf. Acoust., Speech, Signal Processing* (Toronto, Canada), May 1991, pp. 105-108.

# Word Recognition Based on the Combination of a Sequential Neural Network and the GPDM Discriminative Training Algorithm

Wen-Yuan Chen[+*] and Sin-Horng Chen[+]

[+]Department of Electronic Engineering
National Chiao Tung University
[*]Computer & Communication Research Laboratories
Industrial Technology Research Institute
Hsinchu, Taiwan, R.O.C.

### Abstract

This paper proposes an isolated-word recognition method based on the combination of a sequential neural network and a discriminative training algorithm using the Generalized Probabilistic Descent Method (GPDM). The sequential neural network deals with the temporal variation of speech by dynamic programming, and the GPDM discriminative training algorithm is used to discriminate easily confused words by enhancing the distinguishing sounds of them during the scoring procedure. A Mandarin digit database uttered by 100 speakers was used to evaluate the performance of this method. The recognition rates are 99.1% on training data and 96.3% on testing data.

## 1 Introduction

Recently, the connectionist models, particularly the Multi-Layer Perceptrons(MLP), have been widely used for speech recognition. Some experiments[1,2] have directly employed MLPs as discriminators for word recognition. Even though they have achieved encouraging results, there still exist some practical problems on applying an MLP to speech recognition [3]. The most severe constraint lies on the fact that an MLP discriminator can not absorb temporal distortion of speech patterns. On the contrary, many high-performance recognizers based on Dynamic Time Warping (DTW) algorithm and Hidden Markov Model (HMM) can handle the problem of time alignment. Hence, to cope with the temporal variation of speech signal, some hybrid methods which incorporate the concepts of HMM and DTW into MLP discriminators have been reported [3,4,5,6,7].

However, highly confusable words are still difficult to recognize because the distinguishing sounds of them are usually short and low in energy. An effective way to discriminate these easily confused words is to emphasize the distinguishing sounds of them during the scoring procedure. Unfortunately, most recognizers implemented by MLP do not consider this problem and some models [3,5] even lose the discrimination property of MLP. This paper proposes a method for word recognition which is based on the combination of a sequential neural network and the GPDM discriminative training algorithm [8]. The sequential neural network deals with the time alignment problem by employing a sequence of MLPs and using DP to find the optimal state sequences associated with the input utterance. Distortions of these states are weightedly combined for final discrimination. The GPDM discriminative training algorithm is used to learn these weights for emphasizing, in the scoring process, the distinguishing parts of highly confusable words.

The organization of this paper is described as follows. Section 2 introduces the sequential MLP. In Section 3, the incorporating of the GPDM discriminative training algorithm into the
sequential MLP is discussed. A series of experiments are demonstrated in Section 4. Conclusions are given in Section 5.

## 2 The sequential neural network

Fig. 1 shows the block diagram of a sequential neural network composed of $K$ MLPs. Each MLP is regarded as a pattern recognizer of a state, and has $N$ nodes on the output layer to represent $N$ reference words.

### 2.1 Recognition Algorithm

The reference model for each word is represented by a $K-$ state single transition model shown in Fig. 2. The value at the $j-th$ output node of the $i-th$ MLP is regarded as the output value of the $j-th$ reference model at the $i-th$ state. For reference word $w$ represented by the $j-th$ output nodes of the sequential MLP, the target values of MLPs are set to 0.9 for the $j-th$ output nodes, and 0.1 for all others. The accumulated distance $D(w, m)$ between the input utterance $m$ and the reference word $w$ is calculated by

$$D(w,m) = min \sum_{n=1}^{T} \|\hat{a}_n(w, p(n)) - a_n\|^2 \qquad (1)$$

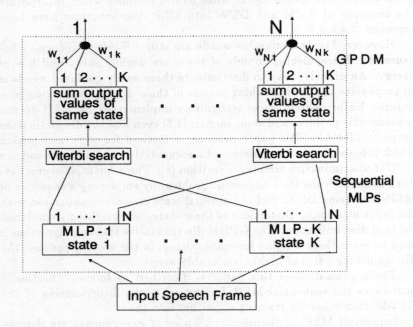

Figure 1: A word recognition model based on the combination of sequential MLPs and a generalized probabilistic descent method.

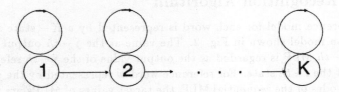

Figure 2: A k-states single transition model.

where T is the length of the input utterance, p(n) $(1 \leq p(n) \leq K)$ is a path that maps the input utterance to the $j - th$ output nodes of the MLP sequence, and $\hat{a}_n(w, p(n))$ and $a_n$ are the output and the target values at the $j - th$ output node of the $p(n) - th$ MLP, respectively. The optimal path can be found by dynamic programming (DP). When only the sequential MLP is used, the word associated to the minimum distortion is taken as the recognized word.

## 2.2 Training Algorithm

The purpose of the training algorithm is to find out a set of MLPs' weights which minimizes the accumulated distance for a training data set. The optimization can be made by an iterative procedure which combines dynamic programming (DP) and the back-propagation (BP) training algorithm. The algorithm is given as follows :

1. Randomly initialize all the network parameters (thresholds and weights) to small values, and normalize them to appropriate values [9]

2. For a training utterance $w$, segment it into $K$ parts of equal length. Data belonging to the $i - th$ segment is used to train the $i - th$ MLP by the BP training method. This procedure is repeated for all training utterances with several iterations.

3. Given the current sequential MLP, use DP to determine, for each training utterance, the optimal segmentations when compared with all reference words.

4. Based on these segmentations, retrain all MLPs by BP training method. Steps 3 and 4 are iterated until a convergence is reached.

We note that, in Steps 3 and 4, multiple segmentations of an utterance are used to train the sequential MLP. This is expected to improve the recognition result.

# 3 Applying a GPDM discriminative training to the sequential MLP

Amari [10] proposed the probabilistic-descent method (PDM) which guaranteed that the discriminant function becomes better on the average. S. Katagiri et al. [8] generalized the PDM to GPDM such that (1) dynamic patterns can be treated and (2) PDM's implementation range can be greatly extended. Here, the GPDM discriminative training method is applied to the sequential MLP for improving its discrimination capability. By taking the output distortions of the sequential MLP as features for discrimination, we first define the following linear discriminant functions

$$g_\alpha(x) = \mathbf{W}_\alpha^t \mathbf{X}, \qquad \alpha = 1, 2, \cdots, m \qquad (2)$$

The decision rule is then becomes: a pattern $x$ is decided to belong to $C_\alpha$ iff $g_\alpha(x) < g_\beta(x)$ for all $\beta(\neq \alpha)$. For applying GPDM to train these weights, a distance is define as

$$d_{\alpha\beta} = g_\alpha(x) - g_\beta(x), \qquad (3)$$

Then, a loss function $l(d)$ is defined to evaluate the cost of decision. The goal of GPDM is to adjust these weights for achieving the minimum cost. By GPDM, a weight $\mathbf{W}$ at the $(k+1)$ iteration is updated by [11]

$$\mathbf{W}(k+1) = \mathbf{W}(k) - \rho_k \nabla_w l(d), \qquad (4)$$

where $\nabla$ is the gradient operator and $\rho_k$ is a positive scaling factor. By substituting (2) and (3) into (4) we obtain [11]:

$$W_\gamma(k+1) = W_\gamma(k) + H_\gamma(k) \qquad (5)$$

where

$$H_\gamma(k) = \begin{cases} +\rho_k l'(d)\mathbf{X}, & \text{if } \gamma \neq \alpha \\ -\rho_k l'(d)\mathbf{X}, & \text{if } \gamma = \alpha \end{cases} \qquad (6)$$

We chose the Gauss-Laplace function $h(d,\nu)$ for $l'(d)$ [11].

$$h(d,\nu) = \frac{1}{\sqrt{2\pi}\nu} e^{-\frac{1}{2}\left(\frac{d}{\nu}\right)^2} \qquad (7)$$

where $\nu$ is a scaling factor. (6) is thus becomes

$$H_\gamma(k) = \begin{cases} +\rho_k h(d,\nu)\mathbf{X}, & \text{if } \gamma \neq \alpha \\ -\rho_k h(d,\nu)\mathbf{X}, & \text{if } \gamma = \alpha \end{cases} \qquad (8)$$

From the definition of $h(d,\nu)$, it is clear that this weight adjusting scheme will response seriously to highly confusable utterances. Some learning algorithms, such as the perceptron learning algorithm, react only on incorrectly classified input utterances. Whereas, (8) reacts on all input utterances (misclassified or not). This will result in safer classification for providing a gap of safeguard.

It is noted that all initial values of weights are set to 1 and all parameters of the sequential MLP are kept fixed when the GPDM is applied.

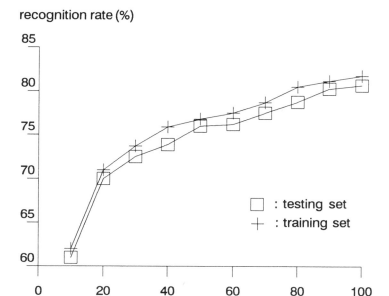

Figure 3: The recognition rate vs. training iteration by sequential MLPs with uniform segmentation.

## 4 Experiments

The Mandarin isolated digit speech database created by Telecommunication Laboratory (TL)[12] is used in our experiments to validate the proposed method. 100 speakers (50 male and 50 female) uttered the ten Mandarin digits two times on different days, one time for training, and one time for testing. The original speech database has been low-pass filtered with 8 kHz and digitized with 16 bits resolution at a sampling rate of 20 kHz. This speech data has been downsampled to 12 kHz. A short-time spectral analysis by 256-point FFT is performed over every 20-ms Hamming-windowed frame padding with 16 zeros at a frame rate of 10 ms. The spectrogram (log) is compressed nonlinearly into 16 frequency channels (mel-scale) with a triangle window according to a model of auditory perceptron. For each frame, energies of 16 channels are normalized to a range between -1.0 and 1.0.

The sequential neural network composed of 7 MLPs was used in the experiments. Every MLP has 16 inputs, 20 hidden units and 10 output units, corresponding to ten Mandarin digits. Fig. 3 shows the recognition rates vs. training iterations of the sequential MLPs learned with uniform segmentation described in Step 2 of Section 2.2. Each itera-

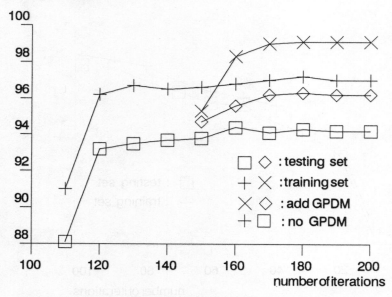

Figure 4: The recognition rate vs. training iteration of sequential MLPs with dynamic programming segmentation.

tion includes 1000 utterances. The recognition rates reach saturation at 81.8% on training data and 80.7% on testing data. After 100 iterations, the DP is used to determine the training data of each MLP which was described in step 3 and 4 of section 2.2. The recognition results are shown in Fig. 4 marked by symbols "+" and "□". It shows that the recognition rates are promoted to 96.2% on training data, and 93.2% on testing data at iteration 120. The GPDM is applied to the sequential MLPs after iteration 150. The initial learning step size $\rho_k$ of (4) is set to 6 and decays linearly with iterations. The scalar $\nu$ of (7) is fixed as 6. These two values are determined experimentally. When the GPDM is applied to the sequential MLPs, the recognition results are shown in Fig. 4 marked by symbols "×" and "◇". We then get the recognition rates of 99.1% on training data and 96.3% on testing data at iteration 180. It should be noted that it is much easier for the parameters of GPDM to be trained than the parameters of the MLPs. Hence, for a sequential MLPs, the GPDM discriminative training algorithm is an effective way to promote the recognition rate with less computation.

## 5 Conclusion

This paper presents a speech recognition method which combines a sequential neural network and the GPDM discriminative training algorithm. The temporal distortion of speech pattern can be alleviated by the sequential neural network, and the GPDM discriminative training algorithm is used to discriminate easily confused words by enhancing the distinguishing sounds of them during the scoring procedure. The training and recognition algorithms are developed, and are based on dynamic-programming, back-propagation, and a GPDM. This method has been used to recognize Mandarin isolated digit words uttered by 100 speakers. The recognition rates are 99.1% on training data and 96.3% on testing data. The results show that the sequential neural network trained with DP can deal with the temporal distortion of speech pattern and promote the recognition rate. The recognition method combined with sequential MLPs and the GPDM discriminative training algorithm is an effective way to discriminate easily confused words and to promptly promote the recognition rate with less computation.

## Acknowledgement

The authors wish to thank the speech groups of the Department of Communication Engineering, National Chiao Tung University and the Computer & Communication Research Laboratories of the Industrial Technology Research Institute for their support.

## References

[1] S. M. Peeling and R. K. Moore,"Isolated Digit Recognition Experiments Using the Multi-Layer Perceptron," Speech Communication, Vol. 7, No. 4, pp.403-409, December 1988.

[2] B. R. Kammerer and W. A. Kupper,"Experiments for lsolated Word Recognition with Single and Two Layer Perceptrons," Neural Networks, Vol. 3, pp.693-706, 1990.

[3] K. Iso and T. Watanabe,"Speaker Independent Word Recognition Using a Neural Prediction Model," Proc. ICASSP-90, pp.441-444, 1990.

[4] H. Sakoe, R. Isotani, K. Yoshida, K. Iso and T. Watanabe,"Speaker-Independent Word Recognition Using Dynamic Programming Neural Networks," Proc. ICASSP-89, pp.29-32, 1989.

[5] J. Tebelskis and A. Waibel,"Large Vocabulary Recognition Using Linked Predictive Neural Networks," Proc. ICASSP-90, pp.437-440,1990.

[6] M. Franzini, K. F. Lee and A. Waibel, "Connectionist Viterbi Training: a New Hybrid Method for Continuous speech Recognition,"Proc. ICASSP-90, pp.425-428, 1990.

[7] W. Luz, Y. Kobayashi and Y. Niimi,"Recognition of Monosyllables Based on the Combination of Neural Network and Dynamic Time Warping," IEICE Technical Report SP90-82, pp.37-43, 1990. (in Japanese)

[8] S. Katagiri, C. H. Lee, and B. H. Juang,"A Generalized Probabilistic Descent Method,"Proc. of Fall Meeting of ASJ, pp.141-142, September 1990.

[9] Q. Jia, N. Toda and S. Usui,"A Study on Initial Value Setting of the Back-Propagation Learning Algorithm," Trans. IEICE D-II Vol. J73-D-II, No. 8, pp.1179-1185, August 1990. (in Japanese)

[10] S. Amari,"A Theory of Adaptive Pattern Classifiers," IEEE Trans. on Electronic Computers, Vol. EC-16, No. 3, pp.299-307, June 1967.

[11] P. A. Devijver and J. Kittler,"Pattern Recognition : A Statistical Approach," Prentice-Hall International, Inc., London, 1982.

[12] J. S. Liou, R. G. Chen, S. M. Yu, J. R. Hwang, and I. C. Jou,"The Speech Database of Telecommunication Laboratory, Ministry of Communications, R.O.C." Proc. of Telecommunications Symposium, pp.128-132, 1990. (in Chinese)

# A SPACE-PERTURBANCE / TIME-DELAY NEURAL NETWORK FOR SPEECH RECOGNITION

Ji Ming  Chen Huihuang and Shen Zhenkang
Department of Electronic Engineering
Changsha Institute of Technology
Changsha  410003, Hunan, P.R.China

Abstract—This paper presents a Space–Perturbance/Time–Delay neural network(SPTDNN), which is a generalization of the time-delay neural network (TDNN) approach. It is shown that by introducing the space–perturbance arrangement, the SPTDNN has the ability to be robust to both temporal and dynamic acoustic variance of speech features, thus, is a potentially competent approach to speaker–independent and/or noisy speech recognition. This paper introduces the architecture, learning algorithm, and theoretical evaluation of the SPTDNN, along with experimental results. Experimental comparisons show that the SPTDNN obtains a performance that improves upon the TDNN for both speaker–dependent/–independent and noisy phoneme recognition.

## INTRODUCTION

One of the major difficulty in automatic speech recognition (ASR) is the variance of speech signals. It is known that the acoustic features of the speech signal, on which the recognition is based, may changes randomly from speaker to speaker, and may also be highly distorted under strong noisy environments. Such acoustic variance makes reliable ASR become very difficult based only upon finite training samples. This problem has been studied for periods. Various techniques have been suggested concerning the training or matching strategies for either speaker-independent or noisy speech recognition.

This paper introduces a new approach to speaker–independent and/or noisy speech recognition using neural network techniques. The approach we present, called the Space–Perturbance / Time–Delay neural network (SPTDNN), is a generalization of the time-delay neural network (TDNN) approach presented by Waibel et al.[1-2]. It was shown by Waibel that, by introducing the time–delay arrangement, the TDNN is enabled to learn acoustic events independent of temporal shifts in the input. Experimental results for speaker-dependent phoneme recognition showed that the TDNN obtained a performance over the HMM approach (according to a more recent literature by Waibel[3], the TDNN has also been applied to multi-speaker phoneme recognition. The recognition was, however, based upon a fully training of the network by all the speakers).

This paper extends the TDNN approach to more general cases, considering speaker-independent and/or noisy phoneme recognition. In these cases where highly acoustic variance associated with speakers or noise may prevail, the net's invariance under phonetic-irrelevant variation in acoustic features becomes of critical importance. In the SPTDNN, a signal-space perturbance arrangement is introduced to provide for such an invariance, incorporating the time-delay arrangement for treatment of time shift problem. A learning algorithm for training the network based on the E-M principle is presented, which can be viewed as an iterative solution of the conventional BP problem. The convergence of this algorithm is discussed.

Further, a theoretical evaluation of the SPTDNN's performance, in terms of the generalization property, is presented. Generalization, a term that describes a network's ability to form universal representation of a problem based only upon finite training samples from the problem, cannot be over-estimated in our considered context – the speech may come either from a speaker for whom the network is not trained, or from a noisy environment in which the noise nature is inexactly known. It is qualitatively shown that, benefiting from the nonlinear augment of the training samples due to the perturbance, an enhanced generalization (in comparison to a comparative network based on the same training set) is achieved for the SPTDNN. Such an enhancement is one of the most important properties of the SPTDNN demonstrated by the experiments.

Experimental comparisons between SPTDNN and TDNN are presented. The experiments are conducted for both speaker-dependent and -independent modes, and for speech degraded by either stationary or nonstationary additive noise. The experimental results demonstrate the potential capability of the SPTDNN for both speaker-independent and noisy speech recognition.

## ARCHITECTURE AND LEARNING ALGORITHM OF THE SPTDNN

### Architecture

The SPTDNN is a three layer feedforward network. The basic unit in the network is shown in Fig. 1. In this figure, $X_i$ (i=1, 2,⋯,I) represent the inputs of the unit ( for the input layer, which are chosen as 16 normalized Melscale spectral coefficients of the input signal). Each input is spatially perturbed by $\triangle_1 \sim \triangle_M$, respectively, and then is time delayed through $D_1 \sim D_N$ delays. For each connection between the input and the unit, two kinds of weights are specified, one for the perturbance (including the zero-perturbance), and another for the delay (including the zero-delay). The unit computes the weighted sum of all the connections, and then passes the sum through a nonlinear function, which is chosen here as a double-side sigmoid function (with output defined in [−1,+1]).

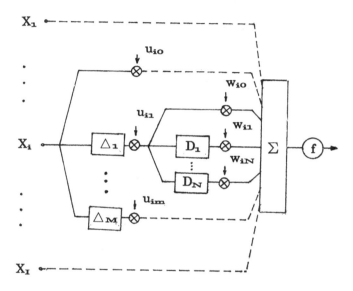

Fig. 1. A Space-Perturbance/Time-Delay Neural Network unit

Thus, the output of the j'th unit in the k'th layer can be written as

$$O_j^{(k)}(t) = f\{\sum_i \sum_{n=0}^{N} \sum_{m=0}^{M} w_{in}u_{im}[O_i^{(k-1)}(t-D_n) + \triangle_m]\} \quad (1)$$

where $D_0 = \triangle_0 = 0$ and $O_i^{(0)}(t)=X_i(t)$, i=1,2,⋯,I. As pointed out by Waibel, with the introduction of the time delays, each input is now measured at several (here for example, (N+1)) different points in time. In this way, the network is enabled to find acoustic events in the input, regardless of when in time they actually occurred. More detailed discussions about this property can be seen in literature[1].

In speech recognition, however, the most difficult problem arises in dealing with the variance in acoustic features. An ideal recognition system should exhibits sufficient robustness to such variance that is phonetic-irrelevant. Here we show that, by introducing the signal-perturbance arrangement, the SPTDNN is enabled to describe such variance directly in terms of its structure.

Considering the input layer of the network. By $X = (X_1, X_2, \cdots, X_I)$ we denote the input of each unit, which is the temporal spectral coefficient vector of the incoming signal. Due to the perturbance arrangement as shown in Fig. 1 or Eq. (1), it sees that the practical input of each unit now is essentially a spectrum set

$$\mathbf{X} = \{ X_i + \triangle_k : i=1,2,\cdots,I; k=0,1,\cdots,M \} \qquad (2)$$

$\mathbf{X}$ includes the original input spectrum X as well as a large class of its nonlinearly distorted versions (in the form of the original spectrum plus different combinations of $\{\triangle_k\}$), with linearly distorted version as a special case. By properly choosing the number and values of $\{\triangle_k\}$, such nonlinearly distorted versions can reasonably represent the temporal variance of the speech spectrum occurred between speakers or in noisy backgrounds.

Now, considering the first hidden layer. We say that the perturbance in this layer describes the dynamic variance (i.e., the variance of the time-varying nature ) of the speech spectrum. Such dynamic variance may usually occur when the uttering manner changes in the course of utterance, or the statistical nature of the noise is time-varying. For illustration, we omit the delay connections and denote by $\triangle^{(k)}$s the perturbance in the k'th layer. Then, it is easy to derive that the perturbance in the first hidden layer can be related to that in the input layer by the relation

$$\triangle_k^{(2)}/f' = \sum_i \sum_m u_{im} \delta (X_i + \triangle_m^{(1)})$$

$$= \sum_i \sum_m u_{im} \delta \triangle_m^{(1)} \qquad (3)$$

where $\delta$ is the variational operator, for the given input $\{X_i\}$, the variation $\delta X_i$ is 0. It sees immediately from (3) that the perturbance in the first hidden layer (i.e., $\triangle_k^{(2)}$s) represents some changes, i.e., $\delta \triangle_m^{(1)}$s, of $\triangle_m^{(1)}$s. If we write the "true" speech spectrum at time t as $X^*(t)=X(t)+\triangle$, where X(t) is the practically measured spectrum and $\triangle$ represents some temporal bias, we see that the dynamic variance of the speech spectrum corresponds to a specific trajectory of $\delta\triangle$, which is supposed can be represented by a specific combination of $\delta \triangle_m^{(1)}$ at different time points. Thus, from (3) it sees that the different combination of $\{\triangle_k^{(2)}\}$, which define a set of possible values of $\delta \triangle_m^{(1)}$ for each time, describes a class of dynamic variance of the speech features. This is an important property, as it makes it possible to use the network for recognition of speech which is degraded by nonstationary noise.

In conclusion, by introducing the perturbance arrangement, the SPTDNN describes in its structure both the temporal and dynamic variance of the input speech. Thus, it is under such variance that the network is forced, guided by the learning rule, to approach some desired output corresponding to specific phonetic information. In this way, to satisfy the classification rule, the network is forced to discover the acoustic features that are independent of the variance described in the network, and then, are only phonetic-relevant. In fact, in performing learning and recognition, the network measures a subset including some neighbourhood of the input acoustic feature, rather than only the feature itself. In this way, the network can learn directly the phonetic-relevant acoustic-class information, less dependent on each specific feature in the class.

### Learning Algorithm

As is shown in (1), as the strength of each connection in the SPTDNN is the multiplication of two weights(corresponding to the perturbance and delay, respectively), conventional BP (back-propagation) algorithms[4] can not be directly applied, due to the derivative of error used for adjusting one weight being always a function of another weight also to be determined. To treat this problem, an iterative BP (IBP) algorithm based on the E-M principle is presented in this paper, and is employed to train the SPTDNN in our experiments. The IBP algorithm iteratively solves the BP learning for each weight subject to some constraints on the weights, and then makes the learning of the SPTDNN computationally tractable. Now, the IBP algorithm is briefly summarized.

Denote by matrices W and U the weight coefficients $\{w_{in}\}$ and $\{u_{im}\}$, respectively, and define the objective function as E(W,U). E(W,U) measures the mean-square error between desired outputs and actual outputs of the network, which is of course a joint function of W and U. The IBP algorithm searches for W and U jointly to minimize E(W,U) with the following steps:

(1) Set initial values for W and U as $W_o$, $U_o$;
(2) For the k'th iteration (k=1,2,···)
 (a) Solve for W using the BP algorithm, which performs

$$\min_{W} E(W_{k-1}, U_{k-1}) \rightarrow W_k$$

 (b) Solve for U using the BP algorithm, which performs

$$\min_{U} E(W_k, U_{k-1}) \rightarrow U_k$$

(3) Set k=k+1, repeat steps (a) and (b) until the error converges.

It is straightforward to show that each step of iteration always makes the error decrease. In fact, since

$$E(W_k, U_{k-1}) < E(W_{k-1}, U_{k-1})$$

and

$$E(W_k, U_k) < E(W_k, U_{k-1})$$

we have

$$E(W_k, U_k) < E(W_{k-1}, U_{k-1})$$

Hence, the IBP algorithm for estimating the weights always

converges to a lower limit of the error function, and the computational complexity is only a linear increase compared to the conventional BP algorithms.

## QUALITATIVE STUDY OF THE SPTDNN'S GENERALIZATION

The problems of multi-speaker or noisy speech recognition can be expressed as a multi-to-single map

$$F : S \longrightarrow p$$

where S is either the subset consisting of acoustic features corresponding to the same phonetic information (indexed by the label p), or the subset containing the possible features of the input speech based on some (inexact) prior knowledge about the noise. In most practical cases, both are involved. The design of a recognizer is essentially to use an algorithm to approximate F, resulting in a F'. Here, in concrete, F' is a neural network. The learning of the network is a process to optimize F' such that F' is as close as possible (in the sense of some priorly defined optimality) to F. When the architecture and learning algorithm of a network have been defined, the optimality of the network is entirely dependent on the values of the weight coefficients, which are in turn dependent on the nature of training sets.

Ideally, if the whole S could be obtained and were used to train the network, the network should be able to classify all inputs correctly. In practice, however, only some subset of S can be obtained. In such cases the generalization property of the network then simply means the network's ability to achieve correct classification for the inputs that are disjoint to the training set, which reflects the robustness of the network to either speakers or background noise.

The essence of the SPTDNN to increase the generalization is by the nonlinear augment of the training set, as is shown in (2). In this way, the performance built upon the fully training with the whole S is approximately obtainable. If we denote the training set as $S_t \in S$, it is known from Section 2 that $S_t$ is augmented by the perturbance, which yields the augmented training set $S_a$. In general

$$S_t \cap S \in S_a \cap S$$

Then, in the sense of coverage degrees $S_a$ is more close to S than $S_t$ is. Considering one simple example. Suppose $S_t$ contains only one point $\{x\}$, whose perturbance results in a set $S_a^x$, and learning makes $F(S_a^x)=p$. Now we have an input $x'$, which deviates from x slightly. Assuming $x' \in S$ and whose perturbance set is denoted by $S_a^{x'}$. Since $S_a^x \cap S_a^{x'}$ may be nonempty, $F'(S_a^{x'})=F'(S_a^x)$ holds with higher probability. However, since $x' \neq x$, $F(x')$ is quite random. The later is the behaviour of most networks, which implies lower robustness.

## EXPERIMENTAL RESULTS

Experiments are conducted for recognizing of three Chinese phonemes "A", "E", "O". 8 speakers (4 male and 4 female) are employed. For each phoneme, a total of $8 \times 1000$ utterances are recorded. Rather than training the network for all speakers, we divided the speakers into two groups (each group containing 2 male and 2 female speakers), one performing the training, and another performing the testing. The TDNN is restored, on which the same training and testing are repeated for comparison.

In the experiments, the same number of units as is used in [1] (i.e., 8, 5, and 3 units for the first hidden, second hidden, and output layers, respectively) is employed for both the TDNN and SPTDNN. The delay arrangement is also chosen as the same as in [1]. Specifically, for the SPTDNN a symmetrical perturbance arrangement is assumed. For each unit in the first and second hidden layers, 6 perturbance connections (corresponding to $\pm \triangle_1$, $\pm \triangle_2$, $\pm \triangle_3$, respectively) are employed, with each $\triangle$ taking a value 0.1 (the input is normalized to lie between –1 and +1). For the output layer, no perturbance is arranged.

Table 1 gives the recognition results for both the trained and untrained speakers. Because the number of training samples used here is much less than that used in [1], the performance of the TDNN for speaker-dependent recognition is not as good as reported in [1]. But, it sees that the SPTDNN approaches the performance of [1] (where the average recognition rate was 98.5%). This result proves the effectiveness of augmenting the training set provided by the perturbance arrangement.

The 3'rd column of Table 1 presents the recognition results of the two networks for the untrained speaker-group, which provide an evaluation of the capability of the two networks for speaker-independent phoneme recognition. It sees that about 15% improvement upon the TDNN is achieved for the SPTDNN, which demonstrates that the perturbance arrangement is an effective means to provide for the robustness against the acoustic variance associated with speakers.

TABLE 1
RECOGNITION RESULTS OF TDNN AND SPTDNN FOR TRAINED AND UNTRAINED SPEAKERS

| network | trained speakers | untrained speakes |
|---|---|---|
| TDNN | 92.7 | 62.8 |
| SPTDNN | 96.6 | 77.6 |

Further, the performance of the SPTDNN for noisy phoneme recognition is evaluated. The experiments are performed in a speaker-dependent mode. Two kinds of white noise, which is added to the trained speaker-group utterances, are considered. The first is a zero-mean stationary process, with an average SNR of 10 dB. The second is a simple nonstationary process. The mean of the process changes sinusoidally, with a changing rate of 5 cycles/one utterance and a SNR of 10 dB. While, the SNR of the white noise (eliminating the mean) is also 10 dB. The recognition results for these two cases are shown in Table 2. It is encouraging to find that for the nonstationary-noisy case the SPTDNN achieves a rate only with a slight decrease in comparison to the stationary-noisy case. Further experiments, involving one kind of colored/nonstationary noise recorded from a real traffic condition, are being undertaken.

TABLE 2
RECOGNITION RESULTS OF SPTDNN FOR TRAINED-SPEAKER SPEECH DEGRADED BY TWO KINDS OF NOISE

| stationary white noise | nonstationary white noise |
|---|---|
| 90.1 | 87.8 |

**CONCLUDING REMARKS**

This paper presents a space-perturbance neural network

architecture to study the problems of speaker-independent and/or noisy speech recognition. The essence of introducing the signal-perturbance arrangement in the neural network is to make the network be able to describe the nature of various phonetic-irrelevant acoustic variance directly in terms of its structure.
Thus, in such a network that describes such variance, the learning of the network is a process to force the network to discover the feature that is independent of such variance, which then makes the feature quite robust and representative.

One problem that should be further studied for the SPTDNN is the selection of perturbance values. In this paper, these values are chosen heuristically. Although a theoretical criterion, the Mini-Max criterion, has been considered, which assumes the acoustic features are subject to a Gaussian $\varepsilon$-mixing distribution, and the perturbance values are chosen such that the maximum cross-entropy using the perturbance points to approximate the distribution is minimized, we didn't find this criterion is applicable for most experiments. To solve this problem, perhaps a joint optimization for both the weight coefficients and the perturbance values should be considered.

## REFERENCES

[1] A. Waibel, T. Hanazawa, G. Hinton, K. Shikano and K.J. Lang, "Phoneme recognition using time-delay neural networks," IEEE Trans. Acoust., Speech, Signal Processing, vol. 37, pp. 328-339, March 1989.

[2] A. Waibel, H. Sawai and K. Shikano, "Modularity and scaling in large phonemic neural networks," IEEE Trans. Acoust., Speech, Signal Processing, vol. 37, pp. 1888-1898, Dec. 1989.

[3] J.B. hampshire and A. Waible, "A novel objective function for improved phoneme recognition using time-delay neural networks," IEEE Trans. Neural Networks, vol. 1, pp.216-228, June 1990.

[4] D.E. Rumelhart, G.E. Hinton, and R.J. Williams,"Learning representations by back-propagation errors," Nature, vol. 323, pp. 533-536, Oct. 1986.

[5] S. Ahmad and G. Tesauro, "Scaling and generalization in neural networks: a case study," in Proceedings of 1988 Connectionist Models Summer School. San Diego, CA: Morgan-Kaufmann, 1988, pp. 3-10.

# Non-linear Prediction of Speech Signals using Memory Neuron Networks

Pinaki Poddar[†]  K. P. Unnikrishnan[‡]
† Tata Institute of Fundamental Research
Bombay 400005, India.
‡ General Motors Research Laboratories
Warren, MI 48090-9055, USA

Abstract – We present a feed-forward neural network architecture that can be used for non-linear autoregressive prediction of multivariate time-series. It uses specialized neurons (called memory neurons) to store past activations of the network in an efficient fashion. The network learns to be a non-linear predictor of the appropriate order to model temporal waveforms of speech signals. Arrays of such networks can be used to build real-time classifiers of speech sounds. Experiments where memory-neuron networks are trained to predict speech waveforms and sequences of spectral frames are described. Performance of the network for prediction of time-series with minimal a priori assumptions of its statistical properties is shown to be better than linear autoregressive models.

## INTRODUCTION

Connectionist network architectures are being studied from various perspectives and several different architectures have been proposed. Multi-Layer Perceptron(MLP) is one of the most widely used architectures and its variants have been applied to solve problems in many different domains with promising performance.

Here, we propose a network design (based on the MLP architecture) that incorporates a non-linear predictor for multivariate time serieses such as speech. The proposed architecture can be organized in a larger framework for classification of temporal pattern sequences, where it is imperative to use the time history of previous states of the network and that of the environment ([1], [2], [3]). This is in direct contrast to majority of applications using MLPs in the domain of 'static' pattern classification where decisions are based only on the current states.

Important qualities of a connectionist design for temporal pattern classification should be (i) an efficient representation scheme for the past information, (ii) a task-independent architecture, (iii) a data-driven decision regarding the effective usage of the past information and (iv) a learning rule that can be formulated using locally available information. With these considerations, the network architecture proposed here incorporates additional neurons (called *memory neurons*) to the conventional MLP architectures, to store the past activations of network neurons in an efficient and meaningful way. As only a single input from the temporal sequence is presented to the network at a time, the network is compact with no arbitrary restrictions imposed on its architecture by the length of given sequences. The effect of past history on the current behavior is not determined *a priori*: the network learns to use these neurons in an optimal fashion for a given task. A learning rule that uses only *locally* available information for adjusting the link weights is developed for this architecture.

In the next section we describe the network architecture, specification of the external environment for operation of the network and the learning rule. The network was trained to build specific models for prediction of different speech waveforms. It was also trained to predict a sequence of multidimensional spectral patterns of continuous speech. Results of these experiments are discussed in the third section.

## DESCRIPTION OF THE NETWORK

This section contains a very brief description of the network and the learning rule. A detailed description can be found in [7].

### Network Architecture

The architecture of a feed-forward multi-layer perceptron with memory neurons is shown in Fig.1. Corresponding to each network neuron, there is a *memory neuron*, which contains information regarding the past activation of its parent network neuron.

The net input to the $j$-th network neuron of the $(l+1)$-th layer at time $t$ is given by,

$$x_j^{(l+1)}(t) = \sum_{i=0}^{N_l} w_{ij}^{(l)} \cdot u_i^{(l)}(t) + \sum_{i=1}^{N_l} f_{ij}^{(l)} \cdot v_i^{(l)}(t) \tag{1}$$

where $u_i^{(l)}(t)$ and $v_i^{(l)}(t)$ denote the output activations of the $i$-th network neuron of the $l$-th layer and its corresponding memory neuron. $w_{ij}^{(l)}$ and $f_{ij}^{(l)}$ are the weights associated with the links that propagate the activations of these units to the $j$-th network neuron of the upper layer.

A sigmoidal non-linearity maps the net input of a network neuron to its output. The activation of a memory neuron is derived from its parent network neuron according to the rule :

$$v_j^{(l)}(t) = \alpha_j^{(l)} \cdot u_j^{(l)}(t-1) + (1 - \alpha_j^{(l)}) \cdot v_j^{(l)}(t-1) \tag{2}$$

where $\alpha_j^{(l)}$ is a variable parameter, referred to as the *memory coefficient*, that controls the rate of decay of the past information. It is straightforward to see that if $0 \leq \alpha \leq 1$, the accumulated activation of the memory neurons will be finite for finite activation of the network neurons and the network will be stable.

Equations (1)-(2) describes the behavior of the network for a given external environment and initial condition. For initial condition, we assume that both $u_j^{(l)}(t)$ and $v_j^{(l)}(t)$ are zero for any $t \leq 0$.

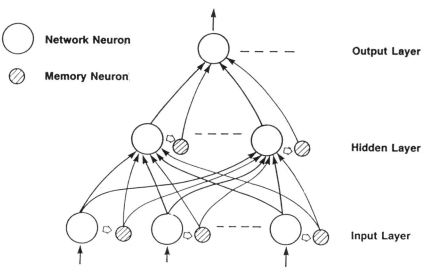

Figure 1 : A schematic of the memory-neuron network architecture. The network neurons are shown as larger open circles and memory neurons as smaller shaded circles. Separate links connect a network neuron and its corresponding memory neuron to another network neuron of upper layer.

**External Environment**

External environment of the network is characterized by a sequence of input vectors $\mathcal{I} = \mathcal{I}(1), \ldots, \mathcal{I}(T)$. At any given time $t$, activation of the network neurons of the lowermost (input) layer is decided by the vector $\mathcal{I}(t)$. The response of the network to the external environment is the activation of network neurons of the topmost (output) layer and should match with a given 'target' signal as specified by the external environment. Specification of the target in temporal pattern classification is a non-trivial issue, because the desired response at any given moment depends on the past outputs of the system. When the network is used as a predictor, the target is the future value of the input itself. Thus, supervised training for a predictor network is completely data-driven and there is no need for an external 'teacher'.

## Learning

The learning process minimizes the difference between the target and actual responses of the network by iterative modification of (i) the connection strengths $w$ between network neurons, (ii) the connection strengths $f$ between memory neurons and network neurons, and (iii) the memory coefficients $\alpha$ for the memory neurons.

The rules for updating the link weights are essentially the same as in the original back-propagation algorithm [4] except for the partial derivative of the total error with respect to the memory coefficient which is computed using the recursive relation,

$$\delta\alpha_j^{(l)}(t) = \frac{1}{\alpha_j^{(l)}}(v_j^{(l)}(t) - v_j^{(l)}(t-1)) + (1 - \alpha_j^{(l)})\delta\alpha_j^{(l)}(t-1) \qquad (3)$$

These derivatives can be computed using only spatially and temporally *local* information. *i.e* various terms involved in the computation are related to the neurons which are either connected by a link (spatial locality) or occured in the previous time step (temporal locality).

The correction term for each variable is obtained by multiplying the corresponding derivative by a *learning rate*. Unlike conventional back-propagation algorithm, the learning rate, here, for each of the variables are decoupled and modified during learning (see [7] for details).

## EXPERIMENTAL RESULTS

In this section we report the results of two experiments that were designed to train the network as predictors of speech signals in different representations. In these experiments, the task is similar to autoregressive(AR) modeling in linear predictive analysis where the prediction of the future sample of a time-series is expressed as a linear weighted sum of the past samples[9]. But this particular connectionist approach differs from autoregressive moving average models in several respects. Firstly, no statistics such as autocorrelation is computed and no assumption is made about the stationarity of the signal. Also, it does not require any *a priori* decision regarding the order of the predictor. And finally, due to the non-linearity introduced through the transfer function of the neurons, the future sample can be modeled as a non-linear function of the past samples.

### Prediction of Speech Waveforms

Speech waveforms (sampled at 16 KHz) of 20 ms duration from the steady portion of the vowels /a/ and /e/ were chosen for this task. Ten examples of each vowel were collected. A single utterance was used for training and the rest were used for testing. Networks with one input unit, two hidden units and one output unit (1 − 2 − 1 network) were trained independently on the waveforms of /a/ and /e/. The input and hidden layers contained memory neurons. Learning was continued for 2000 cycles and afterwards the accuracy of each network was tested using examples of both vowels.

Mean squared error computed over the duration of the entire signal was used as a measure of performance. Table I shows these results for the training and testing data along with those from a linear autoregressive prediction filter. A Wiener filter of 20-th order was calculated by autocorrelation method [10] from the waveforms of each vowel and was used for prediction of both vowels. The tasks for the network and the Wiener filter were similar and number of adjustable parameters were also comparable (there were 14 'learnable' parameters in the network). Hence, performance of this filter provides a fair criterion for comparing the modeling ability of the memory-neuron network for quasi-stationary vowel waveforms.

Table I
Prediction errors by the network and the linear filter

Performance on Training Set

| Trained on | Tested on | | | |
|---|---|---|---|---|
| | Vowel /a/(Train-set) | | Vowel /e/(Train-set) | |
| | Network | Wiener | Network | Wiener |
| vowel /a/ | $0.19 \times 10^{-3}$ | $0.24 \times 10^{-3}$ | $0.75 \times 10^{-2}$ | $0.32 \times 10^{-2}$ |
| vowel /e/ | $0.21 \times 10^{-2}$ | $0.20 \times 10^{-2}$ | $0.47 \times 10^{-3}$ | $0.68 \times 10^{-3}$ |

Performance on Test Set

| Trained on | Tested on | | | |
|---|---|---|---|---|
| | Vowel /a/(Test-set) | | Vowel /e/(Test-set) | |
| | Network | Wiener | Network | Wiener |
| vowel /a/ | $0.24 \times 10^{-3}$ | $0.27 \times 10^{-3}$ | $0.76 \times 10^{-2}$ | $0.26 \times 10^{-2}$ |
| vowel /e/ | $0.30 \times 10^{-2}$ | $0.27 \times 10^{-2}$ | $0.58 \times 10^{-3}$ | $0.63 \times 10^{-3}$ |

From Table I we can see that the performance of the network is consistently better than the autoregressive linear filter. The network trained on one vowel shows a much higher prediction error on a different vowel. *i.e.*, the internal model established by the network for prediction of a particular sequence is distinct. This fact can be useful in organizing such networks into a larger framework for classification of temporal sequences.

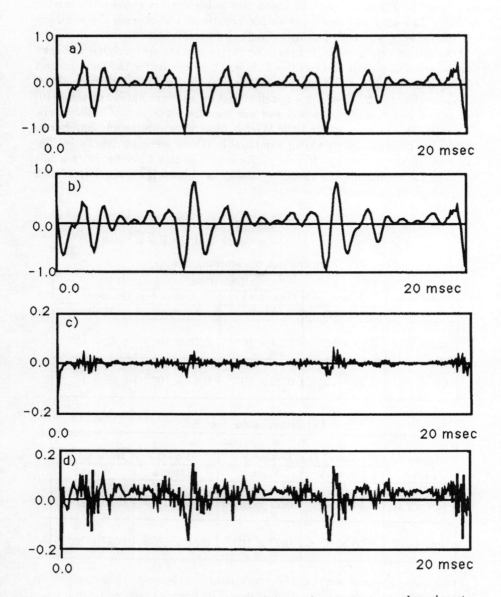

Figure 2 : a) A waveform of vowel /a/ from the test set, used as input. The waveform is sampled at 16 KHz and the figure shows 320 points or 20 ms duration. b) Predicted waveform using the network trained on /a/. c) Prediction error (actual-predicted) for the network trained on /a/. d) Prediction error for the network trained on /e/.

## Prediction of Speech Spectra

The memory-neuron network can also be used for prediction of multivariate time serieses without any explicit computation of correlation. In this experiment the network was used for prediction of a multichannel filter-bank output of continuous speech. Speech signals were preprocessed through a set of filters to generate a 20-dimensional filter-bank output vector at the rate of 300 frames per second (see [5] for details). 20 seconds of speech (6000 vectors) was partitioned into 4000 vectors for training and 2000 vectors for testing. A network with 20 input neurons, 10 hidden neurons and 20 output neurons (with input and hidden layers containing memory neurons) was trained to predict the next vector of the sequence when the current vector is presented as the input. It was trained for 800 learning cycles and then used for predicting the sequence of testing vectors.

Table II
Prediction Error for Spectral Sequence

| Network | MS Error | |
|---|---|---|
| | Train-set | Test-set |
| with memory | $0.24 \times 10^{-3}$ | $0.40 \times 10^{-3}$ |
| without memory | $0.87 \times 10^{-1}$ | $0.98 \times 10^{-1}$ |

Table II shows the mean squared prediction error for the training and test sets. The experiment was then repeated using a $20-10-20$ network without memory neurons. The prediction error in this case differs by more than two orders of magnitude (see Table II). The output neural activity in two cases for a word from the testing set clearly shows the effect of using memory neurons in the prediction task. The predicted output of the memory neuron network smooths out the original spectra but at the same time, captures the essential structure of the input waveform. Hence such prediction can also be useful for segmentation of speech where a high prediction error can be interpreted as a boundary between two phonetic regions [8].

The memory coefficients determine the effective time window for storage of past activations of the network neurons. These coefficients were initially randomized between (0,1). Their values after the training process is shown in Fig. 4. from the figure we can see that the input neurons corresponding to formant positions in the input spectrum tend to have larger memory coefficients (and hence shorter time windows) whereas those corresponding to the higher frequency bands have much smaller memory coefficients (and hence longer time windows). This is expected if we consider the problem from the point of view of an autoregressive modeling. For vocalic regions of the speech sounds, due to high correlation between successive frames, it is enough to consider only the past few frames to predict the future; whereas

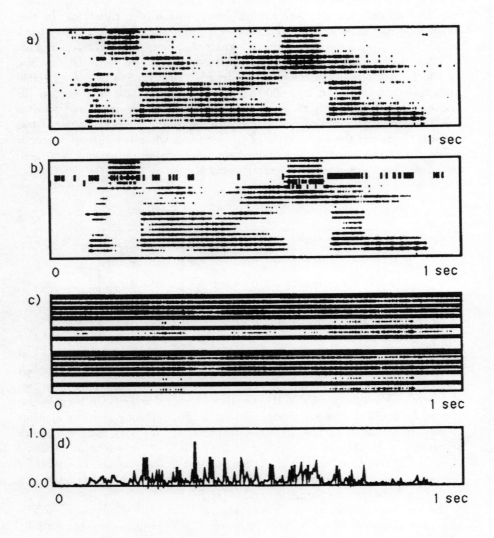

Figure 3 : a) A spectrogram-like display of the actual sequence for a portion of the test set. The spoken word shown here is *RESERVATION*. b) Predicted spectral sequence using a memory-neuron network. c) Predicted spectral sequence using a traditional multi-layer perceptron. d) Euclidean distance between the prediction of the memory-neuron network and the actual spectrum at every frame.

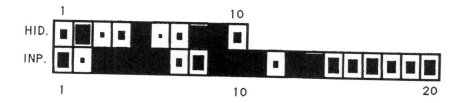

Figure 4 : Memory coefficients of the network that predicts the spectral sequences. Each square corresponds to one neuron of the network. The lower layer is for the input and upper one for the hidden neurons. The size of black squares are proportional to the value of these coefficients. A completely black square represents a value of unity and a blank one represents a value of zero. (see text for further explanation).

for successful prediction in the high frequency regions with low correlations, longer history should be taken into consideration. The coefficients of the ten hidden neurons also have a similar distribution.

## DISCUSSION

We have shown that memory-neuron networks can be used for accurate prediction of quasi-stationary waveforms. The prediction accuracy of these networks are consistently better than comparable linear autoregressive models. Separate networks can be trained to model the dynamic nature of univariate or multivariate time-serieses with minimal *a priori* assumptions about their statistical properties. The models learned by these networks are distinctive enough for simple classification, as shown by the experiments using vowel waveforms. For solving difficult problems involving many classes, independent training of individual networks may not be sufficient. We are currently investigating arrays of memory-neuron networks where each one learns to predict one particular class of sequences, competing with other networks in the array.

## Acknowledgements

Part of this work was done when one of the authors (PP) was at the University of Michigan as a UNDP fellow.

# References

[1] A.Waibel, T.Hanazawa, G.Hinton, K.Shikano and K.Lang, 'Phoneme recognition using time-delay neural networks', IEEE Trans. on ASSP, vol.37, March 1989.

[2] K.P.Unnikrishnan, J.J.Hopfield and D.W.Tank, 'Connected digit speaker-dependent speech recognition using a neural network with time-delayed connections', IEEE Trans. on ASSP, vol.39, March 1991.

[3] F.J.Pineda, 'Recurrent Backpropagation and the dynamical approach to adaptive neural computation', Neural Computation, vol.1, no.1, 1989.

[4] D.E.Rumelhart, G.E.Hinton and R.J.Williams, 'Learning internal representations by error propagation', in Parallel Distributed Processing, vol I, MIT Press, 1986.

[5] P.Poddar and P.V.S.Rao, 'Neural Network based segmentation of continuous speech' presented at International Conference on Spoken Language Processing, Kobe, Japan, December 1990.

[6] Pinaki Poddar and K.P.Unnikrishnan, 'Efficient Real-time Prediction and Recognition of Temporal Patterns' presented at Neural Networks for Computing Conference, Snowbird, Utah, April 1991.

[7] P.Poddar and K.P.Unnikrishnan, 'Efficient Prediction and Recognition of Temporal Patterns using Networks with Memory Neurons' *preprint*, June 1991.

[8] A.Doutriaux and D.Zisper, 'Unsupervised Discovery of Speech Segments using Recurrent Networks', Proceedings of summer school on Connectionist Models, Morgan Kaufmann, 1991.

[9] G.E.Box and G.M.Jenkins, 'Time Series Analysis: Forecasting and Control', Holden-Day, Oakland, California, 1976.

[10] J.D.Markel and A.H.Grey, 'On autocorrelation equations as applied to speech analysis', IEEE Trans. on Audio Electoacoustics, vol.AU-20, April 1973.

# EXPERIMENTS WITH TEMPORAL RESOLUTION FOR CONTINUOUS SPEECH RECOGNITION WITH MULTI-LAYER PERCEPTRONS

Nelson Morgan[†], Chuck Wooters[†], and Hynek Hermansky[††]

[†]International Computer Science Institute, Berkeley, CA 94704, USA
[††]U S WEST Advanced Technologies, Boulder, CO 80303, USA

**ABSTRACT:** Previous work in our group focused on the integration of Multilayer Perceptrons (MLP) into Hidden Markov Models (HMM) and on the use of Perceptual Linear Prediction (PLP) parameters for the feature inputs to such nets. Our system uses the Viterbi algorithm for temporal alignment. This algorithm is a simple and optimal procedure, but it necessitates a frame-based analysis, in which all features have the same implicit time constants. We can provide a range of temporal/spectral resolution choices to a frame-based system by using a layered network to incorporate this information for phonetic discrimination. Thus, we have performed experiments in which we expanded our PLP analysis to include short analysis windows, and in which we trained phonetic classification networks to incorporate this added information. We hypothesized that classification scores would improve, especially for short-duration phonemes. These experiments did not yield the expected improvement.

## BACKGROUND

As shown in [3], the outputs of a classification system trained to minimize the sum of squared errors (or some other criteria, such as the cross entropy [1]) may be considered estimates of the posterior probabilities, that is, the probability of each class given the input. Using Bayes' rule, the likelihoods required for a Viterbi decoding may be derived from these values. The potential advantages to this procedure are the discriminative training inherent to the error-correction method, and the ability to incorporate multiple sources of evidence (temporal context and multiple features) without restrictive assumptions of statistical independence or a particular form of probability density function (e.g., Gaussian).

In previous work by our group, we have applied this approach to the recognition of continuous speech for a speaker-dependent German language database [8], the speaker-dependent portion of the English language Resource Management database [10], and some initial experiments with the speaker-independent English language TIMIT database [9]. In the more recent of these experiments, we have found that we can consistently do better with continuous input features than with discrete vector-quantized features, although networks using the latter can be trained and run with much less computation than networks using the continuous ones. Additionally, it appears that the Perceptual Linear Prediction (PLP) features [5] are more succinct and lead to somewhat higher performance than the mel-cepstral features we were previously using. In essence, the PLP technique computes an auditory spectrum that is then approximated by the spectrum of an all-pole model.

Many practical and research systems have been built with frame-based front ends. The most common approach is to compute spectral features over a 20-25 msec window once every 10 msec. This is a reasonable compromise between bursts that can be a millisecond or less and steady-state sounds that can last for a significant fraction of a second. However, to our knowledge, the optimality of this choice of the frame step and window size has not been extensively studied.

Mammalian auditory systems suggest an alternative to such fixed temporal resolution. Cells at various points along the auditory pathway are provided with signals representing the acoustic input with varying degrees of fidelity to the temporal and spectral transitions. Recent neurophysiological results [4] suggest that (consistent with a filterbank based model of the cochlea) some structures in the early auditory system may have poor temporal resolution at low frequencies, and may track temporal changes more precisely at high frequencies. On the other hand, psychophysical experiments [11] have suggested a narrow range of temporal resolutions, ranging from 8 msec at 1-10 kHz to 13 msec for a 300 Hz band.

The "neurons" used in Artificial Neural Networks (ANNs) are extremely simplified models of neural behavior, and the common architectures, such as the MLP, bear little resemblance to the structure of neural masses. Nonetheless, they are trainable systems that can learn something about the relevance of input features for classification. The work described here is our first attempt at training MLPs to incorporate features with a range of temporal and spectral resolutions. We have hypothesized that the MLP may be able to incorporate variable time resolution to help in phonetic classification, for instance by successfully detect extremely short bursts while maintaining the relatively high spectral resolution to discriminate between steady sonorants.

## METHODS

In previous experiments, we found that our best word recognition performance was obtained with a network that used an input vector consisting of cepstral coefficients calculated from PLP analysis (including log energy), plus an estimate of the instantaneous temporal derivative of these coefficients [9]. The network used a temporal context of 9 frames of these inputs, and had 1024 hidden units and 61 outputs (the latter representing 61 DARPA phones). The resulting nets had 200,000-300,000 connections, and trained in about 12 hours on a 3-board version of the RAP, a 128-MFLOP-per-board machine designed at ICSI [6]. Each frame corresponded to a 20 msec window that was calculated every 10 msec.

In the current experiment, we have augmented these cepstral coefficients with low-order PLP features that have been computed using a relatively narrow window (e.g., 5 or 10 msec). We evaluate these features using a subset of the TIMIT database [1] that we have previously used in [9].

For training of each network, the PLP cepstral coefficients are individually normalized for zero mean and unity variance across all classes. Training is done by an error-backpropagation algorithm [12,13], using an entropy criterion [1] and a cross-validation approach to adaptively set the learning constant and to determine when to stop training [2,7,8]. The recognizer uses the Viterbi algorithm to determine the sequence of single-density phoneme models for which the observed features were most likely. As in previous experiments, within-word transitions are restricted to self-loops and moves to the next state in a phone or word model. The latter is just a concatenation of phone models, since no co-articulation models were used, and no state-skipping was permitted. Transition probabilities between states were assumed to be equally likely for all within-word cases.

For the pilot experiments, training was done using 500 sentences of Resource Management data from speaker dtd, and an additional 100 sentences were used for cross-validation (to determine the end of training). For the final experiments, training was done on a TIMIT subset consisting of 5 sentences from each of 152 speakers, and a test set of 5 sentences from each of 38 speakers. This is the same TIMIT subset that we used in [9], for which our best scores were 53.8% for frame classification and 28.1% for word recognition (using PLP-5 and the first derivatives of each PLP cepstral coefficient).

## RESULTS

Our pilot experiments used PLP-12 cepstral coefficients plus their first temporal derivatives on the Resource Management sample speaker. A variety of techniques were tried, but none of them yielded any significant improvement. Since we could not exclude the possibility that such a result was due to inaccurate frame labeling (which was done by iterative application of Viterbi segmentation on the original fixed temporal resolution data), we proceeded to the use of TIMIT, for which we used energy and PLP-5[2].

(1) For the TIMIT case described above, we augmented the reference input vector (nine 20 msec frames overlapped by 10 msec) with three sets of PLP features that had been calculated over a shorter 10 msec window, with shorter overlap of 5 msec. We trained this network to predict the label

---

[1] We use TIMIT for this purpose because of the prelabeled segmentations, which effectively sidesteps the question of how well our automatic segmentation will work for the higher frame rates. However, we have performed some shorter, pilot runs using the data from a single speaker of the Resource Management task, for which we obtain recognition scores much higher than for TIMIT, since the latter is both speaker-independent and has a very high perplexity.

[2] Earlier tests had shown us that PLP5 gave us a better performance than we could achieve with PLP12 or mel cepstrum 12, probably because the data set included a wide variety of talker types.

of the central 10 msec segment. In spite of a large increase in the number of information carrying features, this experiment resulted in no improvements in classification rate.

(2) We would expect the net to take advantage of short-time information primarily in fast changing sounds such as plosives. Therefore, we built up a subnet that distinguished 8 classes: the 6 basic stops (p, t, k, b, d, g), flaps, and "other". The outputs of this net augmented the reference input vector. Once again, no improvement was seen.

(3) Further, we trained up a network with overlapped 10 msec windows only; this net then had an input vector of 18 overlapped 10-msec windows with 5-msec spacings (so that the speech segment which the net has at its disposal is of the same 90 msec time length as in our previous experiments, but with higher temporal resolution frames). From each subwindow, PLP-5 plus energy was calculated. The resulting net then had 108 features calculated from 90 msec, as for the reference network. There was no overall improvement in phonetic frame classification. To our surprise, while the short-window net did outperform the reference net for some classes, it was uniformly worse for all of the short-duration phonetic classes (e.g. stops).

(4) We then linearly combined the network outputs from this short-window network and the reference network. The weightings were found analytically by linear least squares analysis. Only a nonsignificant improvement was seen over the reference network alone.

All of the above tests were performed by training a network to classify 10-msec frames. TIMIT is presumably marked with better resolution than this, so we finally performed tests with 5-msec frame markings from TIMIT.

(5) First we trained our reference network with a 200 Hz frame rate. To keep the number of parameters as well as the width of the input context constant, we used an input vector consisting of nine 20 msec windows with 10 msec spacing as before. This time, however, each nine-frame overall context window overlapped only 5 msec. This network performed slightly worse than our 100 Hz frame rate reference case. It is interesting to note that once again performance on stops was worse than for the reference case.

(6) With this performance as a reference, we then appended 6 features (PLP5 + energy, calculated from a 10 msec window as before) to the input vector. The performance was roughly the same as for the previous test. Thus, the addition of short-window features essentially had no effect.

## DISCUSSION

We have been unable to substantiate our earlier hypothesis, which was suggested by the physiological observation of fine-time structure at the neural level. The classification of short-duration phonemes was never improved by the addition of short-time features.

The negative outcome of this experiment further stirs our interest in time/frequency tradeoffs for MLP classifiers. We intended to increase the temporal resolution of the analysis by shortening the analysis window and speeding up the analysis frame rate. The extreme of this approach is to use the speech waveform itself as the network input. We have tried this in an experiment with the Resource Management data, and were only able to achieve a 20% frame classification rate, using a net that was roughly the same size as one that had achieved a 70% classification rate from PLP-12 features that were calculated from the same data. This result was consistent over a variety of networks, ranging from 0 to 2 hidden layers of various sizes.

Our experimental results, then, suggest an alternate hypothesis:

*Continuous speech recognition by machine requires constraints on the input features; in particular, coarse representations seem to be preferred. These coarse features might be less sensitive to irrelevant temporal and spectral changes in speech.*

PLP provides us with a gross spectral envelope. We are now satisfied that it provides a speech representation for each frame that is preferable to more detailed measures, such as high-order LPC. Thus, it is possible that, rather than looking into finer temporal resolution analyses, the opposite avenue should be explored. Perhaps we need to investigate how to obtain more global gross temporal pictures of segments of speech. Multiple temporal analysis windows may yet become useful with such a more global view of speech.

## SUMMARY

We have reported some experiments in the use of multiple temporal resolution windows for feature extraction in the phonetic training of a back-propagation neural network. Despite extensive experimentation on close to an hour of hand-marked speech material, no improvement was found. Thus far in our work, only the use of large temporal contexts has helped performance.

## ACKNOWLEDGEMENTS

We note and appreciate the continued support for this work from the International Computer Science Institute (ICSI), US WEST Advanced Technologies, and SRI International (as part of their DARPA contract MDA904-90-C-5253). We continue to appreciate the work of Hervé Bourlard, whose analysis of the probabilistic interpretation of neural networks has formed the basis for our developments. Thanks also to the RAP crew, the ICSI team of James Beck, Jeff Bilmes, and Phillip Kohn who built the RAP hardware and software that made large studies like this possible.

# REFERENCES

[1] E. Baum and F. Wilczek. "Supervised Learning of Probability Distributions by Neural Networks," in *Neural Information Processing Systems, pp.52-61*, American Institute of Physics, 1988

[2] H. Bourlard and N. Morgan. "Connectionist Approaches to the Use of Markov Models for Continuous Speech Recognition," *Advances in Neural Information Processing Systems III*, Morgan Kaufmann, 1991

[3] H. Bourlard, and C.J. Wellekens. "Links Between Markov Models and Multilayer Perceptrons" *Advances in Neural Information Processing Systems I*, Morgan Kaufmann, 1989, pp. 502-510

[4] S. Greenberg. personal communication

[5] H. Hermansky. "Perceptual Linear Predictive (PLP) Analysis of Speech," J. Acoust. Soc. Am. 87 (4), April, 1990

[6] N. Morgan, J. Beck, P. Kohn, J. Bilmes, E. Allman, & J. Beer. "The RAP: a Ring Array Processor for Layered Network Calculations," *Proc. of Intl. Conf. on Application Specific Array Processors*, pp. 296-308, IEEE Computer Society Press, Princeton, N.J., 1990.

[7] N. Morgan and H. Bourlard,. "Generalization and Parameter Estimation in Feedforward Nets: Some Experiments" *Advances in Neural Information Processing Systems II, pp.630-637*, Morgan Kaufmann, 1990

[8] N. Morgan and H. Bourlard. "Continuous Speech Recognition Using Multilayer Perceptrons with Hidden Markov Models," *Proc. IEEE Intl. Conf. on Acoustics, Speech, & Signal Processing*, pp. 413-416, Albuquerque, New Mexico, 1990.

[9] N. Morgan, H. Hermansky, H. Bourlard, P. Kohn, and C. Wooters,. "Continuous Speech Recognition Using PLP Analysis with Multilayer Perceptrons", ICASSP '91, In Press

[10] N. Morgan, C. Wooters, H. Bourlard, and M. Cohen. "Continuous Speech Recognition on the Resource Management Database using Connectionist Probability Estimation," Proceedings of ICSLP-90, 1337-1340, Kobe, Japan, 1990

[11] C. Plack and B.C.J. Moore. "Temporal Window Shape as a Function of Frequency and Level," 117th Meeting of the ASA, Spring 1989, S144, paper GGG8

[12] D.E. Rumelhart, G.E. Hinton, and R.J. Williams. "Learning Internal Representations by Error Propagation," in Parallel Distributed Processing. vol. 1: Foundations, Ed. D.E.Rumelhart and J.L.McClelland, MIT Press, 1986.

[13] P.J. Werbos, 1974. "Beyond Regression: New Tools for Prediction and Analysis in the Behavioral Sciences", Ph.D. thesis, Dept. of Applied Mathematics, Harvard University, 1974

# Neural-Network Architecture for Linear and Nonlinear Predictive Hidden Markov Models: Application to Speech Recognition

*L. Deng, K. Hassanein, and M. Elmasry*
*Department of Electrical and Computer Engineering*
*University of Waterloo, Waterloo, Ont. Canada*

## Abstract

A speech recognizer is developed using a layered neural network to implement speech-frame prediction and using a Markov chain to modulate the network's weight parameters. We postulate that speech recognition accuracy is closely linked to the capability of the predictive model in representing long-term temporal correlations in data. Analytical expressions are obtained for the correlation functions for various types of predictive models (linear, nonlinear, and jointly linear and nonlinear) in order to determine the faithfulness of the models to the actual speech data. The analytical results, computer simulations, and speech recognition experiments suggest that when nonlinear and linear prediction are jointly performed within the same layer of the neural network, the model is better able to capture long-term data correlations and consequently improve speech recognition performance.

## I. Introduction

Speech frames generated by speech preprocessors in automatic speech recognizers typically possess strong correlations over time [5, 8]. The correlations stem, to a large degree, from the complex interactions and overlap patterns among various articulators involved in the dynamic process of speech production [9]. Standard hidden Markov models (HMMs) [1], based on the state-conditioned IID (independent and identical distribution) assumption, are known to be weak in capturing such correlations. The strength of data correlations in the HMM source decays exponentially with time due to the Markov property, while the dependence among speech events does not follow such a fast and regular attenuation.

The linear predictive HMM proposed in [14] and [11] is intended to overcome this weakness but shows no clear evidence of superiority over the standard HMM in speech recognition experiments [11]. This can be understood because the correlation (or the envelop of the correlation function) introduced by the state-dependent linear prediction mechanism decays also in an exponential manner with time lag [2]. This makes the capability of the linear predictive HMM, in dealing with speech-frame correlations, essentially the same as that exhibited by a standard HMM having just a larger number of states.

Nonlinear time series models [15, 16] are believed to be capable of representing the temporal correlation structure of speech frames in a more general and realistic manner. In order to represent the well known nonstationary nature of speech frames, the parameters in the time series models can be made to vary with time. One elegant way of achieving this is to assume that the evolution of the time series model parameters follows a Markov chain.

In this paper we describe an implementation of this idea where three-layered feed-forward neural networks are used as Markov-state-dependent nonlinear autoregressive-type *skeleton* functions (terminology borrowed from [16]) in a time series model. Layered neural networks are ideal tools for implementing mapping functions applicable to speech-frame prediction, an idea originally proposed in [12], for two main reasons. First, it has been proved that a network of just one hidden layer is sufficient to approximate arbitrarily well any continuous function [3, 10]. Thus prediction of highly dynamic and complex speech frames can be potentially made as accurate as possible. Second, the effective back-propagation algorithm is available for network parameter estimation. To understand the properties of predictive models, we carried out detailed analysis on the statistical correlation structures of various first-order predictive models. One principal conclusion drawn from the result of the analysis is that long-term temporal correlations in the modeled data cannot be achieved with only one single predictive term, either linear or nonlinear. However, combinations of linear and nonlinear terms are shown, analytically and by simulation, to be able to produce such signal correlations, which is a desirable property for a speech model. Speech recognition experiments conducted on a speaker-dependent discrete-utterance E-set task with various types of predictive HMMs demonstrate close relationships between the recognition accuracy and the capabilities of the models in handling temporal correlations of speech data.

## II. Correlation Structure in Speech Data and Coarticulation in Speech Dynamics

Speech patterns are known to be highly dynamic and complex in nature [9]. One principal source of this complexity is coarticulation. In articulatory terms, coarticulation results from the fact that several articulators do not always move instantaneously and simultaneously from one targeted articulatory configuration to another. In acoustic terms, coarticulation is related to context dependence, whereby acoustic realization of a sound is strongly affected by the sounds just uttered and to be uttered next. This context dependence makes any IID source model, or the locally IID source model as is the case with the standard HMM, a poor choice for fitting speech data and is a major source of errors in speech recognition [6]. Good models should provide correlation structures rich enough to accommodate the context dependence and other types of temporal dependence in speech data.

## III. Analysis of Correlation Structures for State Conditioned Predictive Models

In this section, we conduct analytical evaluation of the correlation functions for various types of predictive models (linear, nonlinear, and their combination) in order to assess their faithfulness as a speech model for use in speech recognition. For the sake of simplicity in exposition yet without apparent loss of generality, we assume first-order prediction and scalar observations.

### III.1. Linear prediction

The state-conditioned linear predictive source model for speech data $Y_t$'s is chosen to have the following form:

$$Y_{t+1} = \phi Y_t + \epsilon_{t+1}, \qquad t = 0, 1, ..., T. \tag{1}$$

where $\epsilon_t$ is an IID residual random variable with zero mean and variance $\sigma^2$ and the skeleton function is a linear function of the data.

It is well known [2] that when the predictive coefficient $\phi$ is less than one in absolute value, then the process (1) is stationary and its autocovariance function declines exponentially as the time lag $\tau$ with the time constant $-log\phi$.

### III.2. Prediction with a single nonlinear term

The state-conditioned nonlinear predictive source model replaces the linear predictive term in (1) with a symmetric, continuously differentiable but otherwise arbitrary nonlinear skeleton function $f(\cdot)$:

$$Y_{t+1} = f(Y_t) + \epsilon_{t+1}, \qquad t = 0, 1, ..., T. \tag{2}$$

In this section, $f(\cdot)$ is restricted to contain only one single nonlinear term. In actual implementation of the predictive model, we chose $f(\cdot)$ to be a specific nonliner function, such as the tanh function. But otherwise $f(\cdot)$ is not restricted to any specific form when we study the statistical properties of model (2) in this section.

The method we use to derive the correlation function for (2) resembles the perturbation analysis for the study of nonlinear differential equations [13]. To proceed, we construct a family of models which is parameterized by $\alpha$:

$$Y_{t+1}(\alpha) = \alpha f(Y_t(\alpha)) + \epsilon_{t+1}, \tag{3}$$

and model (2) is considered as one model in the family (3) whose statistical properties change continuously with the parameter $\alpha$.

Once the model is parameterized, the autoregression on the data $Y_t$ can be removed by performing power-series expansion of the nonlinear function $f(\cdot)$:

$$Y_1(\alpha) = \epsilon_1 + \alpha f(Y_0(\alpha)),$$
$$Y_2(\alpha) = \epsilon_2 + \alpha f(Y_1(\alpha))$$
$$= \epsilon_2 + \alpha f(\epsilon_1) + \alpha^2 f(Y_0) f'(\epsilon_1) + \frac{1}{2}\alpha^3 f^2(Y_0) f''(\epsilon_1) + \cdots,$$
$$Y_3(\alpha) = \epsilon_3 + \alpha f(Y_2(\alpha))$$
$$= \epsilon_3 + \alpha f(\epsilon_2) + \alpha^2 f(\epsilon_1) f'(\epsilon_2) + \alpha^3 f(Y_0) f'(\epsilon_1) f'(\epsilon_2) + \cdots,$$
$$\vdots$$

and in general,

$$Y_t(\alpha) = \epsilon_t + \alpha f(\epsilon_{t-1}) + \alpha^2 f(\epsilon_{t-2}) f'(\epsilon_{t-1}) + \alpha^3 f(\epsilon_{t-3}) f'(\epsilon_{t-2}) f'(\epsilon_{t-1}) + \cdots. \tag{4}$$

(In the above, $f'(\cdot)$ denotes the derivative of $f(\cdot)$ with respect to its argument.)

From (4) the covariance function for model (3) is calculated to give

$$Cov[Y_t(\alpha), Y_{t+\tau}(\alpha)]$$
$$\approx Cov(\epsilon_t, \epsilon_{t+\tau}) + \alpha Cov[f(\epsilon_t), \epsilon_{t+\tau}] + \alpha Cov[\epsilon_t, f(\epsilon_{t+\tau-1})]$$
$$+ \alpha^2 Cov[f(\epsilon_{t-2}) f'(\epsilon_{t-1}), \epsilon_{t+\tau}] + \alpha^2 Cov[\epsilon_t, f(\epsilon_{t+\tau-2}) f'(\epsilon_{t+\tau-1})]$$
$$+ \alpha^3 Cov[f(\epsilon_{t-2}) f'(\epsilon_{t-1}), f(\epsilon_{t+\tau-1})]$$
$$+ \alpha^3 Cov[f(\epsilon_{t-1}), f(\epsilon_{t+\tau-2}) f'(\epsilon_{t+\tau-1})]$$
$$+ \alpha^4 Cov[f(\epsilon_{t-2}) f'(\epsilon_{t-1}), f(\epsilon_{t+\tau-2}) f'(\epsilon_{t+\tau-1})]. \tag{5}$$

Among the eight terms in (5), the first, second, fourth, and sixth terms are zero for $\tau \geq 0$. This is due to the IID assumption for $\epsilon_t$ and to the fact that $f(\cdot)$ is a static function containing no memory. The fifth term, $Cov[\epsilon_t, f(\epsilon_{t+\tau-2}) f'(\epsilon_{t+\tau-1})]$, is non-zero only for $\tau = 1$ and $\tau = 2$. The seventh and the eighth terms are non-zero only for $\tau = 1$. Likewise, any higher order terms of $\alpha$ in the covariance function which are omitted due to cutoff in the power-series expansion of $Y_t(\alpha)$ would contain non-zero values only for small time lags.

We conclude from the above analysis that prediction of a time series with a single nonlinear term alone does not produce long-term temporal correlations in the model's output.

### III.3. Joint prediction with nonlinear and linear terms

In this section we investigate correlation properties of the data generated from the stationary time series model

$$Y_{t+1} = \phi Y_t + f(Y_t) + \epsilon_{t+1}, \qquad t = 1, 2, ..., T, \tag{6}$$

whose skeleton function has an additional linear predictive term to that of model (2) studied in Section III.2. Although a single nonlinear predictive term just by itself is unable to generate desirable long-term data correlations (Section III.2), we can expect the interaction of the nonlinear term with the additional linear predictive term to produce such desirable properties. The following analysis, and the simulation results shown in Section IV, confirm this expectation.

Following a similar approach to that of Section III.2, the family of models constructed for (6) appropriate for the ensuing perturbation analysis is

$$Y_{t+1}(\alpha) = \phi Y_t(\alpha) + \alpha f(Y_t(\alpha)) + \epsilon_{t+1}. \qquad t = 1, 2, ..., T. \qquad (7)$$

We now decompose the stationary random process $Y_t(\alpha)$ into its stationary component processes by representing it as a power-series expansion on $\alpha$

$$Y_{t+1}(\alpha) = Y_{t+1,0} + \alpha Y_{t+1,1} + \frac{1}{2!}\alpha^2 Y_{t+1,2} + \frac{1}{3!}\alpha^3 Y_{t+1,3} + \cdots. \qquad (8)$$

In order to identify the component processes $Y_{t,i}, i = 0, 1, 2, ...$, we substitute (8) into (7) and approximate the nonlinear function $f(\cdot)$ by truncating its power-series expansion. This gives

$$\begin{aligned}
Y_{t+1}(\alpha) &\approx \phi(Y_{t,0} + \alpha Y_{t,1} + \frac{1}{2!}\alpha^2 Y_{t,2} + \frac{1}{3!}\alpha^3 Y_{t,3}) \\
&\quad + \alpha[f(Y_{t,0}) + f'(Y_{t,0})(\alpha Y_{t,1} + \frac{1}{2!}\alpha^2 Y_{t,2} + \frac{1}{3!}\alpha^3 Y_{t,3})] + \epsilon_{t+1} \\
&= (\phi Y_{t,0} + \epsilon_{t+1}) + \alpha[\phi Y_{t,1} + f(Y_{t,0})] + \alpha^2[\frac{1}{2}\phi Y_{t,2} + f'(Y_{t,0})Y_{t,1}] \\
&\quad + \alpha^3[\frac{1}{6}\phi Y_{t,3} + \frac{1}{2}f'(Y_{t,0})Y_{t,2}] + \cdots. \qquad (9)
\end{aligned}$$

By equating the coefficients of $\alpha^i$ in (8) and in (9), we obtain the following recursive relations among the component processes $Y_{t,k}, k = 0, 1, 2, ...$:

$$\begin{aligned}
Y_{t+1,0} &= \phi Y_{t,0} + \epsilon_{t+1}, \\
Y_{t+1,1} &= \phi Y_{t,1} + f(Y_{t,0}), \\
Y_{t+1,2} &= \phi Y_{t,2} + 2f'(Y_{t,0})Y_{t,1}, \\
Y_{t+1,3} &= \phi Y_{t,3} + 3f'(Y_{t,0})Y_{t,2}, \\
&\vdots
\end{aligned} \qquad (10)$$

According to (10), we can proceed to derive the autocovariance function for $Y_t(\alpha)$ denoted by

$$\gamma = Cov[Y_t(\alpha), Y_{t+\tau}(\alpha)].$$

Using (8) and truncating the expansion up to the first order, we have

$$\gamma \approx Cov[Y_{t,0} + \alpha Y_{t,1}, Y_{t+\tau,0} + \alpha Y_{t+\tau,1}]. \tag{11}$$

Use of the stationarity property of $Y_{t,0}$ and $Y_{t,1}$ leads to

$$\begin{aligned}\gamma &= \phi^2\gamma + \alpha^2 Cov[f(Y_{t-1,0}), f(Y_{t+\tau-1,0})] \\ &+ \phi\alpha Cov[Y_{t-1,0} + \alpha Y_{t-1,1}, f(Y_{t+\tau-1,0})] \\ &+ \phi\alpha Cov[Y_{t+\tau-1,0} + \alpha Y_{t+\tau-1,1}, f(Y_{t-1,0})].\end{aligned}$$

Re-arranging terms and using the stationarity property of $Y_{t,0}$ and $Y_{t,1}$ again give $\gamma$ which is equal to

$$\frac{1}{(1-\phi^2)}\{\alpha^2 Cov[f(Y_{t-1,0}), f(Y_{t+\tau-1,0})] + 2\phi\alpha Cov[Y_{t,0} + \alpha Y_{t,1}, f(Y_{t+\tau,0})]\}. \tag{12}$$

$Y_{t,0}$, the zero-th order expansion of $Y_t(\alpha)$, is a linear process and its properties are well understood (Section III.1). To obtain the desired form for $\gamma$, we need an explicit expression for the component covariance in (12) involving nonlinear process $Y_{t,1}$. Repetitive use of the recursive relations in (10) gives

$$\begin{aligned}Cov[Y_{t,1}, f(Y_{t+\tau,0})] &= Cov[\phi Y_{t-1,1} + f(Y_{t-1,0}), f(Y_{t+\tau,0})] \\ &= \phi Cov[Y_{t-1,1}, f(Y_{t+\tau,0})] + Cov[f(Y_{t-1,0}), f(Y_{t+\tau,0})] \\ &= \phi Cov[\phi Y_{t-2,1} + f(Y_{t-2,0}), f(Y_{t+\tau,0})] + Cov[f(Y_{t-1,0}), f(Y_{t+\tau,0})] \\ &\vdots \\ &= \sum_{i=0}^{t-1} \phi^i Cov[f(Y_{t-i-1,0}), f(Y_{t+\tau,0})]\end{aligned}$$

Substitution of this result into (12) leads finally to

$$\begin{aligned}\gamma &= \frac{1}{(1-\phi^2)}\{\alpha^2 Cov[f(Y_{t-1,0}), f(Y_{t+\tau-1,0})] + 2\phi\alpha Cov[Y_{t,0}, f(Y_{t+\tau,0})] \\ &+ 2\phi\alpha^2 \sum_{i=0}^{t-1} \phi^i Cov[f(Y_{t-i-1,0}), f(Y_{t+\tau,0})]\}.\end{aligned}$$

The first two terms in the above expression are exponentially declining as a function of time lag $\tau$ because the component processes involved are just static functions of linear processes. The remaining summation, however, would in general decay more slowly because of the many contributing terms.

We conclude from the above result that in a model where linear and nonlinear terms are jointly used for prediction, the correlation function tends to decay more slowly than in models utilizing either linear or nonlinear predictive terms separately. In other words, if a model is to be constructed to represent natural data which is known to possess long-term inter-time correlation, such as speech, a model of joint linear and nonlinear predictive terms would be superior to (i.e. more faithful than) that of only one predictive term, either linear or nonlinear.

## IV. Simulation Results on the Predictive Models

Computer simulations were carried out to check the analytical results obtained in Section III, where many approximations were employed to allow for the analysis to be carried out in a closed form. Using a random number generator which produces Gaussian IID residuals $\epsilon_t$ with a zero mean and unit variance, we created artificial "speech" data according to models (1), (2) and (6), respectively. The simulated data consisted of a total of 100,000 points, from which the sampled autocorrelation functions were computed for each model. The autocorrelation functions for models (1), (2) and (6) were superimposed on the same plot for comparison. Figs.1a,b,c,d correspond to four different forms of nonlinear functions $f(\cdot)$ in models (2) and (6): tanh, sigmoid, symmetric square root, and symmetric one-quarter power function, respectively. Parameter $\phi$, interpreted as the neural network weight, is assumed a fixed value less than one (this guarantees stationarity of the modeled processes). It is apparent that regardless of the form of nonlinearities, joint use of linear and nonlinear prediction terms (model 6) produces significantly stronger correlations in the simulated data than the use of separate prediction terms at any $\tau > 0$. This conforms to the analytical results obtained in Section III.

## V. Speech Recognition Experiments

The various predictive HMMs discussed so far were evaluated on a speech recognition task using a database of speaker-dependent speech recorded at the University of Waterloo [7]. The task domain of the recognizer was a total of six CV syllables where C encompasses six stop consonants /p/, /t/, /k/, /b/, /d/, /g/ and V is the vowel /i/. All the syllables were uttered with a short pause in between by native English speakers in a normal office environment. We chose this task for two reasons. First, stop confusion and E-set discrimination are known to be difficult tasks and are of fundamental significance for general speech recognition problems [4, 5]. Second, acoustic realization of stop-vowel syllables exhibits the typical nature of coarticulation and forms special sets of temporally correlated speech data. In order to faithfully represent such speech data, a model would have to be capable of handling long-term temporal correlations. The stop-E-set discrimination task allows us to perform comparative tests on the capability and the effectiveness of various types of predictive models and to assess the practical value of the analytical results obtained in Section III in speech recognition.

Sampled speech data were obtained using a DSP Sona-Graph workstation. Data were collected by digitally sampling the speech signal at 16 kHz. A Hamming window of a 25.6-ms duration was applied every 10 ms. Within each window, a 7-dimensional vector consisting of mel-frequency cepstral coefficients was computed as raw speech data to be fed to the predictive HMMs. These coefficients were appropriately scaled to accommodate the limited dynamic range in the neural network's operation. (We did not use delta cepstral coefficients over time as expanded feature sets since we felt this would violate the principle of consistency; models are considered good only if they can *generate* observations which are statistically consistent with raw data. Our intention was to develop predictive HMMs as data-generator models and we believe that the advantages of using delta coefficients can be coherently embedded in the predictive mechanisms of the models.)

In implementing the neural predictive HMMs, each syllable in the vocabulary was represented by a three-layered, feed-forward fully connected neural network. The network's weight parameters are modulated by a four state left-to-right Markov chain. The network consisted of seven input units, one accepting each scaled cepstral coefficient. Five hidden units were employed which were either all linear, all nonlinear, or two linear and three nonlinear depending on the type of the predictive model considered. Seven output units were all assumed linear, each having the desired value of a corresponding predicted cepstral coefficient one frame ahead. For comparison purposes, we also implemented the standard HMM, which is locally IID and is a degenerated case of the predictive HMM's when the skeleton is fixed at a state-dependent constant (i.e. Gaussian mean). To make a fair comparison, the standard HMM was implemented with an identical structure to the predictive HMM with the mere difference of replacing data prediction with locally IID data generation. The covariance matrices in the standard HMM were assumed identity matrices in keeping with the use of the unweighted least-mean-square error function in the training and testing for the predictive HMMs.

The segmental K-means algorithm was used for training the standard HMM and, in combination with the standard back-propagation algorithm, for training all three types of the predictive HMMs. Details of the training algorithm, as well as the classification algorithm, were described in [12] and omitted here. Training and testing data were obtained from three male speakers. For each speaker, the training set consisted of eight tokens for each of the six syllables in the vocabulary. The test set consisted of 14 tokens for each of the six syllables, giving a total of 84 test tokens for each speaker. Comparative recognition accuracy on the test data for the standard HMM recognizer (locally IID source with a fixed-valued "predictive" term) and for the HMM recognizers using various forms of speech-frame prediction is shown in Table 1. We draw particular attentions to the significantly higher recognition rate obtained with mixed linear and nonlinear hidden units in the neural network architecture compared with other types of recognizers.

## VI. Conclusion

It is concluded from this work that the signal prediction mechanism implemented by carefully structured neural networks is a potentially effective scheme for high-accuracy speech recognition. In the specific task of stop-E-set recognition, use of nonlinear prediction in conjunction with linear prediction was demonstrated to be superior to linear or nonlinear prediction alone, as well as to the standard HMM. This superiority is believed to result from the higher capacity provided by this joint prediction mechanism in representing the inherent long-term correlations between successive speech frames. Analytical evaluation and computer simulations of the correlation functions for various types of simplified predictive models provide strong support for this postulation.

| Speaker | Standard HMM | Linear Pred. HMM | Nonlinear Pred. HMM | Jointly Pred. HMM |
|---|---|---|---|---|
| 1 | 80.9% | 88.1% | 89.3 % | 89.3 % |
| 2 | 85.7% | 84.5% | 92.8 % | 97.6 % |
| 3 | 91.7% | 91.7% | 96.4 % | 100.0 % |
| Ave. | 86.1% | 88.1% | 92.8 % | 95.6% |

Table 1: Comparative recognition accuracy on CV syllables for HMM recognizers using standard HMM and various forms of speech-frame prediction implemented with a layered neural network architecture.

# References

[1] L.E. Baum. "An inequality and associated maximization technique in statistical estimation for probabilistic functions of Markov processes", *Inequalities*, vol. 3, pp. 1–8, 1972.

[2] G.E.P. Box and G.M. Jenkins. *Time Series Analysis—Forecasting and Control*, Holden-Day, San Francisco, CA, pp. 67–72, 1976.

[3] G. Cybenko. "Approximation by superpositions of a Sigmoidal function," *Mathematics of Control, Signals, and Systems*, Vol. 2, June, 1989, pp. 303-314.

[4] L. Deng, P. Kenny, M. Lennig, and P. Mermelstein. "Phonemic hidden Markov models with continuous mixture output densities for large vocabulary word recognition," *IEEE Trans. Signal Processing*, Vol. 39, No. 7, July, 1991, pp. 1677-1681.

[5] L. Deng, M. Lennig, and P. Mermelstein. "Modeling microsegments of stop consonants in a hidden Markov model based word recognizer," *J. Acoust. Soc. Am.*, Vol. 87, June, 1990, pp. 2738-2747.

[6] L. Deng, M. Lennig, F. Seitz and P. Mermelstein. "Large vocabulary word recognition using context-dependent allophonic hidden Markov models," *Computer Speech and Language*, Vol. 4, No. 4, 1990, pp. 345-357.

[7] L. Deng and K. Erler. "Microstructural speech units and their HMM representation for discrete utterance speech recognition," *Proc. IEEE Internat. Conf. Signal Processing*, Toronto, Ontario, Canada, May, 1991, pp. 193-196.

[8] L. Deng, P. Kenny, M. Lennig, and P. Mermelstein. "Modeling acoustic transitions in speech by state-interpolation hidden Markov models," *IEEE Trans. Signal Processing*, scheduled to appear in February, 1992.

[9] G.Fant. *Acoustic Theory of Speech Production*, Mouton, The Hague, 1960.

[10] K. Hornik, M. Stinchcombe, and H. White. "Multilayer feedforward networks are universal approximators," *Neural Networks*, Vol. 2, 1989, pp. 359-366.

[11] P. Kenny, M. Lennig, and P. Mermelstein. "A linear predictive HMM for vector-valued observations with applications to speech recognition," *IEEE Trans. Acoustics, Speech, and Signal Processing*, Vol. 38, No. 2, February 1990, pp.220-225.

[12] E. Levin. "Word recognition using hidden control neural architecture," *Proc. IEEE Internat. Conf. Signal Processing*, Vol. 1, Albuquerque, NM, pp. 433-436, 1990.

[13] N. Minorsky *Nonlinear Oscillations*, Chapter 9, D. Van Nostrand Company Inc., Princeton, N.J., 1962.

[14] A.B. Poritz, "Hidden Markov models: A guided tour," *Proc. IEEE Internat. Conf. Signal Processing*, Vol.1, New York, New York, April 11-14, 1988, pp. 7-13.

[15] M.B. Priestley. *Non-Linear and Non-Stationary Time Series Analysis*, Academic Press, London, England, 1988, pp. 140-174.

[16] H. Tong. *Non-Linear Time Series — A Dynamical System Approach*, Oxford University Press, New York, 1990.

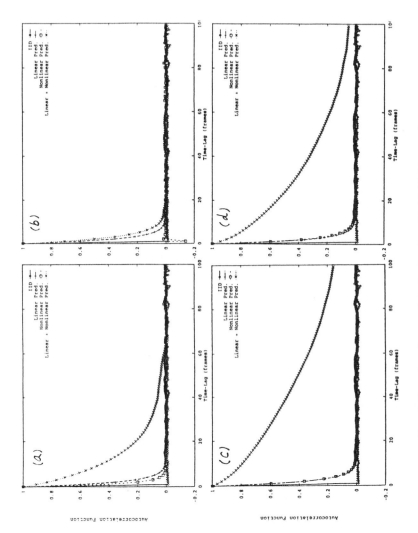

Figure 1: Comparison of autocorrelation functions for models (1), (2), (6) and an IID source. a,b,c and d are for four different forms of nonlinear functions $f(\cdot)$ in models (2) and (6): tanh, sigmoid, symmetric square root, and symmetric one-quarter power function, respectively. Parameter $\phi$ is fixed at 0.6.

# ON ADAPTIVE ACQUISITION OF SPOKEN LANGUAGE

*A. L. Gorin, S. E. Levinson, L. G. Miller and A. N. Gertner*

*AT&T Bell Laboratories*
*Murray Hill, New Jersey*

## INTRODUCTION

At present, automatic speech recognition technology is based upon constructing models of the various levels of linguistic structure assumed to compose spoken language. These models are either constructed manually or automatically trained by example. A major impediment is the cost, or even the feasibility, of producing models of sufficient fidelity to enable the desired level of performance.

The proposed alternative is to build a device capable of acquiring the necessary linguistic skills *during the course of performing its task*. The purpose of this paper is to provide a progress report on our work in this direction, describing some principles and mechanisms upon which such a device might be based, and recounting several rudimentary experiments evaluating their utility.

The basic principles and mechanisms underlying this research program were detailed in [G91], and applied there to a rudimentary text-based task. For completeness, these ideas and results are briefly reviewed here. Since then, however, we have been investigating the application of those ideas to devices with spoken input [G91a], and which are capable of larger and more complex sets of actions [M89][M91][Ge91]. In this paper, we propose some corollaries to those basic principles, thereby motivating extensions of our earlier experimental mechanisms to these more complex devices. We also briefly describe these experimental systems and observe how they demonstrate the utility of our ideas.

## BACKGROUND

This paper reports on our investigation into devices which adaptively acquire the language for their task. There have been two basic principles underlying this work, as exposited in [G91]. A *first principle* is that the primary function of language is to communicate. A consequence of this principle is that language acquisition involves gaining the capability of decoding the message, i.e. of extracting the intended meaning. This is in contrast to much of the research on automated language acquisition, which focuses on discovering syntactic structure, often specifically to the exclusion of meaning. This first principle led us to investigate a language acquisition mechanism based on connectionist methods, in which the network builds associations between input stimuli and meaningful machine responses to them.

A *second principle* is that language is acquired by interacting with a complex environment. A consequence of this principle is that the interaction involves feedback as to the appropriateness of a machine response to a particular input stimulus. Governing learning via such feedback is called *reinforcement learning* [B85][K90] and can be contrasted with *learning by example*. When reinforcement learning occurs at the level of machine action, we've called this *learning by doing*, and the reinforcement a *semantic-level error signal*. This second principle led us to investigate a mechanism for human-machine interaction based on control-theory methods, where the system's input is a message and the error signal is a measure of the appropriateness of the machine's response.

A consequence of these two principles is that the machine *understands* if it learns to respond appropriately to our commands. An underlying paradigm is that the purpose of communication is to effect some transformation in the message's recipient. This transformation may be an immediately and directly observable action (as in our current experiments), an internal change of state which is only indirectly observable, or both as in [L80]. This paradigm can be contrasted with that of Shannon, who considers the goal of communication to be the accurate replication of a transmitted message, specifically without regard for its meaning [S48].

The generic mechanism inspired by these two principles is illustrated in Figure 1, namely a connectionist network embedded in a feedback control system, where connections are strengthened or weakened depending upon feedback as to the appropriateness of the machine's response to input stimuli. In our earliest text-based experiments [G90][G91], we employed a multilayer network as illustrated in Figure 2, where the input nodes correspond to a growing list of vocabulary words, the intermediate layer corresponds to an also growing list of observed word-pairs, the output nodes correspond to a fixed discrete set of possible machine actions, and the connection weights are defined to be the *mutual information* [T69] between words (or phrases) and actions. The mutual information is calculated via smoothed relative frequency estimates, which is a naturally sequential computation. It can be shown that the information-theoretic adaptation step is guaranteed to decrease the single-step error function [G89], although *no* gradient is computed.

In the text-based experiment of [G91], the application scenario was an Inward Call Manager, where the possible machine actions were to transfer a telephone call to an appropriate department. The device was initialized with only the words *no* and *ok,* comprising a prefabricated one bit encoding of the semantic-level error signal. During the course of performing its task, in over 1000 dialogs with 10 users, the device acquired a vocabulary of over 1500 words. Learning was stable in that it retained 99% of what it was taught. More details on that experiment can be found in [G91].

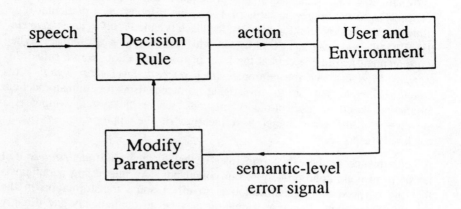

Figure 1: Adaptive Language Acquisition

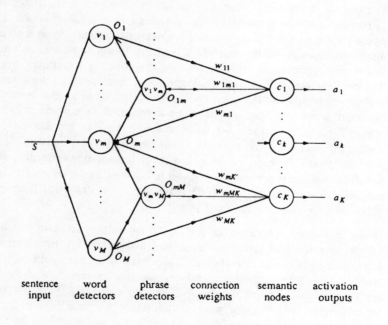

Figure 2: A Multi-Layer Network

## EXTENSION TO SPOKEN INPUT

To acquire spoken language, one needs to address the question *what is a word?* Webster defines a word as *a speech sound ... that ... communicates a meaning without being divisible into smaller units capable of independent use.* In [G91a], we postponed the divisibility issue by restricting the input to isolated word sequences, i.e. spoken words separated by short pauses. We define a *word* as a *cluster* of speech sounds which are both *acoustically* and *semantically* similar. The text-based mechanism of [G91] is thus modified to process speech via incorporating a word-clustering algorithm within the the network training procedure.

In [G91a], we've reported on preliminary experimental results for a small speech database, within the Inward Call Manager application scenario. The vocabulary and grammar are *unconstrained,* and acquired by the device during the course of performing its task. Furthermore, it is unique among speech understanding systems, in that no transcription is involved, i.e. *no* text is involved in either the training or testing of the system.

Transcriptions of two spoken dialogs follow. The first session demonstrates the control-theoretic dialog model, providing error recovery when the device encounters the unknown word *etagere*. At the completion of that first dialog, the network is updated, adding at least one new node to its input layer. The second session illustrates that learning has occurred. Several observations are in order. First, observe that *no* word-level supervision was provided. Second, observe that the device generalized from the singular to the plural of *etagere,* based on the similarity of their spectral representations. Third, observe that in Session One, the device expresses hesitancy in its interpretation, while expresses confidence in Session Two.

### SESSION ONE

HELLO. THIS IS THE J.C. BELL DEPARTMENT STORE. HOW MAY I HELP YOU?

I want to buy an etagere.

MAYBE SHOULD I CONNECT YOU TO THE CLOTHING DEPARTMENT?

No, it's a kind of furniture.

I WILL CONNECT YOU TO THE FURNITURE DEPARTMENT.

HELLO, THIS IS THE FURNITURE DEPARTMENT. HOW MAY I HELP YOU?

# SESSION TWO

HELLO. THIS IS THE J.C. BELL DEPARTMENT STORE.
HOW MAY I HELP YOU?

Do you sell etageres?

I WILL CONNECT YOU TO THE FURNITURE DEPARTMENT.

HELLO, THIS IS THE FURNITURE DEPARTMENT.
HOW MAY I HELP YOU?

## EXTENSION TO LARGE AND COMPLEX TASKS

As a task and language increase in complexity, so does the mapping from message to meaning. It is then reasonable to question the capability of the proposed networks and training procedures to learn such complex mappings. There does indeed exist a body of theory addressing such issues [C89][J88][B88]. Those results, however, tend to be asymptotic in nature, requiring large numbers of examples for learning to occur. In contrast, a striking feature of human language acquisition is our ability to make sweeping generalizations from small numbers of observations. For example, a single occurrence of a new word, in the appropriate context, can be sufficient to acquire its pronunciation, its syntactic role and its semantic associations.

Thus, although a homogeneous network architecture might, given sufficient data, be capable of learning the associations between messages and meaningful responses for complex tasks, we are not satisfied with such asymptotic results. An alternative is to investigate how one might reflect our knowledge of task and language structure in a network architecture, in order to provide improved generalization capability in language acquisition.

We thus propose a *third principle,* that a language acquisition device should be well-matched to its environment and I/O periphery, as measured by its ability to rapidly adapt and generalize. One can anthropomorphize this principle by considering the input and error channels in Figure 1 to comprise the device's sensory periphery, the output channel to comprise its motor periphery, and the network architecture and learning procedure to comprise its innate knowledge. The adaptive training of the network parameters then corresponds to learning during the lifetime of an individual organism. It is intuitively clear that the better matched the innate knowledge of an organism is to its sensory-motor periphery and environment, then the better it will be able to generalize from what it experiences. Indeed, one can take this as a definition of *well matched,* since the ability to generalize implies faster learning and adaptation, which is clearly a critical aspect of intelligence.

Given a device whose range of actions can be formally described, this third principle leads us to investigate methods by which to reflect the structure of that action space in the network architecture. The simplest, yet still interesting, situation occurs when the set of programs which the device can execute are specified by $n$ independent parameters. We view the selection of each parameter as a hidden subaction, denoted a *semantic primitive,* which are then combined to form an observable meaningful action. In this case, the action space can be characterized by a cartesian product of these semantic primitives.

In order to reflect this action space structure in a network architecture, Miller [M89] proposed the concept of a *product network,* comprising $n$ independent networks, one for each semantic primitive. Each is of the form of Figure 2, where the output of the individual networks are combined to form activations for action nodes which correspond to ordered $n$-tuples of semantic primitive values.

In particular, a twofold product network is being experimentally evaluated on a 1000-action Almanac data retrieval task [M91] with text input, responding to queries for some attribute of some state. We denote by $N$ the subnetwork corresponding to the semantic primitive which selects the state parameter, and denote by $A$ the subnetwork which selects the attribute parameter. Abusing notation, we also denote the set of values for the state parameter as $N$, (where the network $N$ has one output node for each possible value), and similarly for A. Thus, one can describe the action space for this task as isomorphic to $N \times A$.

Two sample dialogs follow. First, observe that the device generalizes its acquisition of *50th state* to connote Alaska regardless of the queried attribute. This generalization is enabled by the product network, since the association between words and action is factored through the semantic primitives. Observe also that no word-level supervision is provided, so that each subnetwork must learn which words are relevant to its associated semantic primitive. This task, the product network, and its detailed experimental evaluation are described in [M91].

**SESSION ONE**

HELLO. I KNOW FACTS ABOUT THE FIFTY STATES.
WHAT WOULD YOU LIKE TO KNOW?

Who is the governor of the 50th state?

WOULD YOU LIKE TO KNOW THE GOVERNOR OF ALABAMA?

No, of Alaska.

YOU WANT TO KNOW THE GOVERNOR OF ALASKA.

STEVE COWPER IS THE GOVERNOR OF ALASKA

## SESSION TWO

WHAT WOULD YOU LIKE TO KNOW?

What is the state nickname of the 50th state?

YOU WANT TO KNOW THE NICKNAME OF ALASKA?

ALASKA IS KNOWN AS 'THE LAST FRONTIER'

**Emergent Nonterminals**

Given a task such as the Almanac, let us consider the collection of words and phrases which cause the subnetwork $N$ to fire. Upon inspection, one observes that these mostly comprise place names. Let us hypothesize that this set of words is a *part of speech*. So as to avoid raising a linguistic storm, we will characterize this nonterminal abstractly as $N$, rather than as a *noun*.

Another perspective on this hypothesis is provided via Smyth and Goodman's approach to rule inference [S91]. If one conjectures a production rule $N \to v$, where $v$ is a vocabulary word, they've proposed a priority measure based on the mutual information between the clauses, i.e. $I(v,N)$. This is precisely the output of the information-theoretic network $N$, raising the specter of exploiting meaning to govern the acquisition of nonterminals in a language. This is in contrast to much of the literature on grammatical inference, where meaning plays little or no role.

**Developmental Learning**

People use concepts acquired in simple tasks as building blocks to learn more complex ones. We propose a *fourth principle*, which is that language acquisition should proceed in developmental stages, beginning with mastery of simple tasks, and only then continuing to complex ones. This leads us to consider a mechanism whereby subnetworks are initially trained within a simple task, where their outputs are directly tied to a semantic primitive for that task. These stable subnetworks can then be exploited as nonterminals in higher complexity tasks.

Gertner [Ge91] has investigated the situation of a device whose action space is characterized via an ordered pair of places. The application scenario is a subset of an Airline Information Task [L80], in particular the display of a flight table between some pair of cities. If $C$ denotes the set of possible cities, then the action space is isomorphic to $C \times C$. While again a product space, a little thought shows that the product network mechanism will not suffice to provide rapid adaptation and generalization. We instead proceed developmentally, according to our fourth principle, assuming that a stable subnetwork corresponding to the nonterminal for place names was acquired in an earlier task. (E.g., the subnetwork *N* from the Almanac task.)

A network, analogous to that of Figure 2, is constructed, but where the nodes correspond to nonterminals. The structure is hierarchically self-similar, in that each nonterminal is itself a copy of Figure 2, but with nodes corresponding to vocabulary words (terminals). Thus, the construction of associations between words and actions is factored through these intermediate nonterminals.

As demonstrated in the sample dialogs below, this leads to rapid adaptation and generalization. For the sake of an interesting example, we've chosen the input messages to involve the Nepali language, which uses postpositional rather than prepositional modifiers [Y90]. After Session One, any reasonable network architecture would have memorized the association between that particular sentence and the appropriate action. A product network would even be able to generalize that the phrase *chicago bata* is associated with Chicago being the departure city. One would hope, however, that a network would furthermore learn that these two new words modify the semantics of any city name, not just the ones with which they were observed. One indeed observes this generalization in Session Two, illustrated the power of factoring the word/action associations through intermediate nonterminals. This task, network of nonterminals, and detailed experimental results are described in [Ge91].

## SESSION ONE

I CAN SHOW YOU THE FLIGHT TABLE BETWEEN ANY PAIR OF CITIES.

chicago bata detroit sama

DO YOU WANT TO SEE THE FLIGHTS FROM DETROIT TO CHICAGO?

no, the other way around.

I WILL SHOW YOU THE FLIGHTS FROM CHICAGO TO DETROIT.

## SESSION TWO

ARE YOU INTERESTED IN OTHER FLIGHTS?

washington sama boston bata

I WILL SHOW YOU THE FLIGHTS FROM BOSTON TO WASHINGTON.

## CONCLUSIONS

This paper provides a progress report on our research program in adaptive language acquisition. We've proposed some corollaries to our basic principles, motivating and describing extensions of our earlier experimental mechanisms to more complex devices. We've briefly reported on several rudimentary experiments which demonstrate and evaluate the utility of these ideas, in particular for devices with spoken input, and with larger and more complex text-based tasks.

## REFERENCES

[B85]  A. G. Barto and P. Anadan "Pattern Recognizing Stochastic Learning Automata," *IEEE Trans. on Systems, Man and Cybernetics,* vol. smc-15, no. 3, pp. 360-375, May/June 1985.

[B88]  E. Baum and D. Haussler, "What size of Net Gives Valid Generalization," *Neural Computation* 1, pp. 151-160 (1989)

[C89]  G. Cybenko, "Approximation by Superpositions of Sigmoidal Function," *Mathematics of Control, Signals, and Systems* 2, pp. 303-314, (1989).

[G89]  A. L. Gorin and S. E. Levinson, "A Perceptron with Information-Theoretic Connection Weights," *AT&T Bell Laboratories Technical Memorandum*, Dec. 1989, submitted for publication.

[G90]  A. L. Gorin, S. E. Levinson, L. G. Miller, A. N. Gertner, E. Goldman and A. Ljolje, "On Adaptive Acquisition of Language," *Proc. of ICASSP*, pp. 601-604, April 1990.

[G91]  A. L. Gorin, S. E. Levinson, A. N. Gertner and E. Goldman, "Adaptive Acquisition of Language," *Computer, Speech and Language*, vol. 5, no. 2, pp. 101-132, April 1991.

[G91a]  A. L. Gorin, S. E. Levinson and A. N. Gertner, "Adaptive Acquisition of Spoken Language," *Proceedings of ICASSP 91*, pp. 805-808, May 1991.

[Ge91]  A. N. Gertner and A. L. Gorin, "Adaptive Language Acquisition for an Airline Information Subsystem," *AT&T Bell Laboratories Technical Memorandum*, in preparation.

[J88]  S. Judd, "On the complexity of loading shallow neural networks," *Journal of Complexity* 4, pp. 177-192 (1988)

[K90]  L. P. Kaebling, "Learning in Embedded Systems," Ph.D. Thesis, Computer Science Department, Stanford University, June 1990.

[L80] S. E. Levinson and K. L. Shipley, "A Conversational-Mode Airline Information and Reservation System using Speech Input and Output," *Bell System Technical Journal*, vol. 59, pp. 119-137, 1980.

[M89] L. G. Miller, A. L. Gorin and S. E. Levinson, "Adaptive Language Acquisition for a Data Base Query Task," *AT&T Bell Laboratories Technical Memorandum*, unpublished, Dec. 1989.

[M91] L. G. Miller and A. L. Gorin, "Adaptive Language Acquisition for a Data Retrieval Task," *AT&T Bell Laboratories Technical Memorandum*, unpublished, May 1991.

[S48] C. E. Shannon, "A Mathematical Theory of Communication," *Bell System Technical Journal*, vol. 27, no. 3, pp. 379-423, July 1948.

[S91] P. Smyth and R. M. Goodman, "An Information-Theoretic Approach to Rule Induction from Databases," *IEEE Transactions on Knowledge and Data Engineering*, to appear.

[T69] J. B. Thomas, *Statistical Communication Theory*, Wiley (1969).

[Y90] D. Yarowsky, private communication, Oct. 1990.

# VECTOR QUANTISATION WITH A CODEBOOK-EXCITED NEURAL NETWORK

Lizhong Wu    Frank Fallside
Cambridge University Engineering Department
Trumpington Street, Cambridge CB2 1PZ, UK.

## INTRODUCTION

Assume that $s_i$ is a stationary discrete sequence from a source alphabet $\mathcal{R}$. For a fixed dimension $p$, let $S_i = (s_{pi}, s_{pi+1}, \ldots, s_{p(i+1)-1})^T \in \mathcal{R}^p$. An $r$ bits/vector quantiser consists of a codebook $\mathcal{W} = (W_1, \ldots, W_L)$, where $L = 2^r$, and a mapping $q: \mathcal{R}^p \to \mathcal{W}$ or a partition $\mathcal{P} = (P_1, \ldots, P_L)$ of $\mathcal{R}^p$ such that $q(S_i) = W_k$, if $S_i \in P_k$. Given a distortion measure $d: \mathcal{R}^p \times \mathcal{W} \to [0, \infty)$, and assigning a distortion $d(S_i, W_k)$ to the reproduction of $\mathcal{W}$ for $S_i$, the performance of the quantiser can be measured by the expected distortion:

$$D = \sum_i d(S_i, q(S_i)) = \sum_{k=1}^{L} \sum_{S_i \in P_k} d(S_i, W_k). \tag{1}$$

Two necessary conditions for optimality are [5, 10]:

$$q(S_i) = W_k, \text{if } d(S_i, W_k) \leq d(S_i, W_j), \text{for all } j \neq k; \tag{2}$$

$$\sum_{S_i \in P_k} d(S_i, W_k) = \min_j \sum_{S_i \in P_k} d(S_i, W_j). \tag{3}$$

The basic design scheme for a vector quantiser is the LBG algorithm [10] which satisfies the optimality conditions.

Neural network consists of many simple, neuron-like processing units that interact through weight connections. Any network can be expressed in terms of a transformation function which maps inputs and network weights into network outputs. Assuming that the input vector $X = (x_1, \ldots, x_p) \in \mathcal{R}^p$, the output vector $Y = (y_1, \ldots, y_q) \in \mathcal{R}^q$, and $\mathcal{W}$ is a weight space appropriate to the network architecture, the relationship among $X$, $Y$ and $W$ can be described by a function $F: \mathcal{R}^p \times \mathcal{W} \to \mathcal{R}^q$. Given weights $W$ and input $X$, the output is given as $F(X, W)$ which is compared to the desired output $Y$. A performance measure of the network is the expected cost function:

$$D(W) = E[d(Y, F(X, W))], W \in \mathcal{W}. \tag{4}$$

The goal of network learning is to find a solution to the problem

$$W^* = \arg \min_{W \in \mathcal{W}} D(W). \tag{5}$$

Since the joint probability of $X$ and $Y$ is not known, $W^*$ cannot be solved directly. The learning is usually conducted by optimising performance over sets of samples.

According to rate-distortion theory, if the quantisation performance reaches the rate-distortion bound, the transition probability from source space to codeword space should be [1]:

$$q(Y|X) = \lambda(X)q(Y)e^{sd(Y,X)}. \tag{6}$$

The transition probability of a *neural network-based vector quantiser* is dependent on its connection weights. We denote it as $p(Y|X,W)$ and define

$$p(Y|X,W) = q(Y|X)e^{-d(Y,F(X,W))}. \tag{7}$$

$$p(Y|X,W) = q(Y|X), \text{iff } d(Y, F(X,W)) = 0.$$

By substituting (7) into (5), we obtain

$$\min_{W \in \mathcal{W}} D(W) = \min_{W \in \mathcal{W}} E[d(Y, F(X,W))] = \min_{W \in \mathcal{W}} E\left[\ln \frac{q(Y|X)}{p(Y|X,W)}\right], \tag{8}$$

where $E[\ln \frac{q(Y|X)}{p(Y|X,W)}]$ is the Kullback-Leibler information of $p(Y|X,W)$ relative to $q(Y|X)$. The Kullback-Leibler information is a fundamental theoretical measure of the accuracy of the conditional probability density $p(Y|X,W)$ as an approximation to the desired one, $q(Y|X)$ [16]. From (8), we see that neural network learning for a vector quantiser can be interpreted as modifying the network transition probability by changing the weights so that the *neural network-based quantiser* satisfies the conditions for approaching the rate-distortion bound.

Research on vector quantisation with neural networks is not new, but previous studies concentrated mostly on the following aspects:

- With Kohonen's self-organizing feature map, e.g. [18];

- A set of single-layer nets for linear predictive data as in [3];

- A multi-layer net as front-end as in [2, 9, 11, 14];

- We have studied the performance of a quantiser formed by a set of single-layer neural units and found that it meets the optimal conditions (2) and (3) if an optimal updating gain sequence is employed [19]. We have also studied the use of an associative memory at the output of the set of single-layer neural units. The combined networks form a finite state vector quantiser [22].

A multi-layer neural network possesses more powerful discriminating properties and the possibility of dealing with implicit features of the processed signal. Moreover, contextual information can easily be taken into account by recurrent connections. A lot of research on classification and recognition with a multi-layer neural network has already been reported.

In this paper, we first try to improve the quantisation performance of the coder with a *neural network-based front-end* by sequentially re-optimising the design of the neural network and the quantiser. We then propose an alternative neural network model for source coding and refer to it as a *codebook-excited neural network* (CENN). We study both the feedforward CENN and the recurrent CENN. The multi-stage CENN and the gain-adaptive CENN are also developed to reduce the coding complexity and to adapt the signals with a wide dynamic range. We then compare the CENN with the conventional vector quantiser designed using the LBG algorithm and study the conformal mapping characteristic of the CENN. Finally, as one application of the CENN to speech coding, the Gaussian excitation codebook in the code-excited linear predictive speech coder [13] is replaced by a gain-adaptive CENN. The quality improvement of its reconstructed speech is shown by comparisons of rate-distortion functions and waveforms.

## QUANTISATION WITH NEURAL NETWORK-BASED FRONT-ENDS

A quantiser is inserted into a hidden layer of the neural network. The quantiser encodes the output vectors of the hidden layer into discrete finite-valued code symbols. Both input layers and output layers contain at least one layer of neurons and accomplish the transformation $Y = F_{in}(S)$ and $Z = F_{out}(\hat{Y})$. $F_{in}(\cdot)$ and $F_{out}(\cdot)$ stand for the transformation functions of the input layers and the output layers respectively, $S$ is an encoded vector, $\hat{Y}$ is a reproduction vector of the $Y$ and is selected from the reproduction codebook of the quantiser, and the output vector $Z$ is as the reconstructed vector of the $S$.

The network is trained using the error back-propagation algorithm [12]. This algorithm requires all activation functions to be continuous. Since a quantiser contains winner-takes-all operations, it is impossible to train the quantiser and the neural network jointly. In previous research [9, 11, 14], the quantiser and the neural network were independently designed. When training the network with the error back-propagation algorithm, the quantiser was removed from the network and $\hat{Y} = Y$ was assumed. After the network $F(\cdot) = F_{in}(\cdot)F_{out}(\cdot)$ was established, a data set $\{Y\}$ was generated with $Y = F_{in}(S)$, which was used to design the quantiser $q(\cdot)$. The $q(\cdot)$ was then inserted back between $F_{in}(\cdot)$ and $F_{out}(\cdot)$.

Here we propose a method to improve the design of the coder. After $F(\cdot)$ and $q(\cdot)$ are separately obtained, the input layer network $F_{in}(\cdot)$ and the output layer network $F_{out}(\cdot)$ are respectively trained using the error back-propagation algorithm again. Now the effect of the quantiser is introduced into the training. For the input layer network, its input is $S$, its output is

$F_{in}(S)$ and its desired output is $\hat{Y}$, where $\hat{Y} = q(F_{in}(S))$. For the output layer network, its input is $\hat{Y}$, its output is $F_{out}(\hat{Y})$ and its desired output is $S$. After $F_{in}(\cdot)$ and $F_{out}(\cdot)$ are updated, a new $\{Y\}$ is generated using the new $F_{in}(\cdot)$, which is used to train a new $q(\cdot)$. This procedure for sequentially re-optimising $F_{in}(\cdot)$, $F_{out}(\cdot)$ and $q(\cdot)$ is repeated until the reduction of the distortion between $Z$ and $S$ is smaller than a given threshold.

A simulation [20] with a Gauss-Markov source shows that the SNRs of a 1 bit/sample coder for both the training and test set are improved by about $0.55dB$ after 8 sequential re-optimising iterations, but, even with re-optimising, its performance is still much worse than that of other types of coders. For example, compared with the connectionist vector quantiser studied in [22, 19], the SNR performance of the coder with a neural network-based front-end is about $2dB$ lower at all given dimensions.

## CODEBOOK-EXCITED NEURAL NETWORK

In this section, we develop an alternative model named a *codebook-excited neural network*, or CENN, for source coding, which consists of a multi-layer neural network and a trained excitation codebook. The neural network is driven by a vector selected from the excitation codebook which is formed by training with entries initialized as Gaussian. The encoded signal is applied to the output layer and compared with the network output. For each encoded vector $S$, a vector $X^*$ which leads to the minimum of $d(S, F(X^*))$ is searched in all $X \in \mathcal{X}$. Here, $F(\cdot)$ stands for the network transformation function, $\mathcal{X}$ denotes the excitation codebook, and $d(\cdot)$ is a given distortion measure. The size of $\mathcal{X}$ is determined by the transmission rate $r$, and $\| \mathcal{X} \| = 2^r$. After $X^*$ is found, its index $c$ is transmitted. At a decoding terminal, $S$ is reconstructed by $F(X^*)$. Thus the total coding distortion over an encoded data set $\mathcal{S}$ is:

$$D = \sum_{S \in \mathcal{S}} d(S, F(X^*)) = \min_{X \in \mathcal{X}} \sum_{S \in \mathcal{S}} d(S, F(X)). \qquad (9)$$

For a given $\mathcal{S}$ and an error measure $d(\cdot)$, the quantisation performance of the CENN is determined by $F(\cdot)$ and $\mathcal{X}$.

A feedforward CENN is simply trained using the error back-propagation algorithm. We also study the recurrent CENN where the output from the hidden layer is delayed by $N_h$ samples, which corresponds to the interval between two successive encoded vectors, and fed back into the input terminal of the hidden layer via connectivity $W$. Our training program is based on the work of Williams and Zisper [17].

The excitation codebook is formed by training with Gaussian initial entries. Its optimisation process includes: a) replacing a codeword by a new Gaussian sequence if this codeword has not been chosen after each complete presentation of encoded vectors; b) adjusting the codewords along the descent gradient of the distortion. If we regard the excitation as another layer of the network connected to the input units, in which the weights are $X^*$ and all their inputs are 1, the $\mathcal{X}$ can also be trained using the error back-propagation algorithm.

To reduce the computational complexity of the CENN, we also extend it to multi-stage. For signals with a wide dynamic range, we propose a multi-stage gain-adaptive CENN, which dynamically adjust the gain or amplitude scale of the network output according to the encoded signal level. The quantiser uses a gain estimator to determine a suitable gain factor for each output vector prior to comparison with the encoded vector. The adaptive gain-scaled network extends the dynamic range of its outputs and can more efficiently represent all encoded vectors.

A more detailed description of the architectures and training schemes of these CENNs is described in [20].

## DISCUSSION OF THE SIMULATION RESULTS

The simulations to be discussed in this section are also carried out with a Gauss-Markov source. The first simulation assumes that the number of units in each layer is equal and that the neural network contains no hidden layer. Additional simulations are then carried out by varying the architecture parameters one at a time to observe their effects. For the feedforward CENN, the results are very near to, and for the test set even slightly better than, those obtained by the *connectionist vector quantiser* presented in [22, 19]. We have demonstrated that the latter meets the conditions for an optimal quantiser, so we conjecture that the CENN at least also meets the optimality conditions.

The most encouraging result is the quantisation performance of the recurrent CENN. The performance of the recurrent CENN is close to the asymptotically optimal bound of block quantisers [6, 23] when the dimension $p \geq 2$. We have found that a 2-dimensional quantiser achieves approximately the same generalisation performance as that achieved by an 8-dimensional feedforward CENN. This shows that the recurrent connections accumulate the dependent characteristics over more than one previous input vector and exploit their redundancy. Moreover, the generalisation performance of the recurrent CENN is very near to its design performance. In all given dimensions, the SNRs of the test set are only about $0.2dB$ smaller than those of the training set.

Both effects of the variation of the number of hidden units and input units are similar. Compared to the rectangular-shaped network, the performance reduces when fewer input or hidden units are used, but no obvious improvements result when more input or hidden units are added to the rectangular network. The performance does not improve by adding hidden layers either. We explain this from the point of view of the signal space. The input signal space is partitioned into many small regions by the design. One condition for optimality is that the partitioned regions should be of the Voronoi or Dirichlet type [5]. In general, each Dirichlet region is a bounded polytope and is convex. It is well-known that a one-hidden layer network is sufficient to form convex regions. As explained in the last section, the $n$-hidden layer CENN with an optimised excitation codebook is actually an $(n + 1)$-hidden layer network. Thus the one or more hidden layer CENN does not outperform that with no hidden layer.

If the excitation codebook retains its initial Gaussian entries and is not optimised during training, we find that: a) the zero-hidden layer CENN without an optimised excitation codebook can not form Voronoi regions in the input space, therefore its performance falls off; and b) even with a hidden layer, its performance is worse than that with an optimised excitation codebook, especially for the quantiser with a large excitation codebook. This is because some codewords are rarely chosen if poor initial guesses are given. Such a phenomenon has been discussed by many researchers, e.g. the so called "empty cell" in [10]. Simultaneously optimising the excitation codebook during training in the CENN can alleviate this phenomenon.

Figure 1 shows the performance of the CENN for a Gauss-Markov source. For comparison, the theoretical performance bounds given by the Shannon lower bound to rate-distortion function [1] and the asymptotically optimal block quantiser bound [6, 23] are also plotted in the figure. More other simulation results appear in [20].

Figure 1: Comparison between the quantisation performance of the codebook-excited neural network and some theoretical bounds of the a 1 bit/sample quantiser for a Gauss-Markov source. From top to bottom, the curves respectively correspond to the rate-distortion bound, asymptotic bound, recurrent CENN and feedforward CENN.

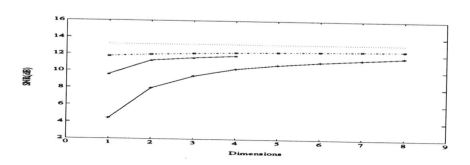

## ADVANTAGES OF THE CENN OVER THAT DESIGNED USING THE LBG ALGORITHM

### Applicability with any Distortion Measure

As described in section 1, the LBG algorithm [10] is well defined if one has a constructive solution to (3). However it becomes difficult to solve (3) for some distortion measures, e.g. the $l_v$ norm distortion measure. Due to their continually-adaptive learning, the CENN does not require direct solving of (3). Therefore, any distortion measure is applicable.

## Uniqueness of Local Minima in CENN for a Gaussian Source

Gray and Karnin [7] demonstrated the existence of multiple distinct local minima for a 1 bit/sample two-dimensional vector quantiser designed using the LBG algorithm for a zero-mean, unit variance, memoryless Gaussian source and a squared-error measure. Three sets of codes were found in their experiments. They were named: (a) two-dimensional product code, (b) one-dimensional code in two dimensions and (c) triangle-dot two-dimensional code. All these codes satisfied optimality conditions but their final average distortions were different.

We present the results obtained with the feedforward CENN designed for this special source, for details see [20]. The following initial codes are applied, (a) very small random value codes, (b) poor one-dimensional local optimal codes and (c) poor triangle-dot local optimal codes. All these three cases converge to the two-dimensional product codes, the best one in all three local optima. The experiment is continued by varying the learning rate sequences and it is shown that the final codes do not change and the distortion is still the same.

For this special example, we see that the CENN is not sensitive to its initial state. Even if the initial states are trapped in the stationary points which are the local minima in the quantiser produced by the LBG algorithm, the CENN can escape and approach a better minimum.

## THE CONFORMAL MAPPING OF THE CENN

The result of vector quantisation with the CENN is that the output vectors of the network tend to describe the distribution density of the encoded vectors. Assume that $r(S)$ is the distribution function of the encoded vectors, $q(Z)$ stands for that of the output vectors, $p(X)$ for that of the excitation vectors and $J(W)$ is the Jacobian of the transformation of the neural network. In the CENN, the training process simultaneously updates $J(W)$ and $p(X)$ so that $q(Z) = \frac{p(X)}{|J(W)|}$ approaches $r(S)$. The output vectors distribute on the encoded signal space and their shape is specified by the entropy of the source, but there is no clear relation between the distribution of output vectors and that of input vectors.

We consider the case where $p(X)$ is given and only $J(W)$ is changeable during learning. Excitation vectors are arranged as uniform meshes, non-uniform meshes, spiral and twinned link patterns. It is demonstrated that the shape formed by the output vectors topologically preserves the pattern structure of the excitation vectors. This phenomenon can be explained by conformal mapping [15]. That is, if the transformation function of the neural network $Z = F(X)$ is analytic and $F'(X) \neq 0$, then the angle between two curves through the point $X$ in the **X** plane is reproduced in magnitude and sense by the angle between the corresponding curves in the **Z** plane.

Several examples have demonstrated another property of the CENN. That is, the shape formed by the codewords can be defined according to applications and the requirement for the quality of reconstructed signals can vary in different signal regions. These properties can be further related to the topology-preserving of Kohonen's self-organising feature maps [8] and the structural vector quantisers, studied for example by Fischer [4]. The structural vector quantisers take advantage of the structure formed by the codevectors, reduce the design complexity and the full search encoding complexity of the vector quantiser based on the LBG algorithm. Though such a quantiser is somewhat suboptimal in a rate-distortion sense, the absolute increment of the distortion itself is still small if the size of the codebook is large enough. Moreover, they can be developed as adaptive vector quantisers.

## CELP SPEECH CODING WITH AN EXCITATION CODEBOOK TRAINED BY A GAIN-ADAPTIVE CENN

The code-excited linear predictive (CELP) method [13] is one of the most efficient speech coding schemes working at around 4.8kbits/s. With this model, the vocal tract is modelled by two cascaded time-varying linear predictive filters driven by a sequence chosen from a codebook formed by Gaussian (or centre-clipped Gaussian) entries. As one application of the CENN to speech coding, the Gaussian excitation codebook in the CELP speech coder is replaced by a gain-adaptive CENN. Some simulation results are shown in table 1. In these simulations, both training and test sets respectively consist of 16 different spoken sentences from the TIMIT speech database. The order of the short term linear predictive filter is set to 12, the order of the long term linear predictive filter to 1, the frame width to 160 samples, its step size to 80 samples and the dimension of the excitation codevectors to 20.

## CONCLUSIONS

An alternative model named a *codebook-excited neural network* has been proposed for source coding or vector quantisation. Two advantages of this model are that the memory information between source frames can easily be taken into account by recurrent connections and that the number of network connections is independent of the transmission rate. The simulations have also shown its good quantisation performance.

The *codebook-excited neural network* is applicable with any distortion measure. For a zero-mean, unit variance, memoryless Gaussian source and a squared-error measure, a 1 bit/sample two-dimensional quantiser with a *codebook-excited feedforward neural network* is found to always escape from the local minima and converge to the best one of the three local minima which

Table 1: Comparison of distortion between the CELP speech coder and the CELP speech coder with the excitation codebook learned by a codebook-excited feedforward neural network (NNCELP).

| Bits/ Ex. Code Vector | Segmental SNR(dB) | | | |
|---|---|---|---|---|
| | Training Set | | Test Set | |
| | CELP | NNCELP | CELP | NNCELP |
| 2 | 3.174 | 4.607 | -0.728 | 4.179 |
| 4 | 5.914 | 6.181 | 5.413 | 5.930 |
| 6 | 7.764 | 8.289 | 7.508 | 8.026 |
| 8 | 8.977 | 9.230 | 8.742 | 9.015 |

are known to exist in the vector quantiser designed using the LBG algorithm. Moreover, due to its conformal mapping characteristic, the *codebook-excited neural network* can be applied to designing the vector quantiser with any required structural form on its codevectors.

We have extended our studies to noisy channels. We derive the optimal gain sequence of the learning process for the joint source and channel coder formed by the Kohonen self-organising feature maps and modify the *codebook-excited neural network* to adapt noisy channels. This study on noise tolerance and robustness of the neural network-based quantisers and coders is reported in [21].

# References

[1] T. Berger. *Rate distortion theory - a mathematical basis for data compression*. Prentice-Hall Inc. Englewood Cliffs, New Jersey, 1971.

[2] G.W. Cottrell, P. Munro, and D. Zipser. Image data compression by back propagation: An example of extensional programming. Technical Report ICS 8702, Institute for Cognitive Science, University of California, San Diego, February 1987.

[3] F. Fallside. Analysis of linear predictive data such as speech and of ARMA processes by a class of single layer connectionist models. In F. Fogelman Soulie and J. Herault, editors, *NATO ASI Series, Vol.F 68 Neurocomputing*. Springer-Verlag, Berlin Heidelberg, 1990.

[4] T. R. Fischer. Geometric source coding and vector quantisation. *IEEE Trans. on Information Theory*, IT-35(1):137–145, 1989.

[5] A. Gersho. Asymptotically optimal block quantisation. *IEEE Trans. on Information Theory*, IT-25(4):373–380, July 1979.

[6] A. Gersho. On the structure of vector quantisers. *IEEE Trans. on Information Theory*, IT-28(2):157–166, March 1982.

[7] R. M. Gray and E. D. Karnin. Multiple local optima in vector quantisers. *IEEE Trans. on Information Theory*, IT-28(2):256–261, March 1982.

[8] T. Kohonen. *Self-organisation and associative memory*. Springer Verlag, New York, Third edition, 1988.

[9] R.M. Kuczewski, M.H. Myers, and W.J. Crawford. Exploration of backward error propagation as a self-organisational structure. In *Proc. Int. Conf. on Neural Networks*, pages II89 – II95, 1987.

[10] Y. Linde, A. Buzo, and R.M. Gray. An algorithm for vector quantiser design. *IEEE Trans. on Communications*, COM-28(1):84–95, January 1980.

[11] A.J. Robinson and F. Fallside. Static and dynamic error propagation networks with application to speech coding. In D.Z. Anderson, editor, *Proceedings of Neural Information Processing Systems*. American Institute of Physics, Denver, 1987.

[12] D.E. Rumelhart, G.E. Hinton, and R.J. Williams. Learning internal representations by error propagation. In D.E. Rumelhart and J.L. McClelland, editors, *Parallel Distributed Processing: Exploration in the microstructure of cognition*, chapter 8, pages 319–362. MIT Press, Cambridge, MA, 1986.

[13] M.R. Schroeder and B.S. Atal. Code-excited linear prediction (CELP): high-quality speech at very low bit rates. In *Proc. Int. Conf. on Acoustics, Speech and Signal Processing*, pages 937–940, 1985.

[14] N. Sonehara, M. Kawato, S. Miyake, and K. Nakane. Image data compression using a neural network model. In *Proc. Int. Joint Conf. on Neural Networks*, pages II35 – II40, 1989.

[15] I. Stewart and D. Tall. *Complex Analysis*. Cambridge University Press, Cambridge, 1983.

[16] H. White. Learning in artificial neural networks: A statistical perspective. *Neural Computation*, 1:425–464, 1989.

[17] R.J. Williams and D. Zipser. A learning algorithm for continually running fully recurrent neural networks. Technical Report ICS 8805, Institute for Cognitive Science, University of California, San Diego, October 1988.

[18] F.H. Wu and K. Ganesan. Comparative study of algorithms for VQ design using conventional and neural-net based approaches. In *Proc. Int. Conf. on Acoustics, Speech and Signal Processing*, pages 751–754, 1989.

[19] L.Z. Wu and F. Fallside. Optimal gain sequence for fastest learning in connectionist vector quantiser design. In *Proc. of International Conference on Spoken Language Processing, Kobe, Japan*, November 1990.

[20] L.Z. Wu and F. Fallside. Source coding and vector quantisation with codebook-excited neural networks. Submitted to *Computer Speech and Language*, December 1990.

[21] L.Z. Wu and F. Fallside. Channel-optimised source coding and vector quantisation with neural networks. To be published, March 1991.

[22] L.Z. Wu and F. Fallside. On the design of connectionist vector quantisers. *Computer Speech and Language* (in press), 1991.

[23] Y. Yamada, S. Tazaki, and R.M. Gray. Asymptotic performance of block quantisers with difference distortion measures. *IEEE Trans. on Information Theory*, IT-26:6–14, 1980.

# SEGMENT-BASED SPEAKER ADAPTATION BY NEURAL NETWORK

Keiji Fukuzawa[†], Hidefumi Sawai[‡], Masahide Sugiyama[†]

[†]ATR Interpreting Telephony Research Laboratories
Sanpeidani, Inuidani, Seika-cho, Soraku-gun, Kyoto 619-02, Japan
[‡]Research and Development Center, Ricoh Co., Ltd.
16-1, Shin'ei-cho, Kohoku-ku, Yokohama 223, Japan

## ABSTRACT

This paper proposes a segment-to-segment speaker adaptation technique using a feed-forward neural network with a time shifted sub-connection architecture. Differences in voice individuality exist in both the spectral and temporal domains. It is generally known that frame based speaker adaptation techniques can not compensate for speaker individuality in the temporal domain. Segment-based speaker adaptation compensates for these spectral and temporal differences. The results of 23 Japanese phoneme recognition experiments using TDNN(Time-Delay Neural Network) show that the recognition rate by segment-based adaptations was 83.7%, 22.8% higher than the rate without adaptation.

## 1 INTRODUCTION

Speaker adaptation is one of the most promising techniques to achieve speaker-independent recognition system. In general, speaker adaptation techniques can be classified into supervised training techniques and unsupervised training techniques. This paper discusses supervised training techniques.

Several frame-based speaker adaptation techniques have been proposed, for example, codebook mapping[1] and neural network based analog mapping [2]. Although these frame-based techniques are effective, the recognition performance with adaptation for an input speaker is still lower than the performance for speaker-dependent recognition. One reason is that frame-based speaker adaptation techniques compensate only for speaker differences in the spectral domain. Nevertheless, voice individuality results in both spectral and temporal structural differences. On the other hand, segment-based approaches allow compensation of both spectral and temporal structural differences.

Several segment-based speaker adaptation techniques have been proposed [3][4]. Generally speaking, training of segment-based adaptation requires considerable speech data for an input speaker. When a neural network is applied to segment-based speaker adaptation, it requires a large number of weighting parameters in the network connection and consequently, considerable speech data. To solve this problem, this paper proposes a **time shifted sub-connection architecture**[5] as shown in Fig.1. 23 Japanese phoneme recognition experiments are performed to evaluate this segment-based adaptation technique.

In the following section, the architecture of the speaker adaptation neural network, the training procedure for the speaker adaptation neural network and a speech recognition system for evaluation are described. Next, the evaluation of these techniques are described.

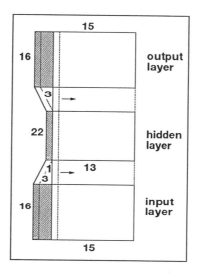

Figure 1: Speaker Adaptation Neural Network Architecture

## 2 SEGMENT-BASED SPEAKER ADAPTATION

### 2.1 Architecture of The Speaker Adaptation Neural Network

A three-layer feed-forward neural network is used for segment mapping. The network has 27,982 connections and its architecture is shown in Fig.1. The input layer, hidden layer and output layer have 15 × 16, 13 × 22 and 15 × 16 units, respectively. The network has a **time shifted sub-connection** architecture: 3-frame × 1-frame × 3-frame type sub-connection shifted 1 frame to maintain segment temporal structure. This architecture reduces the number of weighting parameters. As a result, the network can decrease the mount of speech data for training. Since the power pattern in each segment is normalized over 15 frames, this architecture is different from a simple 3-frame segment mapping architecture. A 4-layer neural network generally provides higher performance than a 3-layer network. Since in a preliminary experiment 3-layer architecture gives almost the same performance as 4-layer architecture in phoneme recognition, in this paper 3-layer architecture is adopted.

## 2.2 Training Procedure for The Speaker Adaptation Neural Network

To train the speaker adaptation neural network, pairs of input and teaching patterns are required. The pairs are generated from the same word utterances of a standard speaker and an input speaker using a DTW(dynamic time warping) technique. Series of segments are generated from the beginning to the end of training words shifted 1 frame. Let $(x_i)$ and $(y_j)$ be series of frames extracted from the training word for the input and the standard speakers. The start and end points of the word are determined by hand labels. $(\hat{x}_i)$ and $(\hat{y}_i)$ mean normalized frames over $K$ frames. The segment $(a_i)$ consists of $K$ frames of $(\hat{x}_i)$ and the segment $(b_j)$ consists of $K$ frames of $(\hat{y}_i)$. A series of segments $(A)$ is generated from the input speaker and a series of segments $(B)$ is generated from the standard speaker. Time alignment between two series of segments for the standard and the input speakers is performed using the DTW technique shown in Fig.2. The segments $(b_{j(i)})$ from the standard speaker are used as the teaching patterns for the corresponding segments $(a_i)$ from the input speaker. The segment adaptation neural network is trained using the BP(Back Propagation)[6] algorithm. The adaptation network training procedure is as follows:

───── Training Procedure (Initialization) ─────

0. $a_i = \hat{x}_i \, \hat{x}_{i+1} \ldots \hat{y}_{i+K-1}, \quad b_i = \hat{y}_j \, \hat{y}_{j+1} \ldots \hat{x}_{j+K-1}. \quad (K=15)$
 $a'_i = a_i.$
$\begin{cases} A' = a'_1 \, a'_2 \ldots a'_i \ldots a'_I, & (1 \leq i \leq I) \\ B = b_1 \, b_2 \ldots b_j \ldots b_J. & (1 \leq j \leq J) \end{cases}$

$L = 1$, $L_e$ : final iteration number, $f_1$ : neural network mapping function with random weighting parameters

## Training Procedure

1. Time alignment between two series, $A'$ and $B$

$$g(i,j) = \min \begin{bmatrix} g(i-2, j-1) + 2d(i-1, j) + d(i,j) \\ g(i-1, j-1) + 2d(i,j) \\ g(i-1, j-2) + 2d(i, j-1) + d(i,j) \end{bmatrix}$$

Alignment window constraint

$\frac{2J}{I}i - J \leq j \leq \frac{2J}{I}i$ & $\frac{J}{2I}i \leq j \leq \frac{J}{2I}i + \frac{J}{2}$.

( $g$ : cost function, $d$ : Euclidean distance measure )

2. Generating training pairs $(a_i, b_{j(i)})$.

3. Training the network

$f_{L+1}(a_i) \longrightarrow b_{j(i)}$.  ( $f_L$ : L-th neural network mapping function )

4. $L = L + 1$, if $L = L_e$ then stop, else $f_L(a_i) = a'_i$, goto 1

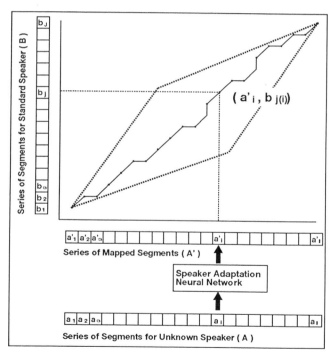

Figure 2: Time Alignment Between Two Series of Segments

## 2.3 Speech Recognition System for Evaluation

Fig.3 shows the speech recognition system for the adaptation neural network evaluation[7]. In this figure the top part is the TDNN(Time Delay Neural Network) aimed at recognizing 23 phonemes[8][9] and the bottom part is the proposed speaker adaptation neural network. The TDNN is trained using 7561 phoneme segment data extracted from 2,620 words uttered by the standard speaker using hand labels.

To recognize an input speaker's utterance, a segment of the input speaker is fed into the input-layer of the adaptation network and mapped to the standard speaker's segment. Then the mapped segment is recognized by the TDNN. Consequently, the input speaker's segment is recognized by the adaptation neural network and the TDNN speech recognition system.

## 3 EVALUATION OF SPEAKER ADAPTATION NEURAL NETWORK

Phoneme recognition experiments for the speaker adaptation neural network were conducted using two male standard speakers and two unknown speakers (one male and one female). The segment consists of 15-frame mel-scaled 16-channel FFT outputs. The speech analysis conditions are shown in Table 1. The recognition system is the TDNN shown in Fig.3. Training the adaptation network was performed using 100 words selected from the 2,620 words which are also used to train the TDNN phoneme recognition system. In these experiments, the final iteration number $L_e$ is set as one. The test data consisted of phoneme segment data extracted from another 2,620 words from the input speakers using hand labels.

Table 1: SPEECH ANALYSIS CONDITIONS

| | |
|---|---|
| Sampling Frequency | 12 kHz |
| Window Function | Hamming |
| Window Length | 21.3ms |
| Analysis Interval | 10ms |
| Analysis Method | 256 point FFT |
| 16 mel-scale Filter-bank Design (Hz) | 0-141; 141-281; 281-462; 462-656; 656-844; 844-1031; 1031-1219; 1219-1406; 1406-1641; 1641-1922; 1922-2250; 2250-2672; 2672-3187; 3187-3797; 3797-4547; 4547-5437 |

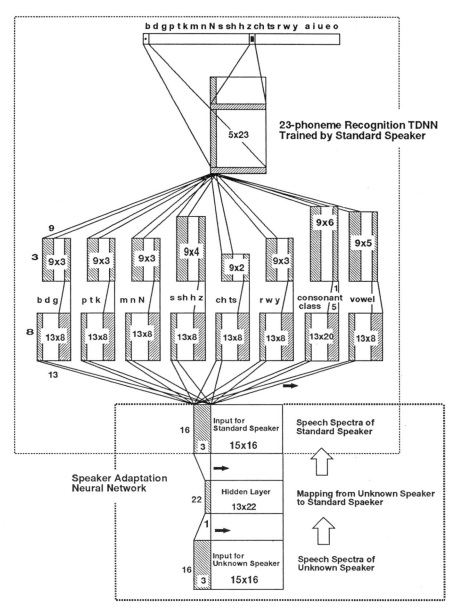

Figure 3: The Architecture of the TDNN 23 Phoneme Recognition System and Adaptation Neural Network

The phoneme recognition performance is shown in Table 2. The averaged recognition rate with speaker adaptation is 83.7%. On the other hand, the recognition rate without adaptation is only 60.9%. Segment-based speaker adaptation thus has a 22.8% higher phoneme recognition rate for input speakers. These results show the effectiveness of the segment-based adaptation neural network.

Table 3 shows the relationship between recognition performance and the number of training words. In this experiment, the adaptation networks were trained using 25, 50, 75 or 100 words. This result shows that the segment-based adaptation works effectively even if 25 words are used for training, and the network constantly works well when it uses more than 50 words for training. This shows that, owing to the time shifted sub-connection architecture, the network does not need a large amount of data for segment mapping training.

Table 2: 23 PHONEME RECOGNITION PERFORMANCE USING NEURAL NETWORK SEGMENT-BASED SPEAKER ADAPTATION

| Unknown ⟶ Standard Speaker    Speaker | Phoneme Recognition Rate ( % ) | | |
|---|---|---|---|
| | Without Adaptation | Adaptation | Speaker Dependent |
| Male2 ⟶ Male1 | 73.9 | 86.2 | 95.6 |
| Female1 ⟶ Male1 | 38.7 | 81.0 | 95.6 |
| Male1 ⟶ Male2 | 78.5 | 83.2 | 96.7 |
| Female1 ⟶ Male2 | 52.3 | 84.3 | 96.7 |
| Average | 60.9 | 83.7 | 96.2 |

Table 3: RELATIONSHIP BETWEEN 23 PHONEME RECOGNITION PERFORMANCE AND NUMBER OF TRAINING WORDS

| Number of Training Words | Number of Phonemes | Recognition Rate ( % ) |
|---|---|---|
| 0 | 0 | 73.9 |
| 25 | 142 | 81.8 |
| 50 | 292 | 85.4 |
| 75 | 441 | 85.8 |
| 100 | 611 | 86.2 |

For comparison, neural network frame-based speaker adaptation was evaluated with one standard speaker and one input speaker. Each segment is mapped frame by frame. The result for 23 phoneme recognition is shown in Table 4. Segment-based adaptation has a 9.1% higher recognition rate than the rate by frame-based adaptation. In another comparison, 7-frame segment-based speaker adaptation by neural network was evaluated with one standard speaker and two input speakers. The task of the experiment is 5 Japanese vowels. The recognition system for the evaluation is a TDNN that has a 7-frame segment input layer. The result for 5 vowel recognition is shown in Table 5. The result shows that the 15-frame segment-based adaptation performance is slightly better than 7-frame segment-based adaptation performance.

Table 4: COMPARISON OF 23 PHONEME RECOGNITION PERFORMANCE BY FRAME-BASED AND SEGMENT-BASED SPEAKER ADAPTATION

| Speaker | Method | Recognition Rate ( % ) |
|---|---|---|
| Male2 → Male1 | Without Adaptation | 73.9 |
| | Frame-based* | 77.1 |
| | Segment-based* | 86.2 |

*number of training words: 100

Table 5: COMPARISON OF 5 VOWEL RECOGNITION PERFORMANCE BY 7-FRAME SEGMENT AND 15-FRAME SEGMENT SPEAKER ADAPTATION

| Speaker | | Phoneme Recognition Rate ( % ) | |
|---|---|---|---|
| | | 7-frame segment | 15-frame segment |
| Male2 → Male1 | Without Adaptation | 93.6 | 91.3 |
| | Adaptation* | 98.0 | 98.1 |
| Female1 → Male1 | Without Adaptation | 67.3 | 62.1 |
| | Adaptation* | 94.8 | 96.4 |
| Speaker Dependent | | 98.5 | 98.7 |

*number of training words: 100

The result for 23 phoneme recognition by VQ-based codebook mapping speaker adaptation in Discrete HMM is shown in Table 6[10]. The recognition rate for the segment-based speaker adaptation neural network is 12.1% higher than the recognition rate for VQ-based codebook mapping. These results cannot be directly compared as they were evaluated by different recognition systems, different training words and different speech feature parameters. The experimental conditions of designing VQ codebook and HMM are shown in Table 7. However, these results suggest the superiority of the segment-based approach.

Table 6: COMPARISON OF 23 PHONEME RECOGNITION PERFORMANCE BY VQ-BASED MAPPING AND NEURAL NETWORK MAPPING

| Unknown → Standard Speaker Speaker | Phoneme Recognition Rate ( % ) | | |
|---|---|---|---|
| | VQ-based Mapping | Neural Network Mapping | |
| | | Frame-based | 15-frame Segment-based |
| Male1 → Male2 | 72.4 | 77.1 | 83.2 |
| Female1 → Male2 | 71.0 | - | 84.3 |
| Average | 71.7 | 77.1 | 83.8 |

Table 7: THE EXPERIMENTAL CONDITIONS OF DESIGNING VQ CODEBOOK AND HMM

| VQ Codebook | Codebooks & Their Sizes | WLR(order 16)[256], normalized power[64], $\Delta$Cep[256] |
|---|---|---|
| | Codebook Generation | Using 100 words of the 216 phonetically balanced Japanese words set |
| | Adaptation | Using 100 words of the 216 phonetically balanced Japanese words set |
| | Fuzzy VQ | fuzziness : $m = 1.6$ number of $k$-nearest neighbors : $k = 6$ |
| HMM | Model (31 models) | 2 states (including the transition), with a loop in the first state for the syllabic nasal and vowels |
| | | 4 states (including the transition), with a loop in the first three states for consonants other than the syllabic nasal |

## 4 CONCLUSION

This paper proposed a segment-based speaker adaptation technique using neural network and reported the evaluation of this technique. It has been shown that this speaker adaptation technique is more effective than frame-based speaker adaptation. We believe that this is because of the ability of the segment-based approach to compensate for the spectral and temporal structural differences occurring between two different speakers. Using the **time shifted sub-connection architecture**, the adaptation network does not require a large amount of speech data for the segment mapping training. This segment-based speaker adaptation technique improves phoneme recognition performance for an input speaker even with less than 100 training words.

These results show the considerable ability of a neural network in segment to segment analog mapping.

This speaker adaptation technique can easily be applied to a segment-based word recognition system or a phrase recognition system. Evaluation by word recognition and phrase recognition task is set for future study.

## ACKNOWLEDGMENTS

The authors would like to thank Dr. A. Kurematsu and Mr. S. Sagayama for their support of this research, Mr. H. Hattori for his useful comments and suggestions and all the members of Speech Processing Department for their discussions and encouragement.

# References

[1] S. Nakamura, et al., Phoneme Recognition Evaluation of HMM Speaker Adaptation Based on Vector Quantization, *Report of Speech Committee*, SP88-106, pp.1-8 (Dec. 1988).

[2] K. Iso, et al., Speaker Adaptation Using Neural Network, *Proc. of Acoust. Soc. of Jpn.*, Spring Meeting, 1-6-16 (Mar. 1989) (in Japanese).

[3] S. Roucos, A. Wilgus, Speaker Normalization Algorithms for Very-low-rate Speech Coding, *Proc. of ICASSP*, 1.1.1-1.1.4 (Apr. 1984).

[4] Y. Shiraki, M. Honda, Speaker Adaptation Algorithms for Segment Vocoder, *Proc. of Acoust. Soc. of Jpn.*, Fall Meeting, 3-6-9 (Oct. 1987) (in Japanese).

[5] K. Fukuzawa, et al., Speaker Adaptation Using Identity Mapping by Neural Networks, *Proc. of Acoust. Soc. of Jpn.*, Fall Meeting, 1-8-16 (Sep. 1990) (in Japanese).

[6] D. E. Rumelhart, et al., Parallel Distributed Processing - Explorations in the Microstructure of Cognition Volume 1: Foundations, Cambridge, MA, MIT Press (1986)

[7] H. Sawai, et al., On Connectionist Approaches to Speaker-independent Recognition,*Proc. of Acoust. Soc. of Jpn.*, Spring Meeting, 1-5-17 (Mar. 1991) (in Japanese).

[8] H. Sawai, et al., Phoneme Recognition by Scaling up Modular Time-Delay Neural Network,*Report of Speech Committee*, SP88-105, pp.73-80 (Dec. 1988).

[9] A. Waibel, et al., Phoneme Recognition Using Time-Delay Neural Networks, *IEEE Transactions on ASSP*, Vol 37, No.3, pp.328-339 (Mar. 1989).

[10] H. Hattori, et al., Supplementation of HMM for Articulatory Variation in Speaker Adaptation, *Proc. of ICASSP*, S3.6, pp.153-156 (Apr. 1990).

# A SIMPLE WORD-RECOGNITION NETWORK WITH THE ABILITY TO CHOOSE ITS OWN DECISION CRITERIA

Kyrill A. Fischer, Hans Werner Strube
Drittes Physikalisches Institut der Universität Göttingen
Bürgerstrasse 42–44, W-3400 Göttingen, Germany
Tel.: +49 551/397731, Fax: +49 551/397720
e-mail: fischer@up3spr1.gwdg.de

Abstract. In the last decades various reliable algorithms for the word classification problem have been developed. All these models are necessarily based on the classification of certain "features" that have to be extracted from the presented word. The general problem in speech recognition is: what kind of features are both *word dependent* as well as *speaker independent* ? The majority of the existing systems requires a feature selection by the designer, so the system cannot choose the features that best fit the above mentioned criterion. We therefore tried to build a neural network that is able to rank all the features (here: the cells of the input layer) according to their functional relevance. This method reduces both the necessity to preselect the features as well as the numerical effort by a stepwise removal of the cells that proved to be unimportant.

## INTRODUCTION

In this paper we will present an artificial neural network model that is able to classify spoken words. The main idea of our approach is the network's ability to distinguish between more and less relevant features. The term "feature" denotes a special kind of stimulus to be extracted from the presented word.

First we will introduce the architecture of the artificial neural network in the next chapter. Then we will present our method to measure the *relevance* of a given feature. Results obtained using these method will be given in a new chapter, followed by a discussion with concluding remarks.

## DESIGN OF THE WORD CLASSIFYING NEURAL NETWORK

### Preprocessing of Speech Data

Our database consisted of 40 words from 10 different speakers (male and female); each word was spoken in 10 slightly different versions. Each utterance is transformed into a 19-channel Bark-spectrogram $x(t,z) \in [0,1]$, where $t = 0 \ldots T$ (word length), $z = 1 \ldots 19$ Bark). We used Hamming windows; window shift interval $= 19.6$ ms, window length $= 25.6$ ms. The final channel-values $x(t,z)$ are calculated from the original power values $p(t,z)$ via $x(t,z) = (p(t,z))^{0.5}$.

## Architecture and Feature Extraction Method

Our network was intended to be as simple as possible: it requires a feature-extracting part that uses the presented utterance to construct a feature vector. This feature vector is then used as an input pattern for the classifying network, consisting of a simple linear perceptron (Fig. 1).

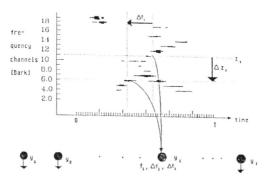

Figure 1:
Architecture and feature-extraction method of the used artificial neural network. A spectrogram of the German word "zwei" /tsvai/ is analyzed using all feature cells. Their activities are calculated according to (1). The activity-pattern of the feature cells is then classified by a linear perceptron.

The number $I$ of input units of the perceptron is given by the number $J$ of features; the number of output cells depends on the number of words to be recognized. The properties of each single feature-cell $i$, $(i = 1 \ldots I)$ are defined by a triple of values $(\Delta t_i, \Delta z_i, z_i) : \Delta t_i = 0 \ldots 200$ ms; $z_i - \Delta z_i, z_i = 1 \ldots 19$ Bark. The activity $y_i$ of a feature-cell $i$ is calculated according to

$$y_{\Delta t_i, \Delta z_i, z_i} = \sum_{t=\Delta t_i}^{T} x(t, z_i) \, x(t - \Delta t_i, z_i - \Delta z_i) \tag{1}$$

using the whole utterance and a constant $(\Delta t_i, \Delta z_i)$-displacement (see Fig. 1). The triple of parameters $(\Delta t_i, \Delta z_i, z_i)$ defines the kind of stimulus (i.e. feature) cell $i$ is sensitive for. The number of feature cells is limited only by the computational power available. The parameters may be chosen arbitrarily,

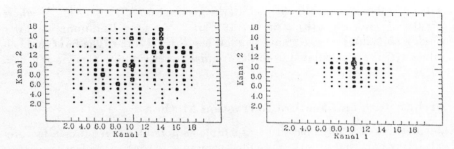

Figure 2:
Positions of highly relevant cells in the $(z, z - \Delta z)$-domain. Relevances indicated by dot-diameters; $\Delta t = 58$ms. Each pair corresponds to a pair of input channels. Details see text.

but repetitions of triples are redundant and therefore should be avoided. In the next chapter it will be shown how irrelevant features (i. e. feature cells) can be detected automatically.

## Classification Method

One of the most important statistical classification methods is the classical discriminant analysis. In a simplified form (see [3]), a pattern $\vec{y}$ is classified into class $j_0$, if and only if $S_{j_0}(\vec{y}) > S_j(\vec{y}) \quad \forall j \neq j_0$, where

$$S_j(\vec{y}) := \vec{\mu}_j^T V^{-1} \left( \vec{y} - \frac{1}{2}\vec{\mu}_j \right) \qquad (2)$$

($\vec{\mu}_j$: mean training vector of word $j$; $V$: covariance matrix of all patterns). Now we will use a *simplification* that will be used throughout this paper: We suppose that

$$V \simeq \text{diag}\left(\sigma_{11}^2, \ldots \sigma_{II}^2\right), \qquad (3)$$

i. e. $V$ can be treated as being diagonal[1]. Then (2) simplifies to

$$\begin{aligned} S_j(\vec{y}) &= c_j + \sum_{i=1}^{I} (\vec{y})_i w_{ij}, \qquad (4)\\ \text{where} \quad w_{ij} &= \frac{(\vec{\mu}_j)_i}{\sigma_{ii}^2}, \\ c_j &= -\frac{1}{2} \sum_i (\vec{\mu}_j)_i w_{ij}. \end{aligned}$$

---

[1] Eqn. (3) holds exactly for features that are statistical independent, i. e. features that are eigenvectors of $V$. This may easily be achieved by a neural principal component analysis (c. f. [4], [5]), but we do not investigate this case here.

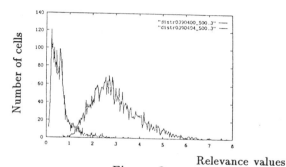

Figure 3: Distribution of the relevance values under different conditions: Left curve: 5 speakers; right curve: 1 speaker. The severe speaker-dependence of the used features can be recognized from the fact that the mean $R$-values drop down with the number of speakers.

Eqn. (4) is a linear expression that can be evaluated by a common linear perceptron. Output cell $j_0$ having maximum activity denotes the recognized word. Essentially, (4) is equivalent to a search for the nearest word representing vector using a variance-weighted Euclidean distance.

# HOW TO MEASURE THE RELEVANCE OF A CELL

Now it has to be decided whether a given feature cell is more or less relevant for the word-classification problem. In order to be able to define a method for determining the relevance value of a cell, we first have to define the meaning of the word 'relevance' in our case. Several definitions and algorithms have been proposed. Some of the proposed definitions are based on a measure of the increment in the averaged output error [1], [6]. Other authors investigated the effects of a pure *random* dilution of the net [8], [9].

In this section we will present a relevance definition that is based on statistical properties of the training ensemble. Then we will show how this relevance definition can be derived from a general statistical approach, using the same consideration (3) that led to (4).

## Definition of the Relevance Function

We follow a very simple but effective idea: For a cell to be relevant, its activity should be as *word-dependent* and *speaker-independent* as possible. Or, more exact: different words should produce different activities ("effectiveness") and same words similar activities ("reliability"). Therefore we tried to measure two basic properties for each feature cell: word-dependence and speaker-independence.

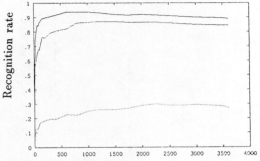

Figure 4: Number of feature cells left

Recognition rate vs. no. of cells used. (40 words, 5 versions, 1 speaker; 1st graph (top): same versions for test and training; 2nd graph: same speaker, new versions; 3rd graph (bottom): unknown speaker)

We defined a relevance function $R_i$ of the form

$$R_i := \frac{\text{Variance of } y_i\text{-activities between different words}}{\text{mean variance between } y_i\text{-training-activities (same word)}} \qquad (5)$$

Each feature cell $i$ gets a relevance $R_i$ according to its behavior during the training. The numerator of (5) describes the average word-dependence of the cell's activity, and the denominator measures the speaker-dependence. Cells showing a high $R$-value are therefore both reliable (denominator $\hat{=}$ speaker-dependence is low) as well as efficient (numerator $\hat{=}$ word-dependence is high).

**Derivation of the Relevance Function**

**The general case.** Due to the fact that our classification is based on a statistical method, we used a corresponding approach to define the relevance function. The *generalized statistical group distance* [2], [3],

$$T^2 = \text{tr}\left(V_W V^{-1}\right) \qquad (6)$$

measures the separability of an arbitrary number of groups ($V_W, V$ being covariance matrices of different-word means and single-word-samples (averaged over all words), respectively). For $J = 2$ groups, (6) reduces to the well known Mahalanobis distance. Removing a cell of the input layer reduces the dimension of the patterns to be classified, and causes a change of $T^2$ too:

$$\begin{aligned} \Delta T_i^2 &:= T_{with\ cell\ i}^2 - T_{without\ cell\ i}^2 \qquad (7) \\ &= \frac{1}{t_{ii}} \sum_{j=1}^{J} b_{ij}^2, \end{aligned}$$

where $t_{ik} := (V^{-1})_{ik}$

$$b_{ij} := \sum_{k=1}^{I} t_{ik} (\vec{\mu}_j - \vec{\mu})_k .$$

Removal of a single cell $i$ requires a subsequent removal of the $i$th row and column of $V$. The new $\Delta T^2$-values have to be calculated using the inverse of this "reduced" matrix.

**Locality from special considerations.** Applying assumption (3) (we already used it for the derivation of the classification method), we get

$$\Delta T_i^2 = \frac{\sum_{j=1}^{J} (\vec{\mu}_j - \vec{\mu})^2}{\sigma_{ii}^2},$$

which is equivalent to (5). Hence, removing a cell with a relevance value $R_i$ simply means to decrease the statistical group-distance by this value. Therefore the definition (5) as a "relevance" seems to be useful. In addition, the above mentioned recalculation of $V^{-1}$ is not necessary under this condition; $R$ becomes a *local* function.

# SIMULATIONS AND RESULTS

## Questions

Our experiments focused on the following three questions:

1. What are the recognition rates of the network under different conditions?

2. How does removal of the low-relevance cells affect the performance?

3. What makes a cell a highly relevant cell?

Since the goal of this paper was not the development of a perfect speech recognition system, questions 2 and 3 have been of more general interest to the authors. (A similar network that maximizes recognition is described in [7] in this issue.)

## Results

**1. Recognition rates.** Our database consisted of 4000 utterances (40 different German words × 10 speakers × 10 versions). The network has relatively poor generalization properties: recognition rates obtained in single-speaker tests are much better than multiple-speaker tests. This is due to the fact, that the features are formed from correlations between *points* of the spectrogram rather than small areas. Tab. 1 gives a good impression of the network's abilities. The model used consisted of $I = 3610$ feature cells that used all possible channel combinations ($z, z - \Delta z \in [1, 19]$) and 10 different time-delays ($\Delta t = 19.6$ms ... $196.0$ms).

Table 1: SAMPLE RECOGNITION RATES

|          | one speaker | 5 speakers |
|----------|-------------|------------|
| 10 words | 96,0%       | 74,8%      |
| 40 words | 92,2%       | 62,5%      |

**2. Removing cells with lowest relevance.** Cell(=feature) removal leads to a characteristic effect: Over a wide range of remaining-cell-percentages, the recognition rates of the "diluted" nets stay nearly constant, until they suddenly drop down. At this moment there are often only 10% to 20% of the initial number of cells left. This behavior shows that the described architecture provides a very high redundance – a typical property of biological networks, too (see Fig. 4).

**3. Parameters of the most relevant feature cells.** After having calculated the relevances, we are able to ask for the parameters that led to relevant cells (see Fig. 2). Distributions of relevance values are shown in Fig. 3. For a single-speaker training, the network adapts very well to this speaker (Fig. 2, left). Therefore, the most relevant features tend to reflect the speakers frequency-characteristics together with the reliability of these features. For a multiple-speaker training (Fig. 2, right) the relevance values decrease; consequently, so does the performance of the net (cf. Table 1). The most relevant cells now tend to reflect the overall speaker-characteristics.

## DISCUSSION AND CONCLUDING REMARKS

The network described is able to recognize a small vocabulary using a simplified statistical classification method. The most important property is its ability to adapt to a unknown task by selecting the best (*a priori* unknown) decision criteria (here: features) from a set of possible ones. A cell's relevance is defined using the statistical group distance rather than by means of the expected change of output error. Removing the unimportant criteria lowers numerical effort and still preserves, or even improves, the network's abilities.

## REFERENCES

[1] M. C. Mozer and P. Smolensky, "Using relevance to reduce network size

automatically", <u>Connection Science</u> vol. 1, Number 1, 1989

[2] J. Hartung and B. Elpelt, <u>Multivariate Statistik</u>, Munich, Vienna: Oldenbourg Verlag, 1982, ch. 4.

[3] A. M. Kshirsagar, <u>Multivariate Statistics</u>, New York, 1972

[4] J. Rubner and P. Tavan, "A self-organizing network for principal-component analysis", <u>Europhysics Letters</u>, vol. 10, pp. 693-698, December 1989.

[5] E. Oja, "A simplified neuron model as a principal component analyzer", J. Math. Biology, vol. 15, pp. 267-273, 1982.

[6] H. Behme and T. Gramss, "Two new methods for word recognition with neural nets", in: J.-P. Ewert and H. Werner (Eds.), <u>Models of brain functions & artificial neuronal nets</u>, Kassel: GhK-University-Edition-Kassel, 1991, pp. 213-246.

[7] T. Gramss, "Word recognition with the Feature Finding Neural Net (FFNN)",
<u>Neural Networks for Signal Processing –Proceedings of the 1991 IEEE-SP Workshop</u>, Princeton, New Jersy, USA, September 29 - October 2, 1991.

[8] R. Kree and A. Zippelius, "Continuous-time dynamics of asymmetrically diluted neural nets", <u>Physical Review A</u>, vol 36, number 9, pp. 4421-4427, November 1987.

[9] A. Rau and D. Sherrington, "Retrieval enhancement due to external stimuli in diluted neural networks", <u>Europhysics letters</u>, vol. 11, number 6, pp. 499-504, March 1990.

# Supervised and Unsupervised Feature Extraction From a Cochlear Model for Speech Recognition

Nathan Intrator*
Center for Neural Science
Box F, Brown University
Providence, RI 02912
nin@brownvm.brown.edu

Gary Tajchman[†]
Cognitive and Ling. Sciences
Box 1978, Brown University
Providence, RI 02912
tajchman@browncog.bitnet

**Abstract**—We explore the application of a novel classification method that combines supervised and unsupervised training, and compare its performance to various more classical methods. We first construct a detailed high dimensional representation of the speech signal using Lyon's cochlear model and then optimally reduce its dimensionality. The resulting low dimensional projection retains the information needed for robust speech recognition.

## INTRODUCTION – SPEECH PREPROCESSING METHODS

Many speech recognition systems, in particular, those based on HMMs, use LPC derived cepstral coefficients as the first step in preprocessing the speech data. These cepstra are then typically passed through vector quantization (VQ), or used directly as input to the HMM. The VQ step discretizes the multidimensional input vectors into a small set of possible inputs. This helps simplify training the system, but also introduces varying degrees of distortion [11]. This limitation is partially overcome by using methods to estimate output parameters for the continuous space defined by the cepstra. These techniques also run into problems when the dimensionality of the input vector gets large. In spite of these potential problems, LPC-based systems have performed well, especially when augmented with energy and time-differenced cepstra [11].

Speech recognition systems using ANNs have employed a much more heterogenous set of preprocessing techniques. Everything from raw speech to LPC-based cepstra has been tried [12]. However, most have used some form

---

*Research supported by NSF, the Army Research Office, and ONR.
[†]Research supported by NSF grant DIR-89-07769.

of preprocessing inspired by the representation produced by the mammalian peripheral auditory system. Examples include Mel scale and bark scale spectra. Other more sophisticated techniques exist that produce more detailed representations.

While there is a tendency for preprocessing based on auditory system constraints to be used with ANNs and preprocessing based on vocal tract constraints to be used with HMMs, this is not always the case. For instance, some current HMM systems include a Mel scale transformation when computing cepstra, and as mentioned above, LPC-based cepstra have been used with ANNs. The differences in preprocessing for HMMs and ANNs can be largely attributed to the fact that ANNs are good at integrating over large dimensional representations, while HMMs do best with much smaller dimensional input.

In this paper we focus on ANN techniques for processing the detailed, high dimensional auditory system representation of speech produced by Lyon's cochlear model [13]. We explore the application of a novel classification method that combines supervised and unsupervised training, and compare its performance to various methods. Our task is feature extraction and classification of voiceless stops extracted from the TIMIT corpus.

**What are features of recognition for speech data**

When moving to a much larger representation of the speech data, many existing techniques such as classifiers, or vector quantizers fail to work, mainly because of the curse of dimensionality [1]. This problem is related to the sparsity of high dimensional spaces, and implies that the amount of training data has to grow exponentially with the dimensionality.

In many cases, it is conceivable to assume that the important information for speech recognition lies in a much smaller dimensional space, and the question becomes, how to find this low dimensional structure, or how to extract the relevant features from the data. This question can be put in a much broader statistical formulation, in which one has a data set that lies in high dimensional space, with a lower dimensional structure and tries to reduce the dimensionality of the data, without losing the important structure. These problems may be addressed using a recent statistical tool called Exploratory Projection Pursuit [3] which has an effective implementation with a biologically motivated neural network [6].

## LYON'S MODEL OF COCHLEAR PROCESSING

We chose to use a fairly sophisticated auditory model to preprocess the speech data for our neural network. One reason for doing this was to assess the feasibility of using such a model as front end for a recognizer. Auditory models typically produce very large output representations in order to retain much of the detail the higher centers in the brain receive from the cochlea.

The auditory model we used to preprocess the speech data was Lyon's cochlear model [13] as implemented by Slaney [18]. For each time slice, 84 channels of output were produced (for data sampled at 16kHz). Time slices were separated by 2 msecs. Therefore, for 56 msec of speech, the model produced 2352 bytes of data. While this is still orders of magnitude smaller than what is transmitted through the auditory nerve to higher centers of the brain, it is much larger than the data representations typically used for speech recognition tasks.

The channels in the model correspond to nerve fibers evenly spaced along the basilar membrane in the cochlea. The center frequencies of the set of channels are logarithmically spaced, giving the lower frequencies a more dense representation than the higher frequencies. Neighboring channels overlap to a large degree. This models the highly redundant representation used by the mammalian auditory nerve. The band pass regions of the channels increase linearly with frequency.

Each channel is implemented as a second order digital filter. The entire filter bank is implemented with a cascade design giving the representation realistic amplitude and group-delay response in addition to making the computation efficient. To model the effects of the inner and outer ear, the signal is passed through a pre-emphasis stage and then processed by the cascade of second order filters. The final stage of processing is preceded by half-wave rectification to model the unidirectional transduction of the basilar membrane movement by the inner hair cells.

The final phase of the cochlear model passes the output of each channel through a series of adaptive gain control (AGC) elements. These AGC elements attempt to keep the output levels of each filter within specific range. Each AGC is coupled with its nearest neighbors to each side. This helps model the masking effects found in real cochlear processing. The resulting rectangular frequency by time representation forms an image of auditory nerve activity and is called a cochleagram.

In sum, much of the detail and character of the representation used by the auditory nerve is retained in the cochleagram representation. The task then becomes how to best use all of this information.

# FEATURE EXTRACTION IN HIGH DIMENSIONAL SPACE – THE BCM MODEL

From a mathematical view point, extracting features from the rectangular representation of the cochleagram is related to dimensionality reduction in high dimensional vector space, in which an $n \times k$ pixel image is considered to be a vector of length $n \times k$. In such high dimensional spaces the *curse of dimensionality* [1] says that it is impossible to base the recognition on the high dimensional vectors, because the number of training patterns needed for training a classifier should increase in an exponential order with the dimensionality, and therefore dimensionality reduction should take place before attempting the classification. Due to the large number of parameters involved, a feature extraction method that uses the class labels of the data, will be biased to the training data [5], which translates to having features with poor generalization or invariance properties. Thus, the feature extraction should be unsupervised. A recent statistical method to address this problem of dimensionality reduction called exploratory projection pursuit (EPP) [3] assumes that features can be constructed from projections of the input space onto a small dimensional space. This method defines interesting features as those projections whose single dimensional projected distribution is far from Gaussian. Since high dimensional clusters translate to low dimensional multi-modal projected distributions, a plausible measure of deviation from normality can be based on a measure of multi-modality of the projected distribution. Intrator [6] has recently shown that a variation of the Bienenstock Cooper and Munro neuron [2] performs exploratory projection pursuit using a projection index that measures multi-modality. A network implementation which can find several projections in parallel is still computationally efficient and therefore may be applicable for extracting features from very high dimensional vector spaces of the type generated by the cochlear model.

The unsupervised feature extraction/classification method is presented in Figure 1. Similar approaches using the RCE and back-propagation network have been carried out by [15], and using the unsupervised charge clustering network by Scofield [17]. Huang and Lippmann [4] described a feature-map classifier for vowel recognition, in which internal nodes compute kernel functions related to the Euclidean distance between the input and cluster centers represented by these nodes. The unsupervised vector quantizer was trained to form the new representation which trained the supervised classifier. Kohonen et al. [10] used a similar approach with LVQ network. Review on various other unsupervised/supervised approaches appears in [12].

Although unsupervised feature extraction has the potential of being less biased to the training data, its result may be suboptimal since it ignores the information contained in the class labels. It is possible for example, that not all the information required for the classification is contained in those directions which are considered interesting by the feature extractor (some trivial examples are discussed in [8]). Therefore, it is possible that a hybrid of unsupervised/supervised feature extractor may yield better performance.

Another way to look at the problem is from the classification side; The performance of the classifier that reduces dimensionality based solely on the class labels, may be improved if an additional measure of the information carried in the projections is added. In the case of a back-propagation classification network, a local penalty term may be added to the energy functional minimized by error back propagation. This penalty which is added only to the hidden layer units, is the projection index defined by the BCM network [6, 9]. Therefore, the modification equations for the hidden layer units are affected by the delta rule [16] and by the BCM modification equations. This method is described in detail in [7].

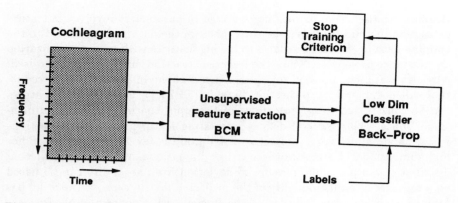

Figure 1: Low dimensional classifier is trained on features extracted from the high dimensional data. Training of the feature extraction network stops when the misclassification rate drops below a predetermined threshold on either the same training data (cross validatory test) or on different testing data.

## METHODS

### Data - Voiceless Stops from TIMIT

In this work we focused on feature extraction and classification of the voiceless stop consonants [p, t, k]. The source of our data was the DARPA TIMIT Acoustic-Phonetic Continuous Speech Corpus (TIMIT). This database contains utterances from many talkers, with coverage of all the major dialect regions in the United States.

All tokens used in these experiments consisted of a stop followed by a vowel. We used only four vowel contexts [aa, ao, er, iy] in the training set. These vowels give a reasonable, but not complete coverage of the vowel space. This restricted set allowed us to test how well the feature extraction generalized to new vowel contexts.

These tokens were drawn from the utterances of 268 different talkers. Multiple talkers and various sentential contexts contribute to a fair degree of variability between tokens of the same CV type. The segment boundaries we used were exactly those provided with TIMIT. We made no attempt to sharpen or correct any misalignments that might exist in the data.

For each CV type, an average over the 25 tokens used for training is presented in the cochleagram matrix shown in Figure 2. The vertical axis is frequency, low to high from top to bottom, and the horizontal axis is time for each cochleagram. Looking at the lower left corner of the images, it can be seen that [p]s have low energy at the high frequencies, [t]s have a sharp burst in the high frequencies, and [k]s have diffuse energy in the high frequencies. These features tend to distinguish between the three voiceless stops for the cochleagram representation.

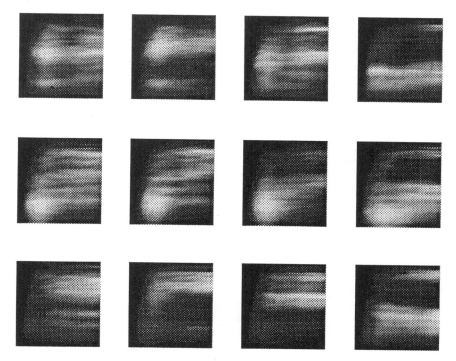

Figure 2: The output of Lyon's cochlear model for the 12 CV pairs. From top to bottom [k, t, p], and from left to right [aa, ao, er, iy]. Each image is the average of 25 tokens from each CV type showing 75msec of speech aligned to burst release. White areas represent high energy.

## Training

In the first experiment features were extracted from the large representation of the speech segment using a BCM network. Here the BCM weights were only affected by the unsupervised modification rule. Classification was accomplished by training a small back-propagation network with the output of the BCM network as shown in Figure 1. An important issue of avoiding over fitting (in either of the nets) was addressed by testing (during training) on a third set of tokens (Pseudo test set).

In the second experiment the modification of the hidden units of a 3 layer back-propagation network, was a combination of the BCM synaptic modification equations, and the error propagated from the top layer. The performance of the networks in the first and second experiments were compared to the performance of a simple back-propagation network.

| Training Method | 4 Vowels Training | 4 Vowels Testing | 7 Vowels Testing |
|---|---|---|---|
| BCM B-P | 81% | 73.8% | 72.7% |
| BCM/B-P | 92.6% | 83.8% | 81.5% |
| B-P | 98.7% | 84.8% | 78.2% |

Table 1: Comparison between classification using (1) projections from BCM unsupervised learning as input to back-propagation; (2) a hybrid of BCM unsupervised learning and supervised learning via error back-propagation; and (3) a plain back-propagation net.

**Testing**

We used two generalization paradigms to test the feature extraction and classification ability of the system. First, the standard type of generalization to new instances of the same class was carried out. For each of the 12 CV types, we tested with 25 novel instances[1]. This kind of generalization requires the system to categorize instances that fall within the region of the input space it has had experience with. Many recognition systems are specifically focused on this kind of generalization. However, the second kind of generalization, where a system trained with a limited set of contexts generalizes well in new contexts, is possibly more important. If a system can transfer to new contexts, or to a region of the input space it has not experienced, the set of abstract features it is using must be capturing highly relevant aspects of the input training space. The ability to discover such features strongly suggests the technique being used is well suited for robust speech recognition. We demonstrate this kind of generalization by training on four vowel contexts [aa, ao, er, iy], and testing with the seven vowel contexts [uh, ih, eh, ae, ah, uw, ow].

## RESULTS AND DISCUSSION

A comparison between the different training methods is shown in Table 1. The low dimensional projections of the cochleagrams discovered with BCM learning, served as input to a small back-propagation network to yield the first set of results. This training method yielded reasonable performance on the training set, and very nearly the same performance on the two test sets. The small difference in generalization to instances of the same-4-vowel-contexts test set and generalization to instances from the new-7-vowel-contexts test set implies the features discovered with this method are good abstractions, and robust. The weight matrices of the eight units used in the BCM network are shown in Figure 3.

Features distinguishing between the different bursts are evident. The

---

[1] There were only 21 new tokens available for [pao]. All other CV groups had 25 tokens.

synaptic weight image on the top row, furthest to the right shows a white area in the high frequencies which corresponds to a distinguishing feature between [t] and [k]. The image directly below is useful for distinguishing [p] from the other two stops.

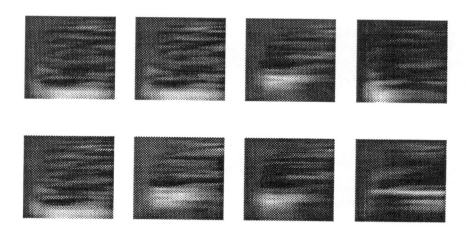

Figure 3: The synaptic weight matrices for 8 units after unsupervised training on 25 tokens of each CV type.

The results of the second training method, in which error back-propagation was modified to incorporate BCM-like constraints, are shown on the second line of Table 1. This novel integration of supervised and unsupervised techniques boosted the performance significantly over the previous training method. However, the pattern of results are very much the same; good and nearly equal performance with both types of generalization.

In contrast, this pattern was not found with the plain back-propagation net. While it did achieve the best performance of the three networks on the training set, it did not transfer its good generalization performance on the same-4-vowel-contexts to the new-7-vowel-contexts test set. Straight back-propagation training only attempts to minimize errors with the training set. It does not necessarily search for abstract features.

At this point, the only comparison we can make with HMM performance is very loose. Niles [14] constructed a baseline HMM system to classify the standard set of 39 phonetic classes in TIMIT. The speech was preprocessed using an order-18 LPC cepstral analysis, and then VQ codebooks for the cepstra, time-differenced cepstra, log energy, and delta log energy were used as input. A three state HMM was trained up for each phoneme. This system classified 82.0 percent correct when tested with just the voiceless stops. While this does give a ballpark indication that the systems we investigated here are doing reasonably well, any further comparison is precluded by methodological differences. For instance, Niles trained the HMMs for voiceless stops with all phonetic contexts, while our tokens always had a following vowel. Also, the HMM system was used as a baseline system, and was not fine tuned.

These preliminary results suggest that BCM training can be beneficially incorporated into a network architecture/training-paradigm for speech recognition. Moreover, the cochleagram input representation produced by Lyon's cochlear model contains details about the speech events that are useful in classifying speech tokens. A set of experiments making specific, quantitative comparisons between the system we have proposed here and current HMM methods is planned.

# References

[1] R. E. Bellman. *Adaptive Control Processes*. Princton University Press, Princton, NJ, 1961.

[2] E. L. Bienenstock, L. N. Cooper, and P. W. Munro. Theory for the development of neuron selectivity: orientation specificity and binocular interaction in visual cortex. *J. Neurosci.*, 2:32–48, 1982.

[3] J. H. Friedman. Exploratory projection pursuit. *Journal of the American Statistical Association*, 82:249–266, 1987.

[4] W. M. Huang and R. P. Lippmann. Neural net and traditional classifiers. In D. Anderson, editor, *Neural Information Processing Systems*, pages 387–396. American Institute of Physics, New York, 1988.

[5] P. J. Huber. Projection pursuit. (with discussion). *The Annals of Statistics*, 13:435–475, 1985.

[6] N. Intrator. Feature extraction using an unsupervised neural network. In D. S. Touretzky, J. L. Ellman, T. J. Sejnowski, and G. E. Hinton, editors, *Proceedings of the 1990 Connectionist Models Summer School*, pages 310–318. Morgan Kaufmann, San Mateo, CA, 1990.

[7] N. Intrator. Combining exploratory projection pursuit with projection pursuit regression, 1991. in preparation.

[8] N. Intrator. Localized exploratory projection pursuit. In Ed Wegman, editor, *Computer Science and Statistics: Proceedings of the 23rd Symposium on the Interface*. Amer. Statist. Assoc., Washington, DC., 1991.

[9] N. Intrator and L. N. Cooper. Objective function formulation of the BCM theory of visual cortical plasticity: Statistical connections, stability conditions, 1991. To appear.

[10] T. Kohonen, G. Barna, and R. Chrisley. Statistical pattern recognition with neural networks: Benchmarking studies. In *IEEE International*

*Conference on Neural Networks*, volume 1, pages 61–68, New York, 1988. (San Diego 1988), IEEE.

[11] K. F. Lee. *Large-Vocabulary Speaker-Independent Continuous Speech Recognition: The SPHINX System*. PhD thesis, Carnigie Mellon University, 1988.

[12] R. P. Lippmann. Review of neural networks for speech recognition. *Neural Computation*, 1(1):1–38, 1989.

[13] R. F. Lyon. A computational model of filtering, detection, and compression in the cochlea. In *Proceedings IEEE International Conference on Acoustics, Speech, and Signal Processing*, Paris, France, May 1982.

[14] L. T. Niles. Timit phoneme recognition using an hmm-derived recurrent neural network. In *Eurospeech91*, September 1991. Genoa, Italy.

[15] D. L. Reilly, C. L. Scofield, L. N. Cooper, and C. Elbaum. Gensep: a multiple neural network with modifiable network topology. In *INNS Conference on Neural Networks*, 1988.

[16] D. E. Rumelhart, G. E. Hinton, and R. J. Williams. Learning internal representations by error propagation. In D. E. Rumelhart and J. L. McClelland, editors, *Parallel Distributed Processing: Explorations in the microstructure of Cognition. Vol. 1: Foundations*, pages 318–362. MIT Press, Cambridge, MA, 1986.

[17] C. Scofield. Learning internal representations in the coulomb energy network. In *Proc. IEEE First Int'l Conf on Neural Networks, San Diego*. 1988.

[18] M. Slaney. Lyon's cochlear model. Technical report, Apple Corporate Library, Cupertino, CA 95014, 1988.

[19] Timit acoustic-phonetic continuous speech corpus. National Institute of Standards and Technology Speech Disc 1-1.1, October 1990. NTIS Order No. PB91-505065.

# Part 4:

# Signal Processing

# A RELAXATION NEURAL NETWORK MODEL FOR OPTIMAL MULTI-LEVEL IMAGE REPRESENTATION BY LOCAL-PARALLEL COMPUTATIONS

Noboru SONEHARA

ATR Auditory and Visual Perception Research Laboratories
Inuidani, Seika-cho, Soraku-gun, Kyoto 619-02, Japan

Abstract — A relaxation neural network model is proposed to solve the multi-level image representation problem by energy minimization in local and parallel computations. This network iteratively minimizes the computational energy defined by the local error in neighboring picture elements. This optimization method can generate high quality binary and multi-level images depending on local features, and can be implemented efficiently on parallel computers.

## 1. INTRODUCTION

Many problems in image processing can be solved by minimizing evaluation functions under complex nonlinear constraints. Analog neural network models proposed by Koch et al.[1] can solve the energy functionals of early vision problems as a surface reconstruction by energy minimization. These models can be categorized into deterministic relaxation neural network models, because they solve large-scale optimization problems by parallel and iterative computations [2] [3]. Because of the deterministic nature, these models converge to the quasi-optimum solution much faster than the stochastic relaxation model [4].
In this paper, a relaxation neural network model is proposed to solve a two-dimensional multi-level image representation problem by energy minimization in local and parallel computations. The author [5] has shown that a relaxation network model, in which each neuron has interactions among its neighbors, can generate a spatial grey-level image representation, and it is also shown that this model is effective for binary representation depending on local features such as edges, by introducing an interaction between the binary process and line process proposed by Geman et al. [4]. Webb [10] has also

developed an accurate algorithm for binary representation using a Gibbs sampler. In this paper, a relaxation neural network model, which is extended to two-dimensional multi-level representation, is discussed.

For signal quantization, Tank et al.[6] have shown that analog-to-digital (A/D) conversion can be considered as a simple optimization problem, and can rapidly be solved by a highly interconnected neural network model. However, this model is designed to provide an optimal scalar quantization, which produces digital codes from instantaneous signals. Since the digital representation of an image should be determined by depending on two-dimensional local features, interactions among neighboring picture elements are introduced to a neural network model for multi-level representation.

The proposed network model iteratively minimizes the total computational energy defined by the summation of the local square error between the grey-level and quantized-level in the neighboring picture elements. This process is a local-parallel one in which the processing of one picture element depends only on the state of the neighboring picture elements. The errors among neighboring picture elements cancel each other out everywhere, so that the energy function is minimized. In this sense, this method can be considered as an optimal and parallel quantization. In Section 2, a neural network model, energy functions, network dynamics and neighboring interactions are described. In Section 3, the multi-level image representation characteristics are evaluated for a geometrical, natural image and a composite image consisting of a photograph and characters, compared with existing serial halftoning algorithms. Processing performance on a SIMD parallel computer is also evaluated.

## 2. A RELAXATION NEURAL NETWORK MODEL

### a. Energy Function and Network Dynamics

In this network model, if an image size is N×N picture elements, there exist K×N×N neurons, each neuron $(k,i,j)$ of which determines the value of the k-th binary digit $b_{kij}$ ($k=1,2,...,K$), either $b_{kij}=0$ or $b_{kij}=1$, for each picture element $(i,j)$. We make the output variable $b_{kij}$ for a neuron have the range ($0 \leq b_{kij} \leq 1$) and be a continuous, monotonic increasing function (sigmoid function) of the internal state variable $q_{kij}$ : $b_{kij}=g(q_{kij})=1/(1+e^{-2\lambda q_{kij}})$, where $\lambda(=1/T)$ is a parameter of temperature (T) determining the slope of the sigmoid function. Let $f_{ij}$ ($0 \leq f_{ij} \leq 1$) denote the intensity-level of a picture element. The quantized-level of $f_{ij}$ is determined as : $f'_{ij}=2^{-K}\sum_{k}b_{kij}2^{k-1}+2^{-(K+1)}$. The proposed method is designed to be a local-parallel one in which the processing of one picture element depends only on the state of the neighboring picture elements. The errors among neighboring picture elements cancel each other out everywhere, so that the following energy function E is minimized:

$$E = \frac{C_Q}{2} \sum_{ij} [\sum_{i'j' \in V_c} f'_{i+i',j+j'} - \sum_{i'j' \in V_c} f_{i+i',j+j'}]^2$$
$$+ \frac{C_V}{2} \frac{1}{2^K} \sum_{ij} \sum_{k} 2^{2(k-1)} b_{kij}(1-b_{kij}) + C_G \sum_{ij} \int_0^{b_{kij}} g^{-1}(b_{kij}) db_{kij} \quad (1).$$

This function consists of three terms: the term $C_Q$ implementing the total error function in some prescribed neighborhood: $V_c$, the binary potential term $C_V$ to favor binary digit representations and to have minimal value when for each $k$, either $b_{kij}=0$ or $b_{kij}=1$, and the gain term $C_G$ to force the binary process $b_{kij}$ inside the hypercube. The dynamics of this network [1] [2] [3] and the update rule for an internal variable using the Euler method are given as follows:

$$\frac{dq_{kij}}{dt} = -\frac{\partial E}{\partial b_{kij}}, \quad q_{kij}(t+\Delta t) = q_{kij}(t) + \Delta t \frac{dq_{kij}}{dt} \quad (2).$$

Let the error signal between the intensity-level and quantized-level denote $\varepsilon_{ij} = f'_{ij} - f_{ij}$. Using the chain rule, we can obtain the following equations:

$$\frac{\partial E}{\partial b_{kij}} = \frac{\partial E}{\partial f'_{ij}} \frac{df'_{ij}}{db_{kij}} = (\frac{\partial E}{\partial \varepsilon_{ij}} \frac{d\varepsilon_{ij}}{df'_{ij}}) \frac{df'_{ij}}{db_{kij}} \quad (3a),$$

$$\frac{dq_{kij}}{dt} = -\frac{2^{(k-1)}}{2^K} C_Q \sum_{i''j'' \in V_c} \sum_{i'j' \in V_c} \varepsilon_{i+i'+i'',j+j'+j''} - \frac{1}{2^{K+1}} C_V 2^{2(k-1)}(1-2b_{kij}) - C_G q_{kij} \quad (3b).$$

The total energy E will always decrease for this update [1] [2] [3]. Note that whether E is a Lyaponov function for the system can easily be checked:

$$\frac{dE}{dt} = \sum_{ij} \sum_{k} \frac{\partial E}{\partial q_{kij}} \frac{\partial q_{kij}}{\partial t} = \sum_{ij} \sum_{k} \frac{\partial E}{\partial q_{kij}} (-\frac{\partial E}{\partial b_{kij}}) = -\sum_{ij} \sum_{k} (\frac{\partial E}{\partial b_{kij}})^2 \frac{db_{kij}}{dq_{kij}} \leq 0.$$

Since $b_{kij}$ is a monotonic increasing function of $q_{kij}$, the right hand side is always negative. Thus, as E always decreases and is bounded below, the system converges to a minimum.

### b. Neighboring Interactions

We define neighboring interactions suitable for local and parallel computations. If there is no interaction among neighboring picture elements, this model correspods to the optimal analog-to-digital converter (scalar quantization) proposed by Tank et al. [6]. In this case, a neighbor $V_c$, and the operator in iterative computation are described as follows:

$$V_c = \{(i',j')\} = \{(i'',j'')\} = \{(0,0)\}, \quad \sum_{i''j'' \in V_c} \sum_{i'j' \in V_c} \varepsilon_{i+i'+i'',j+j'+j''} = \varepsilon_{ij} \quad (4).$$

On the other hand, in the proposed model, interactions among neighboring neurons are introduced for effective representation of an image. As a simple example, the following neighbor $V_c$

is considered, as shown in *Fig. 1* :

$$V_c=\{(i',j')\}=\{(i'',j'')\}=\{(0,0), (1,0), (-1,0), (0,1), (0,-1)\} \quad (5).$$

We can choose the following interaction equation to minimize errors at the neighbor $V_C$. The interactions, and their operator in iterative computation as shown in *Fig. 2* are given as follows:

$$\sum_{i'j'\in V_c}\varepsilon_{i+i',j+j'}=\varepsilon_{i,j}+\varepsilon_{i+1,j}+\varepsilon_{i-1,j}+\varepsilon_{i,j+1}+\varepsilon_{i,j-1}, \sum_{i''j''\in V_c}\sum_{i'j'\in V_c}\varepsilon_{i+i'+i'',j+j'+j''}=5\varepsilon_{i,j}+2(\varepsilon_{i-1,j}+\varepsilon_{i-1,j+1}+\varepsilon_{i-1,j-1}$$

$$+\varepsilon_{i,j+1}+\varepsilon_{i,j-1}+\varepsilon_{i+1,j+1}+\varepsilon_{i+1,j}+\varepsilon_{i+1,j-1})+(\varepsilon_{i-2,j}+\varepsilon_{i+2,j}+\varepsilon_{i,j+2}+\varepsilon_{i,j-2}) \quad (6)$$

By using this network dynamics, deviation of energy function E by a local variable $b_{kij}$, and defining local error interactions ($\sum_{i'j'\in V_c}\varepsilon_{i+i',j+j'}$), the procedure can be done in parallel, and is suitable for implementation on parallel computers. This method can be considered as an optimum and parallel quantization, in the sense that the square error function E is minimized and errors among neighboring picture elements cancel each other out everywhere. We can also introduce another class of neighboring interactions for smoothing or sharpening [9]. For example, we can apply some type of averaging interaction, to minimize the average quantization noise at the neighbor, by introducing the evaluation function:

$$E=\sum_{ij}[(1/N_{V_c})\sum_{i'j'\in V_c} f'_{i+i',j+j'}-f_{i,j}]^2 ,$$

where $N_{V_c}$ is a constant number of neighboring picture elements. Another smoothing example is low-pass filtering interaction such as a Gaussian filter, to maintain the low-passed signal component between the original and quantized image:

$$E=\sum_{ij}[\sum_{i'j'\in V_c} g_{i+i',j+j'}f'_{i+i',j+j'}-\sum_{i'j'\in V_c}g_{i+i',j+j'}f_{i,j}]^2 ,$$

where g is Gaussian filter coefficient. For sharpening, we can consider the neighboring interaction to enhance the high-passed signal component of an original image such as edges:

$$E=\sum_{ij}[\sum_{i'j'\in V_c} l_{i+i',j+j'}f'_{i,j}-\sum_{i'j'\in V_c}l_{i+i',j+j'}f_{i+i',j+j'}]^2 ,$$

where $l$ is Laplacian filter coefficient.

## 3. EXPERIMENTS AND CONSIDERATIONS

Strictly speaking, since $b_{kij}$ is continuous, $b_{kij}$ is converted to discrete $b'_{kij}$ for the evaluation of multi-level image representation:

$$b'_{kij}=Q[b_{kij}]=0 \ (b_{kij}<0.5), \text{ or } 1.0 \ (b_{kij}\geq 0.5), \quad f_{ij}=2^{-K}\sum_{k=1}^{K}b'_{kij}2^{k-1}+2^{-(K+1)} \quad (7).$$

### a. Quantization Characteristics for a Geometrical Image

For easy understanding of the proposed parallel quantization, we show a binary image representation (K=1) for a geometrical image, compared with an optimal scalar quantization (without an interaction). *Fig. 3* shows the result of applying the relaxation neural network model to a quarter-sphere test pattern consisting of the calculated brightness range from 0 to 1.0 (8-bit accuracy) on a background of brightness 0. The experimental conditions are $\Delta t=0.01$, $\lambda=16.0$, $C_V=1.0$, $C_G=0.5$, $C_Q=100.0$ (with no interaction) and $C_Q=0.8$ (with interactions). The quantization characteristics ($f_{ij}-f'_{ij}$) as shown in *Fig. 4* exhibit spatial hysterisis which generates different quantized-levels to the same input-level by minimizing the energy. An actual grey-level is represented in the form of spatial grey-level representation.

## b. Quantization Characteristics for a Natural Image

*Fig. 5* shows binary image representation (K=1) for a natural image, compared with a scalar quantization[6], an ordered dither method[7] and a serial error diffusion method[8]. The proposed method does not have the texture caused by the dither matrix, because the same interactions are applied to all the picture elements. *Fig. 6* shows the result for binary (K=1) using a scanned composite image which consists of the grey scale part (photograph) and what was originally the binary part (characters). The character part of the image is obtained experimentally by Gaussian filtering. An ordered dither and serial error diffusion methods introduce considerable degradation in the resolution of the character parts. As shown in this figure, the proposed method does not introduce such degradation, because the error in the neighbors is minimized. The results for multi-level (K=2,3,4) using a natural, composite image are shown in *Fig. 7*. *Fig. 8* shows quantization characteristics for 16-level(K=4) of the proposed method. In general, finding good energy parameters ($C_G, C_V$, and so on) analytically is hard, so these parameters are determined experimentally so as to be useful practically. The used parameters are $\Delta t=0.01$, $\lambda=16.0$, $C_V=1.0$, $C_G=0.5$ and the maximum iteration number ($N_{itr}$) is 1000, which gives a final stable state image. As shown in these figures, the multi-level image representation is generated in the form of spatial modulation by density of multi-level picture elements, depending on local features. Thus, the optimal multi-level image representation can be realized by a neural network model in local and parallel computations.

## c. Error and Convergence Characteristics

We evaluate the mean square error : $E_{ms}=E[(f_{ij}-f'_{ij})^2]$, where $E[\cdot]$ shows an expectation value. For the discrete binary value of $b_{kij}$, $E_{ms}=E[(f_{ij}-f'_{ij})^2]$ is used. With an increase in the iteration number, the network generates a final stable state image, satisfying constraints. *Fig. 9* shows convergence characteristics for the proposed method (K=4, $\Delta t=0.01$, $\lambda=16.0$, $C_V=1.0$, $C_G=0.5$, $C_Q=3.2$), using a continuous grey-level image along the vertical direction : ($f_{ij}=j/255$, i,j=0,1,..,255). After 200 iterations, the network reaches a stable state. At that time, the errors are $E_{ms}=1.746 \cdot 10^{-4}$,

$E_{ms}^{\ast}=3.783 \cdot 10^{-4}$ for a scalar quantization, and $E_{ms}^{\ast}=2.351 \cdot 10^{-4}$, $E_{ms}^{\ast}=4.384 \cdot 10^{-4}$ for the proposed method. Let a quantization step of **K** bits denote $\Delta h$. Since $\Delta h \cdot 2^K = 1.0$, the variance $\sigma^2$ of the noise is obtained for uniform distribution: $\sigma^2 = \Delta h^2/12 = 1/12 \cdot 2^{2K}$. Since this value is about $3.255 \cdot 10^{-4}$ for **K=4**, this model gives a good approximation from the viewpoint of mean square error. On the other hand, compared with existing halftoning algorithms for a natural image, the errors are $E_{ms}^{\ast}=0.1829$ for an ordered dither method, $E_{ms}^{\ast}=0.1755$ for a serial error diffusion and $E_{ms}^{\ast}=0.0237$ for the proposed method($K=1, C_Q=0.8$). This method can generate an accurate representation for a natural image.

### d. Implementation on a Massively Parallel SIMD Computer

We have implemented this neural model on a SIMD (single-instruction, multiple-data) hypercube parallel computer (**CM-2**: Connection Machine) for the purpose of performance evaluation. The employed parallel computer has 16,384, 1bit-processors, each of which has 32KB of local memory, and 512 floating point processors. The basis of programming for the SIMD parallel computer is to assign data to a processor. In this programming, K neurons for a picture element are assignd to a processor. For 256x256 image data size, we use 65,536 processors in all. The proposed parallel computational model consists of neighboring interactions only, so the neighboring communication function (NEWS operation) provided by the CM-2 system is used for error data exchange. For a 256x256, **K=1** problem, the relaxation processing time is **11.9-13.6** msec using **VP=4** (ratio of the number of virtual processors to physical processors), and **21.3-21.8** msec using **VP=8**.

### 4. SUMMARY

A two-dimensional, parallel relaxation neural network model, which solves the spatial grey level representation problems by energy minimization was proposed. It was shown that this network model can be applied well to multi-level image representation as well as binary image representation of various images involving composite images with less degradation. The proposed relaxation neural network model consists of highly local and parallel computations, and can be implemented efficiently on parallel computers. The proposed relaxation neural network models can be widely applied for various 2-dimensional, local-parallel image signal processing.

**Acknowledgements** The author is grateful to Dr. E. Yodogawa and Dr. K. Nakane of ATR Auditory and Visual Perception Laboratories for encouraging this study. We thank our colleagues, especially Dr. M. Kawato and Dr. M. Sato, for their valuable suggestions. We also thank Mr. T. Yoshikawa for help in programming the neural network model.

**REFERENCES**
[1] C. Koch, J. Marroquin and A. Yuille : "Analog 'Neural'Networks in Early Vision", Proc. Natl. Acad. Sci. USA, 83, pp.4263-4267 (1986)
[2] J. J. Hopfield : "Neurons with Graded Response Have Collective Computational Properties Like Those of Two-state Neurons", Proc. Natl. Acad. Sci. USA, Vol. 81, pp.3088-3092 (1984)
[3] J. J. Hopfield and D. W. Tank : " 'Neural'Computation of Decisions in Optimization Problems", Biol. Cybern. 52, pp.141-152 (1985)
[4] S. Geman and D. Geman : "Stochastic Relaxation, Gibbs Distributions and the Bayesian Restoration of Images", IEEE Trans., PAMI-6, pp. 721-741 (1984)
[5] N. Sonehara : "Binary Representation and Surface Interpolation of the Grey level Image by Relaxation Neural Network Models", The Second IEEE Symposium On PDP, pp. 420-427 (1990)
[6] D. W. Tank, J. J. Hopfield : "Simple 'Neural'Optimization Network : An A/D Converter, Signal Decision Circuit, and a Linear Programming Circuit", IEEE Trans., Vol. CAS-33, No. 5, pp. 533-541 (May 1986)
[7] B. E. Bayer : "An Optimum Method for Two-Level Rendition of Continuous-Tone Pictures", International Conference on Communications, Vol. 1, pp. 26-11-26-15 (1973)
[8] R. Floyd and L. Steinberg : "An Adaptive Algorithm forSpacial Grey Scale", 1975 SID International Symposium Digest of Technical Papers, 4.3, pp. 36-37 (Apr. 1975)
[9] A. Rosenfeld and A. C. Kak : "Digital picture processing", Second edition, Academic press (1982)
[10] J. A. Webb : "Accurate Parallel Digital Halftoning", SID International Symposium (1991)

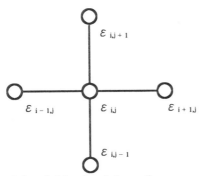

Fig. 1 A neighboring interaction

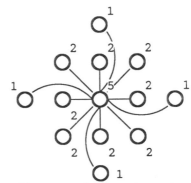

Fig. 2 13 point operator in an iterative computation

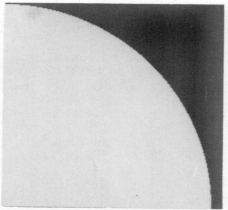

Fig. 3(a) Binary image representation by optimal scalar quantization (with no interaction)

Fig. 3(b) Binary image representation by the proposed quantization (with interactions among neighboring pels)

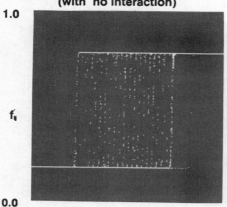

Fig. 4 Quantization characteristics ($f_i$—$f'_i$) of the proposed method

Fig. 5(a) An original image (SIDBA/GIRL:256 × 256 × 8bit)

Fig. 5(b) An optimal scalar quantization (without an interaction, Tank et al.[6])

Fig. 5(c) An odered dither method (matrix size=4 × 4, Bayer[7])

Fig. 5(d) A serial error diffusion method (binary quantization error is apportioned to the 4 neighboring pels, Floyd et al.[8])

Fig. 5(e) The proposed optimal, parallel quantization (with interactions, Sonehara)

Fig. 6(a) An original composite image (character part: Gaussian filtered)

Fig. 6(b) An ordered dither method (matrix size=4 × 4)

Fig. 6(c) A serial error diffusion method (interactions:forward 4 pels)

Fig. 6(d) The proposed optimal, parallel method

Fig. 7(a) Multi-level image representation (K=2, 4-levels, $C_Q$=4.0)

Fig. 7(b) Multi-level image representation (K=3, 8-levels, $C_Q$=4.0)

Fig. 7(c) Multi-level image representation (K=4, 16-levels, $C_Q$=4.8)

Fig. 7(d) Multi-level image representation for a composite image (K=3)

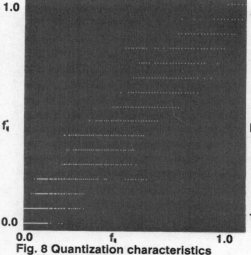
Fig. 8 Quantization characteristics (K=4) of the proposed method

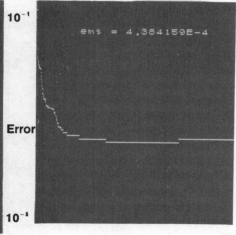
Fig. 9 Convergence characteristics (K=4) of the proposed method

# LITHOFACIES DETERMINATION FROM WIRE-LINE LOG DATA USING A DISTRIBUTED NEURAL NETWORK

Mark Smith
Edinburgh Parallel Computing Centre
University of Edinburgh
Edinburgh, EH9 3JZ
Email: mas@uk.ac.ed.castle

Neil Carmichael, Ian Reid & Colin Bruce
Shell UK (Exploration & Production)
ShellMex House
The Strand
London

## Abstract

A distributed neural network, running on a large transputer-based parallel computer, was trained to identify the presence of the main lithographical facies types in a particular oil well, using only the readings obtained by a log probe. The resulting trained network was then used to analyse a variety of other wells, and showed only a small decrease in accuracy of identification.

Geologists classify well structures using rock and fossil samples in addition to the log data that was given to the network. Results are given here for the accuracy with which the learned network agreed with analyses performed by geologists. The study was then extended into two more areas, firstly to investigate the network's success in predicting physical attributes of the rocks, e.g. porosity and permeability, and secondly to investigate the ability of similar networks to isolate particular geological features.

## INTRODUCTION

Identifying lithofacies types (a description of the constituents of a particular rock) with a minimum of effort has been a major desire of oil companies for many years. There have been recent attempts to use computer technology more extensively for this task; in particular, cluster analysis techniques and neural networks [2] are being investigated. However, the majority of identification is at present performed by hand by geologists. Their work is based upon two sources of information, firstly using data collected by a log probe, this provides readings of a variety of physical attributes of the rock. Secondly, they work from samples of the core rock brought to the surface at regular intervals. By analysing information from both sources, a picture of the main strata divisions and facies types in a particular well can be constructed. While this task is somewhat simplified by the experience of the geologist in identifying common structures between different wells, these relations are somewhat subjective, and their reliance on core samples can make them extremely slow and expensive.

A neural approach to this problem seems very natural since neural networks are an ideal tool for solving problems where we know that a certain set of starting data produces particular results, but we have no knowledge of the actual relations (i.e. a detailed physical model) that connect them. Ideally a network would be trained with a small set of example data sets, using target results based on the geologists classifications. The resulting network could then be continuously improved by presenting new sample data to broaden the network's *knowledge*.

The aim of this project was to investigate the success of a neural network approach to the lithofacies type-determination problem. Following good results from a simple determination between rock that had reservoir potential, and rock which did not, the investigations moved to a more discriminating set of facies types, using the main categories of *Sand, Mud, Coal* and *Cemented* layers. A network was constructed with outputs that corresponded to the probabilities that a certain depth interval contained each of the main rock types. Results were obtained by training a network on a subset of the data from a particular well, and then comparing the definitions for the whole well output by the network to those given by a geologist. Following this, the universality of a network trained on a single well was investigated by using it to identify the facies types present in other, as yet unseen, wells.

Following the success of the first part of the project, it was decided to extend the neural network approach to two further areas of interest. Firstly, the same log probe data was used to predict physical attributes of the rocks such as grain density, permeability and porosity. This type of information is invaluable for accurate reservoir simulations, yet is usually only available through laboratory analysis of core samples. The second extension to the project was to move from oil wells to gas production wells. The network was trained to identify regions of particularly high gas productivity using input data from a logging probe, similar to that used in the oil wells. This classification is of interest since the highly productive areas can be very difficult to identify, even from good core samples.

## THE NEURAL NETWORK

The Edinburgh Parallel Computing Centre (EPCC) has a distributed neural network simulator called *Rhwydwaith* [3]. This is a general purpose Rumelhart-type neural network [4] that runs on a Meiko Computing Surface – a parallel supercomputer based on Inmos transputers.

A Rumelhart network is a feed-forward network containing layers of nodes. Connections exist in only one direction – from the input layer of nodes, through a number of hidden layers, to the output layer. The network uses a learning technique called *back-propagation of error*. Each input pattern (in this case the actual log probe readings for a particular depth) has a target output pattern (the facies types present at that depth), as each input pattern is presented to the input nodes, a forward pass through the network produces a corresponding output state. An error for each pattern is produced by comparing the output to the target pattern, a global error is then computed by summing over all patterns. The weights of all inter-node connections are then iteratively updated until the error for all output nodes is within some predefined tolerance.

## Implementation and Efficiency

Teaching and using neural networks, particularly for complex multi-layer networks with high levels of connectivity between nodes, requires large amounts of calculation. The learning process itself is a gradient technique for solving matrix problems, where the weights in the network constitute the elements of the matrix. A parallel computer can reduce the execution time of the neural network simulator by distributing the weight matrices amongst the available *worker* processors. At each iteration, however, a *master* process must build up a global picture of the network in order to calculate global error values. There must therefore be some method for communicating between processors. For small networks (less than two hundred inter-node connections per layer) a distributed implementation may be inefficient, as communication overheads may cost more execution time than is being saved by the parallelisation of the calculation routines. However as networks become larger the parallelisation of the problem can lead to greatly reduced learning and running times. For the smaller networks used in the initial stages of this work (see Table 1) the parallel implementation gained little in terms of run-time. However, for the larger networks used for the final results, execution times were more than halved through the use of a distributed network.

Rhwydwaith also includes an acceleration strategy to reduce the network's learning time. This strategy, called SuperSAB [5], was developed at Edinburgh as an improvement on a previous strategy SAB [1]. SuperSAB allows the step-size for weight adaption to increase and decrease exponentially once the global error is reducing regularly. This leads to large reductions in learning times for many problems, including those covered by this work. In fact, without the acceleration strategy many of the networks used in the studies covered by this report did not reach an acceptable state in a reasonable time.

## TRAINING THE NETWORK

In order to train a neural network, there must be a store of data from which it can learn any inter-relations that may exist between the input data and the target results. In the case of facies determination there are a vast number of lithographical analyses that have been undertaken in the past. This data consists of a list of *patterns*, each of which contain the log readings for a particular depth in the well, and the corresponding facies type for that depth, as determined by the geologists.

The log probe outputs used in this study consisted of the following measurements:

**Gamma ray** Detects the intensity of gamma radiation emitted from the rock. Typically sand-based rocks contain low quantities of radioactive salts compared to mud-based.

**Neutron** The log probe contains a small neutron source, and makes measurements of the neutron back-scatter produced by the rock. This reading can yield information on the water and carbon content of the rocks.

**Sonic** The sonic log is a measure of the speed of sound through the surrounding rock. In general sound waves travel at decreasing speeds as the rock becomes more porous.

**Density** This measurement is again made using a radioactive source, and relies on the Compton effect as radioactivity is scattered in proportion to the number of scattering centres in the rock. The number of centres is then related to the density of the rock.

It is relatively straightforward to use these readings as inputs to the network, and attempt to obtain the lithofacies types as the output. An identical technique is also used for the later stages of the project with the target outputs being the rock attributes as determined from laboratory measurements, and the level of gas production measured during production.

For the work on lithofacies identification the geological interpretations were split into four main categories: Sand, Mud, Coal and Cemented. Within each of these groups there is a wealth of further subdivisions. The training strategy adopted was to use binary targets for the output nodes to indicate the presence or absence of each of the major rock types. Training was done by selecting a random 10% subset of all patterns and adjusting the network until all outputs came within a certain tolerance of the targets. The complete data set could then be presented to the network and the outputs compared to the targets to give an overall accuracy for the trained network. It is important to note that there was no special selection of training patterns. This allowed the possibility of confusing the network by training with potentially bad examples of each of the major rock types, but we believe that this strengthens our case that the results achieved are very much a lower bound on what may be possible when a neural network approach is coupled with the expert knowledge of a geologist.

When using the network to predict physical properties of the rock from the log data, it was necessary to move to analogue outputs rather than binary as used before. The potential of the output nodes was now used as a direct measure of the porosity, permeability and grain density of the rock at each particular depth. Training was again performed using a random 10% of the patterns for which laboratory core values were known, and the network was trained until the output potentials were all within a chosen tolerance of the target value. For the gas field study the outputs were again binary, indicating the presence or absence of a high production area. Initially randomly selected training data was used, however this produced poor results that were greatly improved after employing some expert geological knowledge to select good examples of high producing rock. This need for selecting training data was exacerbated by the relatively small number of highly productive zones in the complete data set.

## RESULTS

A variety of networks were constructed to identify individual facies types, core values and gas producing regions using only log data inputs. The number of hidden nodes used in each case varied according to the empirically discovered difficulty of each particular determination. The results given below detail the performance of the networks in comparison to either laboratory analysis of rock, or to well interpretations provided by trained geologists. Analyses were performed on both the complete data set from the training well, as well as on other previously unseen wells. The latter case provides some measure of the generality of the network's results between different wells, although it should always be remembered that the network was trained, in this study, to agree with one particular geologist's lithographical interpretation. It is generally accepted in the oil industry that no two geologists will be in complete agreement on all such interpretations.

### Lithofacies Determination

Initially five separate networks were constructed to identify the major rock types individually. These networks therefore had four input nodes, a single output node and a layer of hidden nodes numbering from 10 to 30. Although all training patterns had target outputs with binary values (either that rock-type is present or not), when used for examining complete well data sets the actual output produced was a value between zero and one. It was therefore necessary to introduce a threshold for recognition, and it was decided that any output over 0.6 should be taken as *on*, and any below 0.4 as *off*. The results achieved by these networks are summarised in Table 1. These results correspond to testing the network on the complete data set for well 1, whilst training was performed on a random 10% of that same well. The final column of Table 1 gives the percentage of levels in each well for which the network gave the same interpretation of the rock as the geologist. In other words, the geologists interpretation noted the presence of a certain facies, and the network produced an output for that facies of greater then 0.6, or alternatively the facies was noted as not present and the network's output was less than 0.4. The number of hidden nodes used for each network was not optimised through an exhaustive search of the parameter space. The number of nodes used simply reflects what was required to obtain reasonable accuracy in the predictions.

| Rock-Type | Hidden Nodes | Percentage Recognition |
|---|---|---|
| Reservoir | 10 | 88% |
| Sand | 30 | 90% |
| Mud | 20 | 87% |
| Coal | 10 | 99+% |
| Cemented | 20 | 99+% |

Table 1: Summary of facies-type determinations for the whole of well 1, with training performed on a random 10% sample of that well.

The main obstacle to learning for the networks detailed above lay in the definition of mixed sand and mud rocks, and in particular the definition of which facies dominated the mixture. The results for coal and cemented layers of rock were in comparison much better, however this is not all that surprising as these rocks are markedly different. The main errors in identifying coal and cement lay in the interface regions between strata, where there is both uncertainty in the exact location of the strata boundary, as well as the fact that the logging probe takes readings over a significant volume of rock, and gives an "averaged" result. These same problems occur for the sand and mud results also, but are less significant in comparison to the classification of mixed rocks.

In order to bypass the problem of sand/mud definitions a multi-output network was constructed and trained. This network had four outputs, one for each of the major lithofacies divisions, and contained a large number (between 20 and 60) of hidden nodes, the final production network used 60 hidden nodes. This structure was chosen to allow the concurrent identification of more than one lithofacies type, and thus hopefully provide some definition of mixed rocks and also a measure of the mixing ratios, since output potentials were continuous between zero and one. This network was trained on a subset of the data from a single well, and then tested on the complete data set. Where more than one rock type was indicated as present at a particular depth this was regarded as identifying a mixed rock, and this could be directly compared to the geological description of the well. In order to test the generality of the network's relations, it was then tested on a number of other wells from a similar geological setting. In 3 of the 4 wells tested the results agreed with the geological interpretation in over 90% of outputs, actual results are shown in Table 2. As before agreement between the network and the geologist occurs when the network's output is over 0.6 for an identified facies, and below 0.4 for a facies noted as not present. The vast majority of incorrect outputs occur when the network has an output of between 0.4 and 0.6, and seems uncertain of the presence of a facies, rather than making a complete misidentification.

| Well | Patterns | Result |
|---|---|---|
| 1 | 2988 | 91.4% |
| 2 | 1007 | 83.1% |
| 3 | 1813 | 90.3% |
| 4 | 1400 | 92.5% |

Table 2: Percentage of outputs that match the geological interpretation, training performed on a random 10% of patterns from well 1.

Figure 1 shows the output produced by the trained network when testing a completely unseen well (well 4). It can be seen that there are clear bands of coal and cement in the predictions, and these were shown to correspond to actual coal and cemented strata in the well. Continuous bands of both sand and mud rocks are also clearly identifiable from the graph, and it was very encouraging to find that the regions of overlap between these regions corresponded to sections of the well classified as mixed sand-mud rock.

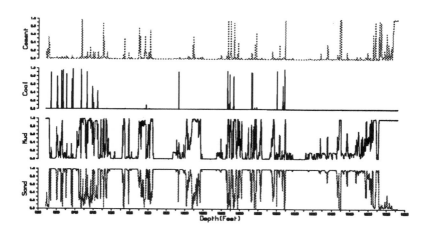

Figure 1: Network outputs for Well 4 showing the predicted rock types present at each level of the well. The vertical scale on each graph ranges from 0.0 to 1.0 and represents the actual output potential of each of the four output nodes.

## Core Values

Following the success of the initial lithofacies investigations a new network was constructed and trained to determine values for core information from the available log data. Specifically the values of the rock's porosity, permeability and grain density were the targets for the network. A single network was used with three outputs, the potential of each representing a measure of the relevant core value for each layer of the well. Training was performed firstly on a small random sample of data sets from well 1, and then on the complete data from that well. The results achieved for well 1 and two similar wells are detailed in Table 3. It can be seen

| Well | Patterns | Tolerance | | |
|---|---|---|---|---|
| | | 25% | 20% | 10% |
| 1 | 591 | 97.8 (97.9) | 95.9 (96.6) | 84.0 (85.3) |
| 2 | 200 | 92.0 (93.7) | 88.1 (91.5) | 73.2 (76.3) |
| 3 | 463 | 92.6 (93.3) | 89.3 (89.8) | 77.7 (78.8) |

Table 3: Percentage of core values correctly predicted to within particular tolerances. Training on a random 10% sample of well 1 and, in parentheses, on the whole of well 1.

that the provision of more training data improves the accuracy of the network for both the training set and in the general case. The results show the percentage of points in each well for which the network's values of porosity, permeability and grain density fall within the indicated tolerance of the laboratory values. It must be remembered when studying these results that measurements of permeability are made on a logarithmic scale, with the majority of each well having particularly low values. Also for most wells the range of grain density values was never more than a few percentage points away from the mean value. However graphical representation of the results shows that the network is successfully identifying the general trends of the core values, and can certainly identify areas with extreme readings.

## Gas Producing Regions

The application of neural network techniques to work in the exploration and production of natural gas fields in the southern North Sea falls into two parts. Firstly the identification of certain lithofacies in gas wells, using techniques identical to those discussed previously, and secondly the attempted prediction of the position of gas producing regions within each well. There is a certain amount of overlap between these topics, since it is most often the case that the particular lithofacies to be identified is exactly the one responsible for the gas production. However the second stage of this exercise also intended to use the network to predict the amount of gas to be output from each level of the well. This work has in the past been performed using statistical techniques such as cluster analysis, however these have been shown to give no more than 75% accuracy when based solely upon log probe data. The results from the neural network approach are at least as good as this in most cases, and often far superior.

Networks were initially trained using randomly selected subsets of completely interpreted gas wells. However, following uninspiring initial results, the help of Shell geologists was obtained in order to construct training data containing good examples of the various types of gas producing rock. It was of particular importance to make this extra effort for this section of the project, since the distinguishing features of the producing rocks are very subtle, and the wells examined were typically of relatively uniform geological formation. The network that was trained on this data contained the same four inputs as for the oil well work, 20 hidden nodes, and had firstly a single binary output to identify the presence of producing rocks, and then an analogue output to provide some prediction of the amount of gas produced. Table 4 shows the results achieved by this network for the six available test wells when training was performed on random data from each of the other wells, and when training took place with the specially manipulated data set. The value of providing good training data can be clearly seen, as can the obvious correlations between the formations of the individual wells.

| Well | Training Set | | | | | | |
|---|---|---|---|---|---|---|---|
| | Special | Well1 | Well2 | Well3 | Well4 | Well5 | Well6 |
| 1 | 93.8 | 85.7 | 81.4 | 72.9 | 72.6 | 59.2 | 81.7 |
| 2 | 70.2 | 83.2 | 75.2 | 68.9 | 58.2 | 53.9 | 81.2 |
| 3 | 84.3 | 85.5 | 72.6 | 82.8 | 71.2 | 62.3 | 76.5 |
| 4 | 94.6 | 83.8 | 81.1 | 81.5 | 91.1 | 68.6 | 73.6 |
| 5 | 88.3 | 64.2 | 70.3 | 78.6 | 80.6 | 72.8 | 47.5 |
| 6 | 92.2 | 85.7 | 71.2 | 62.9 | 53.8 | 55.4 | 88.3 |

Table 4: Percentage agreement with geological interpretations for each of six gas wells when trained on a special training set and on random data from each of the wells.

## CONCLUSION

It is apparent from this work that a neural network can be trained to predict the geologists' interpretation of lithofacies types within an oil well to a high degree of accuracy. This ability is surprising when one notes that the training data set was of a relatively small size and random composition. Therefore the results are probably a lower bound on what could be achieved after collaboration with a trained geologist. However, even with the present accuracy, neural networks provide a possible first filter for the mass of data produced by current oil exploration, by pin-pointing areas of potential interest within large wells.

It is interesting to note that the initial geologists' facies determinations are regarded as being around 80% accurate. Investigations are underway to discover whether the disagreements between geologist and neural network should be attributed to the former or the latter. Work has also been done to reduce the number of hidden nodes in the final large network, while retaining a high level of agreement with geological interpretations. Details of these findings will be included in the full report on this project.

Neural network techniques, such as those described here, are still at an early stage of development. Within the Royal Dutch/Shell group, and the oil industry in general, much more work is required to clarify the relationship between more conventional analysis or interpretation techniques and neural network approaches.

## ACKNOWLEDGMENTS

Edinburgh Parallel Computing Centre is a multidisciplinary project supported by major grants from the Department of Trade and Industry, the Computer Board and the Science and Engineering Research Council. It is a pleasure to acknowledge also substantial support from the University of Edinburgh and from Industrial Affiliates to the EPCC, not least Shell UK (Exploration and Production) Ltd.

## REFERENCES

[1] Devos, M. and Orban G.A., *Proc. NeuroNimes*, Nimes, 1988.

[2] Malferrari, L., Serra, R. and Valastro, G., *An Application of Neural Networks to Oil Well Drilling*, in Proceedings INNC 90, Paris (Kluwer Academic Publishers), 1990

[3] Richards, G.D., *Documentation for Rhwydwaith Version 2.4*, Edinburgh Concurrent Supercomputer Project User Guide **UG-7**, 1989.

[4] Rumelhart, D.E. et al, *Parallel Distributed Processing: Exploration in the Microstructure of Cognition*, MIT Press, Cambridge, MA, 1985.

[5] Tollenaere, T., *SuperSAB: Fast Adaptive Back Propagation with Good Scaling Properties*, Edinburgh Concurrent Supercomputer Project Research Paper, 1990.

# IMPROVED STRUCTURES BASED ON NEURAL NETWORKS FOR IMAGE COMPRESSION

Sergio Carrato, Giovanni Ramponi,
Amedeo Premoli, and Giovanni L. Sicuranza
D.E.E.I., University of Trieste
via A. Valerio, 10, 34127 Trieste, Italy

ABSTRACT — In this paper, the problem of efficient image compression through neural networks (NN's) is addressed. Some theoretical results on the application of 2-layer linear NN's to this problem are given. Two more elaborate structures, based on a set of NN's, are further presented; they are shown to be very efficient while remaining computationally rather simple.

## INTRODUCTION

The problem of image compression has received a great amount of attention in the last years. Many powerful algorithms based on orthogonal transforms have been presented and are now widely used [1]. Recently, nonlinear [2] [3] and linear [4] [5] Neural Networks (NN's) have been employed for data and/or image compression. In general, they are based on a 1- or 2-layer perceptron, in which the first layer performs the compression and the second, when existing, the reconstruction. Linear NN's have been shown to behave as an orthonormal transform, and to perform better than nonlinear ones with respect to the mean square error [6].

In the next section, some theoretical results on the application of linear NN's to this problem are given; they allow us to use a simplified version of the well-known backpropagation algorithm (BP) in the training phase.

In order to obtain even more efficient compression algorithms, two architectures are investigated and presented in this paper. They are based on a set of different linear NN's, which specialize on different types of images.

## LINEAR NN'S FOR IMAGE COMPRESSION

Given a set of images, theoretically the most powerful technique for image compression is the Karhunen-Loeve Transform (KLT) [1]. In this case, a basis for the linear space mapped by the images is found, in which the basis vectors are ordered according to their "importance", so that if the basis is restricted (as in the case of image compression problems) the energy preserved in the remaining coefficients is maximized, i.e. the mean square error due to the basis restriction [1] is minimized.

On the other hand, if a NN is used, a reduced basis may be directly found when the set of images is used as the training set. In [5], a linear single-layer perceptron is used, in which some extra branches connecting output nodes are present. The original images are fed into the input layer and the principal components of the set of images are obtained at the output layer, so that a basis which corresponds to the KLT is found.

As in [2,6], we use a 2-layer linear perceptron as a complete companding structure. More precisely, the original images, fed into the input layer, are compressed in the hidden layer (which has less nodes than the input one) and reconstructed to their original dimension in the output layer (which has obviously the same number of nodes as the input layer). The NN is trained with a suitable set of images, imposing that the desired output coincides with the input. Hence, we do not impose any specification on the signal at the hidden nodes.

The algorithm used for the training is a modification of the well-known BP. It is briefly revised here; details can be found in [6].

## The modified backpropagation algorithm

Let $\mathbf{p}_l$ and $\mathbf{p}'_l$, $(l = 1, 2, \ldots, L)$ indicate the $L$ original and compressed images, arranged into unidimensional arrays, respectively. We may write:

$$\mathbf{p}'_l = \mathbf{W}\mathbf{U}\mathbf{p}_l (l = 1, 2, ..., L), \tag{1}$$

where $\mathbf{U}$ and $\mathbf{W}$ are the $h \times n$ input-to-hidden and $n \times h$ hidden-to-output coefficient matrices, respectively.

In the following we show that, in order to minimize the reconstruction square error due to compression (e.g., $E\{\|\mathbf{p}' - \mathbf{p}\|_E^2\}$, where $\|.\|_E$, whose argument is a matrix, indicates the Euclidean norm), the matrix $\mathbf{U}$ must be the pseudoinverse of the matrix $\mathbf{W}$, i.e., $\mathbf{U} = \mathbf{W}^+ = [\mathbf{W}^T\mathbf{W}]^{-1}\mathbf{W}^T$. Moreover, an orthonormality condition $\mathbf{W}^T\mathbf{W} = \mathbf{I}$, where $\mathbf{I}$ denotes the identity matrix, can be imposed to the matrix $\mathbf{W}$, in which case the pseudoinverse $\mathbf{W}^+$ coincides with the transpose $\mathbf{W}^T$. An alternative approach to the problem may be found in [7].

Let the $n \times L$ matrix $\mathbf{P}$ group the images $\mathbf{p}_l$ columnwise. $\mathbf{U}$ can be regarded as a map from the $n$-dimensional space $S^n$ to a $h$-dimensional subspace $S^h$ ($h < n$ due to compression); similarly, $\mathbf{W}$ maps the subspace $S^h$ in the space $S^n$. Then equation (1) may be rewritten as

$$\mathbf{P}' = \mathbf{W}\mathbf{U}\mathbf{P} \tag{2}$$

where the reconstructed images $\mathbf{p}'_l$ have been embedded columnwise in the matrix $\mathbf{P}'$.

Then the problem consists in finding two matrices, $\mathbf{W}$ and $\mathbf{U}$, which minimize the square of the reconstruction error $\epsilon$ defined as

$$\epsilon^2 = \|\mathbf{P}' - \mathbf{P}\|_E^2 = \|\mathbf{WUP} - \mathbf{P}\|_E^2.$$

We first try to solve the following subproblem:

- if $\mathbf{W}$ is supposed known, find the best $\mathbf{U}$ for a given image $\mathbf{p}_l$.

According to [8], suppose that the columns of $\mathbf{W}$ are linearly independent (otherwise, the rank is not maximum and the solution can not be optimal). Let $\mathbf{x}_l$ be the projection of $\mathbf{p}_l$ onto $S^h$, described in terms of the basis of $S^h$, i.e. $\mathbf{x}_l = \mathbf{Up}_l$. The same vector, described in terms of the basis of $S^n$ and denoted by $\mathbf{p}'_l$, has the form $\mathbf{p}'_l = \mathbf{Wx}_l$ (the columns of $\mathbf{W}$ may be seen as a basis of $S^h$). Now, $\mathbf{p}_l - \mathbf{Wx}_l$ represents the residual vector in the reconstruction; if the projection is chosen to be orthogonal, this vector will be minimum (for a given $\mathbf{W}$). Then we demand that the residual be orthogonal to each vector in $S^h$, or, equivalently, to each vector of its basis, i.e.

$$\mathbf{W}^T(\mathbf{p}_l - \mathbf{Wx}_l) = 0.$$

We obtain

$$\mathbf{x}_l = (\mathbf{W}^T\mathbf{W})^{-1}\mathbf{W}^T\mathbf{p}_l,$$

where the existence of $(\mathbf{W}^T\mathbf{W})^{-1}$ is assured by the full rank of $\mathbf{W}$. Thus the reconstruction $\mathbf{p}'_l$ of $\mathbf{p}_l$ is

$$\mathbf{p}'_l = \mathbf{Wx}_l = \mathbf{W}(\mathbf{W}^T\mathbf{W})^{-1}\mathbf{W}^T\mathbf{p}_l$$

and, by comparing the above formula with eq. (1), we have

$$\mathbf{U} = (\mathbf{W}^T\mathbf{W})^{-1}\mathbf{W}^T, \tag{3}$$

where the right-hand side of the equation may be recognized as the pseudo-inverse [8] of $\mathbf{W}$, usually denoted as $\mathbf{W}^+$.

The result of the above subproblem does not depend on $\mathbf{p}_l$, then it holds true also if we consider all the images, i.e. for the whole matrix $\mathbf{P}$.

It may be easily shown that $\mathbf{P}'$ does not depend on the choice of the basis of $S^h$. Then, without loss of generality we can choose an orthonormal basis, that is a matrix $\mathbf{W}$ satisfying the relation $\mathbf{W}^T\mathbf{W} = \mathbf{I}$. Consequently, we obtain

$$\mathbf{U} = \mathbf{W}^+ = \mathbf{W}^T. \tag{4}$$

The properties of the above subproblem allow one to reduce the original problem to the following:

- find a $n \times h$ matrix $\mathbf{W}$ solving

$$\min_{\mathbf{W}} \|\mathbf{WW}^T\mathbf{P} - \mathbf{P}\|_E^2.$$

It is interesting to note that it is not necessary to impose the constraint $\mathbf{W}^T\mathbf{W} = \mathbf{I}$ explicitely. In fact, let $\hat{\mathbf{W}}$ be the solution found; if $\hat{\mathbf{W}}^T\hat{\mathbf{W}} \neq \mathbf{I}$, being $\hat{\mathbf{U}} = \hat{\mathbf{W}}^T$, we have $\hat{\mathbf{U}} \neq \hat{\mathbf{W}}^+$, in disagreement with eq. 3.

The solution is not unique, as any orthonormal basis of $S^h$ obviously gives the same result.

These results allow us to improve the BP algorithm. During each step of the learning process, we proceed as follows:

- the classical BP is used to update the coefficients of the hidden-to-output matrix $\mathbf{W}$—with the only minor modification due to the absence of the sigmoid in the node transfer function;

- the input-to-hidden matrix $\mathbf{U}$ is set equal to the transpose of $\mathbf{W}$.

This modified procedure has been shown [6] to be more stable and faster than classical BP.

The function which is minimized is the same as the one considered in the basis restriction problem [1], so the NN must converge to a basis which spans the same subspace found by the KLT. The basis is not necessarily the same, this indetermination being taken into account by a suitable rotation matrix.

It is worth noting that, in order to keep the NN's size within reasonable limits, in the actual implementation the input images are partitioned, as usual [2,3], in small blocks, or patterns, which then become the vectors $\mathbf{p}$ considered before.

## IMPROVED LINEAR STRUCTURES

The solution based on a single NN, while rather efficient, is very simple. We have tried to see whether higher compression ratios—or, equivalently, higher Signal-to-Noise (S/N) ratios—are obtainable at the cost of a reasonable increase in architecture complexity. Two possible approaches are presented in the following.

### First proposed structure

The first observation that can be made is that "simple" patterns, i.e. blocks extracted from parts of the images that are smooth or poor in details, need a NN with a smaller number of hidden nodes, with respect to "complex" patterns. The first structure we propose is based on this idea, and is described in the following.

We introduce an "activity" parameter, as in [9], according to which the patterns are subdivided, during both training and test, in four classes (characterized, respectively, by very low, low, high, and very high activity). During the learning phase, four NN's, with increasing number of hidden nodes, are trained each with a pattern subset. In this way, different NN's are obtained, each of which is specialized on a different kind of patterns. During the test

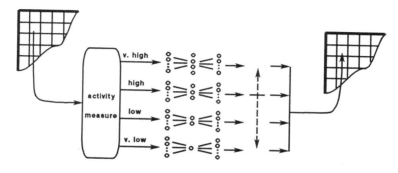

Figure 1: First proposed structure. Each block of the image is treated by a suitable NN, according to its activity.

phase, the patterns taken from the test image are subdivided as before and each of them is elaborated by the appropriate NN. For what concernes the compression ratio, this structure corresponds to a conventional NN having a number of hidden nodes, $h_m$, equal to a weighted mean of the number of hidden nodes used by the four NN's, i.e.

$$h_m = \frac{\sum_{j=1}^{4} h_j b_j}{\sum_{j=1}^{4} b_j},$$

where $h_j$ and $b_j$ are the number of hidden nodes of the $j$-th NN, and the number of blocks coded by the same NN, respectively. A block diagram of the structure is shown in Fig. 1.

## Second proposed structure

In general, a NN is able to specialize on the main features of the patterns belonging to the training set. We then observe that, if the patterns are subdivided according to selected features, a NN can be even more efficient in the coding/decoding operation. The most natural way to divide the patterns is probably based on the preferential direction present in the image details. Four main directions can be considered, i.e. horizontal, vertical, and the two diagonal ones. Of course, this distinction is somehow meaningless for "simple" patterns, their pixel having almost constant gray level, so it can be applied to the "complex" patterns only. Basing on these considerations, and in order to further increase the S/N ratio, a second architecture has been developed and is described below.

In such an architecture, all the patterns are divided according to the above mentioned activity parameter into three classes only. The patterns belonging to the two lowest activity classes are treated exactly as before, and two NN's are trained during learning and used during test. The remaining patterns

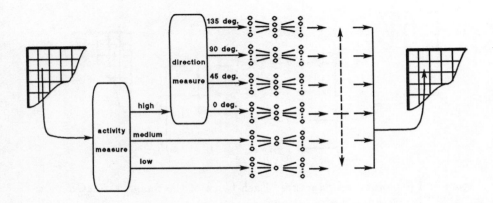

Figure 2: Second proposed structure. The simplest blocks of the images are treated as in Fig. 1, while the most complex ones are further divided according to their preferential direction, and treated separately.

are further subdivided into four subclasses, according to the preferential direction present in each one. This second discrimination is made according to the algorithm proposed in [9]. Four more NN's are then trained by the appropriate patterns taken from the training set, and are successively used during the test phase. A block diagram of the structure, which is then based on six different NN's, is presented in Fig. 2.

For the discrimination of the blocks, we also tested the block classifier algorithm proposed in [10], but we obtained poorer results. This is probably due to the peculiar characteristics of such classifiers, which are more sensitive to edges than to preferential directions; e.g., no discrimination is made between an L-shaped edge segment and a diagonal edge segment.

## SIMULATION RESULTS

For the training of the NN's we use a set of 4096 8×8 non-overlapping patterns extracted from four 256 × 256 images. In order to demonstrate the so called "generalization property" of the NN's, i.e. the capability of a NN to perform well also with data not belonging to the training set, we use a fifth 256 × 256 image ("Lena") to test the NN's. This image is divided into 8 × 8 blocks too.

Image degradation due to compression is evaluated by means of the peak S/N ratio, defined as usual as $\text{PSNR} = 10\log\left(255^2/\overline{e^2}\right)$, where $\overline{e^2}$ is the mean square error.

Results are presented in Fig. 3, in terms of PSNR versus mean number of hidden nodes, for the two proposed structures. As a comparison, data related to a single linear NN are also reported. It may be noticed how the higher specialization of the NN's improves their performance.

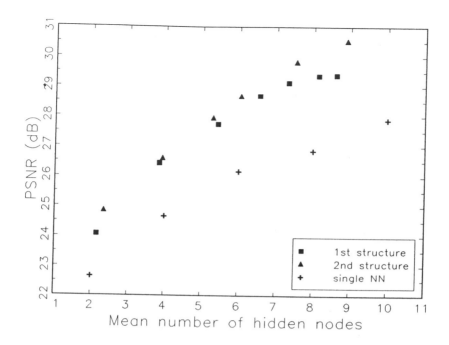

Figure 3: Comparison of the performances of the first proposed structure (based on the activity parameter), the second proposed structure (based both on the activity and the preferential directions), and a single NN.

The difference, in PSNR, between the first and the second structure is rather small. However, we found that the image quality obtained with the second structure is significantly higher. As an example, in Fig. 4, the test image is shown on the left, processed by the first proposed algorithm. The four NN's used have 2, 4, 8, and 12 hidden nodes; the measured values of PSNR and $h_m$ are 27.73 dB and 5.43, respectively. The bases of the four NN's, i.e. the coefficients of each column of the respective matrices $\mathbf{W}$, are presented on the right, shown as gray levels. In Fig. 5, the test image is shown on the left, processed by the second proposed algorithm. The two NN's dedicated to low activity blocks have 3 and 5 nodes, while 8 nodes are used by the four "directional" NN's. The measured values of PSNR and $h_m$ are 27.93 dB and 5.30, respectively. The bases used by the six NN's are presented on the right; it may be seen how the bases of the "directional" NN's are specialized. As a reference, in Fig. 6 the unprocessed test image is shown on the left. On the right, the image processed by a single linear NN having $h = 6$ is also shown; the PSNR is 26.17 dB.

In order to define a transmission rate ($R$, in bit per pixel), quantization at hidden nodes must be taken into account. In a linear network the signal is

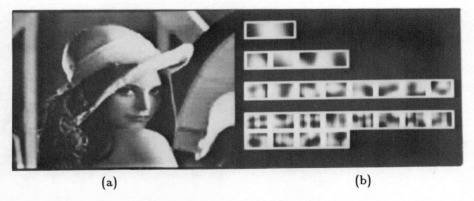

(a)           (b)

Figure 4: a) The test image, processed by the first structure; b) the bases of the four NN's used.

| 1st structure | | 2nd structure | | single NN | |
|---|---|---|---|---|---|
| $R$ (bpp) | PSNR (dB) | $R$ (bpp) | PSNR (dB) | $R$ (bpp) | PSNR (dB) |
| .30 | 26.29 | .31 | 26.20 | .31 | 24.54 |
| .42 | 27.49 | .41 | 27.61 | .47 | 25.98 |
| .64 | 28.92 | .64 | 29.47 | .62 | 26.64 |

Table 1: Comparison of the performances of the two proposed structures, and a single linear NN, in terms of PSNR versus bit rate $R$.

not bounded by the nonlinearity; nevertheless, we have found that a uniform 5-bit quantizer, defined on a suitable and non-critical range, is able to keep the quantization error low while achieving a reasonable compression. The range does not depend on the image being elaborated. In turn, it is slightly different for each NN.

In Tab. 1, a comparison is presented between the two proposed structure and a single linear NN. It may be seen that a minimum increase of 1.5 dB in PSNR is attained by our structures, with respect to single NN's. The data reported on the second line correspond to the structures used for Fig. 4a, 5a, and 6b. The decrease of approximately .3 dB in PSNR is due to quantization. A more accurate calculation should also include the side information which is necessary to specify the NN associated to each pattern. This overhead is very small in our case.

Comparative tests have also been performed with respect to the nonlinear architectures presented in [9], and the superiority of linear structures to nonlinear ones, already shown in [6] for what concerns the conventional NN's, has been confirmed. In fact, an approximate increase of 1 dB in PSNR has been reported.

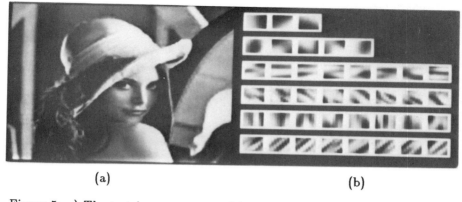

Figure 5: a) The test image, processed by the second structure; b) the bases used by the six NN's used: the first two are dedicated to simple patterns, while the other four are used for the complex ones.

## CONCLUSIONS

As a final remark, it is worth noting that the separation of the patterns on the basis of the activity parameters used within the second architecture relies on the peculiarities of the human visual system, which possesses neurons with a directional sensitivity and therefore is particularly able in discriminating oriented edges and lines. Therefore, coding the patterns with preferential orientations using specialized neural networks is a natural and effective option. During the training phase the specialized networks are able to learn such orientations and then the output image can be reconstructed using well-defined directional patterns as basis functions.

A relevant observation is that the proposed separation of the patterns in an image permits a rough subdivision in the 2-D frequency domain according to the presence of low or high frequency components, and a further subdivision of the highpass information in directional bands. This approach is similar to the multifrequency or the multiresolution techniques, which resort to subband or pyramidal coding. Coding via NN's is at present under consideration for images obtained by means of such decomposition methods.

## References

[1] A. K. Jain, Fundamentals of Digital Image Processing, Prentice-Hall International, Inc., Englewood Cliffs, NJ, 1989.

[2] G. W. Cottrell, P. Munro, and D. Zipser, "Image Compression by Back Propagation: An Example of Extensional Programming", in Models of Cognition: A Review of Cognition Science, N. E. Sharkey, ed., Norwood, NJ, 1989.

(a)  (b)

Figure 6: a) Original test image; b) the same image, processed with a single linear NN having 6 hidden nodes.

[3] G. L. Sicuranza, G. Ramponi, and S. Marsi, "Artificial Neural Network for Image Compression", Electronics Letters, vol. 26, no. 7, March 1990, pp. 477–478.

[4] E. Oja, "A Simplified Neuron Model as a Principal Component Analyzer", J. Math Biology, vol. 15, 1982, pp. 267–273.

[5] S. Y. Kung and K. I. Diamantaras, "A Neural Network Learning Algorithm for Adaptive Principal Component Extraction (APEX)", Proc. Int. Conf. on Acoust., Speech, and Signal Processing, Albuquerque, NM, April 1990, pp. 861–864.

[6] S. Carrato, A. Premoli, G. L. Sicuranza, "Linear and Nonlinear Neural Networks for Image Compression", Proc. 1991 International Conference on Digital Signal Processing, Florence, Italy, September 1991, to be published.

[7] P. Baldi, K. Hornik, "Neural Networks and Principal Component Analysis: Learning from Examples without Local Minima", Neural Networks, vol. 2, pp. 53–58, 1989.

[8] P. Lancaster, M. Tismenetsky, The Theory of Matrices, Academic Press, Orlando, 1985.

[9] S. Marsi, G. Ramponi, and G. L. Sicuranza, "Improved Neural Structures for Image Compression", Proc. Int. Conf. on Acoust., Speech, and Signal Processing, ICASSP-91, Toronto (Canada), May 1991.

[10] B. Ramamurthi and A. Gersho, "Nonlinear space-variant postprocessing of block coded images", IEEE Trans Acoust., Speech, Signal Processing, vol. ASSP-34, no. 5, October 1986, pp. 1258–1268.

# Adaptive Neural Filters

*Lin Yin, Jaakko Astola, and Yrjö Neuvo*
Department of Electrical Engineering
Tampere University of Technology
P. O. Box 527, SF–33101 Tampere, Finland

Abstract-In this paper we introduce a new class of nonlinear filters called neural filters based on the threshold decomposition and neural networks. Neural filters can approximate both linear FIR filters and weighted order statistic (WOS) filters which include median, rank order, and weighted median filters. An adaptive algorithm is derived for determining optimal neural filters under the mean squared error (MSE) criterion. Experimental results demonstrate that if the input signal is corrupted by Gaussian noise adaptive neural filters converge to linear filters and if corrupted by impulsive noise, optimal neural filters become WOS filters.

## 1. INTRODUCTION

The field of adaptive signal processing is closely related to adaptive neural networks in its principles [1]. The two fields have been developing independently but share the feature of an adaptive linear combiner. Adaptive linear filters have been applied to a wide range of problems, including system modeling, statistical prediction, noise canceling, echo canceling, inverse modeling, and channel equalization [2]. Adaptive neural networks, however, had not demonstrated their adaptive filtering abilities until the relationship between neural networks and stack filters was set up recently [11, 12, 13]. It was shown that neural networks have the same ability to suppress noise as stack filters if their neuron functions are unit step functions. As a class of robust filters, stack filters include all rank order filters, weighted median filters, and WOS filters [11, 12, 15]. Using the backpropagation algorithm in neural networks we can find optimal stack filters under either MSE criterion or mean absolute error criterion [11].

It is well known that linear filters can suppress Gaussian noise better than stack filters while stack filters can remove impulsive noise more effectively than linear filters. Based on the threshold decomposition architecture and neural networks with sigmoidal neuron functions, we introduce in this paper a new class of nonlinear filters called neural filters. Neural filters provide a framework which unifies linear FIR filtering and WOS filtering. An adaptive algorithm is derived for determining optimal neural filters under the MSE criterion. This algorithm is analogous to Widrow's LMS algorithm [2] or the back-propagation algorithm [7]. An initial weight vector is chosen arbitrarily, and then the weight vector is adjusted based on the observations of the corrupted and the desired processes. Eventually, the weight vector converges to an optimal solution.

## 2. NEURAL FILTERS AND WOS FILTERS

The definition of neural filters is based on the threshold decomposition architecture [8,9,14,15].

Let $R(t)$ be an arbitrary time continuous bounded signal. Let $R(n)$ denote the sampled and quantized, to $M$ levels, version of $R(t)$. This signal can be decomposed into $M-1$ binary sequences $\{r^m(n)\}, m = 1, 2, ..., M-1$, by thresholding operation called $T^m$.

$$r^m(n) = T^m(R(n)) = \begin{cases} 1, & \text{if } R(n) \geq m; \\ 0, & \text{else.} \end{cases} \quad (1)$$

Note that

$$\{R(n)\} = \sum_{m=1}^{M-1} \{T^m(R(n))\} = \sum_{m=1}^{M-1} \{r^m(n)\} \quad (2)$$

Let a window of width $N$, where $N$ is an odd integer, slide across the input process $R(n)$, and let $\underline{R}(n)$ be the vector containing the $N$ samples in the window at time $n$, i.e.

$$\underline{R}(n) = [R(n-(N-1)/2), ..., R(n), ..., R(n+(N-1)/2)] \quad (3)$$

As shown in Fig. 1, neural filter $F(.)$ is defined by a neuron or neural network as follows,

$$F(\underline{R}(n)) = \sum_{m=1}^{M-1} \sigma(\underline{W}^T \underline{\tilde{r}}^m(n)) \quad (4)$$

where $^T$ denotes the transpose, $\underline{W}$ is the synaptic weight vector,

$$\underline{W} = [W_0, W_1, W_2, ...., W_N]^T \quad (5)$$

$\underline{\tilde{r}}^m(n)$ is an input vector of the neuron,

$$\underline{\tilde{r}}^m(n) = [-1, r^m(n-(N-1)/2), ..., r^m(n), ..., r^m(n+(N-1)/2)]^T \quad (6)$$

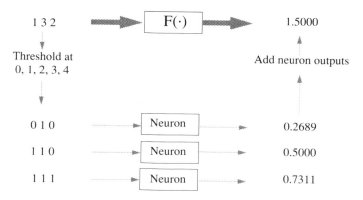

**Fig. 1** A neural filter $F(\cdot)$ of window $N = 3$. The synaptic weight vector of the neurons is $\underline{W} = [2, 1, 1, 1]^T$. The multilevel operation of the filter is shown by wide arrows. The operations performed by the filter when it is in its threshold decomposition architecture are shown by slender arrows.

and $\sigma(.)$ is a neuron activation function. The most frequently used neuron functions are unit step functions and sigmoidal functions. In this paper, sigmoidal functions are used because they can, depending on the situation, approximate either unit step functions or linear functions very well. One intuitive explanation for this is that for large incoming weights, sigmoidal functions are at "high gain" state and can approximate unit step functions and for small weights, however, sigmoidal functions can approximate linear functions. Moreover, the so-called back-propagation technique [7] which is widely used in practice assumes also sigmoidal neuron functions, because a differentiable $\sigma(.)$ is needed for applying gradient descent techniques.

Neural filters can approximate both FIR filters and WOS filters since sigmoidal functions can approximate linear functions and unit step functions. It is intuitive that any neural filter with a linear neuron function reduces to a linear FIR filter. If a unit step function is used and the inputs are binary, the neuron is capable of implementing Boolean functions. That is, neural filters become Boolean filters [13]. Any Boolean filter satisfying the stacking property reduces to a stack filter [13, 15]. The stacking property implies certain attractive properties. Relatively fast algorithms exist for realizing stack filters [15].

Two binary signals, $\underline{u}$ and $\underline{v}$, "stack" if each element of $\underline{u}$ is greater or equal to the corresponding element in $\underline{v}$. Consider $\underline{u}$ and $\underline{v}$ that are filtered with a Boolean filter, the filtering operation being denoted as $y = f(\underline{u})$. Now the Boolean filter $f$ is said to possess the stacking property if and only if

$$f(\underline{u}) \geq f(\underline{v}) \quad \text{whenever} \quad \underline{u} \geq \underline{v} \tag{7}$$

It has been shown that the stacking property of Boolean filters is equivalent to the positivity of the weights of neural networks [6, 11, 12].

*Theorem* Neural networks with positive weights and unit step functions obey the stacking property.

If Boolean functions are performed by neurons with positive weights, the corresponding Boolean filters become WOS filters [11, 12, 13]. WOS filters can be implemented using a sorting operation in the real domain. The output of a WOS filter with integer weights is expressed as

$$F(\underline{R}(n)) = W_0 : th \text{ largest value of the set}$$

$$[W_1 \diamond R(n - (N-1)/2), ..., W_N \diamond R(n + (N-1)/2)] \qquad (8)$$

where

$$K \diamond X = \overbrace{X, ..., X}^{K \text{ times}} \qquad (9)$$

Expression (8) holds also for real valued weights. The output of the WOS filter with real valued positive weights over samples $\underline{R}(n)$ can be calculated as follows. Starting from the higher end of the sorted set add up the corresponding weights until the sum is equal to or greater than $W_0$. The output of the WOS filter is the sample corresponding to the last weight.

WOS filters include many widely used median type filters, such as median, rank order, and weighted median filters. The statistical properties of WOS filters were derived by Yli-Harja *et al* [12]. Optimal WOS filtering algorithms under the mean absolute error criterion have been developed by Yin *et al* [10,11].

## 3. AN ADAPTIVE NEURAL FILTERING ALGORITHM

Like linear filtering, the goal of adaptive neural filtering is to find a neural filter such that the average MSE per time unit between the filter's output and the desired signal is minimized.

Assume $R(n)$ and $S(n)$ are the input signal and the desired output signal of the neural filter, respectively. If $R(n)$ and $S(n)$ are jointly stationary, the MSE to be minimized is

$$J(\underline{W}) = E\left[(S(n) - F(\underline{R}(n)))^2\right]$$

$$= E\left[\left(\sum_{m=1}^{M-1}(s^m(n) - \hat{s}_s^m(n))\right)^2\right] = E\left[\left(\sum_{m=1}^{M-1}(s^m(n) - \sigma_s(\underline{W}^T\widetilde{\underline{r}}^m(n)))\right)^2\right] \qquad (10)$$

where

$$s^m(n) = T^m(S(n)), \quad m = 1, 2, ..., M-1 \qquad (11)$$

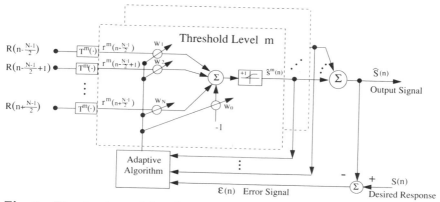

**Fig. 2** The structure of the adaptive neural filtering

$$\hat{s}_s^m(n) = \sigma_s(\underline{W}^T \underline{\tilde{r}}^m(n)), \quad m = 1, 2, ..., M - 1 \qquad (12)$$

and $\sigma_s(.)$ is a sigmoidal function defined as

$$\sigma_s(X) = \frac{1}{1 + e^{-X}} \qquad (13)$$

In analogy to Widrow's LMS algorithm or the back-propagation algorithm, the adaptive LMS neural filtering algorithm can be derived,

$$\underline{W}(n+1) = \underline{W}(n) + 2\mu(S(n) - \hat{S}(n)) \sum_{m=1}^{M-1} \hat{s}_s^m(n)(1 - \hat{s}_s^m(n))\underline{\tilde{r}}^m(n) \qquad (14)$$

where $\mu$ is an adaptation stepsize,

$$\hat{S}(n) = \sum_{m=1}^{M-1} \hat{s}_s^m(n) \qquad (15)$$

The structure of the adaptive LMS neural filtering is shown in Fig. 2.

This algorithm is almost the same as Widrow's LMS algorithm [2]. However, the adaptive LMS neural filters can not be guaranteed to converge to globally optimal solutions because the minima are not unique in optimization problem Eq. (9). Fortunately, experimental results demonstrate that in most situations these solutions are good enough to satisfy our practical requirements.

## 4. EXPERIMENTAL RESULTS

To visualize the differences in the performance of adaptive neural filters and some of the filters which are widely used in practice, we have simulated two experiments.

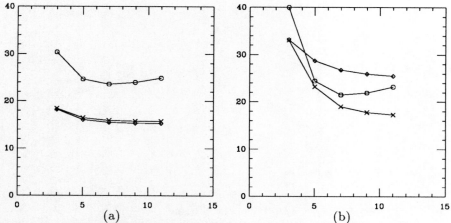

**Fig. 3** Reconstruction MSE as function of the window size. (a) Gaussian noise. (b) Contaminated Gaussian noise. In these figures, three kinds of filters are used, Wiener filter (◇), adaptive neural filter (×), median filter (○).

In the first experiment, the observed sequence is a Markov chain corrupted by additive noise. We considered corrupting noise with two different finite discrete distributions: i) discrete Gaussian, ii) discrete contaminated Gaussian. A comprehensive evaluation is presented in Fig. 3 where the mean squared errors corresponding to three different estimators for the window size from 3 to 11 are compared. For Gaussian noise, the linear filter gives the best result. The performance of the adaptive neural filter is very close to that of the linear filter. For the contaminated Gaussian noise, Fig. 3(b) shows the superiority of the neural filter to the linear filter.

The second experiment gives an application of adaptive neural filters to image processing. Figs. 4(a) and 4(b) show an original image and the image corrupted with impulsive noise. The upper left quarter of the original image and the noisy image were used to train the $3 \times 3$ linear filter and the $3 \times 3$ neural filter. The filters resulting from the training phase were then used to filter the noisy image. The MSE and the MAE values of the restored images are listed in Table 1. Figs. 4(c), 4(d), and 4(e) show the images restored by the adaptive linear filter, the adaptive neural filter, and the $3 \times 3$ median filter. As can be seen by comparing these images, the adaptive neural filter gives better performance than the adaptive linear filter and the median filter.

## 5. CONCLUSION

In this paper, neural networks have been applied for adaptive filtering, which leads to neural filters. Both linear FIR filters and WOS filters can be approximated by neural filters. Experimental results in 1-D and 2-D sig-

Table 1: Results of Impulsive Noise Simulation (3 × 3 window)

| Measured Error | | |
|---|---|---|
| Method | MSE | MAE |
| Adaptive LMS Linear Filter | 414.737488 | 15.081722 |
| Adaptive LMS Neural Filter | 120.233139 | 5.183308 |
| Median Filter | 164.303177 | 7.438873 |

(a)

nal processing demonstrated that adaptive LMS neural filters can effectively remove Gaussian noise, contaminated Gaussian noise, and impulsive noise.

The concept of neural filtering can be generalized by allowing different neural networks to operate at each thresholded level [17]. In this manner, the performance of neural filters can be further improved.

## REFERENCES

[1] B. Widrow and R. Winter, " Neural nets for adaptive filtering and adaptive pattern recognition," *Computer*, pp. 25-39, March 1988.

[2] B.Widrow and, S.D.Stearns, *Adaptive Signal Processing*. Englewood

(b)

(c)

(d)

Cliffs, NJ: Prentice - Hall, 1985.

[3] L. Fu, " Adaptive signal detection in noisy environments," *Journal of Neural Network Computing*, pp. 42-50, Spring 1990.

[4] R. P. Lippmann, " An introduction to computing with neural nets, " *IEEE ASSP magazine*, pp.4-22, April 1987.

[5] S. Tamura and A. Waibel, " Noise reduction using connectionist models," *Proc. of ICASSP 88*, vol.1, pp.553-556, 1988.

[6] S.Muroga, *Threshold Logic and Its Applications*, Wiler Interscience, New York 1971.

[7] D. E. Rumelhart, G. E. Hinton, and R. J. Williams, " Learning internal representations by error propagation ", in *Parallel Distributed Processing: Explorations in the Microstructure of Cognition*, vol.1, Ed. D. E. Rumelhart and J. L. McClelland, MIT Press, pp.318-362, 1986.

[8] E. J. Coyle and J.H. Lin, " Optimal stacking filters and mean absolute error nonlinear filtering," *IEEE Trans. ASSP*, vol.ASSP-36, no.8, pp.1244-1254, Aug. 1988.

[9] J.H.Lin, T.M.Sellke and E.J.Coyle, " Adaptive stack filtering under the mean absolute error criterion," *IEEE Trans. ASSP*, vol. ASSP-38, no.6, pp. 938-954, June 1990.

[10] L.Yin, J.Astola, and Y.Neuvo, " Optimal weighted order statistic filters under the mean absolute error criterion ," *Proc. of ICASSP 91*, vol. 4, pp. 2529-2532, Canada, May 1991.

(e)

**Fig. 4** *(a) Original image, (b) noisy image, (c) adaptive linear filter under the MSE criterion, (d) adaptive neural filter under the MSE criterion, (e) median filter.*

[11] L.Yin, J.Astola, and Y.Neuvo, " Median type filters and perceptrons," *The IEEE Proc. of ISCAS*, Singapore, vol.1, pp.81-84, June 1991.

[12] O. Yli-Harja, J. Astola and Y. Neuvo, " Analysis of the properties of median and weighted median filters using threshold logic and stack filter representation," *IEEE Trans. Signal Processing*, vol.39, no.2, pp.395-410, Feb. 1991.

[13] L. Yin, J. Astola, and Y. Neuvo, "Adaptive Boolean filters," *Proc. of Int. Conf. on Artificial Neural Networks*, Espoo, Finland, vol.1, pp. 759-764, June 24-28, 1991.

[14] J.P.Fitch, E.J. Coyle, and N.C. Gallagher, Jr., " Median filtering by threshold decomposition," *IEEE Trans. on ASSP*, vol.ASSP-32, pp.1183-1188, Dec. 1984.

[15] P. D. Wendt, E. J. Coyle and N. C. Jr. Gallagher, " Stack filter", *IEEE Trans. ASSP*, vol.ASSP-34, no.4, pp.898-911, Aug. 1986.

[16] K.M. Hornik, " Approximation capabilities of multilayer feedforward networks are universal," *Neural Networks*, pp.251-257, 1991.

[17] L. Yin, J. Astola, and Y. Neuvo, " A new class of nonlinear filters - neural filters," submitted to *IEEE Trans. on Signal Processing*, June 1991.

# A Surface Reconstruction Neural Network for Absolute Orientation Problems

Jenq-Neng Hwang,    Hang Li

Information Processing Laboratory
Department of Electrical Engineering, FT-10
University of Washington
Seattle, WA 98195

**Abstract:** This paper proposes a neural network for representation and reconstruction of 2-D curves or 3-D surfaces of complex objects with application to absolute orientation problems of rigid bodies. The surface reconstruction network is trained by a set of roots (the points on the curve or the surface of the object) via forming a very steep cliff between the exterior and interior of the surface, with the training root points lying in the middle of the steep cliff. The Levenberg-Marquardt version of Gauss Newton optimization algorithm was used in the back-propagation learning to overcome the problem of local minima and to speed up the convergence of learning. This representation is then used to estimate the similarity transform parameters (rotation, translation, and scaling), frequently encountered in the absolute orientation problems of rigid bodies.

## 1 Introduction

In computer vision applications, we sometimes encounter the following mathematical problem: given two sets of $m$-dimensional points, $\{\mathbf{x}^{(i)}, i = 1, \ldots, p\}$ and $\{\mathbf{y}^{(i)}, i = 1, \ldots, q\}$, extracted separately from 2-D curve or the 3-D surface of two objects, where one oriented object is a *similar transform* of the other (the reference object). We want to find similarity transformation parameters ($\mathbf{R}$: rotation, $\mathbf{t}$: translation, and $s$: scaling), which can best describe the orientational relationship between these two objects. This problem is sometimes called the *absolute orientation* problem [2]. It is also a very important problem in 3-D motion estimation [5].

Most algorithms proposed so far adopted the *point correspondences* assumption, where the number of root points in each set is equal ($p = q = n$), and each point in one object, say $\mathbf{x}^{(i)}$, has been pre-matched with (corresponded to) one point in the other object, say $\mathbf{y}^{(i)}$. This assumption results in a least squares problem:

$$\min \ \Sigma^2 \equiv \sum_{i=1}^{n} \|\mathbf{x}^{(i)} - s\mathbf{R}\mathbf{y}^{(i)} - \mathbf{t}\|^2 \qquad (1)$$

which can be solved by iterative algorithm [3], or singular value decomposition of a covariance matrix of the data [1, 9].

In reality, in the case of missing points or noisy measurements, there are unequal number of root points ($p \neq q$) in two different sets, and the correspondences of root points from two objects are difficult to determine. By correlating the Fourier transforms, a representation regardless of where the points are sampled, of the functions defined on the coordinates of the root points from two objects, it is possible to estimate the similarity transform parameters with quite limited success [5].

Therefore, it is very important to have a robust algorithm which can efficiently represent the general surface information based on the available points sampled (e.g., via edge detection) from the surfaces of both of the oriented and reference objects. More importantly, this representation should also lead to an efficient mechanism for systematic parameter estimation of similarity transforms.

## 2  Surface Reconstruction Back-Propagation Neural Networks

Representation and reconstruction of 2-D curves or 3-D surfaces of complex objects can be formulated as follows,

**Given:** Points $\mathbf{x}^{(1)}, \ldots, \mathbf{x}^{(n)}$ in $\mathcal{R}^2$ or $\mathcal{R}^3$ are assumed to lie on or close to the curve or surface of an object.

**Goal:** To represent and reconstruct the 2-D curve or 3-D surface of the object.

A useful and general solution to this problem can find applications in a wide variety of fields: e.g., design automation, manufacturing automation, terrain mapping, vehicle guidance, and intelligent robots, etc. Though human can easily understand various shapes, so far machine representation and reconstruction of shapes is still a nontrivial problem.

Our surface reconstruction neural network is an extension from previous research effort using polynomial approximation of an embedded 2-D curve or 3-D surface functions [8], which is based on the concept from *Algebraic Geometry:*, i.e., a planar 2-D curve is the set of roots of a function, $\Psi$, of two (coordinate) variables; a 3-D surface is the set of roots of a function, $\Psi$, of three (coordinate) variables.

Unlike most nonlinear functional approximation ($z = \Psi(\mathbf{x})$) using back-propagation neural network, where samples of distinct training data pairs $\{\mathbf{x}^{(i)}, z^{(i)}\}$ are given for training, the surface reconstruction neural network is only provided with samples of the root points set $\{\mathbf{x}^{(i)}, i = 1, \ldots, n\}$ for training the 2-D curve or 3-D surface, i.e., $\Psi(\mathbf{x}) = 0$. Therefore, a different learning strategy than the standard back-propagation algorithm should be used.

## 2.1 Modification of the Cost Function

A standard 2-layer (one hidden layer) back-propagation neural network is used for the training (see Figure 1). In this network, the activation values of the $k^{th}$ neuron of the $l^{th}$ layer is denoted as $a_k(l)$, $l = 0, 1, 2$, where $l = 0$ specifies the input layer (consists of 2 or 3 input neurons with values equal to $\mathbf{x}$); $l = 1$ specifies the hidden layer (consists of $H$ hidden neurons with commonly used sigmoid functions, $\frac{1}{1+e^{-u}}$); and $l = 2$ specifies the output layer (consists of single output neuron with slightly modified sigmoid function, $\frac{2}{1+e^{-u}} - 1$). The interconnection weight linking the $k^{th}$ neuron of the $l^{th}$ layer with the $j^{th}$ neuron at the $l-1^{th}$ layer is denoted as $w_{kj}(l)$ (with $w_{k0}(l)$ denoting the offset of the neuron). Since only the root points, which all generates the same desired zero outputs, are provided for the training, the standard back-propagation algorithm, which minimizes the mean squared errors between the actual and the desired outputs,

$$E \equiv \frac{1}{2}\sum_{i=1}^{n}(0 - a^{(i)}(2))^2 \equiv \frac{1}{2}\sum_{i=1}^{n}(a^{(i)}(2))^2 \equiv \frac{1}{2}\sum_{i=1}^{n}(\Psi(\mathbf{x}^{(i)}))^2 \qquad (2)$$

will create a network that tells no difference between a root point and a non-root point since none of the non-root points are presented to the network. More specifically, the trained network will basically output zero value no matter what inputs are tested.

Therefore, the cost function to be minimized should reflect some characteristics of the $m$-dimensional root points $\{\mathbf{x}^{(i)}\}$ which are lying on the 2-D curve or the 3-D surface. One potential cost function is to establish a network representation which creates a very large gradient at each root point and small one elsewhere [8]. This can be well interpreted in an $m+1$-dimensional space, where the regions on one side of the curve (or surface) bear positive output values of $\Psi(\mathbf{x})$, the regions on the other side bear negative values of $\Psi(\mathbf{x})$, and the curve (or surface) interpolated from the training root points give zero values of $\Psi(\mathbf{x})$. Therefore the points on the curve (or surface) are actually forced to lie in the middle of the steep cliff with very high gradient values. This results in the following cost function to be minimized:

$$E \equiv \frac{1}{2}\sum_{i=1}^{n}\frac{(\Psi(\mathbf{x}^{(i)}))^2}{||\nabla\Psi(\mathbf{x}^{(i)})||^2} \equiv \frac{1}{2}\sum_{i=1}^{n}\frac{\Psi^2}{\sum_{h=1}^{m}\frac{\partial\Psi}{\partial x_h^{(i)}}} \equiv \frac{1}{2}\sum_{i=1}^{n}\frac{\Psi^2}{||\nabla_i||^2} \qquad (3)$$

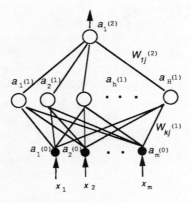

Figure 1: A standard 2-layer back-propagation neural network.

## 2.2 First Order vs. Second Order Optimization

Having chosen the cost function to be minimized, it is now straightforward for us to apply the gradient descent based back-propagation learning (sequential updating) to train the network [7]:

$$w_{kj}(l) \Longleftarrow w_{kj}(l) - \eta \frac{\partial E}{\partial w_{kj}(l)}$$
$$\equiv w_{kj}(l) - \eta \left( \frac{\Psi}{\|\nabla\|^2} \frac{\partial \Psi}{\partial w_{kj}(l)} - \frac{\Psi^2}{\|\nabla\|^4} \sum_{h=1}^{m} \frac{\partial \Psi}{\partial x_h} \frac{\partial^2 \Psi}{\partial x_h \partial w_{kj}(l)} \right) \quad (4)$$

where the second order derivative $\frac{\partial^2 \Psi}{\partial x_h \partial w_{kj}(l)}$ can be computed analytically or approximately. For example:

$$\frac{\partial^2 \Psi}{\partial x_h \partial w_{1j}(2)} = \frac{1}{2}(1 + a_1(2))(1 - a_1(2))\frac{\partial a_j(1)}{\partial x_h} - a_j(1)a_1(2)\frac{\partial a_1(2)}{\partial x_h} \quad (5)$$

where

$$\frac{\partial a_1(2)}{\partial x_h} = \sum_{k=1}^{H} \frac{\partial a_1(2)}{\partial a_k(1)} \frac{\partial a_k(1)}{\partial x_h} = \sum_{k=1}^{H} \frac{\partial a_1(2)}{\partial a_k(1)} a_k(1)(1 - a_k(1))w_{kh}(1) \quad (6)$$

$$\frac{\partial a_j(1)}{\partial x_h} = a_j(1)(1 - a_j(1))w_{jh}(1) \quad (7)$$

$$\frac{\partial a_1(2)}{\partial a_k(1)} = \frac{1}{2}(1 + a_1(2))(1 - a_1(2))w_{1k}(2) \quad (8)$$

Similarly, $\frac{\partial^2 \Psi}{\partial x_h \partial w_{kj}(1)}$ can be calculated. For network with more than one hidden layer, a finite-difference approximation can be used to efficiently compute this second order derivative [4]:

$$\frac{\partial^2 \Psi}{\partial x_h \partial w_{kj}(l)} \approx \frac{1}{\delta} \left( \frac{\partial \Psi(\mathbf{x} + \delta \varrho_h)}{\partial w_{kj}(l)} - \frac{\partial \Psi(\mathbf{x})}{\partial w_{kj}(l)} \right) \tag{9}$$

where $\varrho_h$ is the $h^{th}$ unit vector.

Our experiments on surface reconstruction indicated that the recursive (sequential updating) back-propagation was often stuck in local minima where all weights and offsets reached zero. Therefore, we adopted the Levenberg Marquardt version of the Gauss-Newton method for a batch training of our neural networks [4]. This algorithm takes advantage of the sum of products of the first order Jacobian matrix to explicitly approximate the Hessian matrix of the second order derivatives.

### 2.3 Simulation Results of Surface Reconstruction

For simplicity of illustration, we only presented examples on neural network representation and reconstruction of 2-D planar curves. Figure 2(a) shows 50 root points of a circle for training, and Figure 2(b) shows the reconstructed circle with 5 hidden neurons after 50 training iterations. To test the robustness of the reconstruction upon occlusion, another 50 root points were sampled from 3 quarters of a circle (see Figure 2(c)), and the resulting reconstructed circle with 5 hidden neurons after 100 training iterations is shown in Figure 2(d).

For obviously non-closed curves, the surface reconstruction neural networks try to extrapolate the curves based on the network configuration and the training algorithm used. For example, Figure 2(e) shows 50 root points of a non-closed curve for training, and Figure 2(f) shows the extrapolated curve based on 15 hidden neurons after 200 training iterations.

Simulations also indicated that the surface reconstruction neural networks sometimes generate extra unwanted curves due to either under-fitting or overfitting problems, especially when the given curves are not very smooth or very continuous. Since there is no theoretical support of how many hidden neurons with sigmoid nonlinear basis are required to exactly represent the curves, these extra curves cannot be easily avoided. For example, Figures 2(g) and (h) show such reconstruction effects with 50 root points based on 10 hidden neurons after 400 training iterations. Although this artifact is not desirable, it has little reverse effect on our application of this surface reconstruction network on parameter estimation of similarity transform discussed below.

## 3 Parameter Estimation of Similarity Transform

After trained with the root points $\{\mathbf{x}^{(i)}\}$ of an object, the surface reconstruction neural network establishes an implicit parametric representation of the

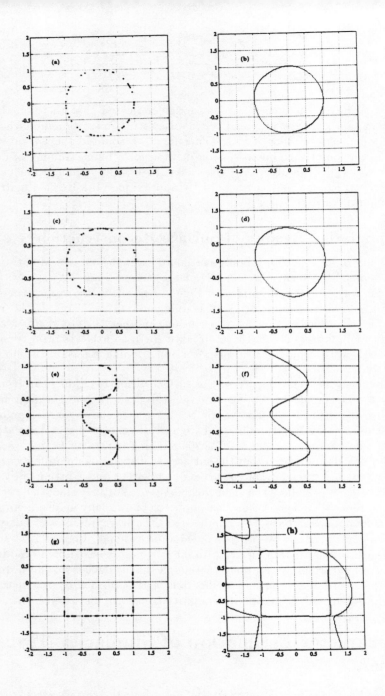

Figure 2: Examples of neural network representation and reconstruction of 2-D planar curves.

curves or surfaces, and can now be used for estimating the similarity transform parameters of the same object with translation/rotation/sclaing described by another set of root points $\{\mathbf{y}^{(i)}\}$ without any *point correspondence*. This task can be achieved by cascading a *linear* similarity transform neural network to the input of the surface reconstruction neural network (see Figure 3). The goal is to allow the similarity transform network update its linear weights and offsets (by keeping constant the weights of the surface reconstruction neural networks) so that the root points $\{\mathbf{y}^{(i)}\}$ can be best transformed (via rotation $\mathbf{R}$, scaling $s$, and translation $\mathbf{t}$) to the same orientation of the reference root points $\{\mathbf{x}^{(i)}\}$, and therefore the surface representation of the newly transformed $\{\mathbf{y}^{(i)}\}$ can best match that of the trained $\{\mathbf{x}^{(i)}\}$ (see Equation 1). Only by using this kind of cascading configuration, we can provide the necessary mismatch information in representation domain from the master surface reconstruction network to the slave similarity transform network in the training of the latter one. This idea is similar in concept to several neural network controller designs [6].

As shown in Figure 3(a), for parameter estimation of 2-D curves, a single layer similarity transform neural network is cascaded to the surface reconstruction network. Upon presenting the new testing root points, this cascaded network can again be trained by the Gauss Newton batch back-propagation algorithm in minimizing the same cost function defined in Equation 3, except that only the weights of the similarity transform neural network are updated, and those of surface reconstruction neural networks are kept constant. The three orientation parameters can thus be derived from the weights of the trained similarity transform neural network:

$$s = \sqrt{w_{11}^2(0) + w_{12}^2(0)} \tag{10}$$

$$\mathbf{R} = \begin{pmatrix} w_{11}(0)/s & w_{12}(0)/s \\ w_{21}(0)/s & w_{22}(0)/s \end{pmatrix} \tag{11}$$

$$\mathbf{t} = \begin{pmatrix} w_{10}(0) \\ w_{20}(0) \end{pmatrix} \tag{12}$$

Note that in order to make $\mathbf{R}$ a valid 2-D rotation matrix, $w_{11}(0)$ has to be forced equal to $w_{22}(0)$, and $w_{12}(0)$ has to be forced equal to $-w_{21}(0)$. This can be done by taking the average of the amount of changes for both weight-pairs at the end of each learning iteration.

Similarly, the parameter estimation of 3-D surfaces can be achieved by cascading a three-layer linear similarity transform neural network to the surface reconstruction network (see Figure 3(b)). This is based on the fact that any 3-D rotation $\mathbf{R}$ can be implemented by three consecutive rotation layers ($\mathbf{R}_x$, $\mathbf{R}_y$ and $\mathbf{R}_z$), each one being a rotation along the specified coordinate axis with special weight connections and weight-pair restriction as in the 2-D rotations. Note also that the offsets for the first two layers of similarity transform network are set to be zero.

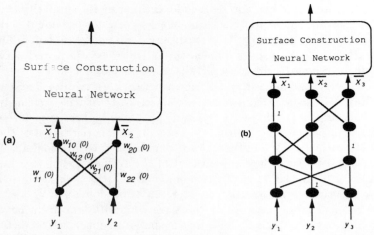

Figure 3: (a) A single layer similarity transform neural network is cascaded to the surface reconstruction network for parameter estimation of 2-D curves. (b) A three-layer similarity transform neural network is cascaded to the surface reconstruction network for parameter estimation of 3-D surfaces.

Simulations for similarity transform parameter estimation of 2-D curves were conducted. Figure 4(a) shows 100 root points $\{\mathbf{x}^{(i)}\}$ sampled from a polygon for training the surface reconstruction network. Figure 4(b) shows the curve reconstruction result based on 10 hidden neurons. Figure 4(c) shows a set of 50 root points $\{\mathbf{y}^{(i)}\}$ sampled from the oriented objects of the same polygon for training the similarity transform network. The resulting oriented and estimated objects (based on the trained similarity transform parameters) coincided perfectly with each other as shown in Figure 4(d).

To further test the robustness of the network, a set of only 25 root points $\{\mathbf{y}^{(i)}\}$ sampled from the oriented objects (see Figure 4(e)) of the same polygon was used for training the similarity transform network. The resulting oriented and estimated objects matched quite well with each other as shown in Figure 4(f).

We also test the noise tolerance capability of the network by adding (small and big) zero mean and white Gaussian noises to the 50 root points shown in Figure 4(c). The small noise corrupted data are shown in Figure 4(g), and the resulting oriented and estimated objects are shown in Figure 4(h). The big noise corrupted data are shown in Figure 4(i), and the resulting oriented and estimated objects are shown in Figure 4(j).

Another important observation worthwhile to mention is that the final value of the cost function $E$ in Eq. 3 is a good indicator of the performance. When $E$ value is bigger than some threshold, the problem can be either that the reconstructed representing surface of the oriented object is very different from that of the trained reference object, or that the search of the weights of the similarity transform network is stuck in the local minimum, which can be easily solved by changing another set of initial weights.

**ACKNOWLEDGEMENT:** We are grateful to Professor Werner Stuetzle of the Department of Statistics at University of Washington for his valuable suggestions and comments on our work.

# References

[1] K. S. Arun, T. S. Huang, S. D. Blostein. Least squares fitting of two 3-D point sets. *IEEE Trans. on PAMI*, 9(5):698-700, Sept. 1987.

[2] B. K. Horn, H. M. Hilden, S. Negahdaripour. Closed-form solution of absolute orientation using orthonormal matrices. *J. Opt. Soc. Am. A*, 5(7):1127-1135, July 1988.

[3] T. S. Huang, S. D. Blostein, E. A. Margerum. Least squares estimation of motion parameters from 3-D point correspondences. In Proc. *IEEE Conf. Computer Vision and Pattern Recognition*, pp. 198-201, Miami Beachd, FL, 1986.

[4] J. N. Hwang and P. S. Lewis. From nonlinear optimization to neural network learning. In Proc. *24th Asilomar Conf. on Signals, Systems, & Computers*, pp. 985-989, Pacific Grove, CA, November 1990.

[5] Z.C. Lin, H. Lee, T. S. Huang. A frequency-domain algorithm for determining motion of a rigid body from range data without correspondences. In Proc. *IEEE Conf. Computer Vision and Pattern Recognition*, pp. 194-198, Miami Beach, FL, 1986.

[6] D. Nguyen and B. Widrow. The truck backer-upper: an example of self-learning in neural networks. In *Int'l Joint Conf. on Neural Networks (IJCNN), Washington D.C.*, pages II357–II363, June 1989.

[7] D. E. Rumelhart, G. E. Hinton, and R. J. Williams. Learning internal representation by error propagation. in *Parallel Distributed Processing: Explorations in the Microstructure of Cognition: Volume 1: Foundations*, Chapter 8. MIT Press, Cambridge, MA, 1986.

[8] G. Taubin. Algebraic nonplanar curve and surface estimation in 3-space with applications to position estimation. Techical Report, Brown University, LEMS-43, Feb. 1988.

[9] S. Umeyama. Least-squares estimation of transformation parameters between two point patterns. *IEEE Trans. on PAMI*, 13(4):376-380, April 1991.

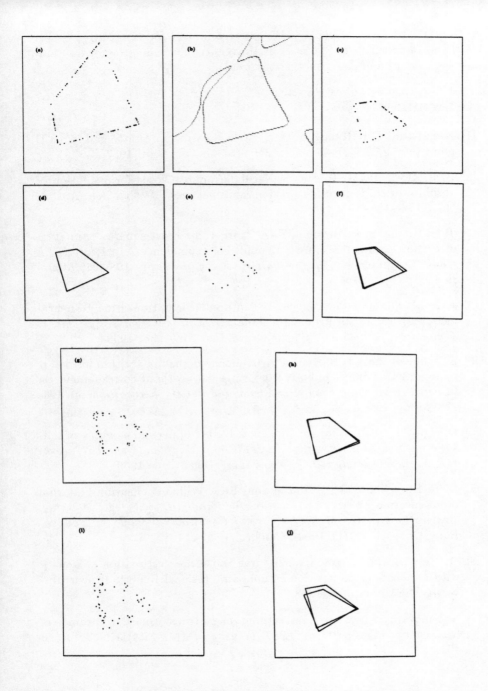

Figure 4: Examples of parameter estimations using similarity transform neural networks for 2-D planar objects.

# RECURSIVE NEURAL NETWORKS FOR SIGNAL PROCESSING AND CONTROL

D. Hush  C. Abdallah  B. Horne
Dept. of Electrical and Computer Engineering
University of New Mexico
Albuquerque, NM 87131 – USA

Abstract – This paper describes a special type of dynamic neural network called the Recursive Neural Network (RNN). The RNN is a single-input single-output nonlinear dynamical system with a nonrecursive subnet and two recursive subnets arranged in the configuration shown in Figure 1. The purpose of this paper is to describe the architecture of the RNN, present a learning algorithm for the network, and provide some examples of its use.

## INTRODUCTION

This paper describes a dynamic neural network structure referred to as the recursive neural network, RNN. This network is particularly well suited for signal processing and control applications. The RNN structure is shown in Figure 1. It consists of three subnets, $A$, $B$, and $C$. All three of the subnets are multi-layer perceptrons with optional high-order connections at the input layer. Subnet $A$ is a *nonrecursive* subnet that processes current and delayed values of the input $d(k)$ while subnets $B$ and $C$ are *recursive* subnets that feedback delayed values from the RNN output, $u(k)$, and the plant output, $y(k)$, respectively. The other components in Figure 1, M and P, are nonlinear dynamical systems which are to be modeled or controlled by the RNN.

This paper is organized as follows. Section 2 describes the operation of the RNN and introduces the notation that will be used. Section 3 presents a learning algorithm for the RNN. Section 4 presents several examples. These examples illustrate the use of the RNN in a variety of scenarios including system identification, nonlinear digital filtering, inverse modeling, control, and nonlinear prediction. Section 5 contains an analysis and summary of the RNN.

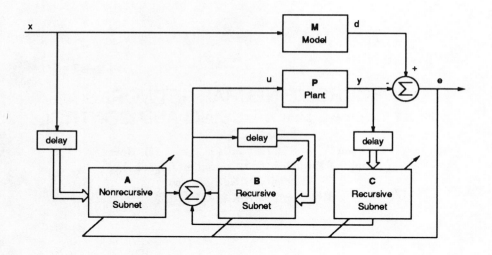

Figure 1: The Recursive Neural Network Structure

## THE RECURSIVE NEURAL NETWORK STRUCTURE

An integral part of the recursive neural network structure shown in Figure 1 is the plant, P. In general we allow P to be a nonlinear dynamical system whose input/output behavior is described by

$$y(k) = g\left(u(k), u(k-1), \ldots, u(k-o_n), y(k-1), \ldots, y(k-o_r)\right)$$

where $o_n$ and $o_r$ are the *nonrecursive* and *recursive* orders of the plant respectively. The plant may be an unknown system, that is we may not know the exact functional form of $g(\cdot)$.

The output of the RNN at time $k$ is the sum of the outputs of three subnets, a nonrecursive subnet $A$ and two recursive subnets $B$ and $C$.

$$u(k) = v_{M_A}^A(k) + v_{M_B}^B(k) + v_{M_C}^C(k)$$

Each of these networks is a multi-layer perceptron with second-order terms at the input layer (SOMLP) [1]. The superscripts $A$, $B$, and $C$ are used to distinguish between parameters in the nonrecursive and the two recursive subnets respectively. $M_A$, $M_B$, and $M_C$ represent the number of layers in subnets $A$, $B$, and $C$ respectively.

If we let $v_l^S(k)$ represent the output vector of the $l^{th}$ layer for the $S^{th}$ subnet ($S \in A, B, C$) then signals are propagated through the networks according to

$$v_l^S(k) = f(W_l^S v_{l-1}^S(k) + w_{l,0}^S)$$

$W_l^S$ represents the weight matrix of subnet $S$ that connects the outputs of layer $l-1$ to the nodes in layer $l$, and $w_{l,0}^S$ are the bias weights for layer $l$.

$f(\cdot)$ is a function which produces a vector result from a vector argument by applying a nonlinear function, $f(\cdot)$ to each component. One of the common choices for the nonlinear function is $f(z_i) = \tanh(z_i)$.

The input vector to subnet $A$, the nonrecursive subnet, consists of current and delayed values of the input signal, and their second order combinations of these values. Thus, $v_0^A(k)$ takes on the form

$$\begin{aligned}v_0^A(k) = [\ &x(k), x(k-1), \ldots, x(k-o_A), \\ &x^2(k), x(k)x(k-1), \ldots, x(k)x(k-o_A), \\ &x^2(k-1), x(k-1)x(k-2), \ldots, x(k-1)x(k-o_A), \\ &x^2(k-2), \ldots, x^2(k-o_A)\ ]\end{aligned}$$

Similarly, the input to subnet $B$ consists of previous RNN outputs (which are the plant inputs) and all second order combinations, and the input to subnet $C$ consists of previous plant outputs and all second order combinations. We will use $o_A$, $o_B$, and $o_C$ to denote the "order" of subnets $A$, $B$, and $C$ respectively (i.e. the number of delayed values that are presented at the inputs to these subnets). Note, all subnets have only one output node so that $\mathbf{W}_{M_A}^A$, $\mathbf{W}_{M_B}^B$, and $\mathbf{W}_{M_C}^C$ are row vectors and $\mathbf{w}_{M_A,0}^A$, $\mathbf{w}_{M_B,0}^B$, and $\mathbf{w}_{M_C,0}^C$ are scalars.

## WEIGHT UPDATE EQUATIONS

The learning algorithm is a gradient search designed to minimize the total sum-of-squared error over the duration of the training signal,

$$E = \frac{1}{2}\sum_{k=1}^{N} e^2(k) = \frac{1}{2}\sum_{k=1}^{N}(d(k) - y(k))^2$$

where $d(k)$ is the output of the model M in Figure 1. The weight updates take on the following form,

$$w_{l,i,j}^S(k+1) = w_{l,i,j}^S(k) + \mu e(k)\beta_{l,i,j}^S(k)$$

where $w_{l,i,j}^S$ is the element of $\mathbf{W}^S$ that corresponds to the weight connecting node $i$ in layer $l-1$ to node $j$ in layer $l$, and

$$\beta_{l,i,j}^S(k) = \frac{\partial y(k)}{\partial w_{l,i,j}^S} = \mathbf{q}_1^T(k)\mathbf{a}_{l,i,j}^S(k) + \mathbf{q}_2^T(k-1)\mathbf{b}_{l,i,j}^S(k-1) \quad (1)$$

where

$$\begin{aligned}\mathbf{q}^T(k) &= \left[\frac{\partial g(\cdot)}{\partial u(k)}, \frac{\partial g(\cdot)}{\partial u(k-1)}, \ldots, \frac{\partial g(\cdot)}{\partial u(k-o_n)}\right] \\ \mathbf{q}_2^T(k-1) &= \left[\frac{\partial g(\cdot)}{\partial y(k-1)}, \frac{\partial g(\cdot)}{\partial y(k-2)}, \ldots, \frac{\partial g(\cdot)}{\partial y(k-o_r)}\right]\end{aligned}$$

are elements of the Jacobian of the plant, and

$$\mathbf{a}_{l,i,j}^{S\,T}(k) = [\alpha_{l,i,j}^S(k), \alpha_{l,i,j}^S(k-1), \ldots, \alpha_{l,i,j}^S(k-o_n)]$$
$$\mathbf{b}_{l,i,j}^{S\,T}(k-1) = [\beta_{l,i,j}^S(k-1), \beta_{l,i,j}^S(k-2), \ldots, \beta_{l,i,j}^S(k-o_r)]$$

If $g(\cdot)$ is known then $\mathbf{q}_1$ and $\mathbf{q}_2$ can be evaluated directly. If $g(\cdot)$ is unknown then $\mathbf{q}_1$ and $\mathbf{q}_2$ must be estimated.

The other vectors in (1) are determined as follows. First we note that the current value of $\beta_{l,i,j}^S(k)$ is a recursive function of previous values through the components of the $\mathbf{b}_{l,i,j}^S$ vector. The components of $\mathbf{a}_{l,i,j}^S$ are given by [2],

$$\alpha_{l,i,j}^S(k) = BP_{l,i,j}^S(k) + FF^B(\mathbf{a}_{l,i,j}^S(k-1)) + FF^C(\mathbf{b}_{l,i,j}^S(k-1))$$

where $BP_{l,i,j}^S(k)$ are the standard backpropagation equations [3] (sometimes called "static backprop") for subnet $S$, and the $FF^S(\cdot)$ operations are given for any vector $\mathbf{c}$ by

$$FF^S(\mathbf{c}) = h_{M_S}^S(k)\mathbf{W}_{M_S}^S \mathbf{H}_{M_S-1}^S(k)\mathbf{W}_{M_S-1}^S \ldots \mathbf{H}_1^S(k)\mathbf{W}_1^S[\nabla_\mathbf{r} \mathbf{v}_0^S(k-1)]\mathbf{c}$$

where,

$$\mathbf{H}_l^S(k) = \mathrm{diag}\left(h_{l,n}^S(k)\right) \qquad n = 1, 2, \ldots, N_l$$

$$h_{l,n}^S(k) = \left.\frac{\partial f(z)}{\partial z}\right|_{z_{l,n}^S(k)}$$

and,

$$\mathbf{r}^T = [u(k-1)\ u(k-2)\ \ldots\ u(k-o_B)] \qquad \text{for subnet } S = B$$
$$\mathbf{r}^T = [y(k-1)\ y(k-2)\ \ldots\ y(k-o_C)] \qquad \text{for subnet } S = C$$

Note that for convenience we have defined $v_{l,0}$, the input to the bias weights, to be 1. Also, $N_l$ represents the number of nodes in layer $l$.

## EXAMPLES

In this section we illustrate the use of the RNN in a variety of applications. These include system identification, nonlinear digital filtering, inverse modeling, control, and nonlinear prediction.

### The RNN in System Identification

In this application subnet $C$ is omitted, P=1, and M represents the unknown nonlinear system to be identified (refer to Figure 1). The two subnets $A$ and $B$ allow the RNN to model both the recursive and nonrecursive behavior of M.

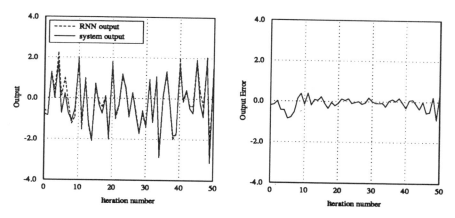

Figure 2: Example of System Modeling Using MLPs

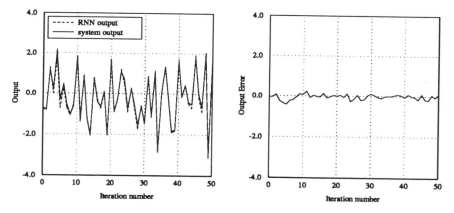

Figure 3: Example of System Modeling Using SOMLPs

One can obtain a linear IIR filter from the RNN by reducing both subnets to a single linear node. If, in addition, subnet $B$ is omitted one obtains a linear FIR filter. Procedures for adapting the weights of both FIR and IIR filters are well known. The gradient learning algorithm which is used to adapt the weights of the RNN reduces to the well known Recursive LMS algorithm when subnets $A$ and $B$ are linear [4]. It is worth mentioning that with the weights of the structure fixed and bounded the RNN is BIBO stable. In contrast, this property does not hold in general for the linear IIR filter.

As an example we have trained the RNN to model the input/output behavior of the nonlinear system described by the following difference equation [5]

$$y(k) = x(k) + \frac{y(k-1)y(k-2)\left(y(k-1)+2.5\right)}{1+y^2(k-1)+y^2(k-2)} \qquad (2)$$

For proper generalization the network should be trained with white noise.

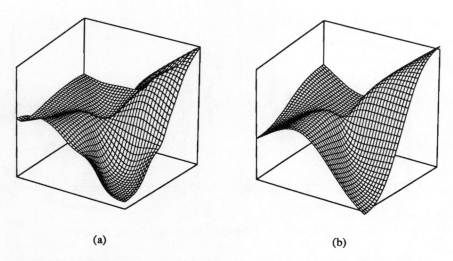

Figure 4: The Mapping (a) of Equation (3), (b) Learned by the RNN.

In our first simulation we used only linear terms in the input layers to the MLPs. The training signal consisted of uniformly distributed white noise over the range [-2,2]. The nonrecursive subnet had two hidden layer nodes and the recursive subnet had thirty. The inputs to the subnets were both of order 2, that is $o_A = o_B = 2$. This yielded a network with a total of 121 weights. Figure 2 compares the output of the RNN (after training) to the output of the actual system when presented with a new white noise.

In our second simulation we used second order terms at the input to the recursive subnet (i.e. a SOMLP is used for subnet $B$). The input signal to the nonlinear system and the RNN in this case was uniform white noise over the interval [-3,3]. Here we found that the RNN was capable of producing a model with the same accuracy as in the first example with only 1 layer (with 1 node) in the nonrecursive subnet, and 2 layers with 6 hidden layer nodes in the recursive subnet. This implementation contains a total of 46 weights. An example of the operation of this network (after training) is shown in Figure 3.

A closer examination of the manner in which the RNN actually forms the model in this second case suggests that the nonrecursive subnet simply passes the input $x(k)$ with a gain of 1, and that the recursive subnet models the recursive part of the system in (2). That is, the recursive subnet attempts to produce a mapping of the form

$$q_3 = \frac{q_1 q_2 (q_1 + 2.5)}{1 + q_1^2 + q_2^2} \tag{3}$$

Actually the recursive subnet is not asked to produce $q_3$ for all possible pairs $[q_1, q_2]$, only those that correspond to valid state trajectories in the nonlinear system. The mapping in (3) is shown in Figure 4 along with the mapping produced by the recursive subnet.

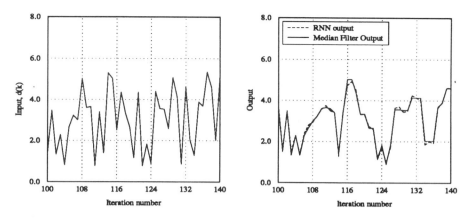

Figure 5: Example of nonlinear filtering.

## The RNN in Nonlinear Filtering

When the RNN is used for nonlinear filtering the network configuration is the same as in the previous section. That is, subnet $C$ is omitted and P=1. The difference in this case is that M represents a known nonlinear filtering operation that we wish to implement with the RNN. In the general case both subnets $A$ and $B$ will be needed. Many popular nonlinear filtering techniques are nonrecursive however, and can be implemented using only subnet $A$. As an example we trained the RNN (subnet $A$ only with linear input terms) to act as a 1-D median filter with a window size of 3 samples (i.e. $o_A = 2$). The subnet had two layers with 12 nodes in the hidden layer. These results are illustrated in Figure 5 which shows the input to the filter, the median filter output, and the RNN output.

## The RNN in Inverse Modeling

In this application the configuration in Figure 1 is such that M=1 and P represents the system that we wish to invert. As in the previous applications subnet $C$ is omitted. Here the RNN is trained to form an inverse model of P so that the transformation from the input of the RNN to the output of P is 1 (possibly with a fixed delay, that is M may be equal to $z^{-\Delta}$). Obviously P must be invertible before the RNN can be applied successfully in this application.

As an example, the RNN was trained to form an inverse model of the plant described by the difference equation in (2). To insure proper generalization the network was trained with uniform white noise in the range [-3,3]. Both subnets in the RNN had two layers, the nonrecursive subnet had 6 hidden layer nodes, and the recursive subnet had 2. Second-order terms were computed at the input to both subnets and $o_A = o_B = 2$. This resulted in an overall network structure with 82 weights. The plots in Figure 6 show the

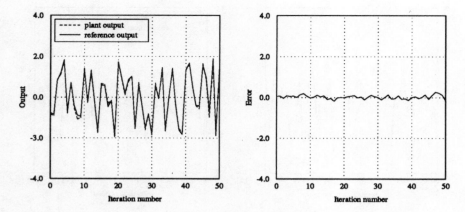

Figure 6: Example of Inverse Modeling

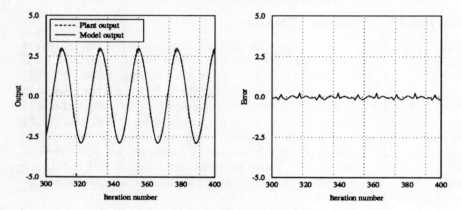

Figure 7: Example of Model Following

output of the plant (after training) with the RNN subjected to a different white noise input.

## The RNN in Control

In this application all three subnets may be used. M is a model that we would like P to follow. Obviously the plant must be controllable before the RNN can be applied successfully in this application. Fortunately, in many control applications, the plant is required to follow the model for specific inputs only. This often makes the training task easier since generalization is not as much an issue.

As an example the RNN was trained to make a plant described by the difference equation in (2) follow a linear model described by [5],

$$d(k) = 0.32d(k-1) + 0.64d(k-2) - 0.5d(k-3) + x(k-1)$$

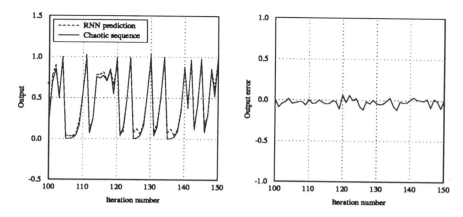

Figure 8: Example of Nonlinear Prediction

The input in this example was a sinusoid of the form $x(k) = \sin(2\pi k/25)$. Subnet $B$ was disabled so that only subnets $A$ and $C$ were used. Subnet $A$ contained a single linear node with $o_A = 2$. Subnet $C$ was a two layer network with $o_C = 2$, linear input terms, and 8 hidden layer nodes for a total of 37 weights. Figure 7 compares the output of the model to the output of the plant after training.

### The RNN in Nonlinear Prediction

When the RNN is used for prediction the structure in Figure 1 is such that P=1, M=1, and subnet $C$ is omitted. This leaves only the RNN with subnets $A$ and $B$. For $\Delta$-step prediction the input to subnet $A$ is delayed by $\Delta$. Thus, the RNN is asked to use past inputs $x(k-\Delta), x(k-\Delta-1)$, etc. to predict the current input, $x(k)$. As an example the RNN was trained to perform 2-step prediction of the chaotic sequence $x(k) = 4.0x(k-1)[1.0 - x(k-1)]$ In this example only subnet $A$ was used. It had three layers with $o_A = 2$, second order input terms, 8 nodes in the first hidden layer, and 4 nodes in the second hidden layer for a total of 89 weights. The results are illustrated in Figure 8 which shows the actual and predicted sequences after training.

### SUMMARY

We have introduced a nonlinear dynamical system called the recursive neural network and presented a learning algorithm for this network based on a gradient search. The network has been shown to be useful in a variety of applications including system modeling, nonlinear filtering, inverse modeling, nonlinear prediction, and control.

The recursive neural network was motivated by problems and concepts from nonlinear filtering and control. It closely resembles the architectures proposed in [5], and can be viewed as a complement to that work.

The RNN is also similar in capability to the recurrent neural networks discussed in [6], in that both networks are nonlinear dynamical structures that contain adjustable weights. In the RNN the states are created by feeding the input and output of the net through an observer composed of the $o_n$ and $o_r$ delays. In recurrent neural networks every node in the network has access to its own state and assumes the ability of directly measuring the state of a system.

It should be cautioned that this paper and the simulated examples are not meant to illustrate the universality or the ease of using the RNN. Indeed, we can point out that the behavior of the closed-loop system is sensitive to many design parameters and modeling assumptions. The fact remains however, that after extensive simulation runs, the RNN presents itself as a useful alternative when analytical methods are lacking.

## REFERENCES

[1] C. Giles and T. Maxwell, "Learning, invariance, and generalization in high-order neural networks," *Applied Optics*, vol. 26, no. 23, pp. 4972–4978, 1987.

[2] D. Hush, C. Abdallah, and B. Horne, "The recursive neural network," Tech. Rep. EECE 91-002, Department of Electrical and Computer Engineering, University of New Mexico, 1991.

[3] D. Rumelhart, G. Hinton, and R. Williams, "Learning internal representations by error propagation," in *Parallel Distributed Processing: Explorations in the Microstructure of Cognition* (D. Rumelhart and J. McClelland, eds.), pp. 318–362, Cambridge, MA: MIT Press, 1986.

[4] B. Widrow and S. Stearns, *Adaptive Signal Processing*. Englewood Cliffs, NJ: Prentice Hall, 1985.

[5] K. Narendra and K. Parthasarathy, "Identification and control of dynamical systems using neural networks," *IEEE Transactions on Neural Networks*, vol. 1, pp. 4–27, March 1990.

[6] F. Pineda, "Generalization of backpropagation to recurrent and higher order neural networks," in *Neural Information Processing Systems* (D. Anderson, ed.), pp. 602–611, American Institute of Physics, 1988.

# A NEURAL ARCHITECTURE FOR NONLINEAR ADAPTIVE FILTERING OF TIME SERIES

Nils Hoffmann and Jan Larsen
The Computational Neural Network Center
Electronics Institute, Building 349
Technical University of Denmark
DK-2800 Lyngby, Denmark

## INTRODUCTION

The need for nonlinear adaptive filtering may arise in different types of filtering tasks such as prediction, system identification and inverse modeling [17]. The problem of predicting chaotic time series has been addressed by several authors [6],[11]. In the latter case a feed-forward neural network was used both for prediction and for identification of a simple, nonlinear transfer function (system identification). These filtering tasks can be solved using a filtering configuration shown in Fig. 1 [17].

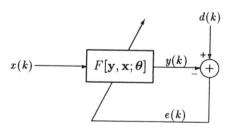

Figure 1: Nonlinear adaptive filtering configuration. $x(k)$ is the input, $y(k)$ the output and $d(k)$ the desired signal. The filter is adapted in order to minimize the cost function: $\sum_{i=0}^{k} \lambda^{k-i} e^2(i)$, where $k = 1, 2, \cdots, N$ and $0 < \lambda \leq 1$ is the forgetting factor [7].

The nonlinear filter may be designed to realize

$$y(k) = F[y(k-1), \cdots, y(k-M), x(k), \cdots, x(k-L+1); \boldsymbol{\theta}] \qquad (1)$$

where $F[\cdot]$ is an unknown nonlinear function parameterized by $\boldsymbol{\theta}$, $k$ is the discrete time index and $L, M$ are filter orders.

The general structure of equation (1) enables one to model any nonlinear, discrete system. $\theta$ is assumed to be slowly time varying and consequently $y(k)$ is quasi stationary. The use of a recursive, nonlinear filter does, however, pose serious difficulties regarding stability. The filter may display limit cycles, chaotic behaviour and unboundedness. The scope of this paper is to implement only the nonrecursive part of (1).

We propose a modularized architecture for the nonlinear filter in Fig. 1 including algorithms for adapting the filter. Further we develop simple guidelines for selecting a specific filter design within the proposed architecture given a priori knowledge of the distribution and origin (type of modeling problem) of the input signal $x(k)$ and the desired response $d(k)$. Finally we present simulations in order to further investigate the nature of the relations between filter design and the statistics of $x$ and $d$.

## NONLINEAR FILTER ARCHITECTURE

The proposed filter architecture is shown in Fig. 2. The filter, which may be viewed as a generalization of the Wiener Model [15], is divided into three partially independent (depending on specific design) sections: A preprocessing unit containing the filter memory, a **m**emoryless, **m**ultidimensional **n**onlinearity (MMNL) and a linear combiner. The structure is selected in order to modularize the modeling problem which ensures a proper and sparse parameterization, and facilitates incorporation of a priori knowledge in contrast to the limited possibilities when using an ordinary feed-forward neural network. The main objective of the preprocessor is to extract the essential

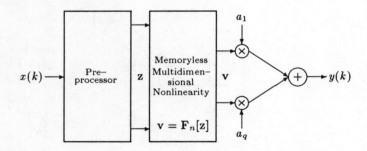

Figure 2: Nonlinear filter architecture.

information contained in $\mathbf{x}_k = [x(k), x(k-1), \cdots, x(k-L+1)]'$ ensuring that $\mathbf{z}$ has a dimension, $p \leq L$. The nonlinearity is memoryless and transforms the vector $\mathbf{z}$ into the vector $\mathbf{v}$, and finally the linear combiner forms a weighted sum $y(k)$ of the terms in $\mathbf{v}$. This corresponds to rewriting (1) as:

$$y(k) = \mathbf{a}' \mathbf{F}_n[\mathbf{F}_p(x(k))] \qquad (2)$$

where $\mathbf{F}_p(\cdot)$ is the preprocessor, $\mathbf{F}_n[\cdot]$ is the MMNL and **a** the weigths of the linear combiner. The filter could be viewed as a heterogeneous multilayer neural network. All sections could be adapted, but this is not always necessary, as we shall see in the next section.

## FILTER DESIGN

### Signal Dependence

It seems reasonable that the specific design of the filter depends on the origin and distribution of the signals $x(k)$ and $d(k)$, and we will summarize some guidelines for choosing an appropriate design as follows:

| Case | x | d | Model |
|------|---|---|-------|
| 1 | Gaussian | Gaussian | $d(k) = \mathbf{a}'\mathbf{x}_k + \epsilon(k)$ |
| 2 | Gaussian | Non-Gaussian | $d(k) = F[\mathbf{x}_k] + \epsilon(k)$ |
| 3 | Non-Gaussian | Gaussian | $d(k) = F[\mathbf{x}_k] + \epsilon(k)$ |
| 4 | Non-Gaussian | Non-Gaussian | $d(k) = F[\mathbf{x}_k] + \epsilon(k)$ |

1. The linear filter is optimal in this case, so $\mathbf{v} = [1\ \mathbf{z}']'$, see e.g. [13, Theorem 14.3].
2. The Wiener model, i.e. a bank of orthogonal linear filters in the preprocessing unit followed by a fixed (non-adaptive) polynomial nonlinearity, provides a filter which can be adapted in a very simple and fast way [15]. The linear filters may be determined using principal component analysis (PCA) on **x**. This case is associated with the problem of nonlinear system identification where $d(k)$ is the output of an unknown system and $x(k)$ the input.
3. In this case there is no obvious choice of filter design. The case arises e.g. in inverse modeling where $d(k)$ is the driving signal of the unknown system and $x(k)$ the resulting output.
4. This case relates to prediction of nonlinear time series where $x(k) = d(k-\tau)$ is, in fact, a delayed version of $d(k)$. Previous simulation studies [11] indicate that the nonlinearity should be constructed from bounded functions (e.g. the commonly used $\tanh(\cdot)$) rather than polynomials, which have the inconvenient property of growing fast towards infinity.

### Preprocessing Methods

We present two possible methods for dimensionality determination. If the unknown system being modeled can be described in the form of a nonlinear differential equation it is possible, in some cases, to determine dimensionality by letting $\mathbf{z} = [x(k), Dx(k), \cdots, D^{p-1}x(k)]'$ where $D$ is a discrete derivative operator. We denote this preprocesor: The **derivative preprocessor (DPP)**.

The following example (the pendulum) illustrates that we often have $p < L$ without loosing information:

$$\frac{d^2x(t)}{dt^2} + \beta \frac{dx(t)}{dt} + \alpha \sin[x(t)] = d(t) \tag{3}$$

$$D^2x(k) + \beta\, Dx(k) + \alpha \sin[x(k)] = F[D^2x(k), Dx(k), x(k)] = d(k) \tag{4}$$

Equation (4) which is a discrete approximation of (3) clearly shows, that $d(k)$ may be expressed as a function of only $p = 3$ variables i.e. the derivatives of $x(k)$. To ensure that the approximation of the derivative operator $D$ is accurate (i.e. approximates the continuous time derivative operator) over a broad range of frequencies it must be implemented using a linear filter with high order. A tapped delay line used without a preprocessing element would necessarily need the same length $L$ to hold the same information so $L$ is obviously greater than $p$ in this example. In practice there exist two major problems with this kind of preprocessing: 1. Derivatives amplify noise (SNR is often low at high frequencies which are amplified the most) and 2. The optimal differentiating filter is non-causal. The first problem may be dealt with by noise reducing lowpass-filtering (see [3] for an optimal approach). The second obstacle may be circumvented by delaying $d(k)$ thus allowing for non-causal filtering i.e. estimating $d(k - r)$ $(r \geq 0)$ using $x(k), x(k-1), \cdots$.

Another method is principal component analysis (PCA) which serves two purposes: 1. PCA makes the components in $z$ mutually uncorrelated (convergence speed-up for certain weight estimation algorithms, e.g. Backpropagation [8]) and 2. It determines the dimensionality which is done by removing the last $L - p$ principal components (PC's) with eigenvalues close to zero. The amount of lost information can be estimated as the total variance of the removed PC's divided by the total variance of $x$, i.e. $L \cdot V\{x(k)\}$. The remaining PC's constitute the optimal *linear* projection (in the mean square sense) of $x$ on the space spanned by the first $p$ eigenvectors of the covariance matrix of $x$. A theoretical wellfounded procedure for on-line estimation of the PC's has recently been described [18]. Other schemes are given in [8, Chap. 8.3].

Further support for the use of PCA can be found in an information theoretical interpretation: Maximize the mutual information $I(\mathbf{x}; \mathbf{z})$ between $\mathbf{x}$ and $\mathbf{z}$. It can be shown that if $\mathbf{z}$ is Gaussian (also valid for certain similar probability density functions): $\max I(\mathbf{x}; \mathbf{z}) = \max H(\mathbf{z}) \Leftrightarrow \max \det V\{\mathbf{z}\}$ where $V\{\cdot\}$ is the variance and $H(\cdot)$ is the entropy. PCA is in fact done by maximizing $\det V\{\mathbf{z}\}$ [10, p. 682] which implicates that dimensionality determination by means of PCA is equivalent to maximizing $I(\mathbf{x}; \mathbf{z})$. When dealing with signals corrupted by noise PCA is not always preferable (especially if the signal to noise ratio is low) because the PC's then will reflect the noise. Furthermore using PCA when the spectral overlap of the signals $x(k)$ and $d(k)$ is small is not reasonable. This is due to the fact that the spectrum of the PC's corresponding to large eigenvalues mainly contains the dominating frequencies in $x(k)$ thus neglecting the frequencies that dominate the spectrum of $d(k)$.

**Memoryless Multidimensional Nonlinearities**

When approximating the MMNL, $\mathbf{F}_n[\mathbf{z}]$, $\mathbf{z} \in \mathcal{I}$ where $\mathcal{I}$ is the input space, we distinguish between *local* and *global* approximation methods [6]. In a local approximation context $\mathcal{I}$ is divided into smaller domains. $F$ is now approximated in each domain by separate nonlinearities. This results in a modularization of the MMNL which ensures a sparse parameterization. By global approximation is meant, that no dividing of $\mathcal{I}$ is done at all. In general there is a trade off between the number of domains and the complexity of the subsequent nonlinearities.

**Global Approximation Methods.** In this case we deal with only one nonlinearity which must have the ability of approximating $F$ arbitraryly accurate. We will discriminate between *fixed* and *adaptive* nonlinearities.

A natural choice of a fixed nonlinearity (FNL) is to let $\mathbf{v}$ contain all possible products of $z_i$, e.g. terms of the form $\prod_{i=1}^{p} z_i^{s_i}$ up to some order $s = \sum_{i=1}^{p} s_i$. When these terms are added by the linear combiner it all amounts to a multidimensional Taylor expansion which combined with the linear filters in the preprocessor defines a discrete Volterra filter. Fréchet showed [15] that any continuous $F(x(t))$ can be represented by a Volterra filter with uniform convergence when $s \to \infty$ for $x(t) \in \mathcal{J}$, $\mathcal{J} \subseteq \mathcal{I}$. A convenient representation can be obtained by using a complete set of orthogonal polynomials $P_i$, where $i$ is the order of the polynomial. If $z_i \in N(0,1)$, $P_i$ are identical to the Hermite polynomials. With these polynomials convergence in mean is assured over a suitable interval $[a;b]$. The generalization to the multidimensional case is done by forming all products of polynomials in different variables e.g. $\prod_{i=1}^{p} P_{s_i}(z_i)$ (see [15] for details). In general the probability density $f_\mathbf{z}(\mathbf{z})$ is unknown which makes it impossible to find the orthogonal polynomials. Instead we propose the use of Chebychev polynomials preceeded by squashing functions that limits the $z_i$ to the interval $]-1;1[$ thereby limiting the $v_i$ to $]-1;1[$.

An obvious choice for an adaptive nonlinearity (ANL) is a layered feed-forward neural network composed of sigmoidal neurons. It is well-known (see e.g. [5]) that a two layer feed-forward network with a linear output neuron (under rather mild conditions on the activation function, $g$) can uniformly approximate any function as the number of neurons in the hidden layer reaches infinity. We suggest that the nonlinearity is composed of to layers. The first layer consists of $p$ neurons which maps $z_i$, $1 \leq i \leq p$ into $g(z_i w_i^1 + w_i^0)$ ($w_i^1$ ensures a proper scaling, see below). The second layer consists of $q$ neurons ($q \to \infty$ for an arbitrarily accurate approximation) and the outputs then form the nonlinear terms $[v_1, \cdots, v_{q-p}]'$. It is further suggested to explicitly model the linear part by letting $[v_{q-p+1}, \cdots, v_q]' = [z_1, \cdots, z_p]'$.

**Local Approximation Methods.** A possible way to divide the input space is to use Localized Receptive Fields [14]. The output from each receptive field is then fed into separate nonlinearities. As above they could be either fixed or adaptive. Note that there is a trade-off between the number of domains in the input space and the complexity of the succeeding nonlinearities. Other local approximation schemes can be found in [6], [16], [15, Chap. 21].

**Scaling of z.** Scaling of z serves two purposes. First, we have to restrict $z_i$ to an interval where the nonlinearity is slowly varying; i.e. neither growing towards infinity (as polynomials for large arguments) nor being constant (like $\tanh(\cdot)$ for large arguments). Secondly, we have to ensure that only the significant amplitude range of $z_i$ (i.e. the interval where $f_{z_i}(z_i) > \varepsilon$, $0 < \varepsilon \ll 1$) is fed into the filter. Otherwise very unlikely values of $z_i$ will be weighted too much in the cost function thus resulting in a poor performance. Scaling with a suitable measure of $z_i$, e.g. 2–3 standard deviations, serves this purpose.

## Weight Estimation Algorithms

The task is to estimate the weights $\theta$ so that the cost function $\sum_{i=0}^{k} \lambda^{k-i} e^2(i)$, $k = 1, 2, \cdots, N$ is minimized [7] where $e$ is the difference between the desired and the actual response and $0 < \lambda \leq 1$ is the forgetting factor.

**Fixed Nonlinearity.** In designs with a FNL it is only necessary to adapt the linear combiner. This is especially simple if **x** is Gaussian and PCA is used as preprocessing making **z** white (independent) and Gaussian. Now if $z_i$ is scaled to unity variance and Hermite polynomials are used in the nonlinearity then the $v_j$ will be uncorrelated and the weights may thus be updated using the crosscorrelation method proposed in [15]: $a_j = C\{v_j d\}/V\{v_j\}$, $1 \leq j \leq q$ where $C\{v_j d\}$ is the covariance between $v_j$ and $d$ and $V\{v_j\}$ is the variance of $v_j$. In most cases, however, **v** is non-white but owing to the fact that $y(k)$ is linear in the weights, adaptive algorithms known from linear adaptive filtering such as the recursive least squares (RLS) [7, p. 385] or the least mean squares (LMS) [17, p. 99] are usable. The latter is perhaps the best choice for large values of $q$ because it needs less computations and memory capacity while the major advantage of the RLS is the much faster convergence for highly correlated inputs ($v_j$).

**Adaptive Nonlinearity.** Designs with ANL implicate that estimation of the weights is a nonlinear optimization task which in general is hard to solve (global optimization) but local optimization schemes have been given, e.g. **Backpropagation (BP)**. BP is known to have very slow convergence [4]. There is therefore a need for development of algorithms with faster convergence. Several second-order algorihms (SOA) have been proposed, see e.g. [4]. A SOA incorporates the information contained in the Hessian (**H**) of the cost function and the weights are updated according to the Newton-Raphson algorithm. In contrast to BP the SOA parameters are given a natural interpretation. $0 < \lambda \leq 1$ is the exponential forgetting factor, $0 < \mu \leq 1$ is the stepsize which normally is non-critical, and $\delta$ ($\mathbf{H}^{-1} = \delta \mathbf{I}$) is initially chosen large. We suggest a further development that takes the problems of nearly singular Hessian matrices into account. This problem arises in "flats" parts of the cost function. It is proposed to use the U-D factorization of $\mathbf{H}^{-1}$ due to Bierman [2].

## SIMULATIONS

### Simulated Systems

In order to compare filter designs when filtering signals with different origin and distribution we study three systems covering the cases 2–4 on p. 3.

| Example | x | d | Model |
|---|---|---|---|
| System Identification (SI) | Band-limited Gaussian noise | Non-Gaussian | Equation (5) |
| Inverse Modelling (IM) | Non-Gaussian | Gaussian lowpass filtered noise | Equation (6) |
| Prediction (P) | Non-gaussian | Non-gaussian | Equation (7) |

**System Identification.** A simple system describing a wave force problem is given by Morison's equation [1, p. 234]. $x(t)$ is the wave velocity and $d(t)$ the wave force. The desired signal $d(k)$ is a discrete version of $d(t)$ sampled with the sampling period $\Delta T = 1$ and the same applies to $x(k)$.

$$d(t) = 0.2\frac{dx(t)}{dt} + 0.8x(t)|x(t)| \tag{5}$$

**Inverse Modelling.** We consider the pendulum where $x(t)$ is the angle deflection and $d(t)$ the force. The desired signal $d(k)$ is a discrete version of $d(t - \tau)$ (sampled with $\Delta T = 0.05$) where $\tau$ is a delay aiming to cancel both the delay between $d(t)$ and $x(t)$ and the delay in the preprocessing unit of the nonlinear filter. $x(k)$ corresponds to the angle $x(t)$.

$$\frac{d^2x(t)}{dt^2} + 0.2\frac{dx(t)}{dt} + 4\pi^2 \sin[x(t)] = d(t) \tag{6}$$

**Prediction.** The signal $x(t)$ is generated by the chaotic Mackey-Glass equation which often is used in a benchmark test for nonlinear predictors [6]. $d(k)$ is a discrete version of $x(t)$ and $x(k)$ a delayed, discrete version of $x(t - \tau)$ where $\tau$ signifies how far ahead we predict. Sampling the signal with $\Delta T = 1$ $\tau$ equals 100 timesteps like in [6],[11].

$$\frac{dx(t)}{dt} = -0.1x(t) + \frac{0.2x(t - 17)}{1 + x(t - 17)^{10}} \tag{7}$$

The systems mentioned above have all been simulated using discrete approximations of the derivatives. These discrete filters are all non-recursive which means that a non-recursive nonlinear adaptive filter is adequate. The actual training and cross validation signals have been obtained by decimating the input and output signals in order to avoid oversampling.

## Numerical Results

In the table below are listed the main results. A measure of the filter performance is given by the error index: $E = \sigma_e/\sigma_d$, where $\sigma_e$, $\sigma_d$ denote the standard deviation of the error and the desired signals respectively (cross validation). The number of parameters $W$ gives an indication of the complexity.

| Ex. | Prep. | L | p | Nonlinearity | | | | | |
|-----|-------|---|---|---|---|---|---|---|---|
|     |       |   |   | Fixed | | Adaptive | | None | |
|     |       |   |   | E | W | E | W | E | W |
| SI  | PCA   | 14 | 4 | 0.263 | 94 | 0.215 | 158 | 0.417 | 5 |
| SI  | DPP   | 19 | 2 | 0.200 | 46 | 0.107 | 128 | 0.389 | 3 |
| SI  | None  | 14 | - | - | - | - | - | 0.382 | 15 |
| IM  | DPP   | 19 | 3 | 0.075 | 152 | 0.116 | 131 | 0.448 | 4 |
| IM  | None  | 19 | - | - | - | - | - | 0.402 | 20 |
| P   | PCA   | 20 | 4 | 0.152 | 170 | - | - | 0.539 | 5 |
| P   | None  | 20 | - | - | - | - | - | 0.539 | 21 |

In all simulations we have used 9000 samples for training and 8000 for cross validation. During training an algorithm based on a statistical test [12], [9] was used to eliminate non-significant weights which accounts for the variations in W. The FNL consisted of bounded Chebychev polynomials and the ANL was implemented using af multilayer neural net. In both cases we used 2. order algorithms for adapting the weights. In general the simulations indicate that the nonlinear filters are clearly superior to the linear with respect to $E$. This improvement is, however, gained at the expense of an increased complexity. The FNL and ANL's seems to show roughly equal perfomance on the selected examples, with the ANL having a better parameterization in the SI example (more significant weights and lower $E$) and vice versa in the IM example. The two preprocessing methods seem to complement each other. In the examples SI,IM, where the equations can be closely approximated with discrete derivative operators of low order, the use of discrete differentiating filters in the preprocessing unit yields a better performance with lower complexity than the use of PCA. This shows that the discrete derivatives are more informative than the PC's in the chosen examples. Using PCA in the example with the pendulum is in fact extremely bad because the PC's mainly reflect the low frequency components in $x(k)$ while the high frequencies carry most of the information about $d(k)$. In contrast, PCA works very well for the example of prediction whereas it is not easy to approximate the Mackey-Glass equation using only low-order derivatives (i.e. the use of a DPP is a bad choice). Finding an appropriate preprocessing method seems thus to require knowledge of an approximate mathematical model for the unknown system. Alternatively a rule of thumb saying that the preprocessor should make the spectrums of the $z_i(k)$ "close" to the spectrum of $d(k)$ could be used. In the prediction example we have used a PCA with $L = 20$ and $p = 4$ which allows us to compare our results with the ones obtained in [11] where a performance of $E = 0.054$ was found using an ANL and a total of 171 weights and $E = 0.28$ using a 6th order fixed polynomial nonlinearity and 210 weights. This indicates, that allthough the ANL still performs better on this example an increase in performance of the FNL has been gained by using bounded polynomials.

## CONCLUSION

In this paper a neural architecture for adaptive filtering which incorporates a modularization principle is proposed. It facilitates a sparse parameterization, i.e. fewer parameters have to be estimated in a supervised training procedure. The main idea is to use a preprocesssor which determine the dimension of the input space and further can be designed independent of the subsequent nonlinearity. Two suggestions for the preprocessor are presented: The derivative preprocessor and the principal component analysis. A novel implementation of fixed Volterra nonlinearities is given. It forces the boundedness of the polynominals by scaling and limiting the inputs signals. The nonlinearity is constructed from Chebychev polynominals. We apply a second-order algorithm for opdating the weights for adaptive nonlinearities based on previous work of Chen et al. [4] and the U-D factorization of Bierman [2]. Finally the simulations indicate that the two kinds of preprocessing tend to complement each other while there is no obvious difference between the performance of the ANL and FNL.

## ACKNOWLEDGEMENTS

We would like to thank Lars Kai Hansen, Peter Koefoed Møller, Klaus Bolding Rasmussen and John E. Aasted Sørensen for helpfull comments on this paper.

## REFERENCES

[1] J.S. Bendat, Nonlinear Systems Analysis and Identification, New York: JOHN WILEY & SONS, 1990.

[2] G.J. Bierman, "Measurement Updating using the U-D Factorization," in Proceedings of the IEEE Int. Conf. on Decision and Control, 1975, pp. 337–346.

[3] B. Carlsson, A. Ahlén & M. Sternad, "Optimal Differentiation Based on Stochastic Signal Models," IEEE Transactions on Signal Processing, vol. 39, no. 2, 341–353, 1991.

[4] S. Chen. , C.F.N. Cowan, S.A. Billings & P.M. Grant, "Parallel Recursive Prediction Error Algorithm for Training Layered Neural Networks," Int. Journal of Control, vol. 51, no. 6, pp. 1215–1228, 1990.

[5] G. Cybenko, "Approximation by Superposition of a Sigmoidal Function," Math. Control Signal Systems, no. 2, pp. 303–314, 1989.

[6] J.D. Farmer & J.J. Sidorowich, "Exploiting Chaos to Predict the Future and Reduce Noise," Technical Report LA-UR-88, Los Alamos National Laboratory, 1988.

[7] S. Haykin, Adaptive filter theory, Englewood Cliffs, New Jersey: PRENTICE-HALL, 1986.

[8] J. Hertz, A. Krogh & R.G. Palmer, Introduction to the Theory of Neural Computation, Redwood City, California: ADDISON-WESLEY PUBLISHING COMPANY, 1991.

[9] N. Hoffmann & J. Larsen, "An Algorithm for Parameter Reduction in Nonlinear Adaptive Filters," in preparation for submission to IEEE International Conference on Acoustics, Speech, and Signal Processing, San Francisco Marriott, California, March 23-26 1992.

[10] P.R. Krishnaiah (ed.), Multivariate Analysis 2, New York: ACADEMIC PRESS, 1969.

[11] A.S. Lapedes & R. Farber, "Nonlinear Signal Processing Using Neural Networks, Prediction and System Modeling," Technical Report LA-UR-87, Los Alamos National Laboratory, 1987.

[12] J. Larsen & L.K. Hansen, "Reduction of Neural Network Complexity," submitted to Neural Information Processing Systems, Denver, Colorado, Dec. 2-5, 1991.

[13] J.M. Mendel, Lessons in Digital Estimation Theory Englewood Cliffs, New Jersey: PRENTICE-HALL, 1987.

[14] J. Moody & C.J. Darken, "Fast Learning in Networks of Locally-Tuned Processing Units," Neural Computation, no. 1, pp. 281–294, 1989.

[15] M. Schetzen, The Volterra and Wiener Theories of Nonlinear Systems, Malabar, Florida: ROBERT E. KRIEGER PUBLISHING COMPANY, 1989.

[16] H. Tong & K.S. Lim, "Threshold Autoregression, Limit Cycles and Cyclical Data," Journal of the Royal Statistical Society, vol. 42, no. 3, pp. 245-292, 1980.

[17] B. Widrow & S.D. Stearns, Adaptive Signal Processing, Englewood Cliffs, New Jersey: PRENTICE-HALL, 1985.

[18] Kai-Bor Yu, "Recursive Updating the Eigenvalue Decomposition of a Covariance Matrix," IEEE Transactions on Signal Processing, vol. 39, no. 5, pp. 1136–1145, 1991.

# ORDERED NEURAL MAPS AND THEIR APPLICATIONS TO DATA COMPRESSION

Eve A. Riskin      Les E. Atlas
Shyh-Rong Lay
Department of Electrical Engineering, FT-10
University of Washington
Seattle, WA 98195

Abstract—The implicit ordering in scalar quantization is used to substantiate the need for explicit ordering in vector quantization and the ordering of Kohonen's neural net vector quantizer is shown to provide a multidimensional analog to this scalar quantization ordering. Ordered vector quantization, using Kohonen's neural net, was successfully applied to image coding and was then shown to be advantageous for progressive transmission. In particular, the intermediate images had a signal-to-noise ratio that was quite close to a standard tree-structured vector quantizer, while the final full-fidelity image from the neural net vector quantizer was superior to the tree-structured vector quantizer. Subsidiary results include a new definition of index of disorder which was empirically found to correlate strongly with the progressive reduction of image signal-to-noise ratio and a hybrid neural net-generalized Lloyd training algorithm which has a high final image signal-to-noise ratio while still maintaining ordering.

# INTRODUCTION

Data compression techniques have become important as large amounts of data need to be stored or transmitted through computer networks. Vector quantization (VQ) [3] is a lossy compression technique that has been used extensively in speech and image compression. VQ is a multidimensional generalization of scalar quantization (SQ) where, in general, VQ can substantially outperform SQ. One simple yet important attribute of SQ is that the indices for the quantization levels (or codewords) are very easily and naturally labeled. This labeling is ordinal; that is, the smallest quantization level is given the lowest label (e.g. 0) and the largest level is given the highest label (e.g. $N-1$). Figure 1 depicts a uniform SQ with ordered labels.

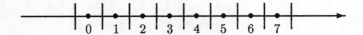

Figure 1: A uniform size $N = 8$ scalar quantizer with ordered labels.

For VQ, the concept of ordered labeling is not as straightforward. For example, should codewords be ordered according to magnitude of the vector or should the codeword indices be vectors themselves? See Figure 2 for a two-dimensional VQ with randomly assigned labels and one possible non-random ordering of the labels.

To the best of our knowledge, there has been no previous work in the coding community in the explicit and beneficial choices of indices for VQ codebooks. While there has been work in the lossless compression of indices (e.g. [9]), there has been no theory for the unified choice of codewords and their indices. We will show how a KNN derived index ordering, much like that on the right-hand side of Figure 2, has significant advantages in the progressive transmission of images. Measures on ordering will also be discussed and applied to explain our empirical results.

# ORDERED NEURAL MAPS

The notion of ordering multidimensional sets of points (i.e. codewords) is related both to the monotonicity requirements in multidimensional scaling (e.g. [10]) and to the unsupervised version of Kohonen's neural net (KNN) [5]. For multidimensional scaling the interrelationship between scaled vectors of measured perceptual responses is constrained to maintain the original ordering of the measurements. KNN training is equivalent to choosing a neighborhood over the indices, and shrinking this neighborhood to effect an ordering.

Both of these techniques simultaneously adjust the choice of codewords and their concomitant indices. The KNN has a straightforward neural interpretation: each codeword is represented by a neuron with its multiplicative input weights and the ordering is inherent in the single or multidimensional arrangements of neurons in an indexed lattice.

Past work in applications of KNN to data compression (e.g. [7, 8]) has compared the resulting distortion for KNN-VQ codebooks to codebooks

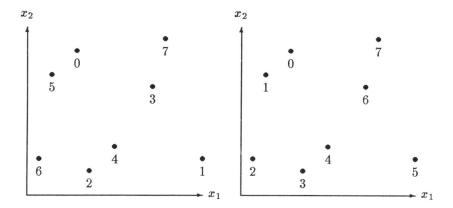

Figure 2: A two-dimensional vector quantizer with randomly assigned labels (left) and the same VQ with ordered labels (right)

generated by the more conventional generalized Lloyd algorithm (GLA) [6]. This work has shown that there is little significant difference in distortion between the two techniques. Additional work by Yair, Zeger, and Gersho incorporated stochastic relaxation into the KNN design to yield lower distortion codebooks than are obtained with GLA [12]. However, these past efforts have not exploited the inherent ordering of the KNN technique. This ordering is manifest for single dimensional indices as a single continuous arc that connects the codewords in order of increasing index. This property of ordered codeword indices is preserved when multidimensional indices are used. For example, for 2-D data vectors, the codebook will form a mesh in the codeword space for 2-D indices.

Kohonen defines the index of disorder $I_D$ as a means to measure the ordering of a neural net topology [5]. This index of disorder is defined as:

$$I_D = (\sum_{i=1}^{N-1} |w_i - w_{i-1}|) - |w_{N-1} - w_1|$$

where $N$ is the number of neurons and $w_i$ is the set of weights which multiply the input to the $i$th neuron. This definition can be easily extended to a VQ, namely $w_i$ is the $i$th codebook entry for a size-$N$ VQ codebook. With this definition, the left side of Figure 2 will clearly have a larger $I_D$ than the right side.

# PROGRESSIVE TRANSMISSION USING AN ORDINALLY MAPPED VQ

In a progressive image transmission system [4, 11], the decoder reconstructs increasingly better reproductions of the transmitted image as bits arrive. This procedure allows for early recognition of the image and has an obvious advantage in telebrowsing (remotely scanning a medical image database) by radiologists: if the wrong image is being received, transmission can be aborted before it is completely sent. The user can thus save both bits and time.

To send an image progressively, rather than send all the bits for each pixel in the image at one time, the encoder scans through the image, sends a small amount of information for each pixel, and then starts the process over again. The received image thus improves in quality as more bits arrive.

By taking advantage of the ordinal mapping feature of the KNN, additional reproduction codewords for the decoder can be designed by clustering neighbor codewords from the codebook for no additional transmission bit cost. Assuming one-dimensional indices, all codewords whose indices have a most significant bit of "1" should lie relatively close together in the input space and likewise for those indices with a most significant bit of "0." When the first bit is sent, the decoder displays the average of all the codewords with the same leading bit. When the second bit arrives for the input vector, the decoder displays the average of all codewords whose indices have the same first two leading bits. For each additional bit, half the number of codewords of the previous transmission will be averaged so the reproduction will be closer to the true input vector. This process continues down to the least significant bit, with the decoder displaying better and better reproductions as more bits arrive. The ordinal mapping feature of the codeword indices ensures that the codeword clusters will be formed of codewords that are near each other in the input signal space.

We modify Kohonen's definition of the index of disorder to one that is more closely linked with our goal of progressive transmission using an ordered codebook. We define the new normalized index of disorder for progressive transmission as

$$I_P = \frac{\sum_{i=1}^{N-1} |w_i - w_{i-1}|}{\sum_{j=0}^{N-1} \sum_{i=j+1}^{N-1} |w_j - w_i|} \times 10^6.$$

This index is the sum of the distances between each codeword and one of its nearest neighbors in the codebook, normalized by the total pairwise distance over the codebook, and multiplied by $10^6$ for scale. The index is used to directly predict how well the first intermediate codebook obtained after clustering will do. A high index of disorder means that

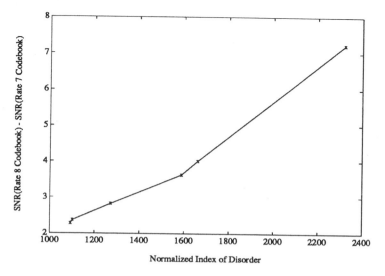

Figure 3: Difference in SNR between rate 8 and rate 7 codebooks vs. the normalized index of disorder, $I_P$.

the nearest neighbors to be clustered are not close (unordered codebook) whereas a low index of disorder corresponds to a more ordered codebook.

Figure 3 shows the relationship between $I_P$ and the first intermediate clustering of nearest neighbors. A typical rate 8 codebook was designed using the KNN algorithm where as the algorithm iterated, the ordering tended to improve to an eventual limit. The index of disorder of the codebook at $5, 10, 20, 50, 100$, and $200$ iterations of the neural network is measured along with the difference in the signal-to-noise-ratio (SNR) between the rate 8 and rate 7 codebook (obtained by pairwise combining nearest neighbors in the codebook). This difference is plotted versus the index of disorder in Figure 3, where we define SNR as

$$\text{SNR} = 10 \log_{10} \frac{EX^2}{MSE}.$$

$EX^2$ is the mean of the squared magnitudes of the training sequence and MSE is the mean-squared error between the inputs and the closest codebook vectors. The graph shows that as the index of disorder increases (the codebook is less ordered), the difference in the SNR between the rate 8 and rate 7 codebooks also increases. Thus, a more ordered codebook (lower index of disorder) suffers substantially less performance loss when the first set of intermediate codewords is formed.

Results for progressive codebook design using a hybrid of KNN and GLA and GLA alone along with results from fixed rate tree-structured

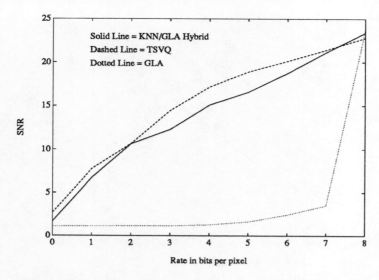

Figure 4: SNR (in dB) vs. bit rate for KNN/GLA hybrid intermediate codebooks, GLA intermediate codebooks, and TSVQ.

vector quantization (TSVQ) [2] appear in Figure 4. The training set was a set of 20 magnetic resonance (MR) images with 4-dimensional vectors. The test set was 5 MR images which were not in the training set. A rate 8 (256 codeword) codebook was designed using the training data for the three different cases. The KNN/GLA hybrid was designed by first running the KNN for 20 iterations to design an ordered codebook (index of disorder was 454). The GLA was then run to ensure that the distortion of the resulting codebook was at a local minimum. This increased the index of disorder to 1061. In contrast, the index of disorder of the pure GLA codebook was 10,315 indicating that the KNN/GLA hybrid still was substantially more ordered than the pure GLA case.

The results on the test data in Figure 4 show that the KNN codebook has the feature that the intermediate codebooks, designed by clustering codewords with similar indices, give relatively good SNR performance at lower bit rates. This is not true for the unordered GLA codebook whose performance severely degrades even at the first set of averages (bit rate of 7 bits per pixel). Thus the KNN codebook would give better image quality in a progressive transmission system at lower bit rates. The TSVQ provides the best intermediate codebook performance but has the worst performance at the highest bit rate which would represent the important final image that is received or stored.

# WORK IN PROGRESS

## Improving Progressive Transmission

Current work is addressing better ways to design the intermediate codebooks than simply by clustering neighboring codewords two at a time. The codewords will instead be combined with their closest neighbors in the space. The number of codewords in a cluster does not necessarily have to be two.

## Ordering Codebooks

The KNN-VQ approach finds low-distortion vectors while simultaneously determining ordered indices. These two steps could also be considered separately. For example, a codebook designed using the GLA could be followed by a form of traveling salesperson heuristic to reorder the codebook indices. Comparative issues to be addressed include computational complexity, final distortion, and intermediate codebook performance.

## Predictive Coding of Codeword Indices

When images are transmitted digitally, they are usually scanned and then transmitted in small contiguous blocks. Because images vary slowly spatially, we expect that significant correlation will exist between neighboring codeword indices of image if the indices are ordinally mapped. As a result, if a predictive code is designed to code the difference between successive codeword indices, the distribution on the differences should be highly skewed toward low magnitude differences. A variable rate code, which results in higher compression on nonuniform distributions, could then be designed on the difference distribution to yield additional compression. We are currently exploring the effect that the dimensionality and the neighborhood definitions of the indices has on the overall amount of compression [1].

# CONCLUSION AND OPEN QUESTIONS

We have applied the ordinal mapping feature of KNN to data compression. The algorithm results in a codebook whose indices are ordered. Initial simulations show that this feature may well be exploited for progressive transmission system. Namely, the final image fidelity was equal to that of non-progressive systems yet the intermediate images were quite close to the quality of the TSVQ progressive transmission technique. We also demonstrated that the distortion of the KNN-VQ technique could be reduced, while still maintaining ordering, by following the KNN iterations with GLA iterations. Lastly, the efficacy of a normalized index of disorder was shown for the first clustering of the highest bit-rate codebook.

There are several open questions about the general application of ordinal neural maps:

- How much does running the GLA after the KNN in the KNN/GLA hybrid destroy the ordering of the KNN codebook?

- Can the normalized index of disorder be extended and generalized for larger sized clusterings of the codebook?

- There is a "natural" ordering for scalar quantization. Does an analogous "natural" definition exist for vector quantization?

- Can ordered indices be advantageously applied to supervised neural network architectures, for example, can the ordering of hidden layers in feedforward neural nets allow for better performance or interpretation of trained nets?

# ACKNOWLEDGEMENTS

This research was supported by the University of Washington Graduate School Research Fund and the Washington Technology Center.

# References

[1] L. E. Atlas, E. A. Riskin, and S.-R. Lay. Using ordered neural maps to reduce the bit rate of image vector quantizers. Submitted for publication, 1991.

[2] A. Buzo, A. H. Gray Jr., R. M. Gray, and J. D. Markel. Speech coding based upon vector quantization. *IEEE Transactions on Acoustics Speech and Signal Processing*, 28:562–574, October 1980.

[3] R. M. Gray. Vector quantization. *IEEE ASSP Magazine*, 1:4–29, April 1984.

[4] K. R. Sloan Jr. and S. L. Tanimoto. Progressive refinement of raster images. *IEEE Transactions on Computers*, 28(11), November 1979.

[5] T. Kohonen. *Self Organization and Associative Memory*. Springer-Verlag, Berlin, 1984.

[6] Y. Linde, A. Buzo, and R. M. Gray. An algorithm for vector quantizer design. *IEEE Transactions on Communications*, 28:84–95, January 1980.

[7] J. McAuliffe, L. Atlas, and C. Rivera. A comparison of the LBG algorithm and Kohonen neural network paradigm for image vector quantization. In *Proceedings of ICASSP*, pages 2293 – 2296. IEEE Acoustics Speech and Signal Processing Society, 1990.

[8] N. M. Nasrabadi and Y. Feng. Vector quantization of images based upon the Kohonen self-organizing feature maps. In *Proceedings of IEEE International Conference on Neural Networks 1988*, pages I-101–I-107, 1988.

[9] D. L. Neuhoff and N. Moayeri. Tree searched vector quantization with interblock noiseless coding. In *Proceedings of 1988 Conference on Information Sciences and Systems*, pages 781–783, March 1988.

[10] R. N. Shepard and J. D. Carroll. *Parametric representation of nonlinear data structures*. Academic Press, NY, 1966.

[11] K-H Tzou. Progressive image transmission: a review and comparison of techniques. *Optical Engineering*, 26(7):581–589, July 1987.

[12] E. Yair, K. Zeger, and A. Gersho. Competitive learning and soft competition for vector quantizer design. *IEEE Transactions on Signal Processing*, February 1992. To appear.

# VECTOR QUANTIZATION OF IMAGES USING NEURAL NETWORKS AND SIMULATED ANNEALING

M. Lech    Y. Hua
Department of Electrical and Electronic Engineering
University of Melbourne
Parkville, Victoria, 3052, Australia.

## INTRODUCTION

Vector Quantization (VQ) has already been established as a very powerful data compression technique [4]. It has been shown to be useful in compressing data that arises in a wide range of applications, including image and speech processing [6]. Specification of the "codebook", which contains the best possible collection of "codewords", effectively representing the variety of source vectors to be encoded is one of the most critical requirements of VQ systems, and belongs, for most applications, to the class of hard optimization problems. A number of new approaches to codebook generation methods using neural networks (NN) and simulated annealing (SA) are presented and compared.

The first part contains discussion of the competitive learning algorithm (CL) and Kohonen's self-organizing feature maps (KSFM). The algorithms are examined using a new training rule and comparisons with the standard rule is included. A new solution to the problem of determining the "closest" neural unit is also proposed.

The second group of methods considered in this paper are all based on simulated annealing (SA). A number of improvements to and alternative constructions of the classical "single path" simulated annealing algorithm are presented to address the problem of suboptimality of VQ codebook generation and provide methods by which solutions closer to the optimum are obtainable for similar computational effort. These involve modifications to cooling schedules (SSA1,SSA2,SSA3), acceptance rules (SA1) and overall structure (MSA1,MSA2). Some of the methods (MSA1) mitigate the need for a lot of experimentation or ad hoc judgment in determining the best combination of the annealing parameters (cooling schedule) for a given problem. Some (SSA1,SSA2,SSA3,MSA2) provide for more "physically integrated" cooling schedules by utilizing feedback from the algorithms as they evolve. And some (SA1,MSA1,MSA2) more effectively explore the search space than classical "single path" SA.

Comparison with neural nets as to minimum average distortion achieved show some of these methods to be highly competitive without the need for some ad hoc estimation in determining gain and neighbourhood parameters.

Examples of practical implementation of the SA and NN methods for coding images with bit rate in the range 0.18 to 0.50 bits per pixel (bpp) are presented.

## VQ OF IMAGES USING NEURAL NETWORKS

## The Classical Learning Rules

The neural network system is built as a number of interconnected nodes (neural units) operating in parallel. Each unit is represented by a weight vector $w_i$ (i=1,....N), where N is the number of units. For vector quantization the weight vectors usually represent the actual codewords. During the training process, input (training) vectors are presented to all units of the neural system. An adaptation algorithm gradually adjusts the weights. The most popular method for adapting the weights is the least mean square algorithm, often called the Widrow-Hoff delta rule [6]. This algorithm minimizes the sum of squares of the linear errors over the training set and alters the weights with each presentation to make an error correction proportional to the error itself.

The error $\epsilon(t)$ is defined to be the vector difference $d(x(k),w(k))$ between the input vector $x(k)$ and the weight vector $w_i(k)$ where k is the presentation number. As learning progresses the weights are updated so that $\Delta w(k) = \alpha(k) \epsilon(k)$, or $w(k+1) = w(k) + \alpha(k) d(x(k),w(k))$, where $\alpha(k)$ (normally in the range $0.1 < \alpha(k) < 1.0$) is a constant which controls stability and speed of convergence. Two basic NN algorithms are considered here each of which require different conditions to be met for the learning (weight updating) to occur. These are: competitive learning (CL) and Kohonnen's self-organizing feature map (KSFM). For CL, the closest neural element is found at each presentation and only that element is updated. For KSFM, a neighbourhood is defined around each computational element and at each iteration the closest node as well as its neighbourhood are updated.

## Determining the "Winning" Neural Unit

Nearly all vector quantization methods based on NN systems require the identification of the closest ("winning") neural unit which involves the problem of fast nearest-neighbour search. The speed of a given NN vector quantization system depends largely on how fast this search can be performed.

All of the VQ systems presented in the paper make use of an optimal nearest-neighbour search based on multidimensional binary tree search [4]. An AVL (balanced) search tree is constructed to achieve optimal search time.

## New Learning Rule

A number of both linear and nonlinear training rules based around the delta rule have been investigated. It was found that some rules are effective in minimizing the mean square error by comparison with the delta rule. For example, the following non-linear rule was found to be effective:

$$w_i(k+1) = w_i(k) + \{\alpha(k) + \beta(k) d[x(k), w_i(k)]\} \, d[x(k), w_i(k)] \tag{1}$$

where $d()$ denotes the scalar distance between two vectors and $\mathbf{d}()$ denotes their vector difference.

## VQ OF IMAGES USING SIMULATED ANNEALING

# The Classical Simulated Annealing Algorithm (SA)

Simulated annealing is an iterative optimization procedure where a given initial codebook is continually refined by random perturbations, which allow the algorithm to escape local minima in its search for the state which represents the global minimum.

The codebook generation algorithm as presented by Flanagan [2] is a fast though suboptimal method. At each iteration the codebook is altered by a trial random perturbation and the change in distortion $\Delta c$ is calculated. If $\Delta c$ is negative the perturbation is always accepted but if $\Delta c$ is positive the change is accepted with probability in proportion to the Boltzmann-Gibbs distribution function :

$$f(\Delta c/T) = \begin{cases} 1 & \Delta c < 0 \\ \exp(-\Delta c/T) & \Delta c >= 0 \end{cases} \quad (2)$$

where T is a temperature parameter.

Initially the temperature is high, but is successively lowered after each perturbation. When T is large almost every perturbation is accepted providing the system with the ability to escape from local minima. When T is small hardly any perturbation that increases the distortion is allowed and the system rarely escapes local minima. The quickness of the algorithm relies on the transformation of the change in Euclidean distortion measure for a single switch of a vector from one set or cluster to another into an efficient form for calculation. According to Connors [1],

$$\sum_{k=0}^{\infty} T_k = \infty \quad (3)$$

is a necessary and sufficient condition to ensure optimality of the algorithm (achievement of the global minimum). This condition however requires a prohibitive amount of computation and so a geometric series may be used to generate temperatures at each iteration allowing a suboptimal solution in finite time. Temperatures are generated according to $T_{k+1} = \alpha T_k$ with $\alpha \in (0,1)$ and best results seem to be achieved with large initial temperatures and large $\alpha$ thereby requiring a large number of computations. This single path descent is nevertheless invariably trapped in some deceptive local minimum due to the sequential irreversible nature of the algorithm and the absence of ideal conditions.

## Cooling Schedule

The parameters determining the cooling schedule are:
(1) the starting value $T_0$ of the control parameter T,
(2) the decrement function $\alpha$ of the control parameter ($T_{k+1} = \alpha(k) T_k$),
(3) the length $L_k$ of the individual sequence of codebooks (the equilibrium condition),
(4) the stopping ( or convergence) criterion to terminate the algorithm.

For an annealing schedule to be problem independent, the above parameters should be determined by the algorithm itself and should not have any predefined values.

For reliable cooling, the temperature should not be decremented until equilibrium is reestablished.

Determination of adequate time-efficient cooling schedules has evolved into an important research topic. The results of implementing various cooling schedules will be presented in the following paragraphs.

## New Versions of the Classical "Single Path" Simulated Annealing Algorithm

The success of the classical SA algorithm might imply the futility of further improvements. However the cooling schedule usually adopted for a suboptimal solution is rather "stiff" and ignores the microscopic processes at work in the evolution of the codebook generation system. A cooling schedule which attempts to explore the "inner workings" of the SA algorithm and exploit this knowledge in the form of feedback can be employed to provide alternatives to the classical algorithm. Three self-adjusted "single path" algorithms (SSA1, SSA2 and SSA3) based on this idea are proposed.

The authors also present another "single path" algorithm (SA1) employing a new acceptance rule deviating from the standard Boltzmann-Gibbs rule.

## Self-Adjusted "Single Path" Simulated Annealing Algorithms (SSA)

**SSA1.** A series of "phase transitions" constitutes the cooling schedule with temperature kept constant within each phase. The sequence of distortion changes within a phase is used to extract statistics as to the **average negative distortion change** $<\Delta c^->$ for that phase. The temperature T applied to the subsequent phase is then derived as a constant of proportionality $\gamma$ multiplying this quantity:

$$T = \gamma <\Delta c^->. \tag{4}$$

**SSA2.** As above a series of "phase transitions" constitutes the cooling schedule with temperature kept constant within each phase. The sequence of distortion changes within a phase is used to extract statistics as to the **standard deviation of negative distortion change** $\Delta c_\sigma^-$ for that phase. The temperature T is then derived as a constant of proportionality $\gamma$ multiplying this quantity:

$$T = \gamma \Delta c_\sigma^-. \tag{5}$$

**SSA3.** A series of "phase transitions" constitutes the cooling schedule with temperature kept constant within each phase. The sequence of distortion changes within a phase is used to extract statistics as to the **average or standard deviation of negative distortion change** $\phi(\Delta c^-)$ for that phase (or any other appropriate statistical measure of the negative distortion change). The temperature T applied to the subsequent phase is then derived as a constant of proportionality $\gamma$ multiplying this quantity:

$$T = \gamma \phi(\Delta c^-). \tag{6}$$

The algorithm differs from the previous algorithms by requiring that for each accepted negative change in distortion (and these are automatically accepted for the algorithms presented) there follows an accepted positive change in distortion whose size is distributed according to the probability of acceptance $f(\Delta c/T)=\exp(-\Delta c/T)$. This has the effect of improving markedly the population of negative distortion changes within a phase when the system moves into a state "in the region of a local minimum" and thereby enhances the stability of the statistical measure used to derive temperature T. The length of each phase may then be substantially reduced without appreciably affecting the stability of the cooling schedule although effective avoidance of local minima may require some finite phase length.

## "Single Path" Simulated Annealing (SA) with New Acceptance Rule

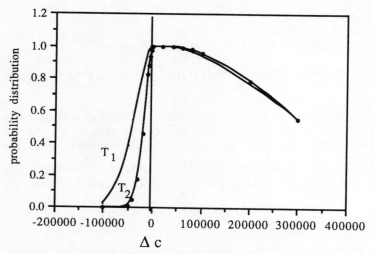

**Fig.1.** Cost function of system at two different temperatures ($T_2 < T_1$).

**SA1.** The classical Simulated Annealing algorithm (SA) is based on a Boltzmann-Gibbs distribution function (6) in analogy to Statistical Physics. The authors have found this choice to be only partially satisfactory and have sought distribution functions which tend to favour large-scale positive "quantum leaps" in distortion ahead of the small-scale ones relative to the Boltzmann-Gibbs rule. Such distribution functions would enable the system to move between states in a more flexible manner and enhance the probability of escape from local minima.

Whilst the cooling schedule adopted is that of classical SA the distribution function for acceptance employed in SA1 is given by :

$$f(\Delta c/T) = \begin{cases} 1 & \Delta c < 0 \\ 1/(1+\Delta c/T) & \Delta c \geq 0 \end{cases} \quad (7)$$

For large positive $\Delta c/T$ the probability of acceptance is $f(\Delta c/T) \approx (\Delta c/T)^{-1}$ and is therefore larger compared to that of the Boltzmann-Gibbs distribution function where the probability of acceptance is $f(\Delta c/T) = \exp(-\Delta c/T)$. Thus large-scale "quantum leaps" to new configurations are more accessible in this system.

The initial temperature $T_0$ is chosen so that when compared to the Boltzmann-Gibbs distribution small-scale perturbations are accepted less frequently to offset the occasional presence of the more frequent large-scale perturbations.

The intuitive arguments expressed above may be further quantified by the following analysis of the probability of acceptance :

Suppose the cost function to be of some form similar to that shown in Fig.1. Let it be descibed by $p(\Delta c)$.

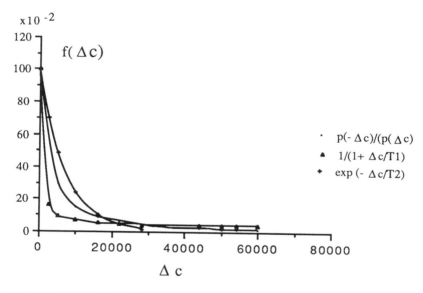

**Fig.2.** A comparison of probabilities of acceptance f () with that for equilibrium for a typical state of the system.

The probability of acceptance of a negative distortion change $\Delta c$ is given by :

$$f(\Delta c)\, p(\Delta c)\, \delta(\Delta c) = p(\Delta c)\, \delta(\Delta c) \qquad \Delta c < 0 \qquad (8)$$

where f () is the imposed probabilty of acceptance for any submitted distortion change and is taken to be unity for negative changes indicating that all negative changes are accepted regardless of size. Note that :

$$P_{accept}(\Delta c, T) = f(\Delta c/T) p(\Delta c) \qquad (9)$$

represents the ultimate probability of acceptance of any distortion change $\Delta c$ at temperature T.

The probability of acceptance of a positive distortion change $\Delta c$ is given by :

$$f(\Delta c) \, p(\Delta c) \, \delta(\Delta c) \qquad\qquad \Delta c \geq 0 \qquad\qquad (10)$$

If these probabilities equate, the imposed probability of acceptance has the form:

$$f(\Delta c) = p(-\Delta c)/p(\Delta c) \qquad\qquad \Delta c \geq 0 \qquad\qquad (11)$$

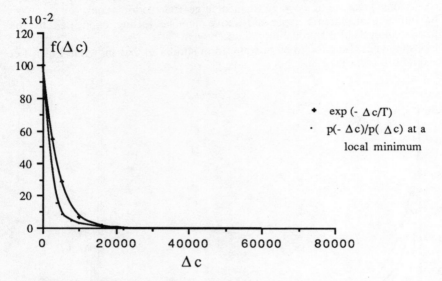

**Fig.3.** Probabilty of acceptance (2) and equilibrium curve (11) upon entering a local minimum

Fig.2 shows the behaviour of the two imposed probabilities of acceptance (2) and (7) in comparison with the above equilibrium curve for a typical state of the system as might be found in the initial phase of the algorithm.

As can be seen from Fig.2, the Boltzmann-Gibbs acceptance rule (2) and the $f() = 1/(1+\Delta c/T)$ acceptance rule allow an imbalance of the probabilities (8) and (10) but in a manner so that:

$$\int_0^\infty (f(\Delta c/T) - p(-\Delta c)/p(\Delta c)) p(\Delta c) \Delta c \, d(\Delta c)$$

$$= \int_{-\infty}^0 p(\Delta c) \Delta c \, d(\Delta c) + \int_0^\infty f(\Delta c/T) p(\Delta c) \Delta c \, d(\Delta c)$$

$$= \int_{-\infty}^\infty f(\Delta c/T) p(\Delta c) \Delta c \, d(\Delta c) = \int_{-\infty}^\infty p_{accept}(\Delta c, T) \Delta c \, d(\Delta c) \lesssim 0 \qquad (12)$$

at least for states of the system represented by Fig.2. In other words statistically there is an overall reduction in cost (per iteration) for the system as a whole with the descent occuring ideally at a minimal rate.

The advantage of the probability of acceptance adopted for SA1 to that of the Boltzmann-Gibbs distribution can now be seen from Fig.3&4. When the state of the system moves from that represented in Fig.2 to some form of local minimum the equilibrium curve alters rapidly in behaviour to something like the distribution illustrated in both Fig.3&4. For sufficiently high temperatures, probabilities of acceptance (2) and (7) provide resistance to the continuance of this state by introducing an overall gain in cost, with (12) now no longer applicable. The previous behaviour of the equilibrium curve tends to be restored. However, as can be seen from the figures, adoption of (7) for the probability of acceptance as opposed to the Boltzmann-Gibbs distibution (2), allows the introduction to the system of a far greater number of large-scale positive distortion changes and also a higher overall gain in cost so as to more effectively escape the barriers of the local minimum.

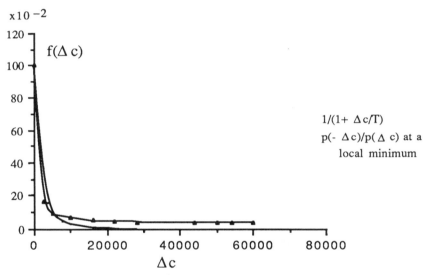

**Fig.4**. Probabilty of acceptance equation (7) and equilibrium curve (11) upon entering a local minimum

## Multipath Simulated Annealing (MSA)

**MSA1.** Another approach which might avoid the suboptimality of the classical SA algorithm as noted above is the notion of a multipath walk through the search space (referred to as Multipath Simulated Annealing, MSA, as opposed to the "single path" SA [2]), each suboptimal path being initiated from some new state or point in the search space generated as the local minimum or endpoint of the previous path whilst recording the best minimum to date. The parameters which constrain the topology of each path are "constants of the path". Here we refer to the initial temperature $T_0$ and the cooling parameter $\alpha$ of the path. $T_0$ is a constant decaying linearly or exponentially with path number whilst α is a constant close to unity. A sufficiently high initial temperature allows escape from the current local minimum and sets up conditions for a descent to a new local minimum. Each path is terminated on recognition of the local minimum (when small progress towards this minimum requires a large effort).

In general better results are obtained with MSA than for the "single path" method for a comparable amount of computational effort. In the case of high dimensionality of the search space the barrier between two low-lying states is typically massively degenerate and the number of ways of getting from one low-lying state to another is large. In other words by adding thermal noise to the system one opens up an enormous variety of paths for escaping from local minima.MSA is well placed to exploit this "variety".

**MSA2.** A variant of the above algorithm is that based on SSA3 with $T_0$ set as above every time a local minimum is encountered (or T might be iteratively incremented by a constant until a condition is met), the latter being identified by the number of attempted perturbations, in the search for a negative distortion change to match the preceding positive distortion change, exceeding a given threshold.

## EXPERIMENTAL RESULTS

Table 1 shows results of coding a one-frame image "Lenna" of size 264x256 pixels with 256 grey levels. The image contains 4224 vectors of length 16 (or blocks of size 4x4 pixels). The initial codebooks were obtained using AVL-tree clustering [4]. MSE denotes mean square error, and bpp denotes the number of bits per pixel.

TABLE 1

| Method | Codebook size [nr of vectors] (Bit Rate [bpp]) | Average Dist. [per pixel] | MSE |
|---|---|---|---|
| AVL-tree clustering (initial codebook) | 8(0.18) | 0.088729 | 532.078 |
| | 16(0.25) | 0.076853 | 399.177 |
| | 256(0.50) | 0.049496 | 165.570 |
| SA | 8(0.18) | 0.085802 | 497.552 |
| | 16(0.25) | 0.071386 | 344.405 |
| | 256(0.50) | 0.045830 | 141.951 |
| SSA1 | 8(0.18) | 0.085873 | 498.376 |
| | 16(0.25) | 0.072240 | 352.695 |
| | 256(0.50) | 0.045879 | 142.256 |
| SSA2 | 8(0.18) | 0.085834 | 497.923 |
| | 16(0.25) | 0.072232 | 352.617 |
| | 256(0.50) | 0.045871 | 142.207 |
| SSA3 | 8(0.18) | 0.085825 | 497.819 |
| | 16(0.25) | 0.072223 | 352.291 |
| | 256(0.50) | 0.045855 | 142.108 |
| SA1 | 8(0.18 ) | 0.085792 | 497.436 |
| | 16(0.25) | 0.071366 | 344.212 |
| | 256(0.50) | 0.043617 | 128.575 |

TABLE 1 (continued)

| Method | Codebook size [nr of vectors] (Bit Rate [bpp]) | Average Dist. [per pixel] | MSE |
|---|---|---|---|
| MSA1 | 8(0.18) | 0.085800 | 497.529 |
|  | 16(0.25) | 0.071376 | 344.309 |
|  | 256(0.50) | 0.044136 | 131.651 |
| MSA2 | 8(0.18) | 0.085798 | 497.506 |
|  | 16(0.25) | 0.071374 | 344.289 |
|  | 256(0.50) | 0.044059 | 131.194 |
| CL | 8(0.18) | 0.086849 | 509.751 |
|  | 16(0.25) | 0.072921 | 359.374 |
|  | 256(0.50) | 0.045420 | 139.427 |
| KSFM | 8(0.18) | 0.085813 | 497.680 |
|  | 16(0.25) | 0.072215 | 352.451 |
|  | 256(0.50) | 0.046603 | 146.782 |
| KSFM with learning rule-(1) | 8(0.18) | 0.085807 | 497.610 |
|  | 16(0.25) | 0.072193 | 352.236 |
|  | 256(0.50) | 0.045824 | 141.916 |

## REFERENCES

[1] D.P.Connors and P.R.Kumar,"Simulated Annealing and Balance of Recurrence Order in Time-Inhomogeneous Markov Chains",Proceedings of the 26th Conference on Decision and Control,1987,pp.2261-2263.

[2] J.K.Flanagan, D.R.Morrell, R.L.Frost, C.J.Read and B.E.Nelson ,"Vector Quantization Codebook Generation Using Simulated Annealing.", IEEE,ICASSP, M7.8,1989,pp.1759-1762.

[3] R.M.Gray, "Vector Quantization",IEEE ASSP Magazine,1984,pp.4-19.

[4] M.Lech, M.F.Rudolph and J.S.Packer,"Vector Quantization Based On Multidimensional Binary Tree Structures", IEEE, ISSPA 90,Gold Coast,August 1990,pp.601-603.

[5] N.M.Nasrabadi and Y.Feng,"Vector Quantizaton of Images Based Upon the Kohonen's Self-Organizing Feature Maps". IEEE International Conference on Neural Networks,1988, pp.1101-1108.

[6] B.Widrow, R.G.Winter, R.Baxter,1987,"Learning Phenomena in Layered Neural Networks",Proceedings of the 1st IEEE International Conference on Neural Networks,vol.2,1987,pp.411-421,San Diego,CA.

# A MULTILAYER PERCEPTRON FEATURE EXTRACTOR FOR READING SEQUENCED DNA AUTORADIOGRAMS

Michael C. Murdock
Advanced Artificial Intelligence Laboratory
Communications Division
Motorola Government Electronics Group
8201 E. McDowell Road, Scottsdale Arizona 85212

Neil E. Cotter
Department of Electrical Engineering
University of Utah
Salt Lake City, Utah 84112

Raymond Gesteland
Howard Hughes Medical Institute
University of Utah Medical Center
Salt Lake City, Utah 84112

Abstract - This paper reports on the application of the three-layer, backward error propagation neural network to the problem of reading sequenced DNA autoradiograms. The network is used for band identification by extracting two features: band intensity level and band intensity gradient. A training set of 16,000 12x12 gray scale patterns is generated. Trained with these patterns, the network successfully learned to identify the degree of presence and absence of these two low level features.

## PROBLEM STATEMENT

Autoradiograms are X-ray films created to image sequenced DNA fragments separated by gel electrophoresis. The DNA sequence appears on the autoradiogram as a ladder of bands distributed over four parallel lanes, where each lane corresponds to one of the four nucleotide bases, adenine, cytosine, guanine, and thymine. The sequence is read by finding the most intense band in the four lanes at each vertical level. This is illustrated in Figure 1.

Alternative efforts [1 - 3] to develop automated sequence readers have considerable difficult with false band identification because of the strong noise component. A typical fragment of an autogradiogram is shown in Figure 2. These readers achieve a high degree of accuracy in identifying bands with error rates at 1%. This has led us to study how human readers proceed in this identification process. The problem, it appears, is solved very easily by the human visual system in experienced readers. The visual system employs extremely complicated dynamics to extract information from the autoradiogram despite the ambiguity and spatial deformities. Human readers are able to observe several different noise corrupted images that share a common feature and invert the differences to extract the feature. We have found that readers acknowledge using a hierarchy of high and low level features to make the identification. Two low level features, band intensity level and band intensity gradient were identified as critical in this process. To automate the reading process and yet maintain high accuracy levels, we wish to mimic the human performance with a trained neural network. This paper describes a multilayer perceptron trained with backward error propagation that identifies these two features and as such can be used in an automated autoradiogram reading system.

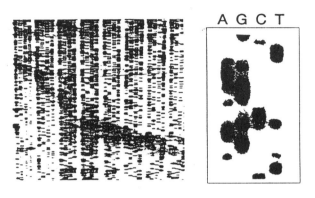

Figure 1. Autoradiogram of sequenced DNA with four lanes enlarged to illustrate how sequence is read. Bases from the top are as follows: GCTTAGAAGA.

Figure 2. Autoradiogram illustrating the effects of noise.

## FEATURE EXTRACTION

Although the operations performed in the human visual system are at present only very superficially understood, it appears that biologicial neural networks do not detect the presence or absence only of a feature, but detect over a range of values. It is this property of the neuron that is exploited for the purpose of extracting feature scores. Multilayer perceptrons have been used for a variety of classification tasks. Handwritten digit recognition [4-5], handwritten signature verification [6], speech recognition [7-8], and sonar target recognition [9] are some of these applications. The neurons in the output layer usually each represent a different class. Thus if neuron c is the most active, the input pattern is classified into class c. In this application, the neurons in the output layer each represent a different feature. This is illustrated in Figure 3. An active neuron indicates the presence of the feature; an inactive neuron indicates the absence of the feature.

In multilayer perceptrons the evolved weight vector and bias determine the characteristics of the decision boundaries. The direction of the boundary is orthogonal to the weight vector; the bias determines the boundary's position. The slope of the boundary is determined by the individual elements of the weight vector. Properly chosen weights can yield a discriminant function that provides a slow sigmoidal sloping boundary separating the two classes. A pattern in the transition region is not assigned directly to one of the classes. The pattern is a member of both classes. This is similar to fuzzy set theory introduced by Zadeh. A fuzzy set allows a continuum of grades of membership. Let $X = \{x\}$ denote the set of patterns to be processed by the neural network. We define the fuzzy sets L and G, corresponding to the intensity Level and Gradient features, as sets of ordered pairs:

$$L = \{x, \mu_L(x)\} \qquad (1)$$

and

$$G = \{x, \mu_G(x)\} \qquad (2)$$

Where $\mu$ is the degree of membership of pattern x in the fuzzy set. In this case, the degree of memberhip applies to the intensity level feature or intensity gradient feature. $\mu$ is a number in the interval [0, 1], with 0 and 1 representing, respectively, complete absence and complete presence of the feature. We define set Y as the set of patterns classified by the neural network:

$$Y = \{x, \mu_L, \mu_G\} \qquad (3)$$

The training set values of $\mu_L$ and $\mu_G$ were obtained and translated appropriately as linguistic descriptions by experienced autoradiogram readers.

## GENERATING THE TRAINING SET

Multilayer perceptrons trained with the backward error propagation algorithm requires a training sample size approximately equal to ten times the number of connections in the network [10]. Feature identification and extraction in image processing require both a large number of network inputs and complex decision boundaries. These two requirements can lead to networks with a large number of connections and hence require a large training set. The input to the network is a 12x12 matrix of gray scale values from digitized autoradiograms. A training set of 16,000 patterns was required. As each pattern must be labeled with a feature score for each of the two features, an automated method was deviced to generate training patterns and label the patterns. This method relies on the use of an autoradiogram degradation model to create realistic training patterns.

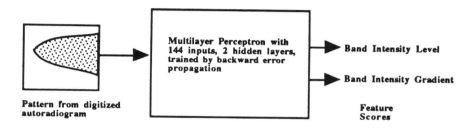

Figure 3. Two features are extracted from each pattern

Autoradiogram clarity is affected by a number of factors including radioactive source crossfire, background, diffusion, contrast variation, surface stress artifacts, film grain, quantum and convolutional noise. To account for these factors a degradation model is proposed that employs the widely used intensity image model for photographic films [11]:

$$g_i(m, n) = \sum_{k, l} h(m, n, k, l) f(k, l) \qquad (4)$$

The band formation process in the film transforms an image function f(m,n), which is a measure of the radioisotope emissions defining the band in the gel, into an image function defining the exposed region of the film. This transformation is the spatial convolution of the image function with h(m,n), the spatial point spread function of the imaging process. The point spread function is determined by the imaging isotope, quantum noise, band diffusion and radioactive source crossfire. The convolution models the smoothing and blurring that takes place in the image formation process. The corruption of the output of this imaging filter is modeled by the following:

$$g(m, n) = N(m, n) [g_i(m, n)]^\tau + \eta(m, n) \qquad (5)$$

where τ is the contrast factor of the film, which is the slope in the linear region of the plot of the optical density of the silver deposit on the film versus the logarithm of the exposure. N(m,n) is the multiplicative film grain noise and η(m,n) is additive artifactual noise. This model accounts for the fundamental sources of degradation in the autoradiogram.

## TRAINING THE NETWORK

Using the autoradiogram degradation model and a seed set of 160 patterns the training set was generated and labeled. The details of this procedure are described in [12]. Backward error propagation training is accomplished by randomly selecting a pattern from the training set and presenting it to the input layer neurons. The output of each neuron, in each layer, is fully connected as input for the following layer neurons. The error between the target (the feature score) $t_{pj}$, and actual output, $o_{pj}$, is computed for each neuron, for each pattern according to:

$$E_p = \sum_j (t_{pj} - o_{pj})^2 \qquad (6)$$

The weights, $w_{ji}$, are modified according to the generalized delta rule:

$$\Delta W_{ji} = \eta\, \delta_{pj}\, o_{pj} \qquad (7)$$

The error signal in the output layer neurons is given by:

$$\delta_{pj} = (t_{pj} - o_{pj})\, o_{pj}\, (1 - o_{pj}) \qquad (8)$$

The error signal in the hidden layers is computed recursively in terms of the error signals of the neurons to which it directly connects:

$$\delta_{pj} = f'(net_j) \sum_k \delta_{pk} W_{kj} \qquad (9)$$

The topology of the networks trained using this rule is specified by the notation w.x.y.z, where w is the number of network inputs, and x, y, and z are the number of neurons in the input, hidden and output layers.

## DISCUSSION

To test its ability to generalize on new patterns, the network was used to extract the two features from 110 patterns from five previously unseen autoradiograms. The method used to quantify the performance of the network was a scatter plot, which compares the human-assigned feature score to the network-assigned feature scores, with exact matches falling on the diagonal. All eight networks, in varying degrees, identified the presence and absence of the two features. Figure 5 shows the performance of the linear 144.8.20.2 network. This is the only network that was able to extract both features and capture the midrange scores. The sigmoid networks performed better than the linear networks but tended to harden the decision regions by pushing the midrange (.2 to .8) feature scores to zero or one. By capturing this midrange behavior, the network appears to have evolved sufficient decision regions without adapting to the peculiarities of the training set.

| Network 1 | 144.8.20.2 |
| --- | --- |
| Network 2 | 144.12.20.2 |
| Network 3 | 144.12.8.2 |
| Network 4 | 144.6.12.2 |
| Learning Rate | .05 linear output, .25 sigmoid output |
| Patterns per Epoch | 16,000 |
| Number of Epochs | 150 |
| Initial Weights | [-.05, .05] |

Figure 4.          Training summary.

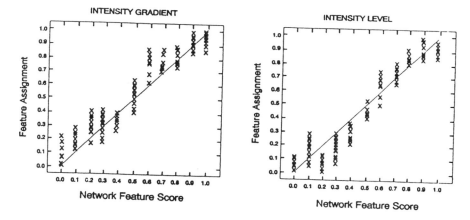

Figure 5.    Scatter diagram illustrating the performance of the 144.8.20.2 linear output network in matching the performance of the human readers in extracting the intensity level and intensity gradient features.

## CONCLUSION

We have described the application of a three-layer, backward error propagation neural network to a non-trivial pattern recognition task. The task is to read the DNA sequence by identifying the bands in an autoradiogram. Figures 1 and 2 illustrate how noise and ambiguity make it difficult to automate the process of reading the bands. As the first stage in this process, we present a feature extractor that identifies two features and the degree to which these features are present.

## REFERENCES

[1] D.Q. Xu, M.K. Tso, W.J. Martin, "Automatic interpretation of digital autoradiograph of DNA sequencing gels," Image Analysis and Processing II, V. Cantoni, V. Di Gesu, Eds., New York: Plenum Press, pp. 501-509, 1988.

[2] J. West, "Automated sequence reading and analysis," Nucleic Acids Research, vol. 16, no. 5, pp. 1847-1856, 1988.

[3] J.K. Elder, E.M. Southern, "Automatic reading of DNA sequencing gel autoradiographs," Nucleic Acid and Protein Sequence Analysis: A Practical Approach, M.J. Bishop, C.J. Rawlings, Eds., Washington DC: IRL Press, pp. 219-229, 1989.

[4] Y. Le Cun, B. Boser, J. Denker, D. Henderson, "Handwritten digit recognition with a back-propagation network," Advances in Neural Information Processing Systems 2, D. Touretzky, Ed., San Mateo, CA: Morgan Kaufmann Publishers, pp. 306-404, 1990.

[5] W. Weidemann, M. Manry, H. Yau, "A comparison of a nearest neighbor classifier and a neural network for numeric handprint character recognition," Proceedings of the International Joint Conference on Neural Networks, vol. I, pp. 117-120, 1989.

[6] D. Mighell, T. Wilkinson, J. Goodman, "Backpropagation and its application to handwritten signature verification," Proceedings of the International Joint Conference on Neural Networks, vol I, pp. 340-347, 1989.

[7] Y. Lee, R. Lippmann, "Practical characteristics of neural networks and conventional pattern classifiers on artificial and speech problems," Advances in Neural Information Processing Systems 2, D. Touretzky, Ed., San Mateo CA: Morgan Kaufmann Publishers, pp. 168-177, 1990.

[8] T. Sejnowski, C. Rosenberg, "Parallel networks that learn to pronounce English text," Complex Systems, vol. 1, pp. 145-168, 1987.

[9]   R. Gorman, T. Sejnowski, "Analysis of hidden units in a layered network trained to classify sonar targets," Neural Networks, vol. 1, pp. 75-89, 1988.

[10]  B. Widrow, "ADALINE and MADALINE," Plenary Speech, vol I. Proceedings of the IEEE 1st International Conference on Neural Networks, San Diego, California, pp. 143 - 158.

[11]  A. Wilsky, Digital Signal Processing and Control and Estimation Theory, Cambridge, MA: MIT Press, 1979.

[12]  M. Murdock, "A DNA autoradiogram degradation model for generating training sets for multilayer perceptrons," Proceedings of the 6th International Conference on Image Analysis and Processing, Singapore: World Scientific Press, 1991.

# CONFIGURING STACK FILTERS BY THE LMS ALGORITHM

Nirwan Ansari and Yuchou Huang
Center for Communications and Signal Processing
Department of Electrical and Computer Engineering
New Jersey Institute of Technology
University Heights
Newark, New Jersey 07102

Jean-hsang Lin
Department of Electrical Engineering
University of Delaware
Newark, Delaware 19716

Abstract — Stack filters are a class of sliding−window nonlinear digital filters that possess the weak superposition property (threshold decomposition) and the ordering property known as the stacking property. They have been demonstrated to be robust in suppressing noise. In this paper, a new method based on the Least Means Squares (LMS) algorithm is developed to adaptively configure a stack filter. Experimental results are presented to demonstrate the effectiveness of the proposed method to noise suppression.

## INTRODUCTION

The subject of adaptive filters [1] has matured to the point where it now constitutes an important part of statistical signal processing. Whenever there is a requirement to process signals that result from operation in an environment of unknown statistics, the use of adaptive filters offers an attractive solution to the problem as it usually provides a significant improvement in performance over the use of a fixed filter designed by conventional methods. Furthermore, it provides new signal processing capabilities that would not be possible otherwise.

All stack filters [2]−[5] obey the threshold decomposition property and the stacking property. The difference between two stack filters lies solely in the Boolean operation performed on each level. A necessary and sufficient condition for a Boolean operation to preserve the stacking property [6] is that the operation must be positive, in which case it has a minimum sum of products representation which is free of complements of any of the variables. Thus, only the logical AND and OR operations are permitted. A brief discussion on stack filters will be presented in Section 2.

Adaptive approaches to configuring stack filters have recently been proposed [7],[8]. In this paper, a new method based on LMS algorithm is developed to configure stack filters. The LMS algorithm [9] along with its properties will be discussed. The incorporation of the LMS learning to configure stack filters will then be developed.

## STACK FILTERS

Median filters [2],[3] and other rank-order operators [4] possess two properties called threshold decomposition property and the stacking property. The first is a limited superposition property; the second is an ordering property which allows an efficient VLSI implementation of the threshold decomposition architecture.

Any filter which possesses the threshold decomposition property and the stacking property is known as a stack filter [5]. Based on the threshold decomposition property and the stacking property, stack filters can be constructed as a "stack" of Positive Boolean functions [6]. Thus, stack filters form a very large class of easily implemented nonlinear filters which include the rank order operators as well as all compositions of morphological operators.

Passing an M-valued discrete time signal through a rank-order filter is equivalent to the following procedure:

(i) Decomposing the M-valued input signal into a set of M-1 binary signals. The $k$th binary signal, where $k$ is an integer in $\{1,2,\ldots,M-1\}$, is obtained by thresholding the input signal at value $k$. That is, it takes a value of 1 whenever the input signal is greater than or equal to $k$, but it is 0 otherwise. Note that summing these M-1 binary signals always provides the original input signal.

(ii) Filtering each binary signal independently with its own rank-order filter. Note that each threshold level is performed in parallel. During the process of filtering, each rank-order filter simply adds the number of 1's in the window and compares the result to an integer r, the desired rank of the filter. The output is 1 when the summation is greater than or equal to r, and 0 when the summation is less than r. For example, if the filter's window is b=2r+1, the filter is a median filter.

(iii) Adding the output of each binary rank-order filter one sample at a time. It is found that the output of the rank-order filter possesses the stacking property. This means that the binary output signals are piled on top of each other according to their threshold levels. It can be seen that a column of 1's always has a column of 0's on top. The desired output value is simply the value of the threshold level where the transition from 1 to 0 takes place.

Consider an M-valued sequence s. The threshold signals $\vec{T}^1, \vec{T}^2, \cdots, \vec{T}^{M-1}$ of the sequence are defined by

$$\vec{T}^j(i) = \begin{cases} 1 & \text{if } s(i) \geq j \\ 0 & \text{if } s(i) < j, \end{cases} \quad (1)$$

where $i$ stands for the $i$th element of the appropriate vector and each element of a threshold vector is binary. Note that these threshold vectors possess the stacking property

$$\vec{T}^1 \geq \vec{T}^2 \geq \cdots \geq \vec{T}^{M-2} \geq \vec{T}^{M-1}, \quad (2)$$

which implies

$$\vec{T}^1(i) \geq \vec{T}^2(i) \geq \cdots \geq \vec{T}^{M-2}(i) \geq \vec{T}^{M-1}(i), \quad (3)$$

where $i$ is the $i$th element of the appropriate vector.

Let $s_1$ and $s_2$ be two binary sequences. A filter defined by a function $F(\cdot)$ is said to have the stacking property if

$$F(s_1) \geq F(s_2) \quad \text{whenever} \quad s_1 \geq s_2. \quad (4)$$

Based on Equations (2), (3) and (4), the output of this filter should have the following relationship

$$F(\vec{T}^1) \geq F(\vec{T}^2) \geq \cdots \geq F(\vec{T}^{M-2}) \geq F(\vec{T}^{M-1}). \quad (5)$$

The function $F(\cdot)$ of a rank-order filter, in fact, is a Boolean function. It means that if a filter defined by the function $F(\cdot)$ has the stacking property, then $F(\cdot)$ must be a positive Boolean function. The operation of these positive Boolean functions is simply the "max" and "min" operations. For example, $F(x_1, x_2, x_3) = x_1 + x_3$, which is a stack filter, is equivalent to $max(x_1, x_3)$.

Stack filters are a generalization of median filters. The structure of a median filter root signal [3] is well known, and the structure and analysis of stack filters are similar. Properties and analysis of stack filters have been covered in great details [5].

## LINEAR DISCRIMINANT FUNCTIONS AND THE LMS ALGORITHM

The problem of finding a linear discriminant function will be formulated as a problem of minimizing a criterion function. A linear discriminant function [10] is defined by

$$g(\mathbf{x}) = \mathbf{a}^t\mathbf{x} + w_0, \quad \text{or} \tag{6}$$

$$g(\mathbf{u}) = \mathbf{w}^t\mathbf{u}, \tag{7}$$

where

$$\mathbf{u} = \begin{bmatrix} 1 \\ \mathbf{x} \end{bmatrix}, \mathbf{w} = \begin{bmatrix} w_0 \\ \mathbf{a} \end{bmatrix}.$$

$\mathbf{x}$ is an input pattern vector, $\mathbf{a}$ is the weight vector, and $w_0$ is the threshold weight. The linear discriminant function is used to classify input pattern $\mathbf{x}$ as one of two possible categories, and the decision rule is: Decide class **A** if $g(\mathbf{x}) > 0$, and class **B** if $g(\mathbf{x}) < 0$. Thus $\mathbf{x}$ is assigned to class **A** if the inner product $\mathbf{a}^t\mathbf{x}$ exceeds the threshold, $-w_0$, and $\mathbf{x}$ is assigned to class **B** if the inner product $\mathbf{a}^t\mathbf{x}$ is less than the threshold, $-w_0$. When $\mathbf{a}^t\mathbf{x}$ is equal to $-w_0$, $\mathbf{x}$ can be assigned to either class **A** or class **B**.

If $g(\mathbf{x}) = 0$, it defines the decision surface that separates points assigned to class **A** from points assigned to class **B**. When $g(\mathbf{x})$ is linear as shown in Equation (6), this decision surface is a *hyperplane*. If $\mathbf{x}_1$ and $\mathbf{x}_2$ are both on the decision surface, then

$$\begin{aligned} \mathbf{a}^t\mathbf{x}_1 + w_0 &= \mathbf{a}^t\mathbf{x}_2 + w_0 \\ \mathbf{a}^t(\mathbf{x}_1 - \mathbf{x}_2) &= 0. \end{aligned} \tag{8}$$

Therefore, $\mathbf{a}$ is normal to the hyperplane. The orientation of the decision surface is decided by $\mathbf{a}$, and the location of surface is determined by $w_0$.

Consider the structure of an adaptive threshold logic element in Fig. 1. This structure has two parts: (1) A transversal filter with adjustable tap weights whose values at time n are denoted $w_1(n), w_2(n), \cdots, w_N(n)$, and (2) a mechanism for adjusting these tap weights in an adaptive manner.

During the filtering process, an additional signal $d(n)$, called the *desired response*, is supplied along with the usual tap input. The desired signal response provides a reference for adjusting the tap weights of the filter. Denote $e_L(n)$ as the estimation error produced during LMS learning. Thus as shown in Fig. 1,

$$e_L(n) = d(n) - \mathbf{w}^t(n)\mathbf{u}(n), \tag{9}$$

where the term $\mathbf{w}^t(n)\mathbf{u}(n)$ is the inner product of the tap weight vector $\mathbf{w}(n)$ and the tap input vector $\mathbf{u}(n)$, and the superscript $t$ stands for vector or matrix transpose. Since there are $N+1$-input which correspond to the $N$ inputs for a filter width of $N$ and a threshold determined by $w_0$, thus the weight vector is $\mathbf{w}^t(n) = [w_0(n), w_1(n), w_2(n), \cdots, w_N(n)]$, and the input vector is $\mathbf{u}^t(n) = [1, u(n), u(n-1), \cdots, u(n-N+1)]$.

Here, only real input data, real weights and real desired output data are considered, and the criterion function is based on the mean squared error,

$$J(n) = E[(d(n) - \mathbf{w}^t(n)\mathbf{u}(n))^2]. \tag{10}$$

If the tap input vector $\mathbf{u}(n)$ and the desired response $d(n)$ are jointly stationary, then the mean squared error $J(n)$ at time n can be written as

$$J(n) = \sigma_d^2 - \mathbf{w}^t(n)\mathbf{p} - \mathbf{p}^t\mathbf{w}(n) + \mathbf{w}^t(n)\mathbf{R}\mathbf{w}(n), \tag{11}$$

where $\sigma_d^2$ is the variance of the desired response $d(n)$, $\mathbf{p}$ is the cross-correlation vector between the tap-input vector $\mathbf{u}(n)$ and the desired response $d(n)$, $\mathbf{R}$ is the autocorrelation matrix of the tap-input vector $\mathbf{u}(n)$,

$$\begin{aligned}\mathbf{p} &= E[\mathbf{u}(n)d(n)], \\ \mathbf{R} &= E[\mathbf{u}(n)\mathbf{u}^t(n)].\end{aligned} \tag{12}$$

The gradient of the criterion function denoted by $\nabla$ is simply the derivative of the mean-squared error $J$ with respect to the tap-weight vector $\mathbf{w}$:

$$\nabla = \frac{dJ(n)}{d\mathbf{w}} = -2\mathbf{p} + 2\mathbf{R}\mathbf{w}(n). \tag{13}$$

An optimal weight vector such that $J(n)$ is minimized can be obtained by letting $J = 0$.

The simplest choice of estimators $\mathbf{R}$ and $\mathbf{p}$ is to use the instantaneous estimates that are based on sample values of the tap-input and desired response, as defined by

$$\begin{aligned}\mathbf{R} &= \mathbf{u}(n)\mathbf{u}^t(n), \\ \mathbf{p} &= \mathbf{u}(n)d(n),\end{aligned} \tag{14}$$

respectively. Correspondingly, the instantaneous estimate of the gradient vector is

$$\nabla(n) = -2\mathbf{u}(n)d(n) + 2\mathbf{u}(n)\mathbf{u}^t(n)\mathbf{w}(n). \tag{15}$$

According to the method of steepest descent [1], the updated value of the tap-weight vector at time $n+1$ is computed by using the simple recursive relation

$$\mathbf{w}(n+1) = \mathbf{w}(n) + \frac{1}{2}\mu[-\nabla(n)], \tag{16}$$

where $\mu$ is a positive real-valued constant.

Substituting Equation (13) into Equation (16), the following updating rule is obtained,

$$\mathbf{w}(n+1) = \mathbf{w}(n) + \mu[\mathbf{p} - \mathbf{R}\mathbf{w}(n)], \tag{17}$$

and substituting Equation (14) into Equation (17), the following LMS learning rule is obtained:

$$\begin{aligned}\mathbf{w}(n+1) &= \mathbf{w}(n) + \mu\mathbf{u}(n)[d(n) - \mathbf{u}^t(n)\mathbf{w}(n)] \\ &= \mathbf{w}(n) + \mu\mathbf{u}(n)e_L(n),\end{aligned} \tag{18}$$

where
$$e_L(n) = d(n) - y(n),$$
$$y(n) = \mathbf{u}^t(n)\mathbf{w}(n). \tag{19}$$

Based on the concept of linear discriminant function, the hardlimiting threshold level should be chosen as follows:

$$y_o = \begin{cases} 1 & \text{if } y > 0 \\ -1 & \text{if } y \le 0. \end{cases} \tag{20}$$

The single layer neuron can be used with both continuous valued and binary inputs. This simple net generates much interest when initially developed because of its ability to learn to recognize simple patterns. For a linear separable case, LMS learning is good enough to classify the input samples. To classify non-linearly separable samples, it is necessary to use multi-layer networks such as multi-layer neurons [11]. On the other hand, the weight vector obtained by the LMS rule [12] cannot be unreasonable if the samples are not separable. LMS learning rule can be generalized to a more general rule by replacing the hardlimiting function by a sigmoid function [9].

## CONFIGURING STACK FILTERS BY THE LMS LEARNING

Denote $s(n)$ as the original signal sequence, $\eta(n)$ as the noise process, and $r(n)$ as the resulting sequence. It is assumed that $s(n)$ is corrupted by additive noise process, $\eta(n)$, and thus the resulting sequence $r(n) = s(n) + \eta(n)$, as shown in Fig. 2.

The problem addressed here is to configure a stack filter $S$ in order to recover the original signal sequence from the corrupted sequence. Since stack filters possess the threshold decomposition and stacking properties, configuring a stack filter is equivalent to first converting the input signal sequence into sequence of binary signals by threshold decomposition, and then finding the appropriate positive Boolean function used for all levels. Now, the input signal sequence is $r(n)$. Assume $r(n)$ is an M-valued sequence, by threshold decomposition the thresholded binary sequences denoted by $\vec{T}^{M-1}, \vec{T}^{M-2}, \cdots, \vec{T}^2, \vec{T}^1$, are obtained where

$$\vec{T}^1 \ge \vec{T}^2 \ge \cdots \ge \vec{T}^{M-2} \ge \vec{T}^{M-1}, \quad \text{and} \tag{21}$$

$$\vec{T}^j(n) = \begin{cases} 1 & \text{if } r(n) \ge j, \\ -1 & \text{if } r(n) < j. \end{cases} \tag{22}$$

Note that 1 and $-1$ rather than 1 and 0 are adopted here.

Let $N$ be the window width of the stack filter. At each threshold level, the input sequence is a binary sequence, and the output is a binary number. Thus, the input-output relationship can be realized by a Boolean function. Recalled from the previous section, some binary Boolean functions can be realized by a linear discriminant function, and thus can be trained by the LMS learning rule discussed in the previous section. However, the Boolean function obtained by training the single neuron may not be a positive Boolean function. Heuristics which will be discussed later are introduced to ensure that the resulting Boolean function is a positive Boolean function.

The general single-neuron structure for configuring stack filters is shown in Fig. 3. The input sequence $r(n)$ is first converted to thresholded binary sequence $\vec{T}^1, \vec{T}^2, \cdots, \vec{T}^{M-1}$. For each window sample of width $N$ of the input sequence $r(n)$, there are (M-1) window samples of width $N$ of the thresholded binary sequences; that is, (M-1) binary input patterns are presented to the single-neuron. Thus, the weights of the neuron are updated by the (M-1) binary input patterns (M-1) times for each window sample of $r(n)$. The serial binary outputs of the neuron are then stacked back into (M-1) levels. Finally, the M-valued filtered output signal is reconstructed, by the stacking property, from the binary outputs by a search for level at which a transition from 1 to $-1$ occurs.

Denote $\vec{r}_i$ as the $i$th window sample of width $N$ of the input sequence $r(n)$, and $\vec{T}_i^1, \vec{T}_i^2, \cdots, \vec{T}_i^{M-1}$ as the (M-1) thresholded binary input patterns that result from the $i$th input sample $\vec{r}_i$. These parallel (M-1) thresholded binary input patterns are transformed into a sequence of binary patterns as follows:

$$\vec{u}_j = \vec{T}_i^k \text{ where } j = (M-1)(i-1) + k. \tag{23}$$

The weights of the neuron are then updated by a learning rule. Using Equation (18) and (19), we have

$$\mathbf{w}_{j+1} = max\{\mathbf{w}_j + \mu \vec{u}_j[d_j - \mathbf{w}_j^t \vec{u}_j]/(N+1), 0\}, \tag{24}$$

where

$$j = (M-1)(i-1) + k.$$

The $max$ operator is to ensure nonnegative weights. Since the elements of $\vec{u}_j$ are either 1 or -1, $\frac{\vec{u}_j}{(N+1)} = \frac{\vec{u}_j}{|\vec{u}_j|^2}$ becomes the normalized input pattern.

Note that during training, negative weights except $w_0$ are set to zero. These heuristics are introduced in order to preserve the stacking property of a stack filter. After training, the final weight vector is used for the remaining inputs. It is easy to show that the above heuristics preserve the stacking property.

Denote $\mathbf{w}^k$ as the weight vector used for the $k$th thresholded level signal. The $k$th level output is:

$$y_o^k = f_H(y^k),$$

where $y^k = (\vec{T}^k)^t \mathbf{w}^k$, and $k = M-1, M-2, \cdots, 1$. Since

$$\vec{T}^{M-1} \leq \vec{T}^{M-2} \leq \cdots \leq \vec{T}^1,$$

and

$$\mathbf{w}^{(M-1)} = \mathbf{w}^{(M-2)} = \cdots = \mathbf{w}^{(1)} \geq \mathbf{0},$$

thus,

$$y_o^{(M-1)} \leq y_o^{(M-2)} \leq \cdots \leq y_o^{(1)}.$$

Hence, the stacking property is preserved.

The LMS method to configure stack filters has been developed and discussed in details above. The effectiveness of the proposed algorithm for noise suppression shall be demonstrated by experimental results. Various types of signals and noise have been considered, but, in this paper, only results for a signal, "Mexican hat," and the $\epsilon$-mixture of Gaussian noise will be presented.

The $\epsilon$-mixture of Gaussian noise has the following probability density function:

$$P(y) = (1-\epsilon)\Phi(\frac{y}{\sigma_1}) + \epsilon\Phi(\frac{y}{\sigma_2}),$$

where $\Phi(y)$ is the probability density function of a Gaussian random variable with zero mean and unit variance. The $\epsilon$-mixture of Gaussian noise is suitable for representing deviations of sample values from a single distribution. With probability

$(1-\epsilon)$, the added noise for a sample, $\eta(n)$, is a Gaussian random variable with standard deviation $\sigma_1$, which represents the background thermal noise; with probability $\epsilon$, $\eta(n)$ is a Gaussian random variable with standard deviation $\sigma_2$, which represents the impulsive noise. In the simulations, the $\epsilon$-mixture of Gaussian noise is generated with $\epsilon = 0.8$, $\sigma_1 = 1$ and $\sigma_2 = 10$.

The "Mexican hat" signal and the $\epsilon$-mixture of Gaussian noise are shown separately in Fig. 4. The filtered output signals obtained by the LMS learning rule with various window widths are shown in Fig. 5-7. In these figures, the first five hundred samples are the results obtained during the training, and the remaining samples are the results after having the trained weights fixed. Fig. 8 shows the mean absolute error between the desired output and the filtered output by the LMS learning rule using various window widths.

From these results, we can draw the following conclusions:

(1) The noise suppression depends on the signal, noise and filter window width.

(2) Filters with larger window widths perform better than those with smaller window widths.

## CONCLUSIONS

A framework for configuring stack filters using the LMS learning rule was established and tested. We have demonstrated through experimental results that the proposed algorithm performs the noise suppression task reasonably well. Other learning rules can be used instead of the LMS algorithm. The current design only makes use of a simple single neuron. Further improvement is expected if a multi-layer network (multi-layer neurons) is employed.

Future research efforts include:
(1) Extend a single-neuron structure to a multi-layer neural network.
(2) To speed up the training by employing a new recently proposed stack filtering architecture [13].

## ACKNOWLEDGEMENTS

The authors thank C. H. Chu for the "Mexican Hat" signal data.

# References

[1] S. Haykin, *Adaptive Filter Theory*, Englewood Cliffs, NJ: Prentice-Hall, 1986.

[2] A. C. Bovik, T. S. Huang and D. C. Munson, "A generalization of median filtering using linear combinations of order statistics," *IEEE Trans. Acoust., Speech, Signal Processing*, vol. ASSP-31, pp. 1342-1349, Dec. 1983.

[3] J. P. Fitch, E. J. Coyle and N. C. Gallagher, "Root properties and convergence rates of median filters," *IEEE Trans. Acoust., Speech, Signal Processing*, vol. ASSP-33, pp. 230-239, Feb. 1985.

[4] J. P. Fitch, E. J. Coyle and N. C. Gallagher, "Threshold decomposition of multidimensional ranked-order Operations," *IEEE Trans. Circuits Syst.*, vol CAS-32, pp. 445-450, May 1985.

[5] P. D. Wendt, E. J. Coyle and N. C. Gallagher, "Stack filters," *IEEE Trans. Acoust., Speech, Signal Processing*, vol. ASSP-34, pp. 898-911, Aug. 1986.

[6] E. N. Gilbert, "Lattice-theoretic properties of frontal switching functions," *J. Math. Phys.*, vol. 33, pp. 57-67, Apr. 1954.

[7] C. H. Chu, "A genetic algorithm approach to the configuration of stack filters," in *Proc. Intl. Conf. on Genetic Algorithms*, George Mason University, June 4-7, 1989, pp. 218-224.

[8] J. H. Lin, T. M. Selike and E. J. Coyle, "Adaptive stack filtering under the mean absolute error criterion," *IEEE Trans. Acoust., Speech, Signal Processing*, vol. ASSP-38, pp. 938-954, Jun. 1990.

[9] S. C. Douglas and T. H. Y. Meng, "Optimum error nonlinearities for LMS adaptation," in *Proc. ICASSP 90*, Albuquerque, New Mexico, April 3-6, 1990.

[10] R. O. Duda and P. E. Hart, *Pattern Classification and Scene Analysis*, Menlo Park, CA: John Wiley & Sons. Inc., 1973.

[11] B. Widrow, R. G. Winter and R. A. Baxter, "Layered neural nets for pattern recognition," *IEEE Trans. Acoust., Speech, Signal Processing*, vol. ASSP-36, pp. 1109-1118, Jul. 1988.

[12] R. Rosenblatt, *Principles of Neurodynamics*, New York: Spartan Books, 1959.

[13] J. H. Lin, "A new architecture and fast algorithm for stack filters," presented at *1991 CISS Conference at John Hopkins University*, March 20, 1991.

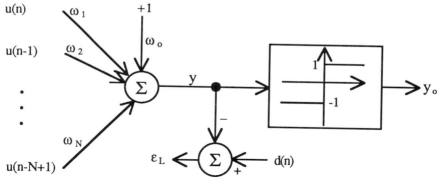

Fig. 1. A neuron: An adaptive threshold logic element — the adaptive filter structure used by the LMS rule.

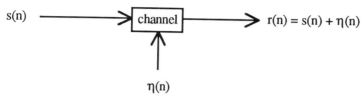

Fig. 2. Additive noise channel.

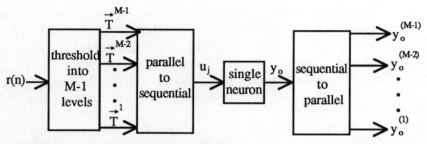

Fig. 3. Single neuron structure for training.

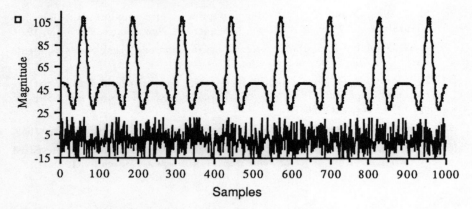

Fig. 4. Original signal ("Mexican hat") and $\varepsilon$-mixture of Gaussian noise shown separately.

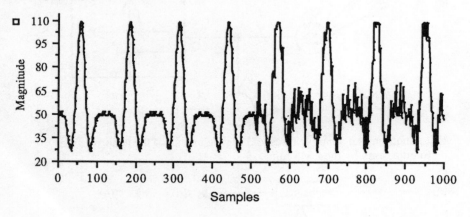

Fig. 5. Output signal obtained by filtering the corrupted signal using the LMS rule with window width=3.

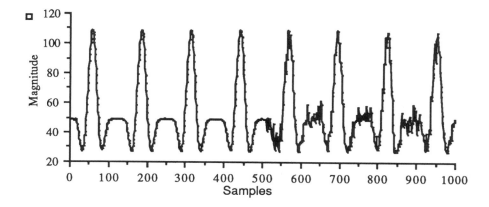

Fig. 6. Output signal obtained by filtering the corrupted signal using the LMS rule with window width=9.

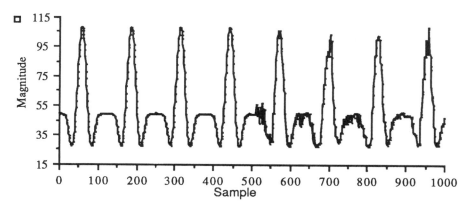

Fig. 7. Output signal obtained by filtering the corrupted signal using the LMS rule with window width=15.

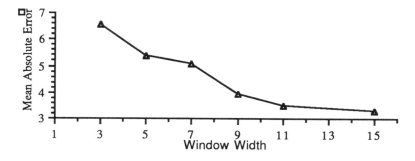

Fig. 8. Mean absolute errors between the original "Mexican hat" signal and the output signals obtained by the LMS rule using various window widths..

# A Neural Network Pre-processor for Multi-tone Detection & Estimation

Sathyanarayan S.Rao    Sriram Sethuraman
Department of Electrical Engineering
Villanova University
Villanova, PA 19085

**Abstract** - A parallel bank of neural networks each trained in a specific band of the spectrum is proposed as a pre-processor for the detection and estimation of multiple sinusoids at low SNRs. A feedforward neural network model in the autoassociative mode, trained using the backpropagation algorithm, is used to construct this sectionized spectrum analyzer. The key concept behind this scheme is that, the network when trained for a certain spectral band, serves as an excellent filter with sharp transition and near complete attenuation in stopband, even at low SNRs. Simulation results to support the advantages of the proposed scheme are presented. Statistical measurements to determine its degree of reliability in detection have been made.

## INTRODUCTION

The ongoing research over the recent years to seek ways of employing neural networks to perform signal processing tasks has been warranted by the ability of these networks to learn and generalize from a small, yet typical training data set and by the prospect of real time processing after training them. The most widely used learning algorithm in such cases is the backpropagation algorithm. This algorithm is well documented[1],[2],[3],[4] and has largely been accepted as a simple yet efficient algorithm for nonlinear optimization. Several interpretations of the way in which the network learns a particular mapping have been proposed in the past. Lapedes & Farber[5] have shown that the network performs an operation similar to modal decomposition obtained by a sum of Walsh functions. The noise reduction property of a neural network has been discussed in [4], [6],[7]&[8]. The noise reduction has been attributed to the semilinear sigmoidal transfer function,[6] which compresses the noise components while preserving the signal components. The concept of a noise reduction filter was introduced in [4] and has been shown to work for noisy EKG data. Thus a backpropagation network in the autoassociative mode can extract the principal components from a noisy data. In this paper, the prospect of such a network as a preprocessor in the detection and estimation of the frequencies of multiple sinusoids in additive white noise has been explored.

## PROBLEM DEFINITION

The detection and estimation of multiple sinusoids in additive white noise may be posed as a problem of finding a transformation that maps the noisy signal to the noise free one. Such a transformation can generally be nonlinear and highly complex. Hence neural networks become a viable alternative to obtain the mapping, compared to linear techniques. Neural networks as noise reduction filters essentially can serve as preprocessors for a spectral estimator.

It has been found that the network has the capability of detecting one sinusoid without any ambiguity up to -3dB[8]. It gives an acceptable performance for the two sinusoids detection case, but the performance degrades rapidly when two or more sinusoids are included.

Simulations performed earlier showed that the network when trained in a particular band, acts like an excellent filter, with high gain in that passband, good attenuation in stopband and a sharp cutoff. It was decided to exploit this fact and thus 'divide and conquer' the problem. Hence the scheme of a parallel bank of neural networks each trained in a specific band of the frequency spectrum was conceived [Fig. 1].

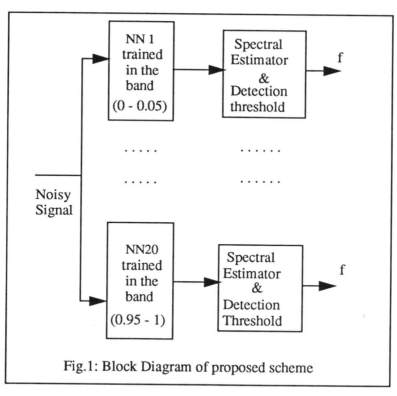

Fig.1: Block Diagram of proposed scheme

This arrangement can prove very useful in multi-tone detection problems under low Signal-to-Noise ratios(SNR). The simulation results for a typical case has been shown. The problems encountered and other performance issues have been discussed.

## NETWORK DESCRIPTION & TRAINING DETAILS

| | |
|---|---|
| Mode | : Auto-associative |
| No. of hidden layers | : 2 |
| No. of hidden nodes | : 16 |
| Learning Algorithm | : Back-propagation |
| Transfer function: | |
| Input layer nodes | : Linear |
| Hidden layer nodes | : Sigmoid |
| Output layer nodes | : Sigmoid |
| Initial weights | : Random |
| Bias | : 1 |
| No. of training patterns | : 200 |
| Error function | : Total squared error |
| Input training vector | : Sinusoids + White noise |
| Desired output vector | : The noise free sinusoid in the band of interest |

The record length of the training vectors was fixed as 32. Some previous detection experiments [9] show that the DFT spectral estimator is quite rugged and performs well for reasonable record lengths and SNRs. They fail only for small record lengths and SNRs below 0dB. The high resolution or AR estimators perform well for small record lengths and high SNRs, but fail for low SNRs. Hence, the choice of 32 for record length has been made in the range where the proposed method may prove to be advantageous compared to the above two methods.

The only change made is, the sigmoidal transfer function has been modified as,

$$f(x) = \frac{1}{1 + exp(-lx)} \qquad - (1)$$

Thus, the derivative becomes,

$$f'(x) = lf(x)(1 - f(x)) \qquad - (2)$$

Hence with decreasing $l$, the slope of sigmoid also decreases. The unsaturated region increases, accommodating a large range of values every time and weights them suitably to minimize the error function rapidly and thus speeds up convergence.

The learning rate is decreased as iterations proceed, to minimize the oscillations. The momentum term correction is either neglected or applied cumulatively at the end of an epoch.

Though the sigmoidal transfer functions stabilize the network by restricting the dynamic range between 0 and 1, they pose the problem of scaling. Hence, to circumvent this problem, the error of an output node is defined as,

$$\delta = d - o = \frac{(s - min(s))}{(max(s) - min(s))} - o \quad \text{- (3)}$$

where, $s$ is the sinusoid in the passband, $d$ is the desired output of that node and $o$ is the actual output.

The output of the network is level shifted during testing to give a zero mean output which is a compressed version of the unit amplitude sinusoid. Thus due to scaling, the information about the amplitude of the sinusoid has been lost.

The additive white noise is varied every epoch, so that the network is exposed to a large number of white noise samples and thus can learn a certain regularity behind the noisy inputs. Even though this results in a lack of convergence for even large no. of iterations, this offers a satisfactory learning of the nature of the noise. Normally, acceptable results are obtained after 500 iterations for an SNR of -3 dB. For the 200 training samples, the time taken for 500 iterations was around 5 hours, on a Sun SPARC Station 1. Though the training time is long, once satisfactory convergence has occurred, the weights will be frozen, and then for any test case the time taken will be in the order of ms. The programs were written and executed in MATLAB, due to the large matrix computations involved.

## TRAINING DATA CREATION

The training data set is very vital for the neural network to learn the desired mapping. It should be representative of the various conditions imposed on the performance of the network. For the problem at hand, the desired performance can be stated as:

(i) It should be able to detect any sinusoid present in the passband.
(ii) It should be capable of rejecting any sinusoid outside the passband.
(iii) It should improve the passband SNR.
(iv) It should limit the false alarm rate.

Keeping these objectives in mind, the data set is created. The spectrum (normalised frequency from 0 - 1) is divided into 20 equal sections each of width 0.05. For each section, the training data is created as follows:

(a) First 15 samples contain a sinusoid in the passband only. This enables the network to learn the passband, by orienting the weights in that direction, from the initial random values.

(b) The next 85 samples contain one sinusoid in passband and three in stopband. The network has to give a noise free sinusoid of the same frequency as the one in the passband. Passband frequencies of the samples are made to sweep the entire passband. Sharp transition may be obtained by having more samples on either sides of the cutoff frequency.

(c) The next 80 samples contain sinusoids only in the stopband. The desired output is 0.5(middle of the sigmoid). This improves the rejection property and keeps the false alarm rate low.

(d) The last 20 samples have one sinusoid in passband of amplitude quite small compared to that of those in stopband. This has been found to keep the false alarm rate low.

## SIMULATION RESULTS

The frequencies of detected sinusoids agree well with the original frequency within a certain tolerance, which depends on the type of spectral estimator used. Since the autoregressive spectral estimators fit an AR model for the observation samples, a nonsinusoidal output will produce a considerably flat spectrum compared to sinusoidal outputs[10]. This results in better detection characteristics. The modified covariance spectra(model order 22) have been plotted for the test case of a network trained in the band, [0.45 , 0.5], for the typical cases discussed in IV[Fig. 2].

The false alarm probability depends on the detection threshold and can virtually be reduced to very low values, but at the risk of increasing the probability of detection failure. The parallel arrangement of a set of filters to cover the entire spectrum has the underlying assumption that the cutoff will be sharp. But this is practically impossible to realize. Experimental results show that there is a large probability of misclassification near cutoff. For the above test case, misclassification occurring in the transition regions is shown in Table I. Except for these transition bands, the classification is found to be satisfactory and up to the expectations. The errors in detection are found to be less than 10%(fig.3). Thus this arrangement will ideally suit applications where the filter sections need not have contiguous frequency bands

It has also been observed that the network when trained for white noise of particular variance can perform well for other variances too (in the sense that it improves detection for lesser variances and shows a degradation for higher variances, which is to be expected).

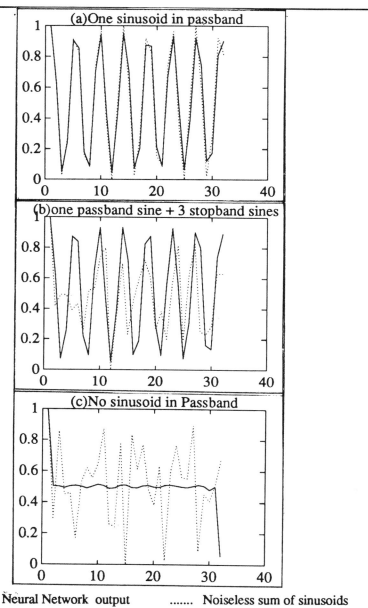

_____ Neural Network output ....... Noiseless sum of sinusoids

Fig.2: plots of the output of a Neural Network trained in the band [0.45-0.5]
for three different cases
(SNR = -3 dB)
a) Neural Network almost reproduces the noiseless sinusoid
b) Neural Network filters out the sinusoid in the passband alone
c) Neural Network rejects the sinusoids & gives a constant output.

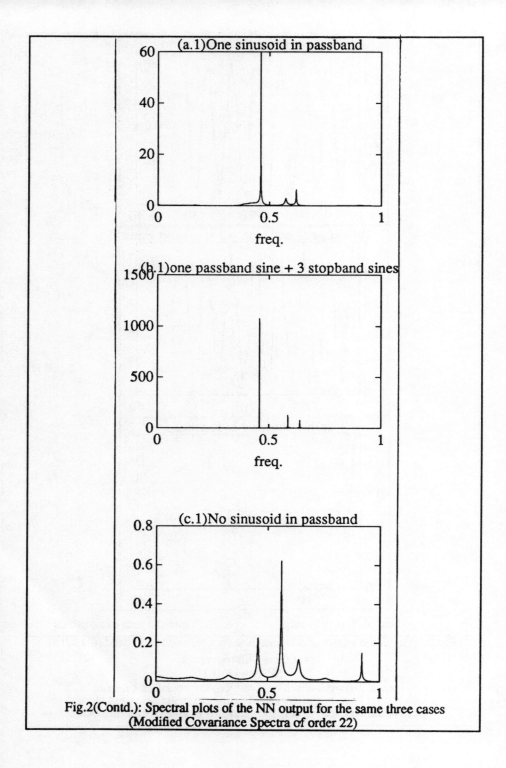

Fig.2(Contd.): Spectral plots of the NN output for the same three cases (Modified Covariance Spectra of order 22)

TABLE I
FALSE ALARM / DETECTION FAILURES IN THE
PASSBAND & NEAR CUTOFF OUT OF 100 TRIALS

| Normalized Frequency | No. of Threshold Exceedings (>8) Out of 100 trials |
|---|---|
| 0.435 | 000 |
| 0.440 | 010 |
| 0.445 | 027 |
| 0.450 | 056 |
| 0.455 | 078 |
| 0.460 | 087 |
| 0.465 | 095 |
| 0.470 | 097 |
| 0.475 | 099 |
| 0.480 | 100 |
| 0.485 | 097 |
| 0.490 | 099 |
| 0.495 | 098 |
| 0.500 | 080 |
| 0.505 | 080 |
| 0.510 | 066 |
| 0.515 | 028 |
| 0.520 | 007 |

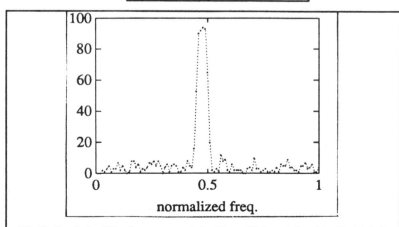

Fig.3: Statistical Performance of the Neural Net trained in [0.45,0.5]
Number of Threshold Exceedings out of 100 trials for sinusoids
of different frequencies plotted against the frequency

## CONCLUSIONS

A neural network preprocessor for spectral estimation has been proposed. Test results and probabilistic analysis show the effectiveness of this arrangement for the multi-tone detection problem. Reasonable frequency estimation has been achieved. Though it has been discussed only for white noise, since no assumption about uncorrelatedness or independence of noise samples has been assumed, the network should perform well for colored noise of the same variance. Yet, a more detailed and rigorous analysis is required to qualify this as a viable alternative for existing schemes.

## REFERENCES

[1] Rumelhart, D.E. & McClelland, J.L., editors, 'PDP: Explorations in the Microstructure of cognition' Volume I, Chapter 8: Learning Internal Representations, MIT Press,1986.

[2] Lippmann, R.P., 'An Introduction to Computing with Neural Nets', pp. 4 - 21, IEEE ASSP Magazine,'87.

[3] Maureen Caudill, 'Neural Networks Primer, Part III', pp. 53 - 59, AI Expert, June '88.

[4] Neural Works Professional II, 'Neural Computing' Volume 1, Networks III, Neural Ware Inc.1989.

[5] Lapedes, A. & Farber, R., 'Nonlinear Signal Processing using Neural Networks:Prediction and Modeling', Los Alamos National Laboratory report, LA-UR-87-2662.

[6] Shin'ichi Tamura, 'An Analysis of a Noise Reduction Neural Network', ATR Interpreting Telephony Research Laboratories, Seika-cho, Souraku-gun, Kyoto, Japan.

[7] Shin'ichi Tamura & Alex Waibel, 'Noise reduction using Connectionist Models', pp.553-556, ICASSP '88.

[8] S.S.Rao & Pradeep M.P., 'A noise reduction neural network as a preprocessing stage in the SVD based method of harmonic retrieval',pp.491-494, Proceedings of the IEEE Circuits & Systems Conference, May '90

[9] Eric K.L.Hung & Robert W.Herring, 'Simulation Experiments to compare the signal detection properties of DFT & MEM Spectra',Communications Research Centre, Ontario, CANADA.

[10] Steven M.Kay, 'Modern Spectral Estimation', Prentice-Hall Signal Processing series,1988.

# FUZZY TRACKING OF MULTIPLE OBJECTS

Leonid I. Perlovsky
Nichols Research Corporation,
251 Edgewater Drive
Wakefield, MA 01880

Existing tracking algorithms have difficulties with multiple objects in heavy clutter [1]. As a number of clutter objects increases, it is becoming increasingly difficult to maintain and especially to initiate tracks. A near optimal algorithm, the Multiple Hypothesis Tracking (MHT) [2], initiates tracks by considering all possible associations between multiple objects and clutter event on multiple frames. This, however, requires combinatorially large amount of computation, which is difficult to handle even for neural networks, when a number of clutter objects is large. A partial solution to this problem is offered by the Joint Probability Density Association (JPDA) tracking algorithm [3], which performs fuzzy associations of objects and tracks, eliminating combinatorial search. However, the JPDA algorithm performs associations only on the last frame using established tracks and is, therefore, unsuitable for track initiation. The problem is becoming even more complicated for imaging, incoherent sensors, when direct measurement of object velocity via the Doppler effect is unavailable.

We have applied a previously developed MLANS neural network [4, 5, 6] to the problem of tracking multiple objects in heavy clutter. In our approach the MLANS performs a fuzzy classification of all objects in multiple frames into multiple classes of tracks and random clutter. This novel approach to tracking using an optimal classification algorithm results in a dramatic improvement of performance: the MLANS tracking combines advantages of both the JPDA and the MHT, it is capable of track initiation by considering multiple frames, and it eliminates combinatorial search via fuzzy associations.

Historically, intrinsical mathematical similarities between tracking and classification problems have not been explored. Tracking problems have been characterized by an overwhelming amount of data available from radar sensors. This has led to the development of suboptimal algorithms amenable to sequential implementation for handling high rate data streams. Such algorithms based on Kalman filters converge to the optimal, Maximum Likelihood (ML) solutions [1], however, their generalization to multiple object tracking has been difficult, as discussed above. Classification and pattern recognition problems, on the other hand, have been characterized by insufficient amount of data for an ambiguous decisions, which has led to the development of Bayes classification algorithms, optimally utilizing all the available information. The MLANS neural network explores mathematical similarity between tracking and classification problems and applies optimal Bayes methods to the problem of tracking multiple objects. The MLANS performs the Bayes association of objects and tracks and the ML estimation of track parameters.

In MLANS tracking, each track is a class characterized by the trajectory state parameters and by the trajectory model. For example, linear trajectories are characterized by their initial positions, velocities, and a linear model relating object coordinates over time to their initial positions. Clutter and maneuvering objects are characterized by appropriate models. The parameters of these models, including the trajectory state parameters are estimated adaptively, in real time. In addition to the trajectory state parameters, each class-track is characterized by its covariance matrix, which can be time dependent or constant and can be adaptively estimated or fixed using prior knowledge about trajectories and sensor accuracy. Unlike the Kalman filter, prediction is not needed for the MLANS track estimation, which performs the optimal estimation of track parameters, utilizing all the available returns, over a wide time window. The covariance matrices in the MLANS are, therefore, determined by sensor accuracy rather than by tracking uncertainties during the Kalman filter initiation and convergence. This optimal utilization of all the available information makes the MLANS ideal for track initiation, as well as for track maintenance in heavy clutter. The MLANS is applicable to both, coherent, radar tracking as well as to incoherent imaging sensor data.

In the example below, the MLANS is applied to tracking three objects in random clutter, using simulated pulse-Doppler search mode radar data. The scattered plot in the Figure shows the distribution of returns from 10 scans in the Doppler-velocity vs range coordinates. About 1000 random clutter returns appear distributed throughout the plot, while three clusters of returns corresponding to the moving objects have approximately 100 returns each. Due to a wide beam employed in a search mode, there is approximately 10 returns from each object per scan. The scatter seen in object returns along the Doppler velocity is determined by the radar accuracy, while the scatter along the range axis is due to both, the radar accuracy and the motion of the objects. Each return is characterized by its time and by its 4 coordinates: range, Doppler velocity, elevation and azimuth angles. The MLANS is using time as a parameter of the trajectory models to project each trajectory to its current position. By clustering these data in the 4-dimensional coordinate space, the MLANS initiates and estimates track state parameters which are shown in the Figure by illustrating 2-$\sigma$ ellipses of the estimated parameters.

This estimation and clustering procedure is performed iteratively after each scan. The MLANS iterates until convergence of neural network weights as described in [5]. This convergence is usually fast, requiring only a few iterations. A more important, fundamental issue in learning efficiency is how accurately track parameters are estimated from a few scans. In the terminology of estimation theory, the question is: what is the efficiency of this neural network as a track parameter estimator? The MLANS performs the Maximum Likelihood estimation, using all the available data from multiple scans. This results in an efficient estimation approaching the Cramer-Rao bound on the best accuracy of track parameter estimation by any algorithm or neural network. The MLANS also has been shown to be close to Cramer-Rao performance bounds on learning efficiency of any algorithm or neural network in several other applications [ 5, 7, 8]

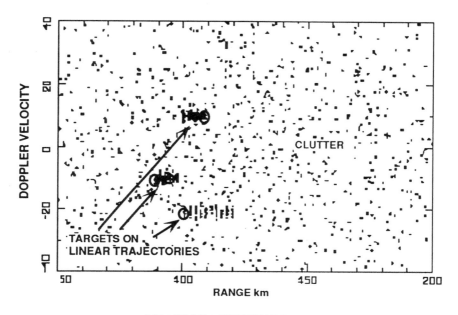

MLANS TRACK BEFORE DETECT

## REFERENCES

[1]   Blackman, S. S. (1986). <u>Multiple Target Tracking with Radar Applications.</u> Artech House, Norwood, MA.

[2]   Singer, R. A., Sea, R. G., and Housewright, R. B. (1974). "Derivation and Evaluation of Improved Tracking Filters for Use in Dense Multitarget Environments" <u>IEEE Transactions on Information Theory</u>, **IT-20**, pp. 423-432

[3]   Bar-Shalom, Y. and Tse, E. (1975) "Tracking in a Cluttered Environment with Probabilistic Data Association". <u>Automatica</u>, **11**, pp. 451-460.

[4]   Perlovsky, L.I. (1987). "Multiple Sensor Fusion and Neural Networks". <u>DARPA Neural Network Study,</u> MIT/Lincoln Laboratory, Lexington, MA.

[5] Perlovsky, L.I. (1988). "Neural Networks for Pattern Recognition and Cramer-Rao Bounds". Boston University Engineering Seminar, Boston, MA.

[6] Perlovsky, L.I. (1988). "Neural Networks for Sensor Fusion and Adaptive Classification". First Annual International Neural Network Society Meeting, Boston, MA.

[7] Perlovsky, L.I. (1990). "Efficient Neural Networks for Transient Signal Processing". XXIII Annual Asilomar Conference on Signals, Systems and Computers, Pacific Grove, CA.

[8] Perlovsky, L. I. & McManus, M.M. (1991). "Maximum Likelihood Artificial Neural System (MLANS) for Adaptive Classification and Sensor Fusion". Neural Networks, 4(1).

# Part 5:

# System Implementation

# NEURAL NETWORKS FOR SIGNAL/IMAGE PROCESSING USING THE PRINCETON ENGINE MULTI-PROCESSOR

N. Binenbaum, L. Dias, P. Hsieh, C.H. Ju, S. Markel, J.C. Pearson, H. Taylor Jr.
David Sarnoff Research Center
and
The National Information Display Laboratory
CN5300
Princeton, NJ 08543-5300

**Abstract:** We describe our modular neural network system for the removal of impulse noise from the composite video signal of television receivers, and our use of the Princeton Engine multi-processor for real-time performance assessment. This system out-performs alternative methods, such as median filters and matched filters. The system uses only eight neurons, and can be economically implemented in VLSI.

## INTRODUCTION

Robert M. Gardner, in his keynote speech at this year's IEEE Symposium on VLSI Technology, emphasized three important points concerning neural networks for signal processing:
(1) Neural networks can solve problems for which conventional methods are unsatisfactory.
(2) Parallel computing technology now permits simulating the real-time operation of neural network systems.
(3) Custom VLSI is needed to achieve practical neural network solutions.

The work presented here exemplifies each of these points: (1) We have developed a neural network system for the detection and removal of impulse noise in television. This has long been considered a difficult signal processing problem for which a solution has never been manufactured and incorporated in television receivers. (2) Much of our computational work, including all neural network training, was done using conventional uni-processor computers. However, we also used a multi-processor SIMD computer, called the Princeton Engine, to simulate the performance of the neural network system in real-time. This testing included side-by-side comparisons with alternative techniques such as median filters and matched filters. In addition, we are currently developing efficient neural network training algorithms for the Princeton Engine. (3) We have determined that the neural network impulse noise reduction system can be implemented in custom VLSI chips at a cost appropriate for inclusion in mass-market television receivers.

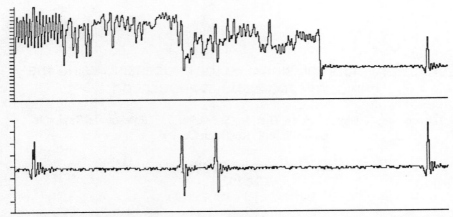

Fig. 1. TOP: Composite video signal (corresponding to 70% of the width of one line) corrupted by 4 impulses. BOTTOM: The impulse waveforms, derived by subtracting the uncorrupted signal from the corrupted signal (re-scaled for clarity). Note also the presence of many impulse-like features in the video signal.

## IMPULSE NOISE IN TV

The color television signaling standard used in the U.S. was adopted in 1953 [1,2]. The video information is first broadcast as an amplitude modulated (AM) radio-frequency (RF) signal, and is then demodulated in the receiver into what is called the composite video signal. The composite signal is comprised of the high-bandwidth luminance (black and white) signal and two low-bandwidth color signals that are encoded in quadrature. This signal is then further decoded into the red, green and blue signals that drive the display. One image "frame" is formed by interdigitating successive "fields" of 262.5 horizontal lines, derived by rasterizing the video signal.

Electric sparks create broad-band RF emissions which are transformed into oscillatory waveforms in the composite video signal, called AM impulses. See Figure 1. These impulses appear on a television screen as short, horizontal, multi-colored streaks which clearly stand out from the picture. There are also echoes of the streak on following lines, due to the way the luminance and chrominance are separated. Impulse producing sparks are commonly created by electric motors. There is generally no spatial (within a frame) or temporal (between frames) correlation between impulses.

## NEURAL NETWORK DETECTION SYSTEM

### System architecture

General considerations suggest a two step approach for the removal of impulses from the video signal — *detect* which samples have been corrupted, and *replace*

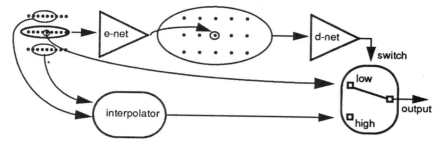

Fig. 2. Schematic of the impulse reduction system. For clarity, the filtering stages after the e-net and d-net are not shown.

them with values derived from their spatio-temporal neighbors. Although impulses are very visually noticeable, they form a small fraction of the data. By altering only those samples detected as corrupted, one avoids blurring the picture. An interpolated average of some sort will generally be a good estimate of impulse-corrupted samples because images are generally smoothly varying in space and time.

For the replacement function, we used a simple average of the samples immediately above and below the current sample. We also designed a neural network interpolator which performed significantly better than the simple averaging method. However, this advantage is only visually apparent if large blocks of the image are replaced. This is not the case for our problem as the impulses are confined to one line and are generally less than 30 samples wide, out of a total of 752 samples per scan line.

For the detection function, we used two small multi-layer perceptrons [3] in series. See Figure 2. The inputs to the first network are nine consecutive samples from the current line centered on the sample of interest. It has three nodes in the first layer, and one output node trained to compute the average of the absolute difference between the clean and noisy signals of the current inputs. It is thus trained to function as a filter for impulse energy, and is termed the e-net. See Figure 3. The output of the e-net is then low-pass filtered and sub-sampled to remove redundant information. This also permits the second network to operate at half the clock rate of the first network, which greatly reduces the complexity and cost of the corresponding chip (discussed further below).

The inputs to the second network are three lines of five consecutive samples each, centered on the sample of interest. This network, like the e-net, has three nodes in the first layer and one output node. It is trained to output one if the sample of interest is contaminated with impulse noise, and zero otherwise. It is thus an impulse detector, and is called the d-net.

The output of the d-net is then fed to a binary switch, which passes through to the final system output either the output of the interpolator or the original signal, depending on whether the d-net output exceeds an adjustable threshold.

Experience showed that the d-net tends to produce narrow spikes to impulse-like features of the image. See Figure 3. To remove this source of false positives, the output of the d-net is averaged over a 19 sample region centered on the sample of

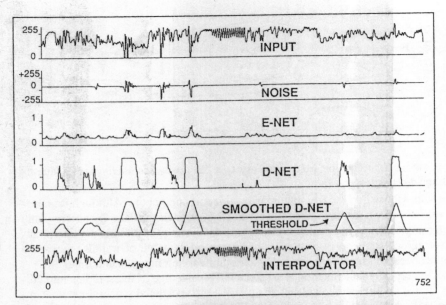

Fig. 3. One scan line of inputs and outputs of the components of the AM impulse removal system.

interest. This reduces the peak amplitude of signals due to impulse-like features (pseudo-impulses) much more than the broad signals produced by true impulses. An impulse is considered to be present if this smoothed signal exceeds a threshold, the level of which is chosen so as to strike a subjectively pleasing balance between low false positive rates (high threshold), and high true positive rates (low threshold). This balance is determined interactively in real-time simulations on the Princeton Engine.

Experience also showed that the fringes of the impulses were not being detected. To compensate for this, all samples within 9 of a suspected corrupted sample are replaced as well.

A single-stage network detector has also been tried, but it has not performed nearly as well as this two-stage detector. A single-stage detector has more information present at its inputs than the d-net, so in principle it should be capable of better performance than the e-net/d-net combination. We presume that the failure of the one-stage network is due not to the inadequacy of its computational architecture or the information at its inputs, but to the inadequacy of the optimization process that determines the network parameters -- the solution is there but it's too hard to find. The two-stage approach breaks the detection problem down into two fairly easy, separable pieces. The e-net removes most of the irrelevant detail, and outputs simple "bump" waveforms to both impulses and pseudo-impulses. The d-net's task then is limited to distinguishing between "bumps" generated by true impulses, and "bumps" generated by pseudo-impulses. This is not too hard because true impulses are rarely vertically correlated, whereas pseudo-impulses often are. Vertically oriented edges and lines are the source of most of the pseudo-impulse waveforms that "fool" the e-net. Some images contain pseudo-impulses that are contained on one line only, such as tennis balls or the sparkle of an eye. These will be falsely detected as impulses by the d-net if they "fool" the e-net.

### Network training

The detection networks were trained on one frame of video containing impulses of five different amplitudes with the largest twenty times the smallest. Visually, these ranged from non-objectionable to brightly colored. The frame of video used (called "album") comprised 6 panels or sub-images, and is widely used as a test pattern within Sarnoff as it contains a representative sampling of important image features. Standard incremental back-propagation and conjugate gradient [4] were the training procedures used. The complexity of the e-net and d-net were reduced in phases. These nets began as three layer nets. After a phase of training, redundant nodes were identified and removed, and training re-started. This process was repeated until there were no redundant nodes.

## PERFORMANCE ANALYSIS

The mean squared error (MSE) is well known to be a poor measure of subjective image quality [6]. A better measure of detection performance is given by the receiver operating characteristic, or ROC [7]. The ROC is a parametric plot of the fraction of corrupted samples correctly detected versus the fraction of clean samples that were falsely detected. In this case, the decision threshold for the smoothed output of the d-net was the parameter varied. Figure 4 (left) shows the neural network detector ROC for five different impulse amplitudes (tested on a video frame that it was not trained on). This quantifies the sharp breakdown in performance observed in real-time simulations at low impulse amplitude. This breakdown is not observed in analysis of the MSE.

Median filters are often suggested for impulse removal tasks, and have been applied to the removal of impulses from FM TV transmission systems [8]. In order to assess the relative merits of the neural network detector, a median detector was designed and analyzed. This detector computes the median of the current sample and its four nearest neighbors (that have the same color sub-carrier phase). A detection is registered if the difference between the median and the current sample is above threshold. The same additional measures were taken to insure that impulse fringes were replaced as were described above for the neural network system. Figure 4 (right) shows both the neural network and median detector ROC's for two different video frames, each of which contained a mixture of all five impulse amplitudes. One frame was used in training the network (TRAIN), and the other was not (TEST). This verifies that the network was not overtrained, and quantifies the superior performance of the network detector observed in real-time simulations. Of course an additional essential criterion is the cost and complexity of hardware im-

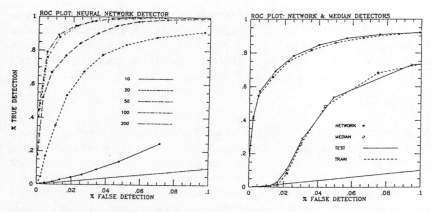

Fig. 4. ROC Analysis of Neural Network and Median Detectors. Straight line marks chance performance

plementation. Median filter chips have been successfully fabricated [9]. The feasibility of constructing a neural network chip is discussed below.

## REAL-TIME SIMULATION ON THE PRINCETON ENGINE

### Architecture Overview

The Princeton Engine (PE) is a SIMD, massively parallel computer. It is scalable in blocks of 64 processors, with a maximum of 2048. They are interconnected in a linear array.

Each processor contains a 16-bit arithmetic logic unit and multiplier, a 64-word triple-port register stack, a 16-bit communication port and up to 128K bytes of 1-cycle and 512K bytes of two-cycle memory. The micro-sequencer has a 144-bit long instruction word. In each instruction time, the processor can simultaneously perform six micro-instructions, such as read from local memory, one ALU operation, and one multiplier operation, etc.

The 16-bit Interprocessor Communications (IPC) bus is used to exchange data between processors. It supports both left/right shifts and broadcast data transfers. In left/right shift operation, the IPC can be set to either normal or bypass mode. In normal mode, each processor's IPC sends or receives data from its nearest neighbors. In bypass mode, each processor's IPC serves as a bridge to connect any two processors. The worst case delay from one processor to any other is 5 cycles (350 ns). In broadcast mode, one processor sends data to all or an arbitrary subset of the others.

The output of each processor is first written to 4 sets of 4x16-bit registers. The Output Timing Sequencer (OTS) selects 1 register set from any one of the processors and sends it to 7, 9-bit digital-to-analog converters. This enables the display of one to four vertically separated windows as well as multiple picture-in-picture displays, and zooming.

**Program Mode**

The Princeton Engine can be programmed in the following three modes:

**Normal Mode:** The complete algorithm operates on one horizontal scan line at a time, with each processor computing the output that corresponds to one of the 910 samples. The instruction clock rate is 14.32 MHz (four times the color sub-carrier frequency). The maximum number of instructions is 910.

**Line Dependent Processing Mode:** There are 64 instruction banks, any one of which can be selected during the processing of each horizontal line. Using this capability, the PE can execute an algorithm which is larger than 910 instructions by applying it to a portion of the picture. For example, to execute an algorithm which requires 1820 instructions, one would process half of the whole picture, using two instruction banks.

**Non Real-time Mode:** The PE behaves more like a general purpose SIMD computer. The micro-sequencer has a 256K, 144-bit, long-word instruction memory. Each processor has local memory of 320K 16-bit words.

**Software Environment**

The PE machine code is created and downloaded using the Apollo host computer. There are two programming methods. In one, a schematic block diagram of the algorithm is graphically constructed, as if it were an actual hardware circuit. The blocks correspond to software modules which can be taken from an extensive signal processing library or created through the GPE (Graphical Programming Editor), which also offers a visual user interface. Micro-instructions are constructed by graphically connecting corresponding registers or gates within the processor. Since each micro-instruction can contain as many as six independent operations, the GPE can be used to produce highly compact code. The Princeton Engine can also be programmed using a high level language, PE-C, which is a subset of C.

The parameters used in the program, such as filter coefficients, thresholds, and algorithm control switches, can be changed during run-time through the Graphical Control Environment (GCE). The GCE reads the symbol table created by the compiler and creates a menu of the run-time parameters.

**Neural Net Simulation**

The neural network system is simulated by using two 910 instruction banks in line-dependent processing mode. The first instruction bank obtains the input from the live video, adds the AM impulses and stores the resulting noisy picture in a frame buffer. See Figure 5. The second instruction bank reads the samples from the

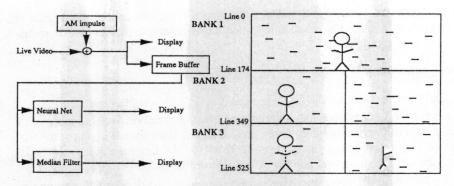

Fig. 5. Illustration of line-dependent processing in the Princeton Engine.

noisy frame buffer, simulates the full neural network system, and outputs the restored picture. The median filter is implemented using a third instruction bank.

The neural network simulations used 16-bit fixed point arithmetic with 8 bits of fraction, and a 10-bit look-up table for the sigmoid function. Comparisons with double-precision simulations on a conventional computer showed no significant reduction in performance.

The output can also be partitioned vertically (using the OTS) to examine different signals in the same instruction bank. For example, the output of the neural network and median detectors can be displayed simultaneously with the corresponding restored pictures (see Fig. 5). Other features that can be controlled at run-time (using the GCE) include: freezing the picture, changing the noise rate and amplitude, adding random noise, and using different thresholds for the median filter and the d-net.

## NEURAL NETWORK HARDWARE FEASIBILITY

With the continued scaling of semiconductor densities it is now possible to conceive of realistic consumer IC based applications of neural networks. A VLSI implementation of an e-net/d-net combination will require a core of about 4000x2000 microns in a CMOS process with a minimum feature size of about 0.8 micron. These estimates are based on a system design that reduces the data rate through filtering and sub-sampling even more than that described above. The e-net takes as input every other set of nine consecutive samples, thus running at half of the data rate. The low-pass filter that follows the e-net takes as input every other set of 3 consecutive outputs from the e-net. The d-net's input is every other set of 5 consecutive outputs from the e-net's low-pass filter (on each of 3 lines), and runs at one eighth of the data rate. The low-pass filter that follows the d-net also takes as input every other set of 3 consecutive d-net outputs. This then finally is the input to the switch.

The direct implementation of the e-net/d-net system topology will be dominated by the number of multiply stages corresponding to the number of connections between nodes in the network. There are a total of 95 multipliers in the e-net and d-nets including FIR filters. However, since most of these multiplies are at a half, a quarter or an eighth of the system clock, it is possible to use serial multipliers for the filter stage of the e-net and all of the d-net, thus making efficient use of available chip area. Basic cell structures for the serial and parallel multipliers in the e-net and d-net were developed and are summarized below.

| Net Stage | #MPY | MPY Bits | Time(ns) | Total Cell Size(um) |
|-----------|------|----------|----------|---------------------|
| E hidden  | 27   | 8x8static | 140     | 3600 x 1200         |
| E output  | 6    | 8x8serial | 280     | 300 x 900           |
| D hidden  | 45   | 8x8serial | 560     | 4000 x 1000         |
| D output  | 6    | 8x8serial | 1120    | 600 x 900           |

Table 1. Summary of the basic cell structures for the e-net and d-net chip.

Because the exact numerical precision and timing requirements of a neural network implementation remains an open research issue some flexibility will be required in the number of bits of precision maintained throughout the network. Cell modules for static multipliers up to 16x16 bits at 28Mhz performance were studied. In addition, a number of serial multiplier configurations have been considered. The basic cell for 8x8 bit serial multiply will require about 250 x 280 microns. A 16x16 bit serial multiplier would require an estimated cell of 500 x 500 depending on the timing.

The second critical component of a direct VLSI implementation of a neural network is the sigmoid function. This will usually involve the use of a PLA or ROM. However, as with the multiplier, there remains considerable interest in both the number of bits and the size of the look-up table which approximates the function. The simulated e-net and d-net used 10-bit tables of 16-bit values. A direct ROM based approach would require a 2K byte ROM for each sigmoid function. We are also evaluating PLA based approaches in order to reduce the size of the table and enable more flexible design trade-offs.

## NEURAL NETWORK TRAINING ON THE PRINCETON ENGINE

An understanding of the effect of limited precision arithmetic on the back-propagation learning algorithm is crucial in the design of a neural network training environment on the Princeton Engine. Given an n bit number, how many bits should be used for the fraction? The decision is between a larger integer range and a smaller minimum step size. For example, given 16 bit numbers with 12 bits of fraction and one bit for the sign, numbers in the range [-8,8) can be represented and the minimum step size is 1/4096. We use m.n to denote a fixed point representation with m total bits and n bits for fraction. Due to the sign bit, m-n-1 bits are used for the integer part.

We compute the sigmoid function by table look-up. All of the following results were obtained using a 10-bit look-up table. The range of the table is [-8,8). We simulated 8, 10, and 12-bit tables, as well as double precision tables, and found that the size of the look-up table had little effect on training. This is also reported by Hoehfeld and Fahlman [10] regarding Cascade-Correlation. To simulate the effects of limited precision, we obtained useful weights by running double precision backpropagation. The performance of the network on different limited precision representations was explored by using these weights. Most development work has been done on Sun workstations with programs designed to mimic the Princeton Engine architecture. In so doing, we have had to work around a problem involving asymmetric truncation which is due to the use of a two's-complement integer multiplier on the PE. For example, small negative numbers are truncated to the smallest possible negative number, while small positive numbers are truncated to zero.

Holt and Baker [11] claim that 8-bit arithmetic (4 bits of integer and 4 bits of fraction) is sufficient for the input/output functioning of feedforward neural networks. Their work deals primarily with binary input and output. Our work on AM impulses deals with continuous input and output. We have found that at least 6 bits are needed for the fraction alone. We believe that at least 12 bits of fraction are necessary for training. Two approaches are being investigated to address the problem of limited precision. One is the use of multiplicative gain factors (Hollis, et al [12] and Kruschke and Movellan [13]), and the other is to increase the number of hidden nodes, thereby increasing the "storage capacity" of the neural network.

A multiplicative gain factor is an extra parameter introduced in the sigmoid function to prevent the weights from reaching the upper bound of the limited precision representation and overflowing. If g is the gain parameter, then the sigmoid becomes $f(x) = 1/(1+e^{-gx})$. Using a gain parameter is equivalent to moving the decimal point of the fixed point number. For example, when using 16.12 arithmetic and a gain of 2, the weights are actually represented in 16.11. Effective use of gain allows us to avoid overflow.

In a preliminary study of the trade off between the number N of hidden nodes and function precision, a 9-N-1 network was trained on two lines of video. Using a varying number of hidden nodes and different arithmetic representations, the network performance, in terms of Mean Squared Error, demonstrated that increasing the number of hidden nodes can compensate for decreasing the numerical precision.

## ACKNOWLEDGEMENTS

This work was supported by Thomson Consumer Electronics, under Erich Geiger and Dietrich Westerkamp. This work was also part of projects for the National Information Display Laboratory. We would also like to acknowledge the help of Jim Gibson, Clay Spence and Ronald Sverdlove at Sarnoff.

## REFERENCES

[1] K. McIlwain and C.E. Dean (eds.); Hazeltine Corporation Staff. Principles of Color Television. John Wiley and Sons. New York, 1956.

[2] D.E. Pearson. <u>Transmission and Display of Pictorial Information</u>. John Wiley and Sons. New York, 1975.

[3] D.E. Rumelhart and J.L. McClelland (eds.). <u>Parallel Distributed Processing: Explorations in the Microstructure of Cognition</u>. MIT Press. Cambridge, 1986

[4] The Numerical Algorithms Group Inc. <u>The NAG Fortran Library Manual, Mark 14</u>. Downers Grove, IL, 1990.

[5] D. Chin, J. Passe, F. Bernard, H. Taylor and S. Knight. "The Princeton Engine: A Real-Time Video System Simulator." IEEE Transactions on Consumer Electronics **34**:2 pp. 285--297, 1988.

[6] J.A. Roufs and H. Bouma. "Towards Linking Perception Research and Image Quality." Proceedings of the SID **21**:3, pp. 247--270, 1980.

[7] D.M. Green and J.A. Swets. <u>Signal Detection Theory and Psychophysics</u>. John Wiley and Sons. New York, 1966. Reprinted with corrections, Krieger. Huntington, N.Y., 1974.

[8] S.S. Perlman, S. Eisenhandler, P.W. Lyons, and M.J. Shumila. "Adaptive Median Filtering for Impulse Noise Elimination in Real-Time TV Signals." IEEE Transactions on Communications **COM-35**:6 p. 646, 1987.

[9] L.A. Christopher, W.T. Mayweather III, and S. Perlman. "A VLSI Median Filter for Impulse Noise Elimination in Composite or Component TV Signals." IEEE Transactions on Consumer Electronics **34**:1 p. 262, 1988.

[10] M. Hoehfeld and S.E. Fahlman. "Learning with Limited Numerical Precision Using the Cascade-Correlation Algorithm." Carnegie-Mellon University Technical Report CMU-CS-91-130, 3 May 1991.

[11] J.L. Holt and T.E. Baker. "Limited Precision Backpropagation Training of Neural Network Benchmarks." 20 September 1990.

[12] P.W. Hollis, J.S. Harper, and J.J. Paulos. "The Effects of Precision Constraints in a Backpropagation Learning Network." Neural Computation 2, pp. 363--373, 1990.

[13] J.K. Kruschke and J.R. Movellan. "Benefits of Gain: Speeded Learning and Minimal Hidden Layers in Back-Propagation Networks." IEEE Transactions on Systems, Man, and Cybernetics 21:1, January 1991.

# DESIGN OF A DIGITAL VLSI NEUROPROCESSOR FOR SIGNAL AND IMAGE PROCESSING

Chia-Fen Chang and Bing J. Sheu
Department of Electrical Engineering
Signal and Image Processing Institute
University of Southern California
Los Angeles, CA 90089-0271

**Abstract**--An efficient processing element for data/image processing has been designed. Detailed communication networks, instruction sets and circuit blocks are created for ring-connected and mesh-connected systolic arrays for the retrieving and learning phases of the neural network operations. 800 processing elements can be implemented in 3.75 cm x 3.75 cm chip by using the 0.5 μm CMOS technology from TRW, Inc. This digital neuroprocessor can also be extended to support fuzzy logic inference.

## I. INTRODUCTION

The techniques for designing a VLSI neuroprocessor have been diversed into two approaches: one is digital design, the other is mixed signal design. Comparatively, a digital design could provide better programmability which is desirable for performing different networks. Moreover, better expandability is also provided to solve large-scale scientific problems.

In the digital design, a systolic system is easy to be implemented because of its regularity and is easy to be reconfigured due to its modularity [1]. A systolic configuration of the connectionist networks can greatly speed up the system throughput in terms of data pipelining. The same architecture is shared between the feedforward propagation phase and the feedback learning phase, which suggests no requirement in reconfigurability [2]. A redundant path can be attached to achieve fault tolerant design.

This paper presents a digital implementation of an efficient processing element (PE). The design is aimed at a wide class of neural network models. Microprogramming technique which results in portability is used. Since the PE is scalable and reconfigurable, multi-chip configuration can be used to increase the network to solve complex problems. The design results can be utilized in the large-volume production and be elastically expanded to Wafer Scale Integration (WSI) to achieve high-performance computing.

## II. COMMUNICATION NETWORKS AND SYSTEM OVERVIEW

A custom-designed PE has been developed. This PE can be directly applied to a ring systolic array [3], a mesh-connected array [4], a cellular neural network [5], or a fuzzy control system in which forward processing and

---

This work was partially supported by DARPA under Contract J-FBI-91-194 and TRW, Inc.

learning operations are expressed in terms of matrix and vector computations. In this writing, the ring and mesh-connected system will serve as two examples for describing the operations of the designed PE.

## A. Ring Systolic Array

The overall system architecture of a ring systolic array used in our system is given in Figure 1. The host computer serves as the interface between the user and the systolic array. It provides the problem-specific parameters, such as input patterns, initial weights, convergence-controlling parameters, etc. The ring controller specifies and monitors the executions in each PE and also performs loading and receiving data to and from each PE. The control line is designed in a broadcast fashion since all commands which are performed in all the PE's are the same during each clock cycle. The control line is operated in a broadcast fashion since all commands which are performed in all the PE's are the same during each clock cycle. To provide an adequate dynamic range for the weights and activation calculations, a word length of 16-bit wide is adopted based on extensive simulations and analysis of the finite word length effects.

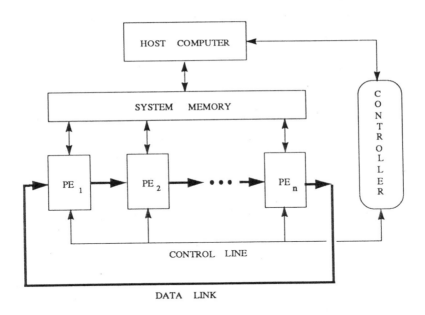

Figure 1  The ring systolic system architecture.

## B. Mesh-Connected Array

A mesh-connected systolic architecture requires two-way communication between each PE and all of its nearest neighbors. In our system as shown in Figure 2, the I/O operations occur only on the right- and left- most columns. The top and bottom rows inside the processor array are connected in a wrap-around fashion. An array controller is used to control all of the operations in mesh-connected processors. The system memory is a two-port memory which is shared by the processor array. It receives data and instructions from the host computer when a host call is received by the array controller. It also loads data into the host computer following a system call. Data are only transmitted to (from) this system memory from (to) the right- and left- most columns.

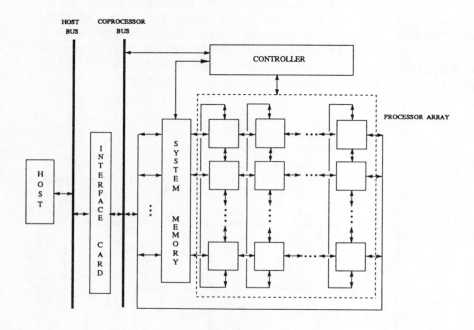

Figure 2 The overall system of a mesh-connected systolic array.

The presented PE is designed to be used either as a neuron processor or a synaptic processor. The mapping of a neural network into a mesh-connected network was discussed by Przytula, Lin and Kumar [4]. Consider a mesh of $N \times N$ processors, two cases must be addressed. In the first one, a neural network of $n$ neurons and $w$ synaptic connections are assigned one-to-one to $n+w$ processors, where $N^2 \geq n+w$. Based on the example of a multilayer perceptron with back-propagation learning, the data routing for a single iteration of the retrieving phase requires $24(N-1)$ elemental shift operations whereas the learning iteration takes $60(N-1)$ shifts. For the second mapping, $N^2 \leq n+w$, the original neural net has to be partitioned to suitably utilize the provided hardware.

Benes network was used for interprocessor routing. All of the information regarding processor assignments, partition strategies and data routing are supplied by the procedure from the host computer. A translator (or compiler) is needed to transform this high-level procedure into microcodes which are recognizable by the array controller.

## III. DETAILED PE DESIGN

The ring systolic architecture is simpler in structure and clearer in data routing. In the following, we will take this architecture for describing the main functions of our custom- designed PE.

Operations in both retrieving and learning phases of the popular error backpropagation can be formulated as matrix-vector multiplication (MVM), outer-product updating (OPU) and vector-matrix multiplication (VMM) problems. The formulas for these three operations are as follows:

MVM:

$$S_i(l + 1) = \sum_{j=1}^{N_l} w_{ij}(l + 1)a_j(l) \quad (1)$$

$$a_i(l + 1) = f_i(S_i(l + 1), \theta_i(l + 1)) \quad (2)$$

where $l$ represents the $l$th layer. The net input value $S_i(l)$, along with the external input $\theta_i(l)$, will determine the new activation value $a_i(l)$ by the nonlinear activation function $f_i(l)$. The weight values $w_{ij}$'s are stored in the on-chip cache memory whereas $a_i(l)$'s are routed through all the processors. After n steps, all the PE's accomplish their jobs simultaneously.

OPU:

$$\Delta w_{ij} = \Delta w_{ij} + g_i(l + 1)a_i(l) \quad (3)$$

$$w_{ij} = w_{ij} - \eta \Delta w_{ij} \quad (4)$$

where

$$g_i(l + 1) = e_i(l + 1)f'(l + 1) \quad (5)$$

the weight value $w_{ij}$, is updated by the value $\Delta w_{ij}$ and the updating rate $\eta$. These $w_{ij}$'s and $\Delta w_{ij}$'s are stored in the on-chip cache memory, $g_i(l)$'s are stored in the registers. whereas $h_i(l)$'s are routed through all the processors.

VMM:

$$e_i(l) = \sum_{j=1}^{N_l} g_j(l + 1)w_{ij} \quad (6)$$

where $e_i$ is the back-propagated corrective signal. The weight values $w_{ij}$'s are stored in the on-chip cache memory while $g_j(l + 1)$'s are routed through all the processors.

The three operations, MVM, OPU and VMM, are basic steps for many signal and image processing applications. For instance, FFT, convolution and Viterbi decoding are mainly composed of these operations. These three matrix and vector computations are supported by the PE.

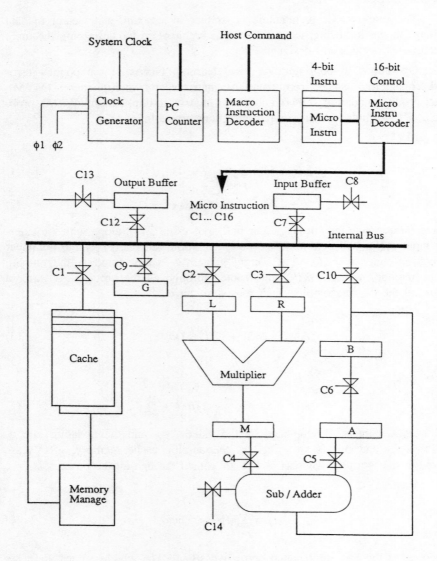

Figure 3  The building blocks inside a PE.

Figure 3 shows the building blocks inside a processing element. Two on-chip cache memories are included in each PE for faster processing. One is instruction cache and the other is data cache. The advantage of separating instruction cache and data cache is that, unlike data, instructions do not change, so the contents of an instruction cache need never be written back to the main memory. Besides the on-chip cache memory inside each PE, there is also a system memory for the PE's. Each PE has its own accessible memory region which is defined initially by the system designer and can then be expanded or shrunk by the user. After the initialization of the systolic array system, the initial parameters (e.g., weights) will be down-loaded into the system memory.

| Step | Micro instru description | c1 | c2 | c3 | c4 | c5 | c6 | c7 | c8 | c9 | c10 | c11 | c12 | c13 | c14 |
|---|---|---|---|---|---|---|---|---|---|---|---|---|---|---|---|
| 1 | load $a_j(1)$ to L from G | 1 | | | | | | | | 1 | | | | | |
| 2 | read $h_j(l+1)$ to buffer | | | | | | | 1 | | | | | | | |
| 3 | $h_j(l+1) \rightarrow$ R,buffer | | | 1 | | | | 1 | | | | 1 | | | |
|   | Do multiplication | | | | | | | | | | | | | | |
| 4 | read $\Delta w_{ij}$ from cache | 1 | | | | | | | | 1 | | | | | |
| 5 | $\Delta w_{ij} \rightarrow$ A | | | | | | 1 | | | | | | | | |
| 6 | Do addition $\rightarrow$ B | | | | 1 | 1 | | | | | | | | | |
| 7 | restore $\Delta w_{ij}$ to cache,R | 1 | | 1 | | | | | | 1 | | | | | |
| 8 | load $\eta$ from G | | 1 | | | | | 1 | | | | | | | |
|   | Do multiplication | | | | | | | | | | | | | | |
| 9 | load $w_{ij}$ from cache | 1 | | | | | | | | 1 | | | | | |
| 10 | $w_{ij} \rightarrow$ A | | | | | | 1 | | | | | | | | |
| 11 | Do addition | | | | 1 | 1 | | | | | | | | | |

(a)

| Step | Micro instru description | c1 | c2 | c3 | c4 | c5 | c6 | c7 | c8 | c9 | c10 | c11 | c12 | c13 | c14 |
|---|---|---|---|---|---|---|---|---|---|---|---|---|---|---|---|
| 1' | Group 1, 5 | | 1 | | | | 1 | | 1 | | | | | | |
| 2' | Group 2, 3, 10 | | | 1 | | | | 1 | 1 | | | 1 | | | |
| 3' | Group 4, 6 | 1 | | | 1 | 1 | | | | 1 | | | | | |
| 4' | Group 7, 11 | 1 | 1 | | | | | | | 1 | | | | | |
| 5' | Step 8 | | 1 | | | | | | | 1 | | | | | |
| 6' | Step 9 | 1 | | | | | | | | | 1 | | | | |

(b)

Figure 4 (a) The original microinstructions for performing OPU operations.
(b) The associative reduced microinstructions.

During the processing, data will flow between the system memory and the cache memory. A DMA controller is used for concurrent I/O and CPU operations.

The system clock is synchronized for each PE in a systolic array. Microprogramming technique is used in the design since it has advantages of being easy to design, maintain, and expand; it also has great portability and compatibility. However, it has the disadvantage in speed when compared with a dedicated-control finite state machine, high speed memory has been used to improve the performance. C1 to C16 in Figure 3 represent the microinstructions. For a particular problem, a systolic procedure is first implemented by a high level language in the host computer, then is sent to the ring controller. Upon receiving the macro commands, the controller decodes them into microinstructions. The microinstructions for performing OPU operations are shown in Figure 4(a). Since there are many overlapping instructions between these steps, they can further be reduced into a compact instruction-step design, which is shown in Figure 4(b). Notice that multiplication and addition have to be separated by at least one clock.

In order to increase the speed of the processor, the internal bus is segmented into multiple local bus lines. By doing so, more steps are overlapped together because they do not occupy the same bus. The speed of the PE is limited by the multiplier. The Wallice multiplier is used. Since a Wallice structure can be separated into 4 x 4 multiplier blocks and Wallice tree blocks, a pipeline procedure can be added here and the speed of the PE can be increased.

The controller includes four major modules: a memory access module, a fault detection and recovery module, an I/O module and a control module. The memory access module is used to down-load data to each local memory. The fault detection and recovery module is used for detecting faulty PE's, correcting and recovering from errors by a spare PE. The I/O module receives data and macrocodes from the host, up-loads data and faulty signals to the host computer. The control module controls operations of each PE, broadcasts and collects data.

A two-level microprogramming is used to speedup the network response because most operations are repetitive and commonly used. Moreover, this two-level control design technique can reduce the total size of the control memory needed, this translates to smaller chip area in the case of one-chip CPU's or PE's [6]. The two-level microprogram and supporting hardware are shown in Figure 5. There are four major operations inside a microprogram:

*Next Address Operation:* next operation is at the next address. Opcode = 00.

*Conditional Branch Operation:* next operation's address depends on the condition in the microcode. Opcode = 01.

*While Do Operation:* could be done by combination of next address and conditional branch operations, and use a dedicated counter to count the iteration number. Opcode = 11.

*Stop Operation:* for the last operation in the nanoprogram, test if repetition, that is, content of the dedicated counter for the dimension (i.e., number of operations) of a macroinstruction from the host, is enough. If yes, the microprogram goes to next operation, else the nanoprogram goes to the first nanooperation and trigger the dimension counter to count down. Opcode = 10.

The implementations of these four operations is shown in Figure 6.

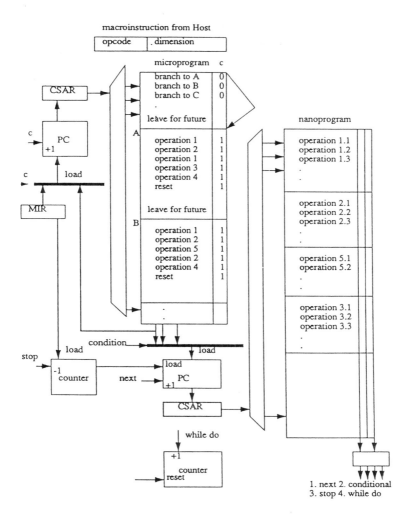

Figure 5  The two-level microprogram.

613

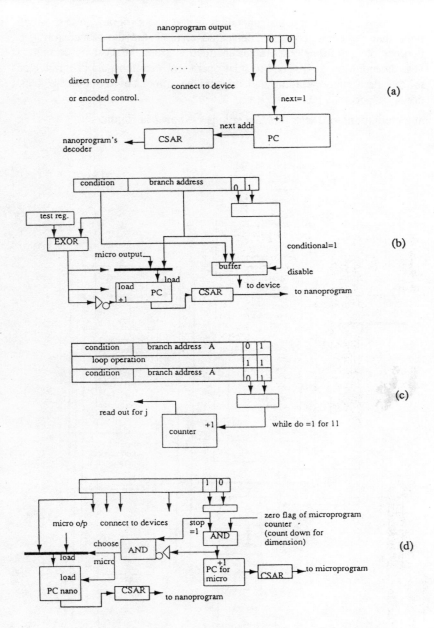

Figure 6 Implementations of the four major operations inside a microprogram.
(a) Next address operation. (b) Conditional branch operation.
(c) While do operation. (d) Stop operation.

From our calculation, 800 processing elements can be implemented in a 3.75 cm x 3.75 cm chip by using the 0.5 µm CMOS technology provided by TRW, Inc. Data buses occupy 20% area of the whole chip. The system memory is supported off-chip.

For the fault tolerant algorithm, the weighted check sum method is used to detect and correct errors [7]. Moreover, parity bits are used to detect one or two adjacent faulty bits in the microprograms and the counters. After completion of error detection and correction, recovery operation can be accomplished by dedicated microprogram.

## IV. CONCLUSION

This microprogram-controlled PE is designed for running various applications in the field of data and image processing. Ring and mesh have been the target communication networks during the design. For a neural network of large size, either partition the network to fit a small hardware size or use multiple chips is achievable.

## V. ACKNOWLEDGEMENT

Valuable discussions with Professors Alvin Despain, Jerry Mendel, Sun-Yuan Kung and Mr. Te-Ho Chen are highly appreciated. James Cable, Advanced Technology Manager, and Mark Miscione, VHSIC Program Manager of TRW, Inc. provided the advanced CMOS technology support.

## REFERENCES

[1]  H. T. Kung, "Why Systolic Architecture?", *IEEE Computer*, 15(1), 37-42, 1982.

[2]  N. Morgan, *Artificial Neural Networks Electronic Implementations*, IEEE Computer Society Press, 1990.

[3]  J.-N. Hwang, J. A. Vlontzos and S.-Y. Kung, "A Systolic Neural Network Architecture for Hidden Markov Models," *IEEE Trans. on ASSP*, vol. 37, no. 12, pp. 1967-1979, Dec. 1989.

[4]  K. W. Przytula, W.-M. Lin and V. K. Prasanna Kumar, "Partitioned Implementation of Neural Networks on Mesh Connected Array Processors," *VLSI Signal Processing, IV*, IEEE Press, pp. 106-115, 1991.

[5]  L. O. Chua and L. Yang, "Cellular Neural Networks: Theory," *IEEE Trans. on Circuits and Systems*, vol. 35, no. 10, pp. 1257-1272, Oct. 1988.

[6]  J. P. Hayes, *Computer Architecture and Organization*, McGraw-Hill, Inc., 1988.

[7]  M. Abramovici, M. A. Breuer and A. D. Friedman, *Digital Systems Testing and Testable Design*, Computer Science Press, 1990.

# Tutorial: Digital Neurocomputing for Signal/Image Processing

S. Y. Kung
Princeton University
Princeton, NJ 08544, USA *

**Abstract** - The requirements on both the computations and storages for neural networks are extremely demanding. Neural information processing would be practical only when efficient and high-speed computing hardware can be made available. In this tutorial, we review several approaches to architecture and implementation of neural networks for signal and image processing. In Section 2, we shall discuss direct design of dedicated neural networks implemented by a variety of hardware technologies (e.g. CMOS, CCD). In Section 3, we introduce an indirect design approach based on matrix-based mapping methodology for systolic/wavefront array processor. The array processors mapping technique presented should be applicable to both programmable neurocomputer and dedictated digital or analog neural processing circuits. In Section 4, we survey several key general-purpose and system-oriented design. Key design examples of existing parallel processing neurocomputers are also discussed.

## 1 Introduction

Neural computing requires a huge number of computations and communications. Therefore, multiprocessors, arrray processors, or massive-parallel-processors are most appealing architecture for neurocomputer design. For hardware implementation, we need to map the neural models onto array architecture. Two kinds of approaches may be adopted: (1) the direct design approach which maps structure of a neural model directly into hardware; (2) the indirect approach which exploits the matrix processing nature of neural models to considerably simplify the hardware requirement. Most existing digital neurocomputers are built with off-the-shelf components. However, the

---

*This research was supported in part by Air Force Office of Scientific Research under Grant AFOSR-89-0501A. The author wishes to acknowledge the invaluable assistances from Mr. W.H. Chou and Mr. T. Miyazaki in the preparation of this tutorial, in particular, Chou's contributions on the array architecture design.

ultimate digital neurocomputer with highest performance will have to be from fully custom VLSI circuits. General-purpose neurocomputers are mostly digital. From an architectural perspective, we shall provide a survey on the possible architectural choice for general-purpose neurocomputers. Examples of digital parallel architectures are [34, 19, 2, 21], [26]. More details can be found in [25].

## 2  Direct Design: Dedicated Neural Processing Circuits

Dedicated neural processing circuits are naturally affected by their implementation technology, such as *electronic processors* (including *analog CMOS, digital CMOS, CCD*) and *optical processors*. Examples of dedicated electronic implementation are [18, 32, 3, 11], [20],[16],[17],[6], [1]. Examples of dedicated optical implementation are [9, 35, 10].

By direct design, the neural computing architecture directly imitates the structure of neural networks. This often involves very straightforward mapping. Consequently, a large number of (mostly global) communications in the connectionist nets are supported directly by hardware, which tends to become a limiting factor of the maximum size of neural networks. *Most existing dedicated neural processors emphasize single chip implementation instead of system building. Before they can be applied effectively to real applications, total system design concept needs to be much enhanced.* Nevertheless, for many preprocessing applications, e.g. early vision processing, dedicated neural circuits are more attractive. Their major characteristics are:

1. Dedicated neural implementation is aimed at high-performance for special applications. Consequently, they are tailored to the architecture of specific neural models.

2. Basic features in special purpose implementation of neural nets are values of neural states, dynamic range of synaptic weights, adaptivity of the synapses (or lack of it), and learning mechanism.

3. The key technologies for neural networks are electronic and optical technologies. So far the most prevailing and practical is the electronic circuits, which include analog CMOS, digital CMOS, CCD, etc.

Due to its maturity, availability, and high-degree of integration, CMOS is by far the most popular technology. (While CCD presents a formidable challenge to the CMOS, optical devices appear to be far from being mature.) They can be built as analog, digital, or hybrid circuits.

**Analog vs. Digital Designs**  Both analog and digital techniques have demonstrated some degree of success in their own areas of applications. In order to select between digital vs. analog circuits for neural nets, many technological issues, e.g. storage and transfer of analog signals, the speed and precision achievable, as well as adaptivity/programmability need to be better understood.

- **Pros and Cons of Analog Circuits**

    For dedicated applications, a neuron can be easily implemented by a differential amplifier, using analog device, and the synaptic weights implemented via resistors. In this way, many neurons can be fit into one single chip. Analog circuits can "process more than 1 bit per transistor" and provide very high speed processing, which compares favorably to digital circuits. For real-time early vision processing; dedicated analog processing chips offer arguably the most appealing alternative. The asynchronous updating properties of analog devices can provide extremely high speed computations which are qualitatively different from those of any digital computer.[18] For examples, analog circuits offer inherent advantages on (1) the computation of the sum of weighted inputs by currents or charge packets; and (2) the nonlinear effects of the devices facilitating realization of sigmoid type functions. In short, analog circuits have a large presence in the existing dedicated chip designs. The circuits will continue to be useful because the integration of analog sensors and neural pre-processing and post-processing is vital to most real-world applications.

    Although the analog circuits are more attractive for the biological type neural networks, its suitability for the connectionist type networks is very questionable. For example, compared with digital circuits, analog circuits are more susceptible to noise, crosstalk, temperature effects, power supply variations, etc. While nonvolatile storage of analog weights provides high synaptic density they are not easily programmable. Since the higher the precision the more chip area will be required, analog precision is usually limited to no more than 8 bits. In resistor-capacitor circuitry, low current consumption calls for high-resistance resistors. In switch-capacitor and resistor-capacitor circuitry, the low-noise constraint limits the minimal transistor surfaces and capacitors. In short, the combined factors of precision, noise, and current consumption lead to larger chip area.

- **Pros and Cons of Digital Circuits**

    For the connectionist networks, digital technology offers some very desirable features such as design flexibility, learning, expandable size, and accuracy. Digital designs have overall advantages in terms of system level performance. Dynamic range and precision is critical for many complex connectionist models. Digital implementation offers much greater flexibility on the choice of precision than its analog counterpart. Designers of digital VLSI circuits also enjoy the mature CAD technology, the convenient building block modular design. In addition, it has clear advantages on its early accessibility of commercial design softwares, silicon fabrication supports, and fast turn-around implementation. The disadvantages of digital circuits are the bulky chip area and (sometimes relatively) slow speeds.

## 2.1 Analog Electronic Circuits

Features of analog design are fast speed, low precision, and small-scale-system. Although analog designs are always continous-valued, they may be implemented in continuous-time ciruits (e.g. RC circuits) or in discrete-time circuits (e.g. analog switch-capacitor circuits, CCD).

**Analog VLSI CMOS Circuits** An excellent discussion on analog design of neural networks can be found in a recent book by Mead[28]. For examples, continuous-time analog electronic real-time retinal processing devices have been developed by Mead and coworkers[32], implemented on 3 $\mu$CMOS VLSI chip. A chip contains an array of pixel cells from 48 × 48 to 88 × 88 cells. Each cell includes a photoreceptor, data processing circuit and a circuit for data transmission out of the chip. The data processing circuits produces the time and spatial derivatives, with the latter generated in a hexagonal resistive network. Some adaptive learning capability has been incorporated into the design of the analog circuits. A continuous-time analog electronic real-time cochlea was developed by Lyon and Mead, which is implemented on a single CMOS VLSI chip. The building blocks are transconductance amplifiers and capacitors. Both are implemented by transistors: the transistors operating above threshold are used as capacitors and below threshold as active devices. Each chip contains a 100-filter stage. Mitsubishi Electric has announced a fast and largest scale learning neurochip. The Mitsubishi chip contains 336 neurons (units) and 28,000 synapses (connections). Synapse weights are controlled by the amount of electricity stored in capacitors, and learning takes place through changes in these amounts. The chip calculates and updates the weights in an analog fashion, as opposed to the more conventional all-digital synapse circuits, making it possible to hold the area needed for a single synapse to 70 square microns. The execution speed is 1 Tera CPS (connections per second) in the retrieving phase (without learning) and 28 Giga CUPS (connection updating per second) in the learning phase.

**Analog CCD Circuits** A CCD works as a delay line, where the charge packets are kept within a potential well for a certain time. By this property, CCD has been very suitable in many signal/image filtering applications. In a neural network, the output of the neurons is a function of the weighted sum on all inputs. By CCD technology, the summation in charge domain can be performed in a single transfer. Thus the CCD device may be used to produce a fast on line analog "multiply and add" operation: The input signal and weights are preloaded into device and the output signal must be immediately retrieved after the operation.

In addition to the advantage of parallel computation of the weighted sum of inputs enjoyed by all analog circuits, CCD offers the key features of (1) reduction of wiring by using transport of charges through silicon device channel; and (2) suitability to support pulse-coded neural models.[16] Moreover, CCD is also suitable for (bit-level) systolic design, making it a very viable candidate for neural processor implementation. [20],[16],[17],[6], [1]

## 2.2 Digital (or Mostly Digital) Electronic Circuits

*Digital VLSI implementation are suitable for dedicated connectionist type neural nets.* By definition, a digital network implies both discrete-time and discrete-valued design. Digital implementation provides a high speed, accurate yet flexible platform for neural network algorithms. A digital design allows a flexible choice of word-length as well as structure. A potentially advantageous option is a hybrid design by properly combine the digital and analog circuits. It is concluded in a comparative study in that digital CMOS appears to be superior to the existing CCD design. [29]

For examples, a mostly digital design is presented by the configurable CMOS neural network chip described by H. P. Graf [2] at Bell Labs. It is a single hybrid chip (using 1.25 $\mu$ CMOS) for synaptic matrix processing. [11, 12] The matrix connects 50 inputs with 24 outputs using 1200 (externally adjustable) synapses. The chip provides 256 blocks each of which consists of 128 binary synapses. The training and calculations are done at a rate of 10 MHZ. The synaptic matrix developed at JPL [33] consists of an array of 32 × 32 interconnections and contains 1024 synapses. The synapses are bistable (on/off) and can be controlled individually by external signals. The setting of the synaptic weights has to be determined a priori. Although limited adaptability is possible, learning algorithms are not implemented in the hardware. Bell Communications Research has designed and implemented a single analog VLSI chip for Boltzmann machine. It incorporates both the simulated annealing technique and a stochastic learning mechanism.

## 2.3 Optical Implementation of Neural Nets

Optical processors, exploiting the global interconnectivity of optical signal flow, have been proposed for the implementations of neural networks[9, 27, 8]. Systems of this type have potential capabilities of massive optical storage, global interconnectivity, and high speed processing [13].

Hybrid VLSI and optical technology represents a very plausible new dimension. A hybrid system can incorporate the advantages of both the complimentary capabilities of optics (efficient linear operations and communications) and electronics (efficient nonlinear operations). Some hybrid designs use systolic arrays to allow parallel processing of digital and analog data. Optical/VLSI pattern-recognition systems also utilize the inherent ability of optical systems to perform transforms for computing high-speed correlations. For a very broad application domain, hybrid systems will offer a significant improvement in cost, size, weight, power consumption, and reliability. Applications will include the areas of pattern recognition, Fourier transform, feature extraction, correlation systems, scene classification, image analysis, machine vision, alphanumeric data analysis, and SAR processing.

# 3 Indirect Design: Mapping Neural Algorithms to Array Structures

By indirect mapping, the original neural network structure may be converted into a simpler computing architecture. Almost invariably, a desirable architecture would have local interconnections only and the computation are performed in pipelined fashion. To achieve such a design, matrix-based mapping methodology is the most mature and effective tool for neural and other information processing applications. The tool can accommodate several key models, such as Hebbian, delta, competitive, and back-propagation learning rules in single layer feedback nets and multilayer feed-forward nets. It is suitable to many useful applications, including image recognition, signal classification, and digital filtering applications.

The main concern in the indirect design approach is that: *given the neural algorithms, how to systematically derive parallel processor which would be cost-effective and yield optimal performance at the same time.* First, we observe that most neural processing algorithms are computationally iterative and intensive, and demand very high throughput. On the other hand, the neural algorithms can be expressed in basic matrix operations (such as innerproduct, outer-product and matrix multiplications), which may in turn be mapped to basic processor arrays.

*This approach matches very well the design principle of VLSI system, which exploits highly regular, parallel, and pipelined architecture, and reduces the communication complexity.* Typical example of VLSI parallel/pipelined architectures are *Systolic* or *wavefront* arrays. They are well suited to VLSI implementation of neural networks, because their properties of modularity, regularity, local interconnection, and pipelining. They have the following key advantages:

1. The exploitation of pipelining is very natural in regular and locally-connected networks. It yields high throughput and simultaneously saves the the cost associated with the *communication*.

2. They provide a good balance between the computations and communication, which is most critical to the effectiveness of array computing.

3. In order to support most of the connectionist models VLSI array architecture appears to be most viable.

In the following, a systematic mapping methodology for deriving systolic arrays is introduced.

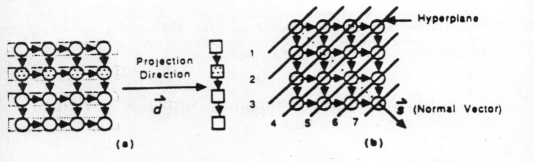

Figure 1: (a) A linear projection with projection vector $\vec{d}$; (b) A linear schedule $\vec{s}$ and its hyperplanes.

**Deriving DGs from Given Algorithms** A DG is a directed graph which specifies the data dependencies of an algorithm. For regular and recursive algorithms, the DGs will also be regular and can be represented by a grid model; therefore, the nodes can be specified by simple indices, such as $(i, j, k)$. *Design of a locally linked DG is a critical step in the design of systolic arrays.*

**Mapping DGs onto Array Structures** The mapping methodology has been adopted to derive systolic/wavefront array architectures for various numerical, signal/image processing, and neural nets algorithms.[24] Two key elements in mapping a DG onto a systolic array are *processor assignment* and *schedule assignment*.[24] It is common to use a *linear projection* for processor assignment, in which nodes of the DG along a straight line are projected (assigned) to a PE in the processor array (see Figure 1(a)). A linear projection is often represented by a *projection vector* $\vec{d}$. A *linear scheduling* is also common for schedule assignment, in which nodes on a hyperplane in the DG are scheduled to be processed at the same time step (see Figure 1(b)). A linear schedule can also be represented by a *schedule vector* $\vec{s}$, which points in the direction normal to the hyperplanes.

For a systolic design, it must satisfy two conditions: $\vec{s}^T \vec{e} > 0$ and $\vec{s}^T \vec{d} \neq 0$, where $\vec{e}$ is any dependence vector in the DG.

## 3.1 Linear Array Design for Data-Adaptive Back Propagation Method

We shall concentrate our discussion on the most popular neural model, i.e. a two-layer Back Propagation ( BP ) network. A two-layer network is labeled an N-K-L network, if there are N nodes on the input layer, K nodes on the hidden layer, and L nodes on the output layer.

The dynamic equations for a two-layer network are

$$\begin{aligned}
\underline{u}_i &= \sum_{j=1}^{N} w_{ij} x_j + \underline{\theta}_i \\
a_i &= f(\underline{u}_i), \quad 1 \le i \le K \\
\overline{u}_i &= \sum_{j=1}^{K} \overline{w}_{ij} a_j + \overline{\theta}_i \\
y_i &= f(\overline{u}_i), \quad 1 \le i \le L
\end{aligned} \quad (1)$$

The dynamics of the retrieving phase can be rewritten in a matrix form:

$$\begin{aligned}
\underline{\mathbf{u}} &= \underline{\mathbf{W}} \mathbf{x} + \underline{\boldsymbol{\theta}} \\
\mathbf{a} &= F[\underline{\mathbf{u}}] \\
\overline{\mathbf{u}} &= \overline{\mathbf{W}} \mathbf{a} + \overline{\boldsymbol{\theta}} \\
\mathbf{y} &= F[\overline{\mathbf{u}}]
\end{aligned} \quad (2)$$

where $\underline{\mathbf{W}}$ and $\overline{\mathbf{W}}$ represent the lower weight matrix and the upper weight matrix respectively, and the operator $F[\cdot]$ performs the nonlinear activation function $f$ on each of the elements of the vector.

As depicted in Figure 2(a), the retrieving phase of two-layer network may be implemented in terms of two consecutive matrix-vector multiplications, interleaved with two stripes of nonlinear processing units.

The back propagation learning rule for a two-layer network is briefly reviewed below.

**Upper Layer** Upon the arrival of the $m$-th training patterns, the updating formula for the upper weight matrix is

$$\begin{aligned}
\Delta \overline{w}_{ij}^{(m)} &= -\eta \frac{\partial E}{\partial \overline{w}_{ij}^{(m)}} \\
&= \eta \, \overline{\delta}_i^{(m)} \, f'(\overline{u}_i^{(m)}) \, a_j^{(m)}
\end{aligned} \quad (3)$$

where $E = \sum_i (t_i^{(m)} - y_i^{(m)})^2$ is the energy function, $\overline{\delta}_i^{(m)}$ is the error signal which can be computed by the BP technique.

For the recursion, the initial value (of the upper layer), $\overline{\delta}_i^{(m)}$, can be easily obtained as follows:

$$\begin{aligned}
\overline{\delta}_i^{(m)} &\equiv -\frac{\partial E}{\partial y_i^{(m)}} \\
&= t_i^{(m)} - y_i^{(m)}
\end{aligned} \quad (4)$$

**Lower Level** The updating formula for the lower weight matrix is

$$\Delta \underline{w}_{ij}^{(m)} = -\eta \frac{\partial E}{\partial \underline{w}_{ij}^{(m)}}$$
$$= \eta \, \underline{\delta}_i^{(m)} \, f'(\underline{u}_i^{(m)}) \, x_j^{(m)} \qquad (5)$$

where the error signal $\underline{\delta}_i^{(m)}$ can be derived as follows:

$$\underline{\delta}_i^{(m)} \equiv -\frac{\partial E}{\partial a_i^{(m)}}$$
$$= -\sum_j \frac{\partial E}{\partial \overline{u}_j^{(m)}} \frac{\partial \overline{u}_j^{(m)}}{\partial a_i^{(m)}}$$
$$= \sum_j \overline{\delta}_j^{(m)} \, f'(\overline{u}_j^{(m)}) \, \overline{w}_{ji}^{(m)} \qquad (6)$$

The overall operations in the learning phase can be divided into three stages: (1) MVM; (2) VMM; and (3) OPU.

- The **MVM step**, i.e. retrieving computation in the forward step, is the same as retrieving phase described in Eq. 2. At the very beginning of the learning phase, it has to be executed to obtain the output response of the input training pattern.

- The **VMM: i.e. the Back-Propagation Step** computes the error signal vectors vectors g and h. The vector g has elements

$$g_i \equiv f'(\overline{u}_i)\overline{\delta}_i$$
$$= f'(f^{-1}(y_i))(t_i - y_i) \qquad (7)$$

The vector h is derived from the error signal according to the back-propagation formula

$$h_j \equiv f'(\underline{u}_j)\underline{\delta}_j$$
$$= f'(f^{-1}(a_j))\underline{\delta}_j \qquad (8)$$

and $\underline{\delta}_j = \sum_i \overline{W}_{ij} g_i$ which may be expressed in form of vector-matrix-multiplication: $\underline{\delta} = \overline{W}^T g$.

- The **OPU step** is the actual updating procedure, which computes the outer-product: $\Delta \overline{W} = \eta g a^T$ and $\Delta \underline{W} = \eta h x^T$

| stages | input | output |
|---|---|---|
| MVM | $x, \underline{W}$ | $a$ |
| VMM | — | — |
| OPU | $h, x, \underline{W}$ | $\underline{W}_{new}$ |

(a)

| stages | input | output |
|---|---|---|
| MVM | $a, \overline{W}$ | $y$ |
| VMM | $t, y, a, \overline{W}$ | $g, h$ |
| OPU | $g, a, \overline{W}$ | $\overline{W}_{new}$ |

(b)

Table 1: The operands and the results in the three stages of the learning phase for the (a) lower-layer and (b) upper-layer.

To facilitate our illustration, the detailed input operands and output results of each of the above three stages of operations for the lower-layer and upper-layer networks are listed in Table 1. In addition to the DG for MVM, the DGs for VMM and OPU are shown in Figure 2(b) and (c) respectively.

In order to facilitate pipeline design, we make use of the following strategies. First, the DG is designed so that the choosen dependency arcs have the same flow direction as shown in Figure 2(a)-(c). To simplify the subsequent mapping, we adopted an **interleaved-DG** design for the VMM/OPU DG here. The interleaved VMM/OPU DG is shown in Figure 3. Also, a NOP block is artificially inserted in Figure 4. Its purpose is to delay the internal data stream long enough to be properly synchronized with the data stream routed from the torus connections.

**Design of Linear Array Structure** Let us use 7-5-3 network as an example. We will use $\vec{d} = [1\ 0]$ and $\vec{s} = [1\ 1]$. The DG and corresponding systolic array design are illustrated in Figure 5.

**Processor Element Design Requirements** The following are the basic design requirements. *Memory*: Each PE should store a row of the lower weight matrix and a column of the upper weight matrix. *Communication*: Data are transmitted uni-directionally between two neighboring PEs. There is no global data bus needed, but some global control lines are useful. A circular link between the first PE and the last PE is necessary to facilitate the pipelining mechanism. *Arithmetic processing*: Each PE should support all the arithmetic processing capabilities, including MAC (multiply-and-accumulate) operation and nonlinearity processing operation. There are two nonlinear functions in the network, f(·) and f'(·). They may be implemented by a table look-up approach.

**Performance** The architectural choice hinges upon several key performance measures, including cost-effectiveness of the number-of-processors, utilization

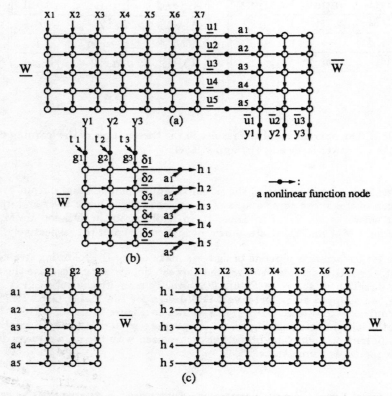

Figure 2: Rectangular DGs for (a) MVM, (b) VMM, and (c) OPU.

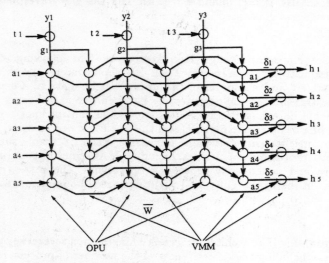

Figure 3: Interleaved VMM/OPU DG.

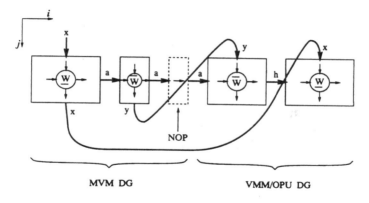

Figure 4: Cascaded Rectangular DGs for MVM DG and the VMM/OPU DG
. Note that the flow directions of the DGs are kept in the same direction to facilitates pipeline design. The natural choice of the projection vector is $\vec{d} = [1\ 0]^T$. The most preferred schedule vector is $\vec{s} = [1\ 1]^T$. If $K > L$, the output of a will arrive earlier than the output of y. To make the schedule feasible, a NOP DG is inserted between the MVM DG and and the VMM/OPU DG for the upper layer. Its purpose is to introduce enough delays such that the schedule vector $\vec{s} = [1\ 1]^T$ becomes permissible.

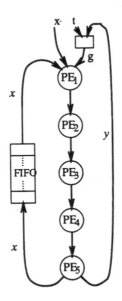

Figure 5: The systolic array design of 7 -5 -3 two-layer BP network.

rate, speed-up-factor, and so on. Here the "speed-up-factor" will be computed as "sequential-time/array-time". The "utilization-rate" will be computed as "speed-up-factor/number-of-processors". We shall use the 7-5-3 network example to illustrate this point. For the linear array, the

$$speed-up-factor = \frac{2NK+3KL}{2N+K+2L} = 115/25 = 4.6$$
$$utilization-rate = \frac{2NK+3KL}{(2N+K+2L)K} = 4.6/5 = 92\%$$

(9)

## 3.2 Rectangular Array Design for Block-Adaptive BP Method

In order to achieve an effective design of rectangular systolic array, we resort to the block-adaptive method - instead of the data-adaptive method discussed previously. The (block) gradient method for a two-layer network is:

$$\Delta \underline{w} = \sum_{m=1}^{M} \Delta \underline{w}^{(m)}$$
$$\Delta \overline{w} = \sum_{m=1}^{M} \Delta \overline{w}^{(m)}$$

(10)

The advantage of block-adaptive method is that the weights are not updated for each training pattern, instead, $\Delta \overline{w}^{(m)}$ are accumulated during the training for the whole set of training patterns. Only after the entire sweep of the training patterns, the weights are updated by the accumulated sum of $\Delta w$.

In Figure 6, we use 3-dimensional DGs to show the block-adaptive computation. The DGs are just the replica of those in Figure 4, with the depth of third dimension being $M$, the number of training patterns. However, it is important to note that the OPU cubic DG is used to compute the values of $\Delta w^{(m)}$, which are accumulated over time to obtain $\Delta w$.

*The main feature distinguishing the block-adaptive method from the data-adaptive one is that the actual updating of the weights is deferred to the end of each block. This implies that there is no need to wait for the OPU DG of the previous pattern to finish before the MVM operations for the new pattern to be started. This feature can be exploited to obtain a much greater degree of parallel/pipeline processing.*

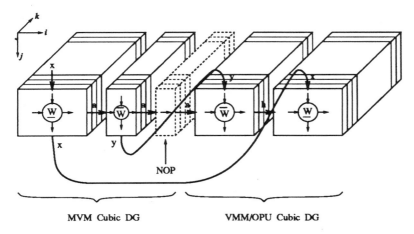

Figure 6: Cascaded Cubic DGs block-adaptive net Note that this DG is obtained by repeating the DGs shown in Figure 4.

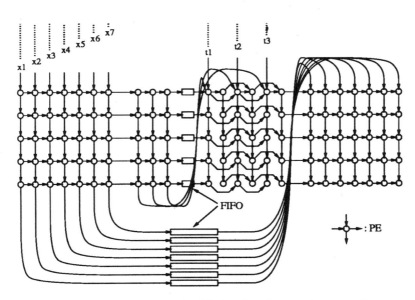

Figure 7: The rectangular arrays for block-adaptive two-layer back propagation ANN

**Design of Rectangular Array Structure** We will use $\vec{d} = [0\ 0\ 1]$ to project the DG in the k direction on a rectangular array. It is easy to verify that schedule vector $\vec{s} = [1\ 1\ 1]$ yields a permissible systolic array design. If the projection direction $\vec{d} = [0\ 0\ 1]$ is used, we will obtain an array processor with four rectangular blocks. The PEs in the upper-layer rectangular array for VMM and OPU operations can perform the VMM and OPU operations concurrently. This is a favorable design since it yields maximal parallel processing capability. Again we shall use an example with a 7-5-3 network and $M = 100$ patterns. The speed-up-factor $= T_{seq}/\text{ATPB} = 11500/121 = 95$.

$$utilization-rate = \frac{(2NK+3KL)M}{(2N+K+L+M-1)(2NK+3KL)}$$
$$= \frac{M}{2N+K+L+M-1} = 95/115 = 82.6\%.$$

Compare with the linear array design, we note that the rectangular array has a clear advantage in the speed-up factor while the linear design has a higher utilization rate. More precisely, with a ratio of 95:4.6, the speed gains by more than 20 times. Moreover, the utilization rate of 82.6 % is considered to be very effective for massively parallel processing.

## 4 General-Purpose Digital Neurocomputers

*The goal of the digital neurocomputer is to provide a high speed, accurate, and yet flexible platform for neural network algorithms using large number of processing nodes, extensive interconnectivity, and complex learning rules.* A neurocomputer must be general-purpose, embracing a broad set of connectionist networks. The development of general-purpose neurocomputers may be viewed as an application-driven approach to massively parallel supercomputers. The intent is to make supercomputing systems which are higher-speed and more fault-tolerant than the conventional supercomputers in various intelligent system applications.

Given the need for multiple sweeps of the training patterns in networks with units numbering in the thousands, a very high processing rate will be required for almost any real-time training performance. On the other hand, higher or as high as processing rates are needed for real time retrieving or recognition. Parallel processing is about the only solution to achieve such processing speeds. Fortunately, exploitation of massive parallel processing is very plausible, since most operations are repetitive in neural net applications.

## 4.1 Basic Design Considerations

### 4.1.1 Balanced Design

For this paradigm, there are extremely high demands on all the digital computing components: *computations, memories,* and *communications*. Neurocomputer design must ensure a balance of these components:

- *Computation:*

    Parallel processing with a large number of processing nodes offers perhaps a realistic solution to the design of real-time neurocomputer. The arithmetic processing units need to support the multiplication, addition, and nonlinear operations with their respectively required precisions.

- *Memory:*

    A proper balance and trade-off must be made between on-chip and off-chip storages, including cache, local memory, global memory. There are many ways to store synaptic weights. For examples, one is to distribute the weights over many neural chips; another is to store the weights off-chip in DRAMs. Unless carefully designed in advance, it may become necessary to reload repeatedly a new set of weights into cache or local memory in the neural chip. However, reloading of weights can be very time-consuming due to the communication bottleneck.[29] As an example, for the Bell Lab. chip [11], it takes 256 clock cycles to reload while the multiplication time (for all the stored weights) takes only one cycle.

    Since the chip area is constant, the trade-off between functionality and storage is inevitable: On one hand, if the storage is vital, then a good fraction of the chip area has to be allocated to on-chip weights storage. On the other hand, if the functionality can not be compromised, then most chip area has to be reserved for arithmetic processing and the weights are stored off-chip in cheaper DRAMs. With the advent of new submicron ULSI technology, it is very hopeful that both the requirements may be simultaneously met in the near future.

- *Communication:*

    The basic requirement of the interconnection network is to ensure that all messages can be successfully fowarded to their destinations as quickly as possible. In practice, constructing direct and physical links for all of the thousands of synaptic weights is not feasible. Various options and trade-offs exist in selecting communication modules and interconnection networks to support parallel neural processing.

### 4.1.2 Overall System Configuration

A possible overall system configuration is depicted in Figure 8, which consists of: host computer, interface system, PE arrays, and interconnection networks.

- The *host computer* should provide batch data storage, management, and data formatting; determine the schedule program that controls the interface system and interconnection network, and generate and load object codes to the PEs. The host provides interaction between the user and the hardware. It initializes or changes network topology and monitors the overall progress. A convenient choice of the host may be a workstation (enhanced with special interface bus) in order to take advantage of its graphics and operating system capabilities.

- The *interface unit*, connected to the host via the host bus, has the functions of down-loading and up-loading data. Based on the schedule program, the control unit monitors the interface system and interconnection network. The functions which are common in interfaces are DMA, buffering, (cache memory, if necessary), handling interrupt, and data and sequence control.

- *Interconnection networks* provide a set of mappings between processors and processors or between processors and memory modules to accommodate certain common global communication needs. Incorporating certain structured interconnections may significantly enhance the speed performance of the parallel processing systems. For data tranfer and message passing in locally interconnected array processors, a special communication unit (i.e. routing hardware) must be included in the processor architecture. In this way, global interconnectivity inherent in some neural networks may be implemented on locally interconnected mesh array.[2]

The following array configurations may be considered for neural information processing.

- One-dimensional architecture:
  (a) linear array;
  (b) ring array.
- Two-dimensional architecture:
  (c) square/rectangular array;
  (d) triangular array;
  (e) hexagoanl array;
  (f) tree array.
- High-dimensional architecture:
  (g) 3-D array;
  (h) hypercube array.

- A *Processor Element (PE) array* comprises a number of processor elements with local memory. The granularity of a PE is closely related to the number of PEs in the array. The arrays are roughly divided into three groups: (1) those with the number of processors on the order of ten, (2) thousands, and (3) millions. The processors in the first and the second groups are usually *single-chip*, based on either custom made or commercially available microprocessors. Processors in the third group are usually *many-per-chip* (and usually bit serial), and are often custom made with VLSI/ULSI technology. This classification is displayed in Figure 8. At the massively parallel end are fine grain machines, such as the Connection Machine and MPP, where bit-serial processors are used. The neurons may be mapped one-to-one onto the processors, but routing units are required to support the desired links. Simple processor primitives are often preferred in many low-precision image processing applications. However, as the level of parallelism increases, and the processing power and memory of processors increases, thus the interconnection and routing problem complicates. Moreover, many neural processing applications require fast multiply-and-accumulate, high-speed RAM, fast coefficient table addressing. In coarser grain PEs, more powerful arithmetic units may be accomodated. TI's TMS320C40 and INMOS' transputer, are examples of a larger PE granularity which belongs to the micro-computer array domain in Figure 8.

### 4.1.3 Synchronous versus Asynchronous Architectures

The array architecture may be synchronous or asynchronous. In a synchronous network, all the neurons change their state in lock-step under the control of a global clock. In asynchronous networks the neuron may be self-clocked. The use of a global clocking signal greatly simplifies the processing element control but it also creates clock distribution problems in large arrays. Given the limited propagation speed of the signals both on a chip and on a board, processors that are far away from the clock source will receive the clock pulses with a few nanoseconds delay relative to processors closer to the source (*clock skew*)[24]. In reality, a global clock can drive approximately up to several tens of processors of FPU (Floating Point Unit) complexity. In order to provide the necessary performance, many machines have to go beyond that range. In addition, any flexible architecture must be expandable. Under such circumstances, clock skew will become a problem. One way to avoid this clock skew is to use multi-stage clock buffer and to minimize the length of the wire. It makes the length of the wire at each stage shorter, and hence reduces the skew. Another alternative involves the use of an *asynchronous* clocking scheme where each processor has its own clock and is synchronized with the others via *handshaking signals*.

### 4.1.4 Implementation Options

There are serveral alternatives of actual implemention of a neurocomputer:

- Implement neural net algorithms in existing parallel processing machines.

    Many neural net algorithms may be ported to an existing machine. The problem is that there are usually some constraints pertaining to a specific machine. As an example, an SIMD neurocomputer such as the Connection Machine can support neural models as long as all the

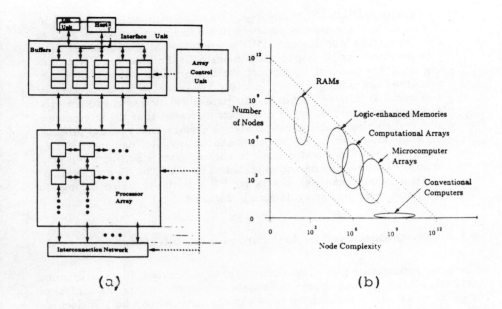

Figure 8: (a) An overall digital neurocomputer system consists of: host, interface system, interconnection networks, and processor array. (b) Different levels of granularity of PE in an array processor system, adapted from [31].

processors can run on the same program. This prompts the search for other system which is more tailored to the desired application.

- Construct a neurocomputer from commercially available chips.

  The use of commercially available components usually keep the cost more affordable, thus this approach is suitable for experimental systems and even some real systems. As an example, the transputer or TMS320C40 chips are designed with enhanced communication capability ready for construction of array processors. Sometimes array processors based on commercial chips have either inadequate FPU performance (e.g. Transputer) or limited interconnectivity (e.g. i860). In this case, a plausible design strategy is to adopt commercial chips for the floating-point arithmetic unit and the memory unit, and attach additional chips for communication unit. Since it is often very difficult to identify well-suited commercial chips for the desired communication unit, custom designed communication chips should be considered.

- Construct new neurocomputer prototypes from custom ASIC chips or gate arrays.

  While it allows optimal design on both the PE and array architectures, it often involves an expensive as well as time-consuming process. Therefore, this should be attempted only after a careful feasibility study of all the specifications pertaining to the given applications.

## 4.2 Processor Element Architecture

The key components in a neural processor element are processing unit, memory unit, and communication unit. The basic arithmetic building blocks are nonlinear processing unit, floating-point/fixed-point MACs, and undeterministic processor. The nonlinear unit is used for activation function and can be built by table-look-up. The MACs are for supporting matrix operations such as MVM, VMM and OPU. The undeterministic processor is meant for stochastic neural models such as Boltzmann machine. The memory units include program memory and data memory. The storage mechanisms include registers, cache, external local memory, or shared memory. Special communication unit will be required to support the array processing environment.

High performance is the most important objective of any PE design. Dynamic logic is in general slower than static logic but requires much less area. Especially in memories such as the microprogram memories in our chip, dynamic logic can provide us with one transistor per bit storage while static logic requires 3-6 transistors per cell. Design time, yield and power consumption are directly influenced by the size of a chip.

The chip should be programmable and independent of the specific algorithm and configuration. It is preferable to use a microprogrammable controller as opposed to one based on random logic in order to be able to change the control of the FPU and also change the functions to suit specific applicational needs. The chip design should support the expandability of the array system. The regularity of operations can be and should be exploited. To anticipate applications requiring more processing nodes than the actual array size, special hardware must be provided to support the partitioning of large problems.

A generic block diagram for the processing element is depicted in Fig.9. It consists of the controller, address generator, three memory banks, the I/O subsystem and the floating point unit containing a multiplier and an adder. Each block is accompanied with a set of registers. In order to provide fast execution, several buses are provided to permit simultaneous data movements from different sources to different destinations. In the following we will detail how this generic architecture may be specialized.

## 4.3 Parallel Array Architectures

Many parallel processing architectures are proposed for neural network implementation. The prominent ones are bus-oriented architectures, SIMD arrays, MIMD multiprocessors, and pipelined (systolic/wavefront) arrays. Comparisons of the different classes will inevitably involve the complex trade-off among various factors, such as programmability, reconfigurability, synchronization, and interprocessor communication. Some neurocomputer design examples are discussed below.

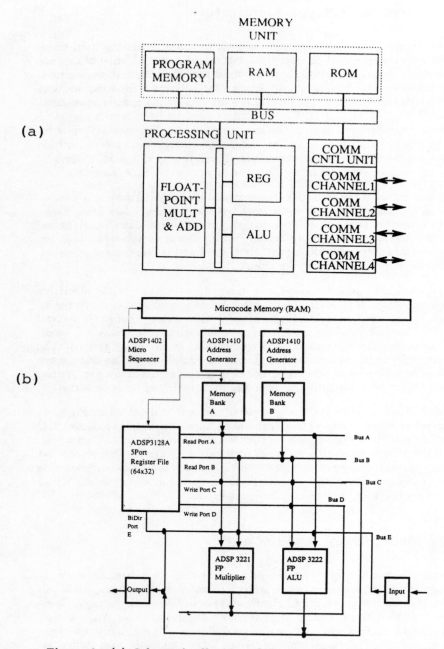

Figure 9: (a) Schematic diagrams for a *complete* PE design. The actual implementation could involve a subset of the components. (b) Board level prototype of Siemens/Princeton PSPS.

## 4.3.1 Bus Oriented Architectures

Single-bus or multiple-bus structures have been widely used for many years in conventional multiprocessor architectures. One of the early examples is the Multimax machine by Encore Computer Corp. which has a speed of 100 Mbyte/s and can be expanded from 2 to 20 microprocessors. Although bus-oriented networks can be made extremely powerful by using very high-speed buses, they are always limited to a certain number of processors by the bandwidth of the buses.

### Example of Bus-oriented Architecture: MARK III and Mark IV .

The Mark III used eight single-board computers. Each of the single-board computers was in the form of a standard VME bus circuit board. The MARK IV is a neurocomputer which is constructed out of signal processing building blocks (memories, multiplier, adders, and barrel shifters), TTL logic parts, and memory parts. This computer is designed to be a node of a much larger neurocomputer (which was never constructed). This larger neurocomputer if constructed would consist of up to 1,000 MARK IV nodes on a fast broadcast bus structure. [14]

### Example of Single Bus Architecture: Hitachi Neurocomputer .

The neurocomputer has about 70 K synapses and 1 K neurons. Each neuron has only 64 synapses, because memory is relatively more expensive in the gate array technology. The key architectural specifications of the Hitachi Neurocomputer are provided below: (More details can be found in [36]. )

- Neuron circuit: Completely digital circuit with learning function.
- Architecture: Time sharing digital bus and dual networks for learning.
- Number of neurons: There are 1152 neurons in the system with 144 neurons per wafer.
- Performance of learning circuits: 2.3 Giga CUPS.

## 4.3.2 SIMD Architectures

A single-instruction-multiple-data (SIMD) machine is a parallel array of arithmetic processors with local memory. An SIMD array has control (instruction) buses and data buses (in lieu of the local instruction codes adopted in the systolic or MIMD arrays). Instructions are broadcast from a host, with all processors executing the same instruction simultaneously [24]. One prominent SIMD architecture for neural implementation is the *Connection machine* [15, 21, 34].

# Example of SIMD Neurocomputer Architecture: Connection Machine

Connection Machine$^{TM}$ is a new type of computing engine which was proposed by W. D. Hillis from Thinking Machines Corporation. It provides a large number of tiny processor/memory cells connected by a programmable communications network. (The assembly-level REL-2 programming language and the higher-level C and LISP programming language are provided.) The current Connection Machine contains 64K processor cells interconnected in a hypertorus structure, each with 4K-bits memory and a simple serial arithmetic logic unit. The basic building blocks are custom CMOS chips, each containing 16 processor cells and one router unit. The router is responsible for routing messages between chips. The system of the connection machine consists of host, processor/memory cells, communications network, and input/output.

A Connection Machine cell is a bit serial processor, comprising a few registers, an ALU, a message buffer, and a finite state machine. Each cell is "intelligent". Based on the incoming message and its internal state, it can execute a sequence of steps such as arithmetic or storage operations on the contents of the message and the registers, adjusting the internal state, and sending out new messages. Very recently, it is reported that the newest Thinking Machines computer CM-200 is capable of 9 Giga operations per second.

# Example of SIMD Neurocomputer Architecture: Princeton Engine

The Princeton Engine is a 29.3 GIPS image processing system capable of simulating video rate signals -including NTSC and HDTV video - in real-time. [7] It consists of a massively-parallel arrangement of up to 2048 processing elements. Each processing element contains a 16-bit arithmetic unit, multiplier, a 64-word triple-port register stack, and 16,000 words of local processor memory. As to the communication supports of Princeton Engine, there are two types of communication. One is *broadcast*, by which one node send and all the others receive. The other is *bypass*, by which one can establish links on one dimensional array. not necessary to be neighbor. (Communication time is 1 cycle when within 64 processor region boundary, with additional 1 cycle per boundary when sent across boundary.) For example, the interprocessor communication bus permits simultaneous exchanges of data between neighboring processors in one instruction cycle.

## 4.3.3 MIMD Architectures

MIMD multiprocessors offer much greater architectural flexibility. The processors communicate either with a shared memory or by a message passing scheme. For the former case, each of the processors has its own local instruction codes and has access to shared memories. For the latter case, the MIMD computer is equiped with special hardware routing communication processor to support asynchronous data distribution. Since the interconnections can be easily time-multiplexed, this technique may help alleviate the problem of the stringent limitation on I/O pin count and other interconnnection hardware. It also allows self-controlled bypassing of defective processor elements - when necessary - and thus enhance the fault-tolerance capability.

Message passing technique in MIMD computer depends on the routing communication processor. The communication processor enables separate processing of communication signals, i.e. the switching communication and the neuro processing tasks can function simultaneously. It executes local and global communication by routing messages through the network of locally interconnected processor chips. A typical message contains a control field, an address field, and some data fields. (Typical data fields are membrane potential, synapse weight, synaptical delay time, neuron threshold, transfer function, etc.)

### Example of MIMD Neurocomputer

In the MIMD architecture proposed in [30], the interconnection tasks are handled by an elaborate communication processor, that allows passing messages through the network. There are three neural processing layers, including preprocessing, feature extraction, and neural classifier layers. All the operations in the preprocessing layer can be executed by a very simple neuron model. For feature extraction and classification more sophisticated models, such as learning or optimization algorithms like back-propagation will be implemented. The use of bidirectional links between the different network layers also allows us to build arbitrary network structures; e.g. feedback loops useful for space-variant operations in contrast enhancement, edge detection and many other applications. Another useful feature is that the neural processors can be re-programmed in less than 2 ms.

1. *Preprocessing Layer*

   A two-dimensional $16 \times 16$ mesh of neural processors forms the first image processing layer of our network, holds an entire $512 \times 512$ pixel grey-tone image and the algorithmic code. Each of neural processors stores at least 1 Kbyte segment of image data, with the segment size $(32 \times 32)$. The neural processors are coded to carry out some fixed-weight preprocessing operations such as median filtering, edge detection.

2. *Feature Extraction Layer*

   The second layer of the network performs feature extraction operations on the preprocessed data. This is a data compression phase. The lower data rate will facilitate more sophisticated neuron processing models for the next layer.

3. *Neural Classifier Layer*

   The third lower layer of neural processors implements trainable and self-organizing classifiers, to be used for different neural recognition experiments.

### 4.3.4 Array Processor Architectures

VLSI pipelined array processors which possess the properties of modularity, regularity, local interconnection, and pipelining. Typical examples are systolic and wavefront architectures. A systolic array is a network of processors which rhythmically compute and pass data through the system [23, 22]. A wavefront array is an asynchronous, self-timed, data-driven computation array. These arrays have local instruction codes and external data are piped into the array concurrently with the processing.

Wavefront processing utilizes the localities of both data flow and control flow inherent in many signal processing algorithms. Since there is no need to synchronize the entire array, a wavefront array is truly architecturally scalable. The wavefront array processor possesses most of the advantages of the systolic array processor, such as extensive pipelining and multiprocessing, regularity and modularity. More significantly, it also possesses the asynchronous data-driven capability of dataflow machines and can, therefore, accommodate the critical problem of timing uncertainty in VLSI array systems. For many applications, it would offer a most attractive solution for a very high speed and cost-effective neural net implementation.

The ring array requires far less complicated routing and a smaller number of processors thus satisfying the economy requirements as well. Ring architectures have been shown to be a good match to a variety of signal processing and connectionist algorithms. The simple communication ring topology is common to several other proposed and realized machines. Two examples are Siemens/Princeton PSPS and Intel's iWARP.

**Example of Pipelined Array Processors: PSPS -SIEMENS .**

The PSPS-Siemens board level processing element is based on microprogrammable, commercially available building blocks in order to provide a platform where hardware and software features of our array can be tested. To achieve this goal the design emphasizes flexibility and performance but in doing so, its cost, size and power consumption is increased making the construction of a large system very difficult.

The block diagram of the processing element is shown in Fig.9 and the functional blocks are briefly discussed below.

**Example of Pipelined Array Processors: iWarp .**

As an example, backpropagation algorithms have been implemented on WARP - a programmable linear systolic array consisting of 10 processors. A more advanced version of WARP is the iWarp VLSI chip by Intel.[4] The iWarp is a scalable parallel computing system, aimed at supporting both systolic (in a wavefront processing fashion) and message-passing applications. The iWarp component consists of two essentially independent agents: a computation agent and a communication agent. The communication agent consists of 4 bidirectional pathways, each 8-bits (+control) wide and 40 Mbytes/sec per bus (320 Mbytes/sec aggregate). This allows 20 simultaneous connections through/to/from the cell. The peak performance of iWarp is 20 MFLOPs, 20 MIPs, 160 Mbytes/sec memory operations, 80 Mbytes/sec data transfer (send), and 80 Mbytes/sec data transfer (receive).

For example, the NETtalk neural network benchmark runs at 16.5 million connection per second and 70 MFLOPS on a 10 cell Warp array. For a comparison, the same benchmark runs at 36 million connections per second and 153 MFLOPS on an iWarp array of the same number of cells. [5].

### 4.3.5 System Design Considerations

Generally speaking, in an array system design, one seeks to maximize the following performance indicators: effective array configuration, programmability for different networks, flexibility on problem partitioning, fault-tolerance to improve system reliability, word-length-effect in fixed/floating point arithmetics, and efficient memory utilization [24].

**Partitioning and Scalability** An ideal parallel system should *scale* properly, meaning that its speedup factor should remain roughly constant with an increasing number of processors. In other words, the complexity of the system should not introduce additional communication and synchronization overhead.

**Fault Tolerance and Reconfigurability** Actual array processor hardware will be very large scale and additional attention should be taken to achieve an acceptable system reliability. A fault at any node of the ring array architecture results in disconnecting the communication of the ring, therefore, redundant paths have to be established if a degree of fault tolerance is to be expected [37].

It is sometimes necessary for a digital implementation to be reconfigurable, possibly "on the fly". In order to defend the array against run time failures, concurrent error detection schemes should be incorporated inside each PE for fault detection, and (stand-by) spare PEs should be supplied for array reconfiguration.

**Software Environment** Software problem is in general a demanding task for a programmable neural architecture or general purpose computers attached by neural-oriented hardware accelerator. The more the system architecture differs from the inherent structures of the neural nets, the more important are the software supports. In a user-friendly environment, the user should be allowed to concentrate on the problem at hand instead of worrying about managing the hardware resources of a parallel system. Some operating system support is therefore necessary (e.g. languages, memory management, communications).

## 5 Concluding Remarks

When compared with their analog counterparts, digital implementations of neural networks suffers from the drawbacks of larger silicon area, relatively slower speed and the greater cost of interconnecting processing units, when compared with their analog counterparts. These difficulties can be overcome by design innovations. Digital techniques neural networks, when fully developed, can offer many unique and appealing advantages not matchable by their analog counterparts.

## References

[1] A. J. Agranat and et al. A ccd based neural network integrated circuit with 64k analog programmable synapses. In *Proc. IJCNN-90*, volume II, pages 551–555. IJCNN, June 1990.

[2] L.A. Akers, M.R. Walker, D. K. Ferry, and R. O. Grondin. Limited interconnectivity in synthetic systems. In R. Eckmiller and C. V. D. Malsburg, editors, *Neural Computers, Computer and Systems Science Series*, pages 407–416. Springer-Verlag Inc., 1988.

[3] J. Alspector and R. B. Allen. A neuromorphic VLSI learning systems. In P. Losleben, editor, *Advanced Research on VLSI*, pages 313–349. MIT Press, 1987.

[4] B. Baxter and et. al. Building blocks for a new generation of application-specific computing systems. In S.Y. Kung and et. al, editors, *Application-Specific Array Processors*, pages 190–201. IEEE Computer Society Press, 1990.

[5] S. Borkar and et al. iwarp: An integrated solution to high speed parallel processing. In *Proceedings, Supercomputing '88*, pages 300–339. IEEE Computer Society Press, 1988.

[6] A. Chiang, R. Mountain, J. Reinold, J. LaFranchise, and G. Lincoln. A programmable ccd signal processor. In *Proceedings, IEEE International Solid-State Circuits Conference, ISSCC90*, pages 146–148. IEEE/ISSCC, 1990.

[7] D. Chin and et. al. The princeton engine: A real-time video system simulator. In *IEEE Trans. on Consumer Electronics*, volume 34, pages 285–298, 1988.

[8] G. Eichmann and H. J. Gaulfield. Optical learning (inference) machines, vol. 24. *Applied Optics*, 1985.

[9] N. H. Farhat, D. Psaltis, A. Prata, and E. Paek. Optical implementation of the Hopfield model. *Applied Optics*, 24:1469–1475, May 1985.

[10] A. D. Fisher and J. N. Lee. Optical associative processing elements with versatile adaptive learning capability. In *Proc. IEEE, COMPCOM Meeting*, volume 137-140, 1985.

[11] H. P. Graf and P deVegvar. A CMOS implementation of a neural network model. In P. Losleben, editor, *Advanced Research on VLSI*, pages 351–367. MIT Press, 1987.

[12] H. P. Graf, W. Hubbard, L. D. Jackel, and P. G. N. DeVegvar. A CMOS associative memory chip. In *Proc. IEEE, 1st Intl' Conf. on Neural Networks, San Diego*, pages III461 – III468, 1987.

[13] R. Hecht-Nielsen. Performance limits of optics, electro-optics, and electronic neurocomputers. In *Proc. SPIE, Optical and Hybrid Computing*, volume 634, 1987.

[14] R. Hecht-Nielsen. *Neurocomputing*. Addison Wesley, 1990.

[15] W. D. Hillis. *The Connection Machine*. Massachusetts Institute Technology Press, 1985.

[16] J. Hoekstra. (junction) charge-coupled device technology for artificial neural networks. In U. Ruckert U. Ramacher, editor, *VLSI Design of Neural Networks*, chapter 2, pages 19–45. Kluwer Academic Publishers, 1991.

[17] M. Holler and et al. An electrically trainable artifical neural network (etann) with 10240 floating gate synapses. In *Proceedings of the IJCNN-89*, pages 11–191. IJCNN, Washington, D. C., June 1989.

[18] J. J. Hopfield and D. W. Tank. Neural computation of decision in optimization problems. *Biological Cybernetics*, 52:141–152, 1985.

[19] K. Hwang and Joydeep Ghosh. Hypernets for parallel processing with connectionist architectures. Technical Report CRI-87-03, University of Southern California, January 1987.

[20] R. S. Withers J. P. Sage, K. Thompson. An artificial neural network integrated circuit based on mnos/ccd principles. In *AIP Conf. Proc.*, pages 151, 381. Morgan-Kaufmann, 1986.

[21] N. H. Brown Jr. Neural network implementation approaches for the connection machine. In *Proc. IEEE, Conf. on Neural Information Processing Systems – Natural and Synthetic, Denver*, November 1987.

[22] H.T. Kung. Why systolic architectures? *IEEE, Computer*, 15(1), January 1982.

[23] H.T. Kung and C.E. Leiserson. Systolic arrays (for VLSI). In *Sparse Matrix Symposium*, pages 256–282. SIAM, 1978.

[24] S. Y. Kung. *VLSI Array Processors*. Prentice-Hall, 1988.

[25] S. Y. Kung. *Digital Neurocomputing*. Prentice Hall, Englewood Cliffs, NJ, 1992.

[26] S. Y. Kung and J. N. Hwang. Parallel architectures for artificial neural nets. In *IEEE, Int'l Conf. on Neural Networks, ICNN'88, San Diego*, pages Vol.2: 165–172, July 1988.

[27] H. Mada. Architecture for optical computing using holographic associative memories. *Applied Optics*, 24, 1985.

[28] C. Mead. *Analog VLSI and Neural Systems*. Addison Wesley, 1989.

[29] U. Ramacher. Guide lines to vlsi design of neural nets. In U. Ruckert U. Ramacher, editor, *VLSI Design of Neural Networks*, chapter 1, pages 1–17. Kluwer Academic Publishers, 1991.

[30] P. Richert and et al. Digital neural network architecture and implementation. In U. Ruckert U. Ramacher, editor, *VLSI Design of Neural Networks*, chapter 7, pages 125–152. Kluwer Academic Publishers, 1991.

[31] C. Seitz. Concurrent VLSI architectures. *Invited paper, IEEE Transactions on Computer*, C-33, December 1984.

[32] M. A. Sivilotti, M. A. Mahowald, and C. A. Mead. Real-time visual computations using analog CMOS processing arrays. In P. Losleben, editor, *Advanced Research on VLSI*, pages 295–312. MIT Press, 1987.

[33] A. P. Thakoor. Content-addressable, high density memories based on neural network models. Technical Report, Jet Propulsion Laboratory JPL D-4166, March 1987.

[34] S. Tomboulian. Introduction to a system for implementing neural net connections on SIMD architectures. Technical Report, NASA Langley Research Center, Hampton, NASA Contractor Report 181612, January 1988.

[35] K. Wagner and D. Psaltis. Multilayer optical learning networks. *Applied Optics*, 26:5061–5076, December 1987.

[36] M. Yasunaga and et.al. Design, fabrication and evaluation of a 5-inch wafer scale neural network lsi composed of 576 digital neurons. In *Proc. IJCNN'90(San Diego)*, pages II-527, 1990.

[37] P. Zafiropulo. Performance evaluation of reliability improvement techniques for single loop communications systems. *IEEE Trans. on Communications*, pages 742–751, June 1974.

# Conference Author Index

## A

Abdallah, C. 523
Aghajan, H. K. 188
Aikawa, K. 337
Ansari, N. 570
Ardalan, S. H. 151
Astola, J. 503
Atlas, L. E. 543
Austin, S. 347

## B

Beck, S. 21
Binenbaum, N. 595
Bourlard, H. 309
Bruce, C. 483

## C

Carmichael, N. 483
Carrato, S. 493
Chakravarthy, S. V. 21
Chang, C. F. 606
Chen, H. 40
Chen, H. H. 319, 385
Chen, M. S. 70
Chen, S. H. 376
Chen, W. Y. 376
Cotter, N. 562

## D

Deng, L. 411
Deuser, L. 21
De Vries, B. 101
Diamantaras, K. I. 50
Dias, L. 595
Drucker, H. 198
Duren, R. 236

## E

Elmasry, M. 411

## F

Fallside, F. 432
Fischer, K. A. 452
Fukuzawa, K. 442

## G

Geiger, D. 60
Gertner, A. N. 422
Gesteland, R. 562
Ghosh, J. 21
Gorin, A. L. 422
Gramss, T. 289
Guedes de Oliveira, P. 101

## H

Hassanein, K. 411
Hermansky, H. 405
Hirata, Y. 256
Hoffmann, N. 533
Horne, B. 523
Hsieh, P. 595
Hua, Y. 552
Huang, X. D. 357
Huang, Y. 570
Hush, D. 523
Hwang, J. N. 513

## I

Intrator, N. 460

## J

Ji, M. 385
Ju, J. 595
Juang, B. H. 11, 299

## K

Kai, A. 327
Kailath, T. 188
Katagiri, S. 11, 299
Kohonen, T. 279
Kung, S. Y. 50, 616

## L

Larsen, J. 533
Lau, W. H. 226
Lay, S. R. 543
Le Cun, Y. 198
LeBlanc, M. J. 208
Lech, M. 552
Lee, C. H. 11, 299
Lee, K. F. 357
Lee, Y. C. 319

Lenz, R. 121
Leung, S. H. 226
Leung, W. F. 226
Levinson, S. E. 422
Li, H. 513
Lin, J. H. 570
Lippmann, R. 266
Liu, R. 40
Liu, Y. D. 319
Luk, A. 226

## M

Makhoul, J. 173, 347
Mammone, R. 90
Manolakos, E. S. 208
Manry, M. T. 70, 246
Markel, S. 595
Matsuyama, Y. 141
Mendel, J. M. 80
Miller, L. G. 422
Miyamura, T. 131
Moody, J. E. 1
Morgan, N. 309, 405
Murdock, M. 562

## N

Nakagawa, S. 256
Neuvo, Y. 503
Nobakht, R. A. 151

## O

Ono, Y. 256
Osterberg, M. 121

## P

Paliwal, K. K. 367
Papadakis, I. N. M. 217
Pearson, J. C. 595
Peikari, B. 236
Pereira, R. A. M. 60
Perlovsky, L. I. 589
Poddar, P. 395
Premoli, A. 493
Principe, J. C. 101

## R

Ramponi, G. 493
Rao, S. S. 580

Reid, I. 483
Renals, S. 309
Riskin, E. A. 543
Rohani, K. 70
Rossen, M. L. 111
Russo, L. E. 161

## S

Sagayama, S. 327
Sakaniwa, K. 131
Sankar, A. 90
Sawai, H. 442
Schaper, C. D. 188
Schwartz, R. 347
Sethuraman, S. 580
Shen, Z. K. 385
Sheu, B. 606
Sibul, L. H. 30
Sicuranza, G. L. 493
Smith, M. 483

Sonehara, N. 473
Strube, H. W. 452
Sugiyama, M. 442
Sun, G. Z. 319

## T

Tajchman, G. 460
Takami, J. 327
Taylor, H. , Jr. 595
Tutwiler, R. L. 30

## U

Unnikrishnan, K. P. 395

## V

Van den Bout, D. E. 151

## W

Waibel, A. 357
Wang, L. X. 80
Wooters, C. 405
Wu, L. 432

## Y

Yamada, I. 131
Yau, H. C. 246
Yin, L. 503

## Z

Zavaliagkos, G. 347